MOTION
AND
TIME STUDY

Improving Productivity

PRENTICE-HALL INTERNATIONAL SERIES
IN INDUSTRIAL AND SYSTEMS ENGINEERING

W. J. Fabrycky and J. H. Mize, Editors

BLANCHARD   *Logistics Engineering and Management*
BROWN   *Systems Analysis and Design for Safety*
BUSSEY   *The Economic Analysis of Industrial Projects*
FABRYCKY, GHARE, AND TORGERSEN   *Industrial Operations Research*
FRANCIS AND WHITE   *Facility Layout and Location: An Analytical Approach*
GOTTFRIED AND WEISMAN   *Introduction to Optimization Theory*
KIRKPATRICK   *Introductory Statistics and Probability for Engineering, Science, and
   Technology*
MIZE, WHITE, AND BROOKS   *Operations Planning and Control*
MUNDEL   *Motion and Time Study, 5th edition*
OSTWALD   *Cost Estimating for Engineering and Management*
SIVAZLIAN AND STANFEL   *Analysis of Systems in Operations Research*
SIVAZLIAN AND STANFEL   *Optimization Techniques in Operations Research*
THUESEN, FABRYCKY, AND THUESEN   *Engineering Economy, 5th edition*
TURNER, MIZE, AND CASE   *Introduction to Industrial and Systems Engineering*
WHITEHOUSE   *Systems Analysis and Design Using Network Techniques*

MOTION AND TIME STUDY

Improving Productivity

5th edition

MARVIN E. MUNDEL, P.E.

M. E. Mundel and Associates

PRENTICE-HALL, INC., *Englewood Cliffs, New Jersey* 07632

Library of Congress Cataloging in Publication Data

Mundel, Marvin Everett
   Motion and time study.

   Bibliography: p.
   Includes index.
   1. Motion study. 2. Time study. I. Title.
T60.7.M86 1978   658.5'42   77–21141
ISBN 0–13–602987–6

10 9 8 7 6 5 4 3 2 1

Printed in the United States of America

PRENTICE-HALL INTERNATIONAL, INC., *London*
PRENTICE-HALL OF AUSTRALIA PTY. LIMITED, *Sydney*
PRENTICE-HALL OF CANADA, LTD., *Toronto*
PRENTICE-HALL OF INDIA PRIVATE LIMITED, *New Delhi*
PRENTICE-HALL OF JAPAN, INC., *Tokyo*
PRENTICE-HALL OF SOUTHEAST ASIA PTE. LTD., *Singapore*
WHITEHALL BOOKS LIMITED, *Wellington, New Zealand*

CONTENTS

v

PREFACE

This book is intended to provide a systematic, practical, and yet scientifically correct treatment of present-day motion and time study in a readily understandable form. My aim has been to present the basic principles underlying successful work. The development of these principles is examined with the aid of carefully selected examples, many of which are new to this edition, so that the principles and procedures may be understood and later applied naturally rather than as clerical procedures. For each technique, at least one example is given in detail so that the highly important reasoning processes involved in the application of the procedures are discernible. The relationship of motion and time study to the industrial engineering techniques involving mathematical models has been indicated where appropriate.

A generalized theory of managerial control has been introduced early in the book to provide a framework for relating the material of this book to the whole of management activity. A concept of work-units and a work-unit structure is also introduced early in the book to make possible the relating of subsequent chapters to service activities and indirect work, as well as to direct manufacturing and processing. These materials have been thoroughly evaluated in numerous, actual working situations to determine the form in which to include them.

Although the text is directed at the *how* of motion and time study, as seems appropriate, great care has been taken to keep the *why, when,* and *where* carefully in focus. The material is oriented to the serving of objectives; techniques are treated solely as a flexible means of achieving these objectives.

The scope of application of the examples chosen includes not only direct and indirect work, agricultural work, and office activities, but also the broad variety of activities that characterize all service and government organizations. The illustrations appear in the form in which they would appear in practice, so that they may provide working guides and examples rather than unrealistic examples of the graphic arts. Care has been taken to provide broad coverage.

The chapters on work measurement reflect the greatly increased variety and scope of applicability of modern techniques. Many of the newer techniques have already been applied so widely that I have been able to select specific and important illustrations.

The text has been completely rewritten and reformatted. First, the introductory chapters introduce the whole of the field. Second, to increase readibility, the outlines used to write the chapters have been visibly incorporated into the book in the form of numerous center and side headings. Third, the sequence of presentation has been made more logical.

Finally, this book has also been redesigned to facilitate the use of the same material by a large variety of people whose individual interest in motion and time study may be highly diverse. In this manner, within the limits of their interest, they will all have a common ground for communication and understanding. To this end, the chapters that discuss general aspects of the field have carefully been kept separated from those presenting a discussion of techniques and the specifics of application. As was noted, these introductory chapters all appear together at the front of the book. These are followed by a series of chapters discussing techniques, which have been made as short and concise as appeared consonant with an adequate explanation, and contain, with a few exceptions, only a single illustration of relative simplicity. Each of these chapters is followed by a set of appropriate problems. The problems are of a wide variety, to provide an instructor with a maximum number of models for problems that he or she may wish to generate to replace those given, after the book has been in use for an extended time.

These chapters are followed by a series of chapters presenting more complex ramifications of the techniques for more serious students or practitioners who may wish to pursue the subject more fully or who are seeking guidance in more difficult situations.

Hence, the generalist who wishes only a brief presentation, the person who wishes a general familiarity with all the techniques, and the person who wishes to pursue some or all of the techniques to a more exhaustive level of detail may all work from the same book. For instance, the generalist may confine his reading to the general subject chapters, Chapters 1 through 6.

The person who wishes to acquire a general familiarity with the totality of the subject may extend his reading to Chapters 7 through 27. The person who wishes to pursue the subject (or selected topics) more fully will have recourse to Chapters 28 through 46. The person who is seeking some specific information will be aided by the variety of ways provided in the index for locating the appropriate sections.

During the past thirty years the traditional approaches as well as the new approaches described in this book have been thoroughly tested by a wide variety of applications to the direct and indirect work of light and heavy industry and to a wide variety of government activities, ranging from the Secret Service of the U.S. Department of the Treasury to the Animal Health Division of the U.S. Department of Agriculture, as well as organizations of foreign governments. They have also been used with thousands of college students and industrial and government seminar participants in many countries.

In particular, during the ten years that I taught at Purdue University, I also served under the Fulbright Program as visiting lecturer at the University of Birmingham, England. There, through the good offices of the late T. U. Matthew, I met numerous British industrial engineers and toured many British plants. The opportunities for study and discussion that the visit made possible are reflected in this book.

Also, I undertook to organize and operate on a national basis a management engineering training program for the U.S. Army Material Command, which employs over 160,000 civilians. The techniques of motion and time study play an important part in this program, and the opportunity to work with these techniques on such a wide scale has influenced the thinking throughout this book.

During the next ten years, when I worked as a consultant engineer, I also had the opportunity to serve as seminar team leader of groups assigned to introduce industrial engineering into Japan, followed by a period during which I served as a visiting professor at Keio University in Tokyo and as a consultant to a broad variety of industries. These experiences occasioned a careful scrutiny of many of the basic assumptions previously associated with much motion and time study technology.

Subsequently, as Principal Staff Officer for Industrial Engineering in the Executive Office of the President, U.S. Bureau of the Budget, I had a unique opportunity to work with a broad variety of government service activities.

Most recently, again working as a consultant engineer with industrial concerns in America and abroad, as well as with a wide variety of government agencies, I have been faced with an enormous range of problems.

In this book I have included those approaches that were basic to successful work throughout these broad areas of activity.

Acknowledgments

In writing this book I have drawn freely on the background provided by the pioneer work of Frank B. and Lillian M. Gilbreth and on the teaching of the man who started me in this field, David B. Porter of New York University. Many other individuals and groups have helped make this book possible.

I owe much to the cooperation given my activities over the years by L. P. Persing of the General Electric Co.; Dean Parsons of the Perfect Circle Co.; J. F. Rittenhouse, Jr., and F. Savage of C. G. Conn Ltd.; L. J. Fletcher of the Caterpillar Tractor Co.; John Harder of R. G. LeTourneau Inc.; Claude Campbell and Guy Avery of Acme Steel Co.; Guy Bates of General Motors Corp.; and Charles Winkleman of Belden Manufacturing Co. I am also indebted to American Steel Foundries, L. S. Ayres & Co., City Ice and Fuel Co., Sayco Valve Corp., and the Hotel Sherman. I owe much to C. J. Allen of Patrick Cudahy, Inc., who has so magnificently co-operated in maintaining for six years a thorough statistical analysis of the results of an industrial application of the objective time study technique. I am also indebted to L. Piel of Johnson Service Co.; C. DeWitt and G. Kaplan of Western Printing and Lithographing Co.; R. Amstutz, R. Solger, A. Mandele, and M. Ottow of the S & C Electric Company; the industrial engineering staff of S. C. Johnson Co.; F. Pleva of Herbst Shoe Co.; E. Schendel of Trostel Leather Co.; A. Erickson of Barber-Colman Co.; J. Ruedebusch, G. Ploger, and Allen Drobke of Mirro Aluminum Co.; A. T. Tseng and J. Carpenter of the Diamond-Gardner Corp.; R. Klockzim of Inland Steel Co.; Lacey Randolph of the American Steel Foundries; R. Belt of Hamilton Manufacturing Co.; and Evan Scheele of Evco Inc., for their cooperation on applications that are reflected either in examples or in the approach to various problems treated in this book.

W. Spooner and Commander Ginn of the Navy Bureau of Ships, John Blake of the Long Beach Naval Shipyard, M. Peal of Benecia Arsenal, and Colonel J. MacGruder of the Air Force Advanced Logistics School have all influenced the making of the book.

To B. Carlberg, L. Fahlin, L. Rask, R. Kristensson, and Yngve Svensson of Sweden, I am deeply indebted for their assistance, cooperation, and advice on much presented in this book. Likewise to Winston Fear, Anne Shaw (Mrs. J. Pirie), R. M. Currie, P. B. R. Gibson, R. Connelly, and W. Rodgers of Great Britain; A. Takanaka, N. Isogai, T. Harada, K. Yoneda, and S. Sakai of Japan; and H. H. Hilf of West Germany.

Further, particularly with respect to the sections dealing with time standards for indirect work, I am deeply indebted to Louis Mayne and Donald Houston of the Meat and Poultry Inspection Program of the U.S. Department of Agriculture. When Mr. Mayne was with that activity, he set the stage for the first wide-scale testing and use of the new techniques; Dr. Houston provided the veterinarian knowledge, the arduous leg work, and the devoted attention required for seeing the techniques fully developed and put to their proper use. In connection with the continuation of this effort, special thanks are due to R. Sommers, W. Caplinger, J. Stein, H. Steinmetz, V. Berry, A. Geisman, Clyde Smithson, Clyde Jenner, and Ken Duff, all of the Meat and Poultry Inspection Program, for their substantial and continuous support. Their help has been invaluable in this effort, which was one of the pilot projects of extending motion and time study on a broad scale into a varied service-type activity.

I also want to thank the numerous individuals in the Department of Health, Education, and Welfare who have been my colleagues in a wide-ranging application of the newer techniques. Particular thanks are due to Michael A. Nigro, Martin Skutnik, and Justin Shook, and again to Clyde Smithson for his contribution while with DHEW. Special thanks are due to Anne Kaufman for her concept of the "reconstructed log."

Also, I want to thank Kojiro Yamaoka, of the Yanmar Diesel Engine Company of Osaka, Japan, and the numerous engineers of that company who have worked with me in making so many industrial applications of the newer techniques included in this book. The products of the Yanmar Diesel Engine Company range from small ultra-high-precision fuel-injection pumps to custom-built 3,000-horsepower diesel-engine generator and pump sets. The activities range from light and heavy foundry through light, heavy, and ultra-high-precision machining and include continuous and custom assembly. The support and overhead activities extend over the typical wide range of a modern diversified and multiplant industrial organization. An ideal test ground was provided for the development, testing, and application of the enormous range of motion and time study technology needed to support managerial control over such a broad range of activities. My debt really extends to the whole of the Yanmar organization.

I should also give special thanks to Tsuneo Ono, formerly of the Japan Management Association, and to Ziro Yamauti of Keio University in Tokyo, both of whom made feasible the large-scale testing of the techniques in this book in a wide variety of Japanese industries; I am also deeply indebted to Takeshi Kawase, who worked with me on so many projects. Further, with respect to my work in Japan, I should like to thank T. Ono; K. Shibata, who worked with me on Asian Productivity Organization projects; the management of the Nippon Electric Company, the Mitsubishi Heavy Industries, Inc., the Kawasaki Dockyard Company, the Japan Steel Tube Company, the Toyo Rayon Company, the Kanebo Spinning Company, the Matsushita Electric Company; and the engineers at those companies for the assistance I have received. Particular thanks are due those who worked with me on projects in Australia, particularly Kevin A. McDonald and G. McCracken.

I also have a debt to the various people in Korea, China, The Philippines, Singapore, and Hong Kong who have been coworkers on a wide variety of projects.

I have a large and continuing debt to the personnel of the Army Management Engineering Training Agency, particularly Lynn Bryant, the Director; James Jensen, the Deputy Director; William Shallman, the Chief of Industrial Engineering; and Vincent Poulin for the steady aid they have given to my work.

Acknowledgment should also be made of the assistance received from Art Jebbins, A. Weinberg, Donald Clark, Edith Mahon, and Andrew Latvala of the U.S. Department of the Interior.

Work on the new approaches has also been greatly aided by E. Saulman, G. Wise, R. Omohundro, J. Atwell, R. Brown, and W. McCallon of the Animal Health Division of the Agricultural Research Service, U.S. Department of Agriculture. Mention should also be made of the superb support work in making applications provided by Horace Robinson and Charles Hendricks of the Administrative Services unit of that organization.

Thanks are also due to my former colleagues in the Executive Office of the President, Bureau of the Budget, for the many thought-provoking discussions in which we engaged. Also, acknowledgment should be made of the suggestions and advice received from R. Thelwell and B. Usilaner of that organization.

The list, if extended to include all those whose effect is discernible anywhere in the book, would become much longer. I have attempted to single out those whose contribution is large, but the others were also important. I hope no one feels slighted by being left out; no slight is intended.

Finally, I would like to thank my patient book typist, Beatrice Levine, who so graciously provided the enormous effort that took this work from initial draft to finished copy, and San Panasethaned, who assisted with the complete reworking of the illustrations.

Silver Spring, Maryland   MARVIN E. MUNDEL

LIST OF TABLES

part one

INTRODUCTION TO
MOTION AND TIME STUDY

MOTION AND TIME STUDY
IN MANAGEMENT

General Definitions

The term *motion and time study* refers to a broad branch of knowledge dealing with the systematic determination of preferable work methods, with the appraisal, in terms of time, of the value of work involving human activity, and with the development of material required to make practical use of these data.

Motion study. The motion study aspect consists of a wide variety of procedures for the description, systematic analysis, and improvement of work methods, considering (1) the raw materials; (2) the design of the outputs (products or services); (3) the process or order of work; (4) the tools, work-place, and equipment for each step in the process; and (5) the human activity used to perform each step. The aim is to determine (or design) a preferable work method. The criterion of preference is usually economy of money, but effectiveness of the activity, ease or economy of human effort, economy of time, or economy of material—as well as other criteria, discussed in later chapters—frequently may take precedence. The terms *methods study* and *work design* appear in writings on the subject. *Value analysis* is also used to refer to the analyses applied to the raw material selection and to the consideration of the alternatives of output design.

Time study. The time study aspect consists of a wide variety of procedures for determining the amount of time required, under certain standard conditions of measurement, for tasks involving some human activity. (The term *work measurement* is also commonly used.)

Interrelationship. It is difficult to separate completely these two aspects, inasmuch as a specified method, frequently in the form of a written standard practice employing one of the techniques of motion study, is one of the conditions of time measurement. Also, time measurements are often a part of the basis on which alternative methods are compared. In addition, method determination and time appraisal complement each other's utility in application. The combined term—motion and time study—is used to denote all three phases of activity: method determination, time appraisal, and the development of material for the application of these data.

Field of Application

In any activity or occupation, motion and time study can be usefully employed to help find a preferred way of doing the work and assist in effectively managing or controlling the activity. The approach of motion and time study fits equally well when applied to heavy or light factory, office, production, maintenance, staff, or supervisory work. It is equally applicable to farm work, housework, surgery, cafeteria work, department store or hotel work, the whole range of government activities, battle activities, or any other human activity. What is accomplished by human activity will vary from job to job. The nature of the raw material will vary widely. In one case it may be information; in another it may be some simple or complex substantive material. The outputs may be services, responses to another group's actions, or any one of the almost infinite variety of products found in a modern society. The varieties of process are almost limitless. The varieties of tools, equipment, and workplaces (using these terms in the broadest sense to refer to things worked with and places worked at) is enormous. However, the requisite human efforts will in all cases be composed of the same basic acts, and the information relating to the economical use of human effort will be universally applicable.

Basic problems needing solutions. The problem of determining a feasible and preferable method of accomplishing the work will always be present. The problem of determining the required amount of human work time will be a normal concomitant. Regardless of the variation in accomplishment or field of knowledge some procedure must be employed to design the work and to determine the amount of time required to perform it. Suitable, analytical procedures must be selected from among the broad variety of available motion and time procedures to effectively assist in applying available knowledge to the solution of the design and measurement problems.

Productivity. *Productivity* is the term used to describe the ratio of units of output per unit of input. Productivity indexes are the value of such ratios divided by a similar ratio for a base year.[1] It should be obvious that the techniques of motion and time study (using the term in the broadest sense) play an important role in efforts to enhance productivity. When the effort is to enhance labor productivity the techniques of work measurement are vital in developing a procedure for meaningfully adding diverse outputs into an aggregated measure of output.

Effectiveness. *Effectiveness* refers to the success of an organization in achieving its objectives. With respect to economic aspects of objectives, with manufacturing, service, and government organizations, motion study and time study obviously are of great use in improving effectiveness. With respect to measuring other aspects of effectiveness, other technologies come into play.

Managerial uses. Motion and time study may well be used to provide a means for communication and cooperative activity among the various divisions of an organization in selecting, planning or designing, and controlling the proper integration of materials, design of product or work achieved, process, tools, workplaces and equipment, and human activity. Motion and time study techniques are aids for systematically performing certain managerial tasks. This is not the same as saying that motion and time study includes all management, any more than one would say that accounting embraces all phases of business operation because it is used in most of them. Both motion study and time study, and accounting are tools that are used in the majority of organizations to assist in solving certain problems.

---

[1] M. E. Mundel, "Measures of Productivity," *Industrial Engineering*, Vol. 8, No. 5, pp. 24–26; W. G. Hines, "Guidelines for Implementing Productivity Measurement," *Industrial Engineering*, Vol. 8, No. 6, pp. 40–43.

Management Defined

Management, whether it is supervisory management, middle management, or top management, in industry, service, or government, may be described as a task that can be represented by the following nine steps. These steps are continually repeated, essentially in a cyclic fashion, with a feedback (a flow of information) taking place between and among the various steps as well as between the first and final steps.[2]

1. Determine quantitative objectives for a given span of time in conformance with guiding principles or assigned criteria, considering known limitations and freedoms of action.

2. Plan a series of programs (outputs to be produced in accord with a time-phased plan) for achieving the desired objectives.

3. Determine the workload (necessary activities that must be performed to accomplish the programs).

4. Determine the resources required to perform the workload, such as man–job time use, equipment time use, and material.

5. Acquire authority to employ these resources to perform the workload.

6. Use the resources to accomplish the workload, constantly constraining and improving the manner of use.

7. Determine the workload accomplished and the resources consumed, compare with the workload planned and the resources estimated as required, and take corrective action, as necessary, to cause performance to conform to plan.

8. Determine the amount of program accomplished, compare with the program planned, and take corrective action, as necessary, to cause conformance.

9. Determine the degree of accomplishment of objectives, compare with original objectives, and take corrective action, as necessary.

The cycle of managerial control described above is shown graphically in Figure 1.1, with only the major feedbacks indicated.

Human values. The defining of the task of management in such mechanistic terms is neither to deny the importance of such problems as motivating or leading people nor to deny the need to give weight to human values. The

---

[2]The definition of management and cyclic diagram are a slight modification of material extracted from M. E. Mundel, *A Conceptual Framework for the Management Sciences.* New York: McGraw-Hill Book Company, 1967, pp. 159–172.

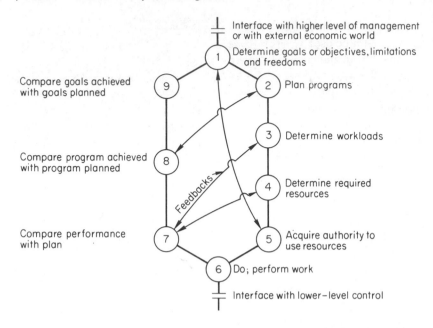

Figure 1.1. Cycle of managerial control.

mechanistic description is merely the definition of the matrix in which these problems occur. Much more will be said about human problems in Chapter 2.

Where motion and time study apply. It is in facilitating or assisting in the decision-making process in the performance of various of these steps in the cycle of managerial control that motion and time study procedures have their application. For instance, in performing step 2 of managerial control, *plan programs*, the problems of designing the outputs, selecting the materials, designing or selecting the tools, equipment, workplaces, and methods must be solved. Various motion study techniques may be usefully employed to assist in systematically solving the problems associated with performing this second step of managerial control. Of course, which techniques are of value will be affected by the nature of the outputs in question, the level of managerial control, and the degree of detail being handled at that level.

Similarly, in proceeding from step 2, *plan programs*, to step 3, *determine workload* (the setting forth of a list of the quantities of activities to be performed), the techniques of motion study will be of assistance. Again, which specific techniques are of value will be affected by the same factors mentioned with respect to step 2.

In proceeding from step 3, *determine workload*, to step 4, *determine required resources*, the techniques of time study are employed in determining the amount of man-time (and the use-time of any adjuncts to the work) required to perform the workload. These time study (or work measurement) data will again be used to assist in the performance of step 6, *do*, to constrain the actual performance of the work.

Subsequently, steps 7, 8, and 9, the steps of managerial control wherein the plans are compared to performance, will be assisted by information flow systems. The techniques of motion and time study will not only be useful in designing these information flow systems, but the time study data used to convert the data of step 3 to that needed for step 4 will be used to evaluate the new data carried by these information flow systems.

Summary. In short, motion and time study techniques are neither the managerial process nor a substitute for it, but are a series of techniques that may be usefully employed to assist in the performance of many of the steps in the managerial process. Naturally, in assisting the higher levels of managerial control, the techniques which are suitable will usually differ from those which are suitable for lower levels of managerial control wherein the attention is focused on smaller details. Also, the nature of the outputs will have an effect upon the suitability of any particular technique.

Simple Examples

It seems appropriate at this point to introduce a few brief examples of the results of the application of motion and time study to give further substance to the subject of this discussion.

Enveloping a plant magazine

ORIGINAL METHOD

Preparing the plant magazine shown in Figure 1.2 for mailing involved stuffing the magazine into a preaddressed envelope and turning in the flap of the envelope. The face of the envelope bore the sender's name in the upper left-hand corner, the second-class postmark in the upper right-hand corner, the address in the center, and postmaster's instructions in the lower left-hand corner. Thirty-one thousand of these magazines were stuffed into envelopes each week.

IMPROVED METHOD

The magazine was redesigned, doubled in size, and folded an additional time. One-half of the back page was used for all the information that had previously appeared upon the envelope. The improved magazine is shown in Figure 1.3. This enabled the company to put out a magazine of twice the size for less money than was involved in the original method.

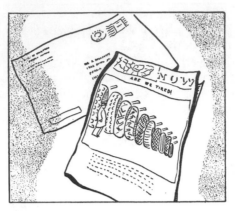

Figure 1.2. Original plant magazine and envelope.

Figure 1.3. Improved magazine format.

Assembling the bearing and oil seals in a heavy steel roller wheel

ORIGINAL METHOD

This roller wheel is used on the tailgate of heavy earth-grading equipment. In the original method, one of the oil seals was inserted first. Because there was nothing under this oil seal, it had to be hammered in edgewise and then turned into position with the thumb. This procedure caused the oil seal to enter farther into the bore than was required. After inserting the first oil seal, the operator turned this heavy steel roller wheel over and dropped the bearing into the bore. Since the first oil seal was farther in the bore than was necessary, it was necessary to take a ram and lead mallet and hammer the first oil seal back to its desired location, as shown in Figure 1.4. (For clarity of the illustration, only a few parts are shown on the bench.) The ram was

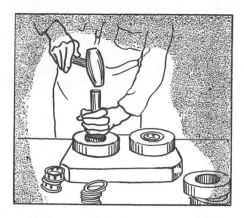

Figure 1.4. Original workplace for assembling roller wheels.

used to distribute the pressure over the entire surface of the bearing and prevent damaging it. After the oil seal was hammered back, it was possible to put in the second oil seal. Inasmuch as the bearing provided a support for this oil seal, it was easier to assemble than the first, even though it was identical.

IMPROVED METHOD

Two steel spacers were welded to the bench, as shown in Figure 1.5. (Again, for clarity of illustration, only a few parts are shown on the bench.) These spacers were the exact thickness of an oil seal. The operator first positioned the roller wheel over these spacers and then dropped the roller bearing into place. The spacer on the workplace held the bearing up in its

Figure 1.5. Improved workplace for assembling roller wheels.

proper position, permitting the insertion of the first oil seal in the same easy manner as was used on the second in the original method. After the first oil seal was in place, the roller wheel was turned over and the second oil seal was added as easily as was the first. This method was not only faster, in that an operator could achieve two and a quarter times as much production with the same effort, but easier, since the heavy work of using the lead mallet and ram was eliminated.

Sorting day-old chicks

ORIGINAL METHOD

As soon as the operator finished inspecting a box of chicks, he placed the lid on the box of inspected chicks, which was on the left side of the table. He then carried this box back to the rack of boxes of chicks that had been brought from the incubator. He then returned to the table, obtained the cover and the box in which the chicks had come (in the last inspection), and moved them across the table, placing the box on the left side of the table and the cover alongside the left of the table. Next, he again walked over to the chick rack and obtained a fresh box of chicks for inspection. He lifted the cover off this box and placed it alongside the table on the right. He lifted the chicks out of the box, four or five at a time, as shown in Figure 1.6, placed them on the table, and inspected them. He placed the first-grade

Figure 1.6. Original method of inspecting chicks.

chicks in the box to his left and, when necessary, stepped back a foot and turned around to deposit the second-grade chicks in the box behind him. Rejects were thrown in a can under the table. He continued this until the hundred-odd chicks in the box were inspected. The cycle was repeated until all the boxes on the rack were processed.

IMPROVED METHOD

The operator was seated on a stool. The rack of chicks had been brought up close to him. It was now possible without getting up to turn around in the chair and pick out a fresh box of chicks or replace a box of inspected chicks in the rack. Using the new method he placed the box of chicks for inspection on the table directly in front of him and inverted the cover over the box at an angle, as shown in Figure 1.7. He placed a piece of paper over

Figure 1.7. Improved method of inspecting chicks.

the holes in the cover of the box and, with two scoops of his hands, lifted the entire 25 chicks (there are usually more than the nominal 25) out of one of the four compartments in the box and put them on the inverted cover. Both hands were engaged in sorting the chicks. After the 25 were placed on the lid in front of the inspector, he picked out the obviously good chicks and placed them in the compartment from which he had withdrawn them; the obviously second-grade chicks he put into the box to his right. Those which were doubtful he left for further scrutiny before placing them in the proper container. If the number in the box was too low, he added chicks from the box to his left. By performing the operation in this new manner, the operator was able to inspect 95 per cent more chicks than with the original method, and the inspection was better.

Handling work to sandblast room

ORIGINAL METHOD

As shown in Figure 1.8, the materials handler worked at a chipping bench until the sandblaster signaled a need for more work. The handler then went to the unloading spot, where the painters had left an empty skid,

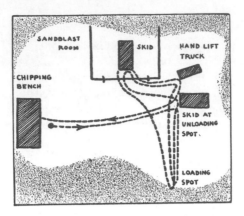

Figure 1.8. Plan view of factory floor showing path of materials handler with original method.

dragged the skid to the loading spot, where the process conveyor ended, and loaded the skid with parts that were ready for sandblasting. Next, the worker went to get the hand lift truck and pulled it to the sandblast room. The material handler opened the door of the sandblast room, placed the lift truck under the skid, and lifted the skid of blasted parts. This skid was moved to the unloading spot and lowered. The lift truck was then taken to the freshly loaded skid, used to lift it, move it into the sandblast room, and lower it. Finally, the lift truck was removed (which would be damaged if sandblasted), the door of the sandblast room lowered, and the truck stored. The worker then returned to chipping, traveling 244 feet to do the total task.

IMPROVED METHOD

As shown in Figure 1.9, the skids and lift truck were replaced with special wheeled buggies. On the signal from the sandblaster the materials handler went from the chipping bench to the unloading spot, rolled the empty buggy to the loading spot and loaded it. When it was loaded, it was rolled to the sandblasting room, the door opened, the buggy of blasted parts rolled out, and the new buggy rolled in. The material handler then stepped out, closed the door, rolled the buggy of blasted parts to the unloading spot, and returned to chipping. The new method saved 47 per cent of the material handler's time and required only 128 feet of movement.

Manufacturing a transformer core

ORIGINAL METHOD

Flat rings were punched from soft iron stock. These rings were sprayed on both sides and bound into the core shown in Figure 1.10. In the illustration, a few rings are turned up to show the construction. This process involved

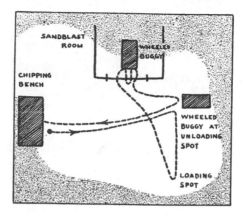

Figure 1.9. Plan view of factory floor showing path of worker with wheeled buggies.

Figure 1.10. Original version of transformer core.

several handlings, punch presses, assembly, and considerable scrap in the form of punch-press skeleton.

Improved Method

The core was constructed by winding a long narrow strip of slightly different metal, with laminations parallel to the axis rather than perpendicular as in the original design. The new core is shown in Figure 1.11. The end is loose to show the construction. The new material could be sprayed in continuous strip form. By eliminating several handlings, punch-press work, and scrap, this new method saved $131,000 per year.

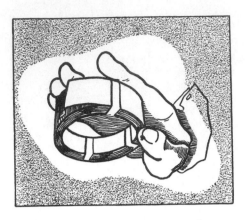

Figure 1.11. New design of transformer core.

Citrus picking.[3]   Citrus fruit was chosen[4] because it is Israel's leading export. The investigation was carried out at the request of the grove owners, the agricultural workers' union, the Citrus Board, and the Ministry of Agriculture.

Scores of workers in selected and varied groves were studied for some weeks by observers who recorded the methods used and kept a close tally of the output and, with the aid of stopwatches, of the time spent on picking, resting, and delays caused by shortage of equipment (boxes, ladders, etc.). Experiments were made with various methods of picking, and one method was finally selected as superior. Production norms based on time studies were set for various types of groves. The data were so extensive, so well collated, and so thoroughly analyzed that at one sitting the results were accepted by both the association of grove owners and the agricultural workers' union.

In cooperation with the adult vocational training division of the Ministry of Labor, courses were held for over 250 grove foremen of a total of 600, showing them the selected method and training them to instruct workers. Colorful instructional posters were displayed at all groves. A simple, profusely illustrated brochure describing the selected method and containing other work aids was printed in three languages (Hebrew, Yiddish, and Arabic) and distributed to all the workers, most of whom were new immigrants with little or no literacy.

Although all groves did not make all the necessary changes, the industry as a whole employed during the season 7,000 pickers instead of the 9,000

[3]H. Fish "Raising Productivity in Israel," *International Labor Review* (Geneva, Switzerland), Vol. 68, Nos. 4–5.
[4]By the Israel Productivity Office.

employed the season before, and 40 per cent more exportable fruit was picked.[5] Thus, there was not only a substantial increase in productivity but also a decrease of spoiled fruit. Where payment by results was introduced, worker's earnings were increased by 20 to 40 per cent.

Motion and Time Study Applied to the Manufacture of a Whole Product

The preceding illustrations have been primarily confined to three types: (1) improvements on a single operation on a product, (2) improvements on some of the steps on a product, and (3) improvement of a component of a larger product. Despite the magnitude of the typical results, the reader may be wondering what the net result might be if the techniques were brought to bear upon the whole of the manufacturing of a more complex product or a larger product, or the whole of a manufacturing activity or the whole of a governmental operation. The next five examples, therefore, instead of illustrating the details of the changes, consist primarily of a presentation of the net results obtained when various motion and time study techniques were applied on a larger scale than in the preceding examples.

Diesel-engine fuel-pump manufacturing plant. Various motion and time study techniques were applied to assist in improving the product design, process, equipment, labor utilization, and product flow. The largest change was with respect to the machining sequence. The production of the plant was raised from 20,000 units per month to 60,000 units per month without changing the number of employees and without extensive automation.

Shipyard. Various motion and time study techniques were used to develop time standards and controls so that the number of labor-hours could be determined in advance of the work and labor effectively dispatched. The labor required to erect a ship was reduced by 45 per cent.

Office of the Solicitor, U.S. Department of the Interior. Various motion and time study techniques were applied to assist in forecasting the workload that could be anticipated and in converting this into a statement of the required number of lawyers. Subsequently, procedures were designed for using these data for monitoring the output and backlog of the approximately 200 lawyers working in 29 offices in the United States. These data were used

---

[5]There were no dismissals, since picking is a seasonal occupation and the normal turnover of labor from season to season exceeds 75 per cent [Fish's footnote].

as a guide in the shifting of lawyers among the locations in order to keep abreast of workload. Over a 6-year period productivity increased by 25 per cent.[6]

Livestock Market News Service, U.S. Department of Agriculture. Time study technology was applied to provide a basis for determining the staffing needed for each of the 42 cities in which the Livestock Market News Service maintained an office. These data permit the staffing to be rationally adjusted when the workload in any market area shows significant changes. These data also assist in the determination of the additional cost of extending the Service to an additional market area or the determination of which services must be curtailed when funds are short. They also permit the identification of offices whose workload is such that they are convenient places for training new employees.

Large manufacturer of automotive electrical components. Joseph Lucas Ltd., the largest producer of automotive electrical equipment in the United Kingdom, applied motion and time study techniques as a fundamental tool in an effort to increase productivity and reduce costs. Goals are set and progress measured in terms of indices. Three of the indices used are as follows:

1. Technical index $= \dfrac{\text{rationalized time}}{\text{current standard time to produce product}} \times 100$

The rationalized time is the estimated standard time considering possible improvements in design and production methods. This index reflects motion study effects.

2. Labor index $= \dfrac{\text{current standard time}}{\text{current performance time}} \times 100$

The current performance time reflects shortcomings such as waiting for materials, machine breakdown, or failure on the part of the operators. This index reflects time study measures and their uses.

3. Productivity index $=$ labor index $\times$ technical index

This index combines the effects of motion study and time study.

It is to be noted that all these indices are designed so that the value of the index when the goal is achieved is 100 per cent. The progress of Joseph

---

[6]M. E. Mundel, *Measuring and Enhancing the Productivity of Service and Government Organizations.* Tokyo: Asian Productivity Organization, 1975, pp. 113–132.

Lucas Ltd. on one product, as measured with these indices, with an intensive motion and time study program, is shown in Figure 1.12.

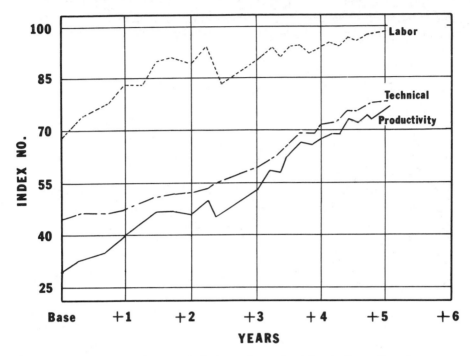

Figure 1.12. Labor, technical, and productivity indices, Joseph Lucas Ltd., over a 7-year period.

Discussion of the Examples

Numerous other illustrations could be cited. Motion and time study has been applied to improving or designing the pattern for activities ranging from the loading of naval weapons to the cleaning of cow barns. However, the foregoing detailed examples indicate a fair variety of applications.

For the purpose of showing the effect of using the methodology of motion and time study, in all cases, existing jobs that were improved were used as illustrations. However, it should be understood that the techniques of motion and time study are most useful, even if less dramatic, when they are employed to design an effective method prior to the outset of an activity.

The examples and later chapters. As the various techniques of motion study are examined in later chapters, the procedures used to develop many of the improved methods described in the illustrations will be given in detail to demonstrate how almost anyone familiar with these techniques could have arrived at similar results. These procedures are equally effective for the initial design of efficient methods. In the sections on time study, the methods used to evaluate the magnitude of the change will be given. In the section on application, the means of seeing the new methods adopted as practice will be discussed.

PROBLEMS

1.1. Determine what applications of motion and time study have been made in an organization you are familiar with.

1.2. Give a short résumé of the lives and work of F. B. and L. M. Gilbreth.

1.3. Describe F. W. Taylor's experiments at the Midvale Steel Company.

1.4. For what accomplishments is the Gilbreth Medal awarded by the Society for Advancement of Management?

1.5. Describe the accomplishments of one recipient of the Gilbreth Medal.

1.6. For what accomplishments is the Frank B. and Lillian Gilbreth Award made by the American Institute of Industrial Engineers?

1.7. Describe the accomplishments of one recipient of the Frank B. and Lillian Gilbreth Award.

1.8. Taking size and color into account, as well as other variables, how many products appear in:
(a) A Sears Roebuck catalog?
(b) A Montgomery Ward catolog?
(c) A Spiegel catalog?
(d) An A.S.M.E. mechanical catalog?
(e) Any catalog (as assigned)?

1.9. What are the quantitative objectives, limitations, and freedoms of:
(a) Your present or past employer?
(b) Your university?
(c) Your immediate or previous organization?

1.10. What means are used to determine the achievement of objectives by the organization cited in answering Problem 1.9?

1.11. Suggest and describe, from your own knowledge, a situation where method changes would appear appropriate.

1.12. Suggest and describe, from your own knowledge, a situation where the manpower resources employed appear:
(a) Inadequate.
(b) Excessive.

1.13. In what professional societies would we find people whose major interest is in motion and time study work?

1.14. Find and summarize an article in the periodical literature (within assigned dates) to describe the application of motion study to a situation.

1.15. Through what channels should a proposal for a method change be made in an organization you are fimiliar with?

1.16. Describe and evaluate the procedure used by the postal employee the last time you:
(a) Purchased stamps.
(b) Sent a registered letter.
(c) Sent an insured parcel.

1.17. Describe and evaluate the procedure used to gas a car at a gas station.

2   THE HUMAN FACTOR

Introduction

The major part of this book is devoted to a systematic discussion of techniques and procedures for the development of improved work methods and for the setting of standards for human performance. As indicated in Chapter 1, these techniques play a vital role in the managerial process so necessary for the effective control of human undertakings. These techniques have had a vital role in increasing the productivity of undertakings in the industrial, service, and government sectors of the modern world. The continued growth of these undertakings—growth in size, complexity, and variety—increases rather than decreases the need to apply these techniques. The growth of enterprises also creates a need for new motion and time study techniques. Beginning with Chapter 6, in keeping with the subject of this book, the emphasis will be on the techniques of motion and time study rather than on what is called the "human factor," although there is no intention of minimizing its importance.

20

People and Motion and Time Study

In that people are a basic element of any enterprise, any discussion of either the old or new techniques must include an examination of the interaction between these techniques and people. However, rather than scattering the material relating to the human factor throughout the book or repeating it in each chapter, it has been gathered together in this chapter, which should be considered as a prologue to each of the chapters that follow.

Variety of effects of motion and time study. In any activity the application of the techniques of motion and time study has profound effects upon the conditions and wages of employment. These effects may be classified as long- or short-run effects; the actual impact may be economic, emotional (psychological or sociological), or mechanical (physiological), or more likely a combination of these. However, it should be noted that people do not react to this sort of impact as purely economic, emotional, or mechanical men, but to the combination of these aspects as perceived by them; their reactions are influenced by past experience, mental outlooks, social relations, general economic conditions, and so forth.

Impact of innovation. In many cases, the very introduction of an innovation or change in the way of doing work, whether the change is in the work method or in the managerial control procedure, substitutes the unknown for the known. The unknown creates a generalized fear often difficult to isolate. In other cases, ego problems created by the obsolescence of a skill or by a change in managerial control procedures which reduces or appears to reduce the freedom of action of the employee may enter into the main problem. Attitudes may also be affected by the real or imagined economic aspects of such a change. Externally imposed change also implies criticism of past practice, and people resist the change to avoid the acceptance of the implications. All these aspects may be confused in a problem of innovation; motion and time study techniques are essentially systematic ways of finding innovations. Conflict may ensue unless proper measures are taken.

The Hawthorne Study

The complexity of the interaction of all the factors affecting the reaction of a person to his job (and innovations) was superbly illustrated by a classic experiment performed at the Hawthorne Works of the Western Electric Co. in the late 1920s.[1]

---

[1] G. A. Pennock, "Investigation of Rest Periods, Working Conditions and Other Influences," *Personal Journal*, Vol. 8, No. 5, 1930.

This experiment was performed to obtain answers to the following questions:

1. Do employees actually get tired out?
2. Are rest pauses desirable?
3. Is a shorter working day desirable?
4. What is the attitude of employees toward their work and toward the company?
5. What is the effect of changing the type of working equipment?
6. Why does production fall off in the afternoon?[2]

Experimental procedure. In this experiment, six employees (five assemblers and one material handler) were selected from a group of 100 assembling telephone relays. Previous to the selection, it had not been customary to keep individual production records and the entire group of 100 was paid on the basis of the output of the group. For a period of 2 weeks a record was kept of the individual production of the six selected employees to provide a base record. This record was kept without their knowledge.

The six employees were then told of the purposes of the experiment and, after agreeing to participate, were moved into a test room physically separated from the large group. They worked in this test room for 5 weeks, without any other changes in the conditions of work, with no change in the basis of computing pay, in order to establish a base record for the new conditions.

Changes introduced. For the next 8 weeks they were paid on a new basis, a group rate based on the production of the five assemblers in the test room rather than on the basis of the whole group of 100 as previously. Production increased by about 4 per cent.

For the next 4 weeks, the group was given 10-minute rest pauses twice a day, and again daily and weekly output rose.

For the next 4 weeks, the group was given six 5-minute rest pauses, further reducing working time. The employees expressed some dislike of this constant interruption of the flow of work, and the output curves showed a small decrease.

For the next 11 weeks, the group was given a 15-minute rest pause in the morning with refreshments supplied by the company, and a 10-minute break in the afternoon. Despite the loss of production time, output returned to the highest level previously obtained in the experiment.

For the next 7 weeks, the group stopped a half-hour earlier each day. This was accompanied by a spectacular rise in both daily and weekly output. (At this point in the experiment, unforeseen circumstances necessitated replacing two of the operators.)

---

[2]Ibid., p. 297.

For the next 4 weeks, the working day was shortened by an additional half-hour. The hourly output rose so that the daily and weekly output diminished only slightly.

For the next 12 weeks the group returned to working the full day but retained the 15-minute midmorning refreshment break and the 10-minute afternoon break. Daily and weekly output reached a new high.

For the next 12 weeks, the half-day of work on Saturday was eliminated. Daily output increased. Weekly output fell somewhat but remained above all but two of the previous highs.

For the next 12 weeks, the workers returned to all but one of the original conditions of work; no rest pauses, no refreshments, no shortened hours, but they were paid as a group of six. The daily and weekly outputs rose to record highs and continued to rise throughout the entire 12 weeks.

For the next 31 weeks, the group returned to rest pauses with the company supplying the morning-break beverage and the employees supplying their own food. The output rose again, establishing a new high.

Conclusions. It became obvious that more was happening than was attributable directly and mechanically to the experimental changes. Mr. Pennock, in charge of the investigation, explained, "The results are mainly due to changes in the mental attitude."[3]

Elton Mayo further explained the increases in productivity as follows: "The supervisor took a personal interest in the employees and their achievements; he showed pride in the record of the group. He helped the group to feel that its duty was to set its own conditions of work (oriented with the organization's objectives); he helped the workers find the 'freedom' of which they so frequently spoke in the course of the experiment."[4]

Communication and Its Effects

The interpolation inserted in the preceding quotation has great importance. Implicit in Mayo's remarks is the assumption that the supervisor (representing the organization) and the workers understood each other and *communicated successfully*.

> Successful communication between individuals depends upon something more than a common language, a common set of words. People and groups with different experiences and social places, although having in common many of the same words, may vary widely in mental attitudes. These differences in modes of thought and ways of viewing

---

[3]Elton Mayo, *The Human Problems of an Industrial Civilization.* New York: Macmillan Publishing Co., Inc., 1933, p. 70.

[4]Ibid., p. 71.

things may make communication in some instances almost impossible. The trained expert with his precise and logical vocabulary has difficulty in communicating with the layman. The customary ways of thinking of the skilled toolmaker, for example, are quite different from those of the non-machine minded unskilled worker. They differ also from those of the engineer, the accountant, the marketing expert, the executive, or the administrator. As it is commonly expressed, people with different ways of thinking do not "get" each other.[5]

In brief, the Hawthorne experiment resulted in increases in productivity because of mutual understanding and acceptance of the changes as compatible with the objectives of both parties.

Basic Problem Areas

Indeed, it may be taken as axiomatic that the members of a group will not cooperate willingly in an activity they do not understand.[6] It should be obvious that the application of motion and time study has not only problems of communication, but that the communication must result in motivation, that is, an active acceptance of the change.

Those applying the techniques of motion and time study may have one objective in mind; those to whom it is applied may "understand" this somewhat differently. Even if the understanding is correct, the evaluation of the suitability of the objective may differ, and motivation to a state of active participation may become a difficult problem.

In substance, whether the changes resulting from motion and time study will really assist the worker or be an advantage to him is not the only problem. Of importance also is what the worker thinks the changes will do to him. The problem is further complicated because in the past, in all too many cases, motion and time study activities have been used in an unscientific fashion as cover for programs requiring excessive effort from employees. (Even the determination of whether effort is excessive or not is not a simple problem, as will be seen in subsequent chapters.) Also, in some cases motion and time study activities have been allowed to cause some unemployment; this fear is frequently uppermost in the worker's mind. In the particular area of work within the comprehension of the worker he may know that there is much excess manpower long before management begins to apply motion and time study techniques to determine the amount of manpower really

---

[5]F. J. Roethlisberger, *Management and Morale*. Cambridge, Mass.: Harvard University Press, 1941, p. 62.

[6]Mayo, op. cit., pp. 119–120; see also A. Zander, "Resistance to Change—Its Analysis and Prevention," *Advanced Management*, Vol. 15, No. 1, pp. 9–11.

required. The fear of potential unemployment may have a real basis. Even if management has established a reputation for moving excess workers to other work rather than discharging or laying them off, there is still the fear of change unless active steps are undertaken to minimize it.

Innovating in New Areas of Application

When motion and time study techniques are applied to work other than direct work, one often encounters a deep-rooted skepticism with respect to the useful potential for change. In government activities one hears the remark, "But we don't make shoes!" With other indirect groups the common retorts are, "Our work can't be planned!"; "You can't measure us with a stopwatch!" Some of these remarks may be true, although they may not have any bearing on the suitability of applying motion and time study techniques. They all reflect a variety of fears; fears that undue regimentation will ensue, fears that undue effort will be required. These fears may not be related to a fear of unemployment or to a preference for an easy way of working. Indeed, the more the true attitude reflects dedicated effort, the more they may fear the consequences of motion and time study. They fear the unknown.

General Philosophy

Many will agree that any activity that does not benefit society deserves to fall by the wayside. Where the general effects are desirable but the local effects entail hardship (whether it is real or imaginary makes no difference) for some individuals, the activity should continue but a means of lessening or removing the undesirable results must be found. Only the most socially blind persons will point to the long-run effects of motion and time study and disregard the few individuals who may suffer, or expect these few individuals to submit willingly to hardship "for the good of the group." Calling unemployment "temporary technological unemployment" neither makes it more pleasant nor assists in communicating in a manner that will gain mutual cooperation in effecting changes. In many plants, trades, and organizations that are backward in method and procedure, this fear of unemployment often forces retention of needlessly uneconomical ways of doing work.

Social needs. It must, however, be borne in mind that unless working methods continually improve, unless each working hour continually becomes more productive, a society will become economically static, and the standard of living will no longer rise. Indeed, such a condition includes factors that will inevitably depress the standard of living.

The employment of the techniques of motion and time study must not be considered a mere mechanical problem but must be so carried on and applied that it relates properly to the people who are affected.

A Simple Example

A crisis was occasioned in one plant with 1,300 employees by the introduction of a new method that made it possible for 15 employees easily to perform the work previously requiring 50 persons. The plant management averted a crisis by guaranteeing employment for the displaced workers, including carrying them on the payrolls until the normal turnover and expansion of production made vacancies for them in the plant. This was accompanied by an extensive education program concerning the need for the change and the consequences of not making the change. It was not sheer altruism, of course, that prompted this action. A trained worker, adjusted to the plant organization, represents considerable value. This action, which postponed only temporarily the obtaining of the benefits of the new method, set the stage for willing cooperation in plant-wide methods improvement.

A More Complex Example

In another plant a crisis arose during contract negotiations. Although the plant did not have a piecework system, they did have time standards and expected the direct production workers to perform at some reasonable per cent of these. The union, during the contract negotiations, charged that the standards were capricious; that some were relatively easy to attain; that others were impossible. For this reason the union requested that all time standards be abolished and that all time study work cease. The company felt that time standards were so vital to their continuing in business that they could not yield on this point. They had the standards evaluated and found that the charge of excess variation was valid. They offered to restudy all tasks and set more consistent standards. The argument was protracted. A federal mediator failed to bring the two sides into some sort of agreement.

The conflict situation. A strike ensued. After 6 weeks the union agreed to go back to work, with the understanding that the company and union representatives would later work out a mutually agreeable clause with respect to time standards and time study. These negotiations were protracted. The union would not budge from its position that time standards and time study must be abolished. The company was adamant in its need for such standards. During the negotiations the company exerted a large effort to review and correct all time standards. At the same time they examined all cases of failures to attain a reasonable performance with a cause-seeking rather than a fault-finding attitude. No disciplinary actions were taken. The plant remained sullen; time study observers were reviled; negotiations continued without progress. This state of affairs continued for 6 months.

A new approach; successful communication. Finally, the company tried a new approach. They set up a group of meetings with the employees. The 1,100 employees were taken in groups of 50 during paid-for working time. The plant manager, personnel manager, the chief industrial engineer, the area time study observer, and the appropriate supervisor attended each meeting. Each meeting lasted 1 hour. During the first 30 minutes the employees were shown a short series of 20 slides accompanied by a tape-recorded talk explaining the time study methods used and the reasons time standards were considered so vital to the company. During the second 30 minutes the management representatives answered all questions. If unanswered questions remained at the end of the assigned hour, the employees were promised an opportunity to attend another meeting, and this meeting was held at the end of the series of meetings.[7] The results were immediate and profound. The whole attitude of the plant brightened; time study observers were greeted and given cooperation; the union withdrew its request for further negotiations; the efforts of the company to make the standards more consistent were accepted as the correct approach.

A Successful Program in a Service Activity

The Consumer and Marketing Service of the U.S. Department of Agriculture has had an extremely successful motion and time study program. In this case the success may be traced directly to the involvement of line managers and operating people in the studies rather than having them done by technicians.

Generalizations

In all cases, some solution is possible, although it may differ from the ones just given. It surely seems foolish to continue to burden ourselves with

---

[7]The slides and tape, of course, had been shown to all managers (including foremen) before the start of the meetings with the employees.

the hard way of doing work when more economical methods are feasible, but are not adopted because we have not learned to install them cooperatively. Toil is certainly not that sweet. Also, pride in work must be expanded to include not only pride in what is done but also pride in doing it as economically possible.

Hawthorne. It should be noted that the results at Hawthorne, referred to earlier, were greatly influenced by the group's feeling that it was setting its own conditions of doing work. This, unfortunately, does not point the way toward an easy means of effectively employing motion and time study techniques in all applications. As will be seen in the next chapter, the most effective methods may require changes in factors, many of which are far beyond the scope or knowledge of the worker. Hence, it is not always possible to give the worker direct participation in developing the innovations.

Application prior to the inception of work. At this point the advantage of designing effective methods prior to the outset of an activity should be obvious. Problems of resistance to change, worker displacement, and retraining may largely be avoided. Only through the use of the techniques of motion and time study, as a design tool, may such a goal be approached. However, there is no limit to the perfection of methods. Therefore, designing effective methods prior to beginning work only minimizes the problem of introducing change; it does not eliminate it.

Management-originated problems. While the preceding discussion has emphasized worker problems, this is not the whole picture. Management resistance to change is often as great. For motion and time study to achieve its fullest and most fruitful use, it must be thought of as a series of techniques applicable to problems affected by all the functions of all the individuals of an organization: from workers to sales and design engineers in a plant; from buyers to delivery persons in a department store, and so forth.

Basic requirements for successful use. For all these functions to be properly integrated into an economical operating pattern, supervisors at all levels must be educationally equipped to cooperate intelligently in the motion and time study approach, even to the extent of actually applying motion and time study as a normal part of their activity. They must also be emotionally susceptible to change and innovation and have adequate executive direction, the latter greatly conditioning the former. For any executive adequately to direct or participate in an activity, the executive usually must be well acquainted with the procedures used, their capacities and limitations. To aid subordinates in performing their work properly, the executive should have a good grasp of how their work should be performed. Consequently, we may say that the top executive attitude toward change and the procedures used to bring about change is one of the important human factors affecting the success of the motion and time study work. This holds true with respect to both the aspect of improving methods and that of designing methods.

Reporting requirements. Furthermore, a motion and time study department must actually justify its existence in an organization by reporting on its achievements. Because a well-operated motion and time study group usually cooperates in achieving a desired result with other staff or line groups, these reports pose a real problem in that they must not confuse the true role played by each group or individual concerned. Also, an effective motion and time study group will have its effect "before the fact"; they will participate in developing effective methods prior to the start of work. Their achievements will be "cost avoidance" or "excess manpower avoidance" rather than cost or manpower reductions; the improvement may merely make it possible to accomplish more without increased cost. A reasonable evaluation of the group's efforts in such circumstances poses an even more complicated problem.

Summary

In summary, it is suggested that, because of the "human factor," a motion and time study group or an individual engaged in performing such functions must not only execute the necessary technical activities in a sound, accurate fashion, but also actively take part in furthering the integration of motion time study into the organization by:

1. Disseminating motion and time study information throughout the whole organization, not only to overcome the normal resistance of people affected by it, but also to aid everyone in the organization to cooperate in finding better ways of doing work. Communication must be maintained with each working group, to permit their objectives and problems to be understood and integrated into the design of effective applications. Formal training courses, supervisor conferences, information via the house organ, and person-to-person discussions of each issue are desirable. The ideal situation occurs when innovations are originated, or at least participated in, by persons as close to the point of application as possible.

2. Submitting adequate financial and narrative reports of its activities to management, properly crediting cooperating, participating, or originating individuals or groups.

3. Actively seeking equitable solutions for possible hardships connected with technological changes.

Also, because of the "human factor," all members of the organization should actively undertake to understand the procedures and techniques of motion and time study. In the typical manufacturing organization practically all members of the "production" group are able to understand the drawings used to describe the product. For motion and time study to be of greatest value, all members of any organization must similarly be able to read the job designs, process designs, and control system designs described with the procedures and techniques of motion and time study. Thus, they may equip themselves to take part in cooperative action for finding equitable ways of raising the standard of living through more effective production, without increased effort, and without hardship.

PROBLEMS

2.1. Job methods training (JMT) was an important phase of the Training Within Industry program of the War Manpower Commission during World War II. What features of JMT made it as effective as it was?

2.2. If you knew of a better way to perform a common household task, describe in detail the procedure you would use to persuade someone to adopt it. (Do not detail the improved method.)

2.3. Try the approach and report on the effects and results.

2.4. Report on the effects after a lapse of 1 week.

2.5. Interview an industrial production worker and report on his or her concept of motion and time study. Analyze and evaluate the validity of any conclusions the worker may have come to.

2.6. Report from your experience on the reactions that accompanied a change of method on a job.

2.7. Interview an industrial engineer or an industrial engineering technician and describe his or her place in the organization.

2.8. Report on a pertinent reading from the literature in sociology or industrial psychology.

2.9. Report on the changes in the past 50 years with respect to:
(a) Employment in the manufacturing industries.
(b) Employment in the service industries.
(c) Employment in the government service:
    1) Municipal.
    2) State.
    3) Federal.
    4) Total.
(d) Employment in the agricultural sector.
(e) Horsepower per employee in the manufacturing industries.
(f) Purchasing power of money.

(g) Value of output per employee-year in the manufacturing industries:
    1) Gross, unweighted.
    2) In terms of a constant-value dollar.
(h) Life span of human beings.
(i) Population of:
    1) Your country.
    2) The world.

2.10. With respect to the use of motion and time study, of what importance are the changes noted in the examination of the assigned item from Problem 2.9?

2.11. Examine the news media (of an assigned period) and report on a labor dispute related to method or workload and assigned manpower.

2.12. Suggest the solution you would have proposed for the situation described in Problem 2.11.

2.13. Report on your attitude toward potential changes in your workload. Explain your rationale.

2.14. A group of 15 executives have been in the habit of dictating letters to their secretaries, who record them by shorthand. It is proposed to use tape-recording and transcribing equipment (suitably designed). Prepare the speech you would use to make this proposal to:
(a) The executives.
(b) The secretaries.

2.15. List the possible objections to the proposal of Problem 2.14 by:
(a) The executives.
(b) The secretaries.

2.16. List your responses to the objections of:
(a) Problem 2.15a.
(b) Problem 2.15b.

2.17. In addition to the changes of Problem 2.14, it is further proposed to create a pool of secretaries with a pool supervisor. This pool would serve all the executives rather than any one secretary responding to one executive. What do you think of this proposal? How would you effect this change?

2.18. *The incentive problem.*[8] A junior time and methods worker has set a method and rate (standard time for a job) in the finishing department which has created considerable ill feeling. He has asked for 320 pieces per hour per operator. Since the work is on wage incentives (100% standard hour plan), and since none of the operators have come up to standard, they are a little disappointed, to say the least. On your way up through the department (you are head of the motion and time study staff) to talk with the supervisor of the aforementioned department, who has asked for a consultation with you, one of the operators says to you, "This is a — of a rate Block set!" (Block is the junior.) You tell the operator that the rate will be looked into.

---

[8]From M. E. Mundel and L. C. Pigage, *Case Problems in Industrial Organization and Management.* Ann Arbor, Mich.: Edwards Bros., p. 43.

The supervisor first practically cries on your shoulder, asking, "How can I get my department up to standard when I get such nutty rates as this? I can't even honestly ask them to do it for me." The supervisor turns to the assistant and says, "Moon, do you think it is a fair rate?"

"Heck, no," says the assistant, "and you guys (turning to you) usually do a good job. Now you take that job we are doing for Smith and Co. It is almost like this one and you don't call for anywhere near as much production. Do you?"

"You make our job tough," says the foreman.

You soothe both of them and tell them that you will restudy the job yourself but ask to reserve discussion and decision until then. You promise to give it your unbiased attention.

The assistant asks: "Does that mean you are just going to look at it and then swear that your department was right the first time?" This is said in a joking manner and not offensively. Once before, a question similar to this came up and you upheld the rate. However, at that time the shortness of the production run precluded the chance of a real test. They think that perhaps the restudy was just a blind.

You tell the two of them that you will fully consider the job. The assistant is really worried about his crew and in earnestness assures you that instead of being 320 per hour, the standard should be 285 per hour. You tell him that you will look into the job but that you cannot commit yourself now.

You restudy the job carefully and take a new stopwatch time study. After the inclusion of all normal delays, personal time, and allowing for the tiringness of the task you find that your figures ask for 287 per hour. Block missed several highly intermittent but important parts of the operation. Since your policy requires production standards in multiples of five when over 200 per hour, this figure of yours would reduce to 285 per hour.

(a) How intelligent is the policy of using multiples of five for production over 200 per hour? Why?

(b) Is there any potential trouble in this situation? Explain.

(c) Would you merely issue the rate change and say nothing? Explain.

(d) Would you issue the rate change? If so, as of the date or retroactive?

(e) Exactly what would you do? (Make a scientific analysis of the problem.)

2.19. *The typewriter.* A friend who works in the same organization you do has been bringing his reports to you for comments and editing before making a final copy. He works on them at home at night. Your comments are often extensive but are always welcomed and accepted. The last report required so many corrections to the equations and suggestions concerning changes in phrasing that, in some places, you could hardly squeeze them in between the double-spaced typed lines. When you were through discussing the notes with him you suggested that on the next report it would be advantageous if he would type with triple spacing. He replies, "I can't. My typewriter only has a 1, $1\frac{1}{2}$, and 2 throw return lever."

(a) Is it possible that he is really as stupid as he sounds?

(b) Explain what you think has really happened.

2.20. In order to save costs connected with paying the salaried personnel in your plant, you, the comptroller, are planning to pay these workers twice a month instead of weekly, as is now done. You plan to issue checks on either the 15th of the month, or the first workday following it, and on the last working day of the month. Describe, in detail, your approach to gaining acceptance of your plan.

2.21. The secretaries in your department have been using standard typewriters. You plan to introduce IBM Selectric typewriters with a moving element and a nonmoving paper carriage. Explain, in detail, the procedure you would use to effect the change.

2.22. The scientists in your laboratory have been, in your opinion, abusing the concept of compensatory time (time off from work to compensate for work done at home or on weekends). You are the chief. Explain, in detail, the approach you would use in reducing compensatory time.

2.23. The Bloodgood Meat Packing Company employs approximately 1,200 people on a one-shift operation. They kill between 400 to 600 head of hogs, 40 head of cattle, and 150 lambs per hour, although the workload fluctuates to some extent with livestock and meat-market prices. The company makes a full line of meat products, such as canned ham, bacon, bologna, liverwurst, knockwurst, spiced beef, and pork sausage, as well as selling meat.

On a potential order from the U.S. government for bologna the Company bid on the basis of $\frac{1}{4}$ of 1-cent profit per pound and was aghast when their largest competitor, noted for good, profitable marketing practices, received the order at a price below the cost of raw materials for Bloodgood bologna.

Bloodgood had been making bologna from 40 per cent beef chuck, 30 per cent D pork, 20 per cent C pork, 10 per cent A veal, and 10 per cent ice. An alternative formulation used by the company would have resulted in a higher price. It was known that their competitor paid approximately the same price for raw material as did Bloodgood. The sausage department assured the manager that the two formulas used by Bloodgood were the only possible formulations for bologna.

Your subsequent research into bologna making revealed that:

(a) Bologna could be made from many mixtures of many different cuts of beef, pork, and veal, provided that the resultant mixture had not more than 38 per cent fat and that the moisture did not exceed four times the protein plus 10 per cent.

(b) The price of each cut or grade of beef, pork, and veal varied daily, independent of the price of the whole animal.

(c) A series of equations could be written for the most economical mixture, taking into account the analysis and price of all potential cuts of all meats with respect to:
   1) Fat.
   2) Protein.
   3) Moisture.
   4) Texture.
   5) Color.
   6) Cost of the mixture.

(d) These equations could be solved simultaneously each day to obtain the minimum-price mixture for a grade A bologna typical of Bloodgood output and usually produced a formula that made them more competitive than their competitor.

Using this method of formulation, the company made and profitably sold a greatly increased quantity of bologna and satisfied all customers, but the chief of the sausage department maintained that: "The mathematically formulated bologna is inferior and I can tell the difference! So can anyone who knows good bologna!" On this basis he continually argued, at every opportunity, with respect to the cost-reduction program.

If you were the industrial engineer in this plant, what would you do about this situation?

3

THE SCOPE OF
MOTION AND TIME STUDY

Recapitulation

The improved methods described in Chapter 1 were the results of the use of motion study techniques, as defined there; the evaluation of the magnitude of the changes was the result of the use of time study techniques. The actual application of these data required the development of some type of instructional material from which the people were trained to do the job in the desired manner; also some procedure for planning and controlling their efforts. As was noted earlier, this book makes use of the term *motion and time study* to describe the entire field of activity. The terms *work simplification, work study, work standardization, work measurement, time study, standards work, motion study, methods research, value analysis, work design,* as well as many others, are in common use either for parts of this field or to describe the entire field. Although this is not an important issue, this book, for clarity, will use only those terms suggested in the first chapter, in the sense in which they were suggested.

Organization of This Book

This book is divided into five parts, as follows:

Part One.   General overview, background material, and introduction to motion and time study: Chapters 1 through 6.

Part Two.   The motion study techniques: Chapters 7 through 19.

Part Three.   The time study (or work measurement) techniques: Chapters 20 through 26.

Part Four.   The application of motion and time study data: Chapter 27.

Part Five.   Additional details, for serious students and practitioners, related to more complex aspects of the materials in Parts Two through Four: Chapters 28 through 46.

This is only a general breakdown because, as was noted in the first chapter, all of these aspects are interrelated.

Fundamental Philosophy

The fundamental philosophy of motion and time study involves three assumptions:

1. There are usually numerous ways to perform any task, but with the knowledge obtainable at any one time, one method is usually superior to the others.
2. The scientific method of solving problems is more productive of better work methods than is undisciplined ingenuity.
3. A standard of performance, or a time value for work, may be determined so as to permit dimensioning the required manpower inputs into any organization, thus permitting the creation of a true managerial design.

In other words, motion and time study procedures may be said to be based on the assumptions that for every job there is always, in the light of present knowledge, a "one best way,"[1] that a scientific method is the surest

---

[1] This phrase was coined by Frank B. and Lillian M. Gilbreth, the originators of much of our motion study. Some psychologists have objected to this attitude because, as they have well established, individual differences in people point to the desirability of individualized motion patterns. However, the differences between various alternative methods are usually so gross and the requirements of individuals so minute in contrast, that the one

way of determining this one best way, and that the time value of work may be measured in consistent units. It should be noted that the concept of the one best way does not rule out the determination of a still better method when our basis of preference, our knowledge, or our ability is altered.

The scientific method. The scientific method of solving problems involving the determination of a preferred way of doing a job, a preferred method of production, or a preferred method of doing new work requires the application of a logical procedure consisting of the following steps[2]:

1. *Aim*—determination of objective in terms of area of job to be changed, and establishment of criteria for evaluating the preferability or success of solutions.

2. *Analysis*—separating the whole of the work method into subdivisions or steps, pertinent to the job, appropriate to its scope, possessing known characteristics, or concerning whose performance information is already available.

3. *Criticism*—application to the analysis of basic data, or checklists of desirable arrangements of the steps into a preferable work pattern and of information concerning desirable ways of performing each of the steps.

4. *Innovation*—formulation (or synthesis) of a new suggested procedure for performing the work.

5. *Test*—evaluation, by means of the data previously used in step 3, of the desirability of the method formulated in step 4, with respect to the objectives set up in step 1.

6. *Trial*—sample application of the method tested in step 5 to ascertain the completeness with which all variables have been taken into account.

7. *Application*—final standardization, installation, evaluation, and use of the improved work method.

Areas of use. This general procedure is usually the same whether the work is in an office, in a light or heavy industry, on the farm, in a hospital, laundry, service station, library, or home, or in any government activity. It is merely the details of performing the various steps that change with the nature and requirements of the problem.

---

best way as selected is usually preferable for all the workers involved. Indeed, the question of individual differences is not even pertinent in cases concerning the flow of material through a plant, or the design of an output.

[2]These steps, which merely represent a pattern of thinking, could be divided into a greater or lesser number. The breakdown chosen seemed convenient.

Main Body of Knowledge

The main body of knowledge in motion and time study consists of procedures for each of the seven steps of the scientific method and may be grouped as follows:

1. Systematic means of selecting a feasible-appearing objective.

2. Procedures for analyzing work into appropriate units possessing known characteristics.

3. Checklists of desirable and undesirable features applicable to these analyses.

4. Means of describing the new method prior to performance.

5. Means of checking, prior to performance, the desirability of the new method.

6. Means of observing the initial application of a method and evaluating such observations.

7. Procedures for setting down work methods in such a manner as to aid in training people to follow the desired procedures; means of determining the time that should be required for the job and means of controlling the performance of the workforce.

Additional requirements. Before a systematic means of selecting a feasible-appearing objective can be discussed, we must first examine and divide the total area of application of motion and time study into reasonable sections. Such a division will provide additional terms for clearly describing our objectives.

Different Kinds of Changes

To improve[3] a work method it may be necessary to begin by introducing innovations or changes in any one of the five areas that affect its performance. These areas are:

1. *Human activity.* The hand and body motions or the perceptive or cognitive activity or their sequence may be changed to ease or improve the task.

---

[3]The use of the word "improve" does not necessarily imply an existing working situation. The method under study may well be in the planning stage, but it is still subject to "improvement."

2. *Work station* (*tools, workplace layout, or equipment*). The design of any single work station or the equipment used for any part of the task may be modified.

3. *Process or work sequence.* The order or condition in whch the various work stations receive the in-process output may require change or the number of work stations may be modified.

4. *Output design.* The product design or the form of goods sold or the material sent out (in the case of a substantive product) or the nature of the completed service (in the case of a service type output) may require either a slight or a drastic change in order to facilitate the attainment of the objectives of improvement.

5. *Inputs.* The incoming supplies or raw materials (for substantive products) or the incoming information (for service outputs) brought into the organization may require a change with respect to the form, condition, specification, or timing of the arrival to allow the desired improvements to be made.

Interrelationships among kinds of changes. A change in any one of these areas, other than area 1, usually involves changes in other areas with lower numbers. Indeed, the reason for introducing a change in output design may be to effect economies in one of the preceding areas; a change in inputs may be advantageous despite the increased cost of what is obtained if this increased cost is more than offset by better equipment usage, labor savings, or an improved service.

The situation, with substantive outputs, may be represented by a diagram, as in Figure 3.1. Essentially the same situation exists with service outputs, although minor changes in the words selected to describe the situation help clarify the concept. With service outputs, the "supplier" supplies

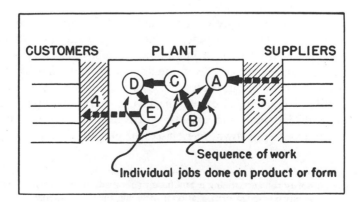

Figure 3.1. Areas affected by motion and time study.

information or a request. The "customer" is the one who receives the service; with government activities this person may not be a customer in the usual sense of the word. Further, with government services the supplier and customer may be the same person. However, the same five separable parts of the situation will exist and must be recognized as potential areas for change.

Classification of changes. To facilitate, in actual cases, the discussion of potential changes let us designate innovations involving a change in the manner of performing the work at any single location, such as A, B, C, D, or E in Figure 3.1, provided that the change of equipment would not cost more than $100,[4] as *class 1 changes* to correspond to the list of areas of change.

Class 2 changes are those introduced by a change costing more than $100 in the equipment used at any single location, such as A or B of Figure 3.1. Area 1 changes usually accompany changes of this type for maximum gain.

Class 3 changes are those introduced by a change of the sequence of jobs A, B, C, etc., brought about by adding, eliminating, combining, or changing the order of jobs. This frequently involves area 2 or area 1 changes as well, for maximum improvement.

Class 4 changes are those introduced by a change in the form or nature of the output (product design) moving across area 4 of Figure 3.1. Such changes will usually involve all lower areas of change as well.

Class 5 changes are those introduced by a change in some characteristic of the inputs flowing across area 5 of Figure 3.1. Such changes will usually also involve all lower areas of change.

Factors Affecting Class of Change Sought

The class of change sought on a particular job is a function of how far the analyst thinks it desirable, or feels able, to carry the improvement. In the examples given in Chapter 1, the amount of change varied greatly. In the case of the transformer coil, even the specification of the raw material and the design of the final product were changed; in the case of the chick sorting, nothing was changed except the layout of the workplace and the pattern of hand and body movements. These are the two extreme cases. Altogether, as has been shown, five classes of possible changes exist, with

---

[4]The $100 limit is arbitrary and represents a common level at which a supervisor's or foreman's authority for authorizing a purchase or requisition on a service department ends. If the limit in any particular organization is different, the line of demarcation between class 1 and class 2 changes should also be shifted.

Class of Change	Hand and Body Motions	Tools, Workplace, and Equipment	Process	Product	Raw Material
1	New	Minor changes	Same	Same	Same
2	New	New (Radical)	Same	Same	Same
3	New	New	New	Same	Same
4	New	New	New	Modified	Same
5	New	New	New	Modified	New

Note: The word *modified* appears in the "Product" column to emphasize that this is only a modification rather than a different product for a different purpose. Also, in any row, the item farthest to the right undergoing change, indicates where the change must usually be initiated.

Figure 3.2. Outline of the five classes of change possible with motion and time study.

the characteristics summarized in Figure 3.2. It should be noted that the higher the number of change, the more items usually changed, although there will be many exceptions to this general rule. Also, the higher the class of change, the more authority required, and the more people affected.

Organization levels and classes of change. For instance, a class 1 change may be initiated by almost anyone in an organization. A class 2 change may involve the superior of a first-line supervisor for authority to purchase, or the tool superintendent[5] for design of the jigs or tools. Indeed, a class 2 change may be initiated by the tool division. A class 3 change may involve all groups mentioned in the preceding classes, plus the process division; a class 4 change may require the additional cooperation of the engineering (or design) and sales division; and a class 5 change may include all the previously mentioned groups, plus the purchasing department. In service organizations equivalent groups will be involved, but the titles will vary over such a wide range that a single example would not appear to be of particular value.

Classification of changes as a guide. These five classes of change may be considered as a guide to areas in which changes are possible or as a guide to "organized brainstorming" as opposed to random hunting.

---

[5]The occupational titles given here are samples and may vary from plant to plant. They certainly will be different in service organizations.

Motion Study Technique Selection

Which analysis technique is to be used will be a function of the class of change sought and the physical characteristics of the subject of the methods scrutiny, together with some economic considerations. The techniques will be summarized in Chapter 4. Each technique will be treated later, in detail, in a separate chapter.

Time Study Technique Selection

At some stage of the analysis a suitable work measurement technique must also be selected. The technique chosen will depend upon factors such as the nature of the work, the time for each repetition of the work, the uses to be made of the time standard, and so forth. Time study, or the appraisal of work in terms of time, will be treated in summary fashion in Chapter 5. The detailed techniques will be examined, later in this book, following the motion study techniques. Time measurement is vital in the modern managerial process. Unless one can determine, in advance, how long work should take, the managerial functions are essentially based on "rule of thumb" rather than on fact. The newer and more sophisticated mathematical techniques of scheduling, or the planning of output mix, by means of linear or dynamic programming, frequently depend upon time study to give accurate coefficients for the variables in the required systems of equations. Time study in this manner influences many managerial decisions.

Changes May Be Extensive

Figure 1.12 showed the effects of an extensive motion and time study program on one product at Joseph Lucas Ltd. During this program raw material specifications were changed, designs altered, process sequence and factory layout changed, tooling redesigned, and operators trained to perform new jobs with new methods. Time standards were used as guides for supervisors, for manpower planning, work scheduling, and for payment by means of an incentive pay plan. The entire fabric of the industrial plant was subjected to change as a result of the motion and time study program.

Obtaining Cooperation

It should be obvious at this point that a motion and time study analysis may affect many different levels in an organization. If the techniques that follow are understood by all concerned, cooperative action for selecting economical operating procedures may take place much more easily than if all the initiative and action originate in a single staff department. The staff department may still be necessary to coordinate the work and to carry the burden on extensive projects, but motion and time study affects, and should be thought of as being part of, everybody's job.

PROBLEMS

3.1. List the class of change involved in each of the first six cases described in Chapter 1. Give the reason for your classification in terms of what was actually achieved in each case.

3.2. Find an application of motion study in the periodical literature (within assigned dates); make a brief résumé of the improvement; and indicate the class of change that was involved.

3.3. Use the scientific method in the following problem situations:

(a) *Gulliver's travels:* In "Gulliver's Voyage to Laputa," Swift's satire on Oxford and Cambridge, he pokes ridicule at the astronomers by describing how the savants at the University of Laputa claim to have observed a planet, spinning like the earth, but with two moons, one of which rises in the East and sets in the West, while the other moon rises in the West and sets in the East, despite the fact that both moons are moving around the planet in the same direction. The astronomers at one of the universities retorted that two of Jupiter's moons do behave in this fashion. Swift then pointed out that they were really Laputans, his terms for utter nitwits, to propose such a nonsensical situation.

1) Who was "off base"?

2) Defend your position.

(b) *The Iwakaze spinning mill.* The product of the mill is fine cotton and blended thread. There are 160 "sets" of spinning machines. Each machine has 60 spinning heads; each head spins a thread (pulls and twists a loose sliver of fiber into a thread) independently, and winds it onto a spool called a bobbin. Hence, each machine has 60 bobbins. The machine stops automatically when any bobbin is full. Each machine in a set of machines is spinning the same thread as the other three machines in a set. However, owing to the manner in which production of the various sizes of thread is changed from time to time, all sets spinning a given size of thread are not in the same general area. Some small changes are made to the schedule once each month. The machines are of various types; they are of German, English, American, or Japanese origin. All the machines in any one set are alike. All machines can spin all types of thread.

There are crews of doffers and spinners. Doffers remove all the bobbins from a set of machines when the machines have stopped automatically, put on empty bobbins, and restart the machines. The spinners patrol the operating machines and repair broken threads and replace the sliver supply cans. When a thread breaks, the head stops spinning. The other 59 heads on the machine continue. The only loss is the loss of production from the idle head until the thread is tied by the spinner, and a "short" bobbin which causes some inconvenience in subsequent processing. If left unattended, from 3 to 10 per cent of the heads would stop each hour. Breakage is a function of thread type and fiber mixture.

There are 18 doffing crews per shift; 10 crews have a manning of four experienced doffers; eight crews consist of one experienced doffer and four trainees. There are 72 spinners per shift. All work an 8-hour shift with $\frac{1}{2}$ hour out for lunch. Spinners are the more experienced group, having been promoted from doffer and given additional training for spinning. During the $\frac{1}{2}$-hour lunchtime the machines are shut down. The cycle of work is such that no 2 days' schedules of doffings are the same despite the fact that all products are made continuously. There are two shifts (see the table).

MACHINE CYCLES

Item	Machine Type	Number of Machines	Cycles/Day/Set
SB 20	TN 6″	104	14.08
SB 16	TN 7″	88	12.18
SB 21	TN 7″	80	9.37
SB 30	TN 6″	36	7.50
RB 30	TN 6″	20	8.50
RB 30	OM 6″	24	8.50
RB 60	TN 7″	56	3.75
RB 60	OM 6″	48	4.69
NR 30	TN 7″	24	7.50
NR 38	TN 7″	56	4.69
NR 40	TN 7″	56	4.69
NR 42	NM 7″	48	4.69

The method of operation is for each spinner to patrol a given area of machines on a patrol path. The doffers doff the finished set of bobbins from a set of machines, as a group. A whole set is always doffed at the same time. A crew of doffers is dispatched to doff a set of machines when the bobbins on that machine are full and the machines have stopped. The short bobbins are also doffed at that time. The supervisor who does the dispatching sends any idle crew of doffers that he can find. The spinners on patrol pass the machines which the doffers are working on.

Under this procedure the most conscientious crew of doffers gets the largest workload, although all doffers are paid alike. It was decided that a schedule should be developed and crew assignments made to achieve a more equitable distribution of the workload. However, owing to the fact that each doffing on a set of machines occurs at a different time each day, the computation of the daily schedule has involved much work.

You have been asked to "Devise a method of reducing the complication of computing the daily schedule for doffers and spinners."

1) What do you do?

2) Explain your rationale.

3.4. List all the similar and all the different aspects of the items in the following list:

(a) Mr. Puccini's spaghetti.

(b) Mr. Chow Kim's Kimchi.

(c) Mr. Front-de-beouf's roast beef.

(d) Mr. Frost's curry.

(e) Mr. Jackson's zither.

(f) The Swiss navy.

4

INTRODUCTION TO
MOTION STUDY TECHNIQUES

Categorizing the Techniques

Motion study techniques may be divided, with respect to purpose of use, into three major categories:

1. Used to assist in determining the most feasible-appearing class of change.
2. Used to delineate the units of output, as a preliminary to the use of either category 1 motion study techniques or the use of time study techniques.
3. Used to assist in examining, in appropriate detail, the manner of performing the work.

Multiuse techniques. Some of the techniques have sufficient flexibility of use that they may be used to assist with several of the varieties of purpose given.

46

Sequence of the Use of the Techniques

With substantive outputs. In general, a study of work related to the production of substantive outputs starts with the use of a technique chosen from those in category 1. The purpose is to assist in selecting the most feasible class of change. In most cases this is followed by the use of a technique from category 3 to examine the details of the work as a preliminary to improvement.

With service outputs. A study of service outputs, in contrast, usually starts with the use of a technique chosen from category 2; a technique to assist in delineating the outputs. This is usually followed by the use of some time study technique and then finally with the use of a technique chosen from category 3, the work analysis techniques.

CATEGORY 1, FOR SELECTING THE CLASS OF CHANGE   *– Substantive output*

1. Preliminary possibility guide.
2. Detailed possibility guide.
3. Work activity analysis.
4. Work sampling.
5. Memomotion study.

CATEGORY 2, FOR DELINEATING THE OUTPUTS   *— Service output*

1. Work-unit analysis.
2. Work activity analysis.

CATEGORY 3, FOR EXAMINING THE APPROPRIATE DETAILS
OF THE WORK

1. Work activity analysis.
2. Work sampling.
3. Process chart—product analysis.
4. Horizontal time bar chart.
5. Network diagram.
6. Process chart—man analysis.
7. Information flow analysis chart.
8. Operation chart.
9. Multiple-activity analysis chart.
   a. Man and machine chart.
   b. Multiman chart.
10. Micromotion analysis.
11. Memomotion analysis.

Selecting a Technique to Use

In most cases, several techniques will be used in sequence. After all, techniques are "tools." Tools are usually used in sequence. For instance, if one wished to make some object, first he would make a drawing of it. A drawing is a "tool." Next, if the object was to be made of wood, he would use a saw to cut, a drill to make holes, a hammer to drive nails, and so forth. The choice of tool is affected by sequence and type of assistance needed. The selection of motion study techniques involves similar considerations.

Basis of selecting from category 1. If the work involves substantive outputs, the techniques of category 1 will be considered first. In general, the more detailed knowledge of the work situation possessed by the investigator, the more likely the study will start with the use of possibility guides. When more detailed information is needed to meaningfully examine the possibilities of change, the study may start with work activity analysis or with work sampling.

Basis of selecting from category 2. In general, if the study is of a service activity, this category will be used to provide an initial technique. If the activity is such that a meaningful method for counting the outputs is not readily apparent, work-unit analysis will be the first step. If the service activity is well organized, from an overall point of view, work activity analysis may be chosen as a starting point.

Basis of selecting from category 3. In the case of substantive outputs, techniques in this category will be considered after the appropriate technique from category 1 has been applied. In the case of many service outputs the use of category 3 techniques may be preceded by the use of time study techniques whose use may follow the application of the category 2 technique. In any case the choice of technique from category 3 will be a function of two major factors:

1. The class of change sought.
2. The nature of the work.

As was noted in Chapter 3 with respect to the class of change sought, there are numerous subfactors to be considered. The nature of the work has an even more complex effect. Work varies with respect to physical scope. For example, compare the work of Figures 1.6 and 1.7 with the work of Figures 1.8 and 1.9. The range of physical scope (area used to do the work) of these two jobs is only a small fraction of the whole range of scope of work.

Work also varies with respect to the time to perform repetitions of the work. Some jobs are completed in seconds; others take minutes, hours, days, weeks, or months. Work also varies with respect to the demands upon the worker with respect to physical or cognitive skills. This is only a partial list. Each different situation may require a different technique for analysis of the appropriate details of the work.

Material that follows in this book. In each of the chapters dealing with the details of technique, information will be included with respect to the nature of the work to which the technique is applicable. The purpose is to assist with respect to choice of technique. Also, as was noted earlier, some of the techniques of motion study are of such a nature that they may be used in different ways for different purposes. Hence, in the chapters dealing with these techniques, the differences in the manner of use, depending on purpose, will also be carefully indicated.

Conclusion

It should be apparent that to usefully employ the techniques of motion study, one needs to be familiar with:

1. Which techniques serve which purposes.
2. In what sequence the techniques should be used.
3. To what kind of work each technique may be usefully applied.
4. The details of the techniques.

As a consequence, the chapters describing the techniques are organized to serve these four needs.

PROBLEMS

4.1. Provide a list of 10 tasks, representing a range from:
   (a) A very short time per repetition to a long time per repetition, arrayed.
   (b) Tasks that are performed within a small area (workbench) to those covering large physical areas, arrayed.
   (c) Tasks that require little physical skill to those requiring great skill, arrayed.
   (d) Tasks that require little or no cognitive effort to those that require great cognitive effort, arrayed.
   Note: Each of these lists should be independent and separate.

4.2. Describe a task you are familiar with. Give details with respect to:
  (a) Time per repetition.
  (b) Area covered.
  (c) Physical skills required.
  (d) Cognitive effort required.
  (e) Knowledge required to do the task.
  (f) Equipment involved.
  (g) Importance of accuracy of work.
  (h) Hazards involved.
  (i) Seriousness of errors.
  (j) Time required to evaluate results.

5

INTRODUCTION TO
TIME STUDY TECHNIQUES

Purpose of This Chapter

In Chapter 1 the relationships between the cycle of managerial control and the techniques of motion and time study were discussed. Subsequently, the nature of the motion study techniques were examined. The discussion included, in Chapter 4, an examination of the factors that affected the choosing of a specific technique in any particular situation.

This chapter will examine in greater detail the objectives for using the time study or work measurement techniques. The words *time study* and *work measurement* will be used synonymously. The relationship to the cycle of managerial control will be examined. This chapter will also examine some basic measurement problems and some basic facts relating to these problems as preliminary to a later study of the actual techniques.

51

Definitions

Time study (or work measurement) was defined in Chapter 1 as *a set of procedures for determining the amount of time required, under certain standard conditions of measurement, for tasks involving some human activity.* The result of such a measurement is called a *standard time.*

Basic Uses of Standard Times

The first and fundamental use of a standard time is to assist in the operation of the management cycle, as the cycle was described in Chapter 1. The standard time is used as *a numerical coefficient for converting a quantitative statement of the workload* (*step 3*) *to a quantitative statement of the required manpower resources* (*step 4*). If the time use of other resources is concomitant with the use of manpower, the amount of these other required resources may also be computed from the standard time. Planned or standard costs will be computed from the same data. Standard times, as will be seen in later chapters, also play a role in the design of methods. Standard times also provide a basis for the comparison of the actual and the planned use of manpower (step 7 of the cycle of managerial control).

Detailed Uses of Standard Times

1. To determine labor and equipment requirements. Any managerial plan for the production of outputs must be tested for feasibility with respect to available resources. Feasibility is examined by converting the desired quantity of outputs to a statement of required resources and determining whether this amount of resources is within the allowable limitations. (In industrial enterprises such limitations are economic, that is, outputs must be profitable; in government enterprises they are substantive, for example, number of men, space, money.) If the plan is not feasible, either the amount of the desired outputs must be altered or the factors affecting the need for resources must be altered. (This second area of change is the area of impact of the motion study techniques.) When labor and equipment requirements, expressed as their money cost, are added to materials and overhead costs we have *standard costs.*

The need for standard times for these basic uses in the cycle of managerial control is so fundamental that some sort of standard time must be generated, no matter how informal are the techniques used. Even in situations where managers may deny the feasibility of work measurement, some alternative of technique is being used. It is one of the purposes of this book to describe procedures which are more effective than these informal techniques so as to improve the results obtained by the exercise of managerial control. Standard times, for these basic uses, must either reflect expectable performance, or the ratio for converting to such performance must be known.

2. *To assist in developing effective methods.* (As was noted in Chapter 1, the motion study aspect and the time study aspect are not fully separable; this is an area of great overlap.)

a. To Determine the Number of Pieces of Equipment a Person May Run

Time values for the human parts of the cycle are important factors in setting up the job method for such work. Here, the time standards should be such that they are attainable by most of the workers if good machine utilization is to be attained.

b. To Balance the Work of Crews, Coordinate or in Sequence

Efficient crew work demands an even distribution of work among the members of the crew. It is the crew member with the longest job who determines the output of the crew. Office processing sequences, assembly lines, and most crew activities usually achieve higher production and lower cost than individuals doing complete operations. Lower cost is achieved because of the greater automaticity possible with the smaller tasks, the specialized tooling and workplaces possible, and the reduction in training time and cost. However, an unequal distribution of work among the crew members can more than offset these gains. With an 11-man work sequence, or production line, if the task of one worker requires 10 per cent more work than the next longest job, the work of at least one whole worker may well be lost. (The relative ability of the worker with the longest task may increase or decrease this figure.) Whether the standard times are easy or difficult to achieve does not matter as long as all standard times are of equal ease of difficulty. Consistency of the standard times is the sole requirement for this application.

c. To Compare Methods

As can easily be seen, a standard of consistent difficulty is required to provide an unchanging yardstick for the comparison of two or more methods of performing the same work. In this case the relationship of the standard time to possible performance is immaterial.

3. To constrain the use of manpower

a. To Set Schedules

Production (or work accomplishment) schedules are a vital necessity for any organization. They are used to control the rate of using resources. They also serve as a basis for planning sales programs for profit-motivated organizations; for planning accomplishment programs for government or service organizations. They should permit proper coordination of staff departments, operations, purchases, and delivery or sales, and so forth. Production schedules, if they are to be reliable guides, must be based on measures bearing a known relationship to the expected rate of output. Any standard time that is greater or less than that which may be actually expected as typical performance needs a known correction factor for use in schedule making, but is still quite usable. Scheduling is greatly facilitated if the correction factor for computing typical performance is the same for all standard times in use in the organization.

b. To Set Labor Standards

This does not necessarily refer to wage incentives. Labor standards can be the levels of individual or group production deemed satisfactory, and may be applied without financial incentives. The standard times used for this purpose should be readily attainable, by the type of worker who is expected to be average for the job, to avoid either making "substandard performance" typical or creating a frustrated feeling on the part of the workers. The importance of this can hardly be overstressed. Labor standards, properly determined and properly understood, are an asset to both management and labor, since they fix a level of satisfactory activity and protect the interests of both groups.

c. To Determine Supervisory Objectives

A foreman or supervisor is supplied with a mix of men, materials, space, machines, tools, and methods. The mix will vary greatly, depending upon the nature of the outputs. It is his job to supervise the coordination of this mix to achieve an expected result. Time standards for this use should indicate the rate at which he is expected to coordinate his facilities in order to meet schedules and produce outputs within the standard costs. Such time standards will also help the foreman or supervisor to select workers who need additional training, who are misplaced, or who have unusual aptitudes or apply themselves with unusual diligence. Time standards for these uses should indicate typical expected performance, or some known proportion of it, so that an individual's performance may be evaluated against such standards.

d. To Provide a Basis for the Setting of Piece Prices
   or Incentive Wages

Incentive wages are a means of automatic financial supervision for both labor and management. They tend to reward the more productive workers in proportion to their output. They also give rise to worker insistence on management's keeping a steady flow of work during the working day and on eliminating sources of work stoppage such as poor maintenance. To keep production up to the desired level, the standards at which incentives begin must be more than attainable and worth exceeding. They must also be consistent so as to provide equal incentive opportunities from job to job and thus avoid inequities within labor grades or the upset of established job hierarchies. There must also be a sufficient difference between base pay and the pay at a reasonably possible level of performance so as to provide an incentive that will stimulate production. Hence, if the wage incentive is calculated on the basis of the work done in excess of standard performance, the standard time should be greater than that needed by the average worker, as well as being expressed in relatively consistent units.

4. To assist in comparing performance with plans with respect to workload and resource usage. An organization, if profit-motivated, usually prices its merchandise prior to manufacture. To do this it must predict how much labor or production-center time will be expended on each phase of the work, and must have a means of continuously comparing actual performance to predicted performance. For performance predictions, as with schedule setting, the standard times are usable if they bear a known relationship to the time that will actually be required. Standard times for each operation are used for detailed cost checks on operations and in the determination of the exact places for the application of corrective action. They assist in "pinpointing" jobs that are not being performed as expected. For nonprofit activities such as government agencies, the same continuous comparison of plans with performance must take place so that corrective actions may be taken, when needed, to obtain the maximum possible conformance of events with plans. For this use, the time standards must have the same basic characteristics as given above.

Conclusion

From the foregoing, one conclusion may be reached; the concept of standard time may vary somewhat with the intended use. Hence, the particular technique chosen for a particular problem will be a function of (among other factors) the use to be made of the standard time.

To What Does a Standard Time Relate?

A standard time, it was noted, is a numerical coefficient for converting a quantitative statement of workload to a quantitative statement of the required manpower resources. This statement raises two questions:

1. In what units is workload or output expressed?
2. In what units are manpower resources expressed?

These two questions are worth examining in that practically all the difficulty that has been encountered in attempting to serve managerial needs with standard times can be traced to a failure to satisfactorily answer these two questions, a failure that is often caused by a lack of realization of the full implications and ramifications of these questions.

Units of Workload

With substantive outputs. In industrial organizations, where the output is substantive, such as home appliances, shoes, clothing, and so forth, the problem of quantifying the workload with respect to direct labor has been looked upon, quite properly, as a minor problem. Little difficulty has been encountered.

With nonsubstantive outputs. With the indirect work of manufacturing organizations and with service-type organizations or with government agencies, poor methods of quantifying the workload have frequently frustrated work measurement efforts, leading many to conclude that real work measurement and, consequently, real managerial control is not possible. For instance: "The [Army] Corps [of Engineers] had, over a period of years, attempted to develop a comprehensive work measurement system. Based on the frustrations of these attempts, the agency contended that its heterogeneous mix of work had, and would, defeat such efforts."[1]

---

[1]U.S. Department of Army, Corps of Engineers, "Chronology of Actions Taken by the Corps of Engineers on Work Measurement." Report prepared by the Manpower Management Division, Office of Personnel Administration, Office of the Chief of Engineers, August 1963, as described in John R. Hadd, "An Investigation of the Feasibility of Agency-wide Work Measurement in the United States Army Corps of Engineers," M.A. thesis, George Washington University.

Differences Between Substantive and Service Outputs

Almost any service-type output has an important difference as compared with substantive outputs. Substantive outputs can be represented by an engineering drawing; service-type outputs cannot be so represented. Because of this, substantive outputs are easily converted to a quantitative statement of workload, but service-type outputs present a more difficult problem. Methods of defining such outputs are feasible and will be described later in this book.

Units of Manpower

The second basic question was: In what units are manpower resources expressed? To answer this by saying "man-hours" or "man-years" is to miss many important aspects. Implicit is the need to state what *kind of a person* (abilitywise), and *how fully is the person to exert the ability.* We need to state the *rate* of work input implicit in the time use of manpower embodied in the standard time. We need to state to the extent appropriate, the method, conditions of work, equipment, type of individual, and degree of exertion of such an individual's capabilities.

Basic requirements for defining standard. As with physical measurements, to avoid chaotic results a standard must be defined in such a manner that consistent, reliable measurements may be made, sufficient for the purpose for which the measurement is made. If, in a machine shop, each worker's micrometer had different graduations on it, product control would be extremely difficult and inefficient, if not impossible. Similarly, if no adequate standard of human performance is defined, managerial control will lack effectiveness. However, garments are not made to micrometer-type measurements. Such meticulous accuracy would be inappropriate. In many cases of human performance, non-micrometer-type measures may also be appropriate.

Consistency of measurement. In many work measurement problems an adequate narrative description of the unit of measurement is not enough. It must also, for the most effective use, be embodied when possible in some physical form, as are other standards. For example, the metric unit of length, the meter, was originally defined verbally as 1 ten-millionth of the distance on the earth's surface from the pole to the equator. Imagine trying to apply this standard consistently with this statement as the sole guide! Before the meter was of any use as a measuring unit it had to be reduced to physical forms, such as rulers or gage blocks. The fact that the meter, as thus finally defined by an objective embodiment, was somewhat at varience with its verbal definition did not greatly hinder its ultimate usefulness. Such a discrepancy between verbal and objective definition would not be entirely inconsequential in the case of time study, but it is of less consequence than the need for consistency. However, the need for adequate objective representation of the standard is frequently just as great with work measurement as with other types of measures.

Human factors. In view of the foregoing it is desirable at this point to examine some fundamental facts concerning the factors relating to the "rate of work input." These data will provide a background for understanding some of the procedures examined later.

With any human physical attribute (and the ability of a person to do work is, in the final analysis, a function of physical attributes), the ratio between the best and the worst, excluding highly unusual people, is seldom greater than 2 to 1, with few cases of these extremes. Ralph Presgrave, analyzing data developed by Hull,[2] Wechsler,[3] and Barnes,[4] suggests that for practical purposes range of human capacity may be taken as being between 1:2 and 1:2.5. He suggests the reasonable dependability of a 1:2.5 range, although he states, "In fact a range of 1:2 in practice is extremely unusual and very often includes examples that any competent observer would refuse to take into account, even though, as a matter of courtesy, or in an attempt to assist poor operators, he might take a study."[5] Undoubtedly, the actual range of 1:2.5 is reduced to the limits of 1:2 by even rudimentary worker selection.

The nature of the distribution of capacities is such that with each trait the greatest number of people tend to possess an average amount, with the number diminishing as the extremes are approached. This is what we call the "normal" distribution.

Industrial implications. Consequently, for any job on which a large number of people work, or on a group of jobs with consistent time standards,

---

[2]C. L. Hull, *Aptitude Testing*. Yonkers, N.Y.: World Book Company.

[3]D. Wechsler, *The Range of Human Capacities*. Baltimore Md.: Williams & Wilkins Co.

[4]R. M. Barnes, *Motion and Time Study*, 2nd ed. New York: John Wiley & Sons, p. 271.

[5]R. Presgrave, *The Dynamics of Time Study*. Toronto: University of Toronto Press, p. 108.

the distribution of workers and performance (if all workers worked at their jobs at the maximum pace they could maintain on such jobs without harmful exertion) would be as shown in Figure 5.1.

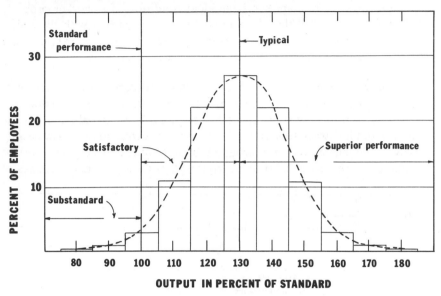

Figure 5.1. Distribution of performance of a large group of workers with typical performance denoted as 130 per cent of standard. (Designating typical performance as 130 per cent of standard is common industrial practice.)

It is worth noting that a change in the ratio between standard time and time at typical performance will alter the scale of the abscissa shown in Figure 5.1, change the scale value at the midpoint, and shift the position of the line indicating the division between substandard and below-average workers. With the upper end of the curve shown in Figure 5.1 always bearing a value equal to twice the lower end, it can be seen that when the ratio of standard time to time at typical performance reaches 100/100, the line below which we find substandard workers lies at the midpoint and half the workers are "substandard," but when the proportion is 150/100, the substandard line lies at what is, for all general purposes, the bottom of the curve, and we have (excluding unusual cases) no "substandard" workers.

Workers who cannot achieve acceptable performance. It should also be apparent that, in most cases, there will be workers who will not, even with the proper training, reach the level of standard performance. They are on the wrong job and, in most cases, the sooner they are shifted to other tasks for which they are suited, the more advantageous it will be for the management and the worker. There will also be workers who will greatly exceed the standard level of performance. With the variation in human capacity, this is to be expected. This represents the ideal condition of a worker properly fitted to a job, provided that the standard time is consistent with the definition it embodies, although, our mores being what they are, an extreme case may cause considerable disturbance with a wage incentive system.

An actual example. It is to be noted that well-performed worker selection, pacing by limits of equipment, and group work will tend to reduce the spread of the curve shown in Figure 5.1. Figure 5.2, an actual curve for the hours worked in 1 week by 1,100 workers at a plant with an incentive system, displays some increased central grouping because of extensive group work at this plant. On the other hand, it clearly displays the basic characteristics of

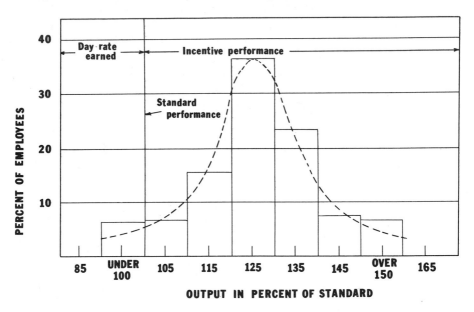

Figure 5.2. Per cent of hours worked on incentive versus per cent performance for 1 week. (Courtesy of C. J. Allen, Chief Industrial Engineer, Patrick Cudahy, Inc., Cudahy, Wis.)

the theoretical curve of Figure 5.1. The incentive system in this plant is based on standard times designed so that the average worker should make 25 per cent incentive.

Problems to be recognized. It may be seen that both the selection and application of a rate of work input for use in the standard time must take into account the uses intended and the social aspects of the situation. These are matters that require careful consideration and frequently introduce difficult problems. However, the vital necessity of adequate work measurement to the effective management of an enterprise should be implicit in the uses made of standard times; hence, also the need to solve such problems.

Conclusions from Preceding Discussions

We may now reach some conclusions with respect to time study or work measurement.

Time study or work measurement was defined, in two compatible ways, as being:

1. A set of procedures for determining the amount of time required, under certain standard conditions of measurement, for tasks involving a human activity.
2. A set of procedures for developing numerical coefficients for converting a quantitative statement of the workload (step 3 of the cycle of managerial control) to a quantitative statement of the required manpower resources (step 4).

Also, it was shown that the relationship of a "standard time" to real-world expectancy could vary, depending upon the use to be made of the standard time.

The Basic Nature of a Standard Time

All time study techniques are procedures for obtaining four kinds of data. These are:

1. WT = work time.
2. WC = work count associated with WT.
3. M = modifier used to adjust the meaning of WT with respect to the real world.
4. A = additive used to adjust the ST to real people.

These kinds of data are used to compute a standard time as follows:

$$ST = \frac{WT}{WC} \times M + A$$

where ST = standard time.

Categories of Time Study Techniques

The techniques of time study may be divided, in terms of general source of the necessary data, into five categories. It should be noted that each technique listed should be thought of as a subgroup; there are many different ways, with respect to details, of applying each technique.

The five categories

1. REQUIRING DIRECT OBSERVATION
   a. Direct time study—intensive sampling.
   b. Direct time study—extensive sampling.

2. USING PAST PERFORMANCE RECORDS
   a. Simple mathematical techniques using historical data.
   b. Complex mathematical techniques employing historical data and linear programming or multiple regression.

3. USING PREVIOUS TIME STUDY DATA
   a. Predetermined time systems.
   b. Standard data systems.

4. IMPLICIT IN THE NATURE OF THE WORK
   a. Time standards by fiat.

5. INVOLVING THE WORKER IN THE DATA COLLECTION
   a. Fractioned professional estimates.
   b. Self-reporting.

Criteria for Choice of Time Study Technique

With respect to categories. The basis of categorization given differs from that used for the motion study techniques; those related to sequence of use. The categories of time study techniques are divided on the basis of data-gathering methods. Not all can be applied in all situations. For a simple illustration, a cognitive task can seldom be studied by the direct observational techniques; one cannot *observe* the thinking process.

Other criteria. In general, in addition to the feasibility of using the data-gathering procedure implicit in each technique, the different techniques vary with respect to:

1. Accuracy (this varies with the nature of the work; work situation and technique frequently interact with respect to accuracy).
2. Ease of studying cycles of various lengths.
3. Staff power required to set standard times.

Conclusion

It should be apparent that to usefully employ the techniques of time study, one needs to be familiar with:

1. Which techniques can be used under what circumstances.
2. The details of the techniques.

As a consequence, the chapters describing the techniques are organized to serve these two needs.

PROBLEMS

5.1. Describe the extent of use and the uses made of time standards in a plant or an organization in which you have worked, or one with which you are familiar.

5.2. If the range of performance of a random group of operators is roughly 2:1, and the average performance is 130 per cent of standard production, what approximate percentage of the operators will probably:
(a) Fall below 90 per cent?
(b) Fall below 100 per cent?
(c) Exceed 100 per cent?
(d) Exceed 130 per cent?
(e) Exceed 150 per cent?
(f) Exceed 160 per cent?

5.3. From the literature, summarize an application of time study with special reference to the benefits gained.

5.4. From the literature, report on a criticism of time study.

5.5. Report on the time study contribution of:
(a) F. W. Taylor.
(b) Carl Barth.
(c) Henry Babbage.
(d) Charles Fayol.
(e) Charles Bedaux.

5.6. Describe a task for which one could obtain the necessary work measurement data:

(a) By observation.

(b) From "historical" data.

5.7. Describe a task for which direct observations would reveal little about the relationship between the work and time.

5.8. Make a histogram, as in Figure 5.1, of:

(a) Student grade averages.

(b) Student scores on a test.

(c) Scores from 20 students throwing five darts at a dartboard.

5.9. (a) If a histogram, as in Figure 5.2, were to be made of the golfers in an important tournament, would the results differ from Figure 5.1?

(b) If different, why? If the same, why?

5.10. Where is the reference standard of the meter kept? In what form is it?

5.11. What kind of projects does the Army Corps of Engineers undertake?

5.12. It was stated that an extremely productive worker, producing much more than normally expected (even taking the 1:2 range of capability into account), would strain the mores if a wage incentive system were in use. Why would this happen?

6 APPLYING MOTION AND TIME STUDY

Motion and Time Study and the Objectives of Organizations

Motion and time study activities are only of value to an organization when they are carried on in a manner designed to serve the organization's objectives. The objectives of an organization are established by top management.

It is worth noting that the performance of this step of setting objectives is the primary differentiating characteristic between the top management of an industrial enterprise operating in a free economy and either an industrial enterprise operating in a state economy or the top management of a government organization. The top management of an enterprise in a free economy has economic objectives; there is considerable freedom of choice in developing substantive programs to achieve these objectives. An enterprise in a state-owned economy usually has substantively stated goals assigned to it. It may have some freedom of choice with respect to the economic aspects of achieving them, but in stringent economies, such as represented by communist or fascist states, these economic choices may be either severely delimited or nonexistent.

A governmental organization, even in a free economy, usually has substantively stated goals, which are either assigned or derived from more general statements of the nature of desired achievements; in addition, there are usually numerous economic restrictions concerning the allowable nature of programs to be followed in achieving the goals.[1]

Measures of effectiveness. The usual measure of the effectiveness of the management of an industrial enterprise is the amount of profit that it makes. Indeed, the need for profit is overwhelming; without a profit the organization cannot endure. The usual measure of the effectiveness of either an industrial enterprise operating in a state economy or a government organization is whether (or to what extent) it achieves its substantive goals. Indeed, the need for some reasonable achievement is overwhelming; the manager of a government organization who cannot "deliver" usually will be replaced. Hence, it is easy to see why industrial enterprises are generally money-motivated whereas government enterprises or organizations are primarily program-motivated.

Use of motion and time study with military bases. The difference caused by government objectives is most evident with a military base. The commander may consider using time study but the objective is to determine the amount of resources that he needs to meet the maximum workload that may be imposed upon the base during an emergency; the use of motion study would be to increase the capability to take on more workload during an emergency. There will not necessarily be any savings, in the same sense as in the industrial enterprise.

Differences in priorities between the private sector and government. Motion and time study may be used in any cycle of managerial control; the priorities will differ when the motivations of the organizations differ. Failure to understand these differences can cause much frustration and failure. The industrial organization usually needs to employ time study to determine costs and motion study to reduce them. The government organization usually wishes to determine costs only so that the economic limitations set for its objectives are reasonable; it wants to use motion study to assist in making the achievements of objectives feasible or in increasing the substantive achievements.

---

[1] M. E. Mundel, *A Conceptual Framework for the Management Sciences.* New York: McGraw-Hill Book Company, 1967, p. 274. (A few phrases have been deleted.)

Hence, in both industrial enterprises and in government organizations we find pressure for the effective use of techniques and procedures as presented in this book. Private industry has economic pressure; government agencies have pressure from their Chief Executive through the budget process.

Basic Manner of Employing Motion and Time Study

None of the foregoing alters the essential nature of the manner in which the techniques of motion and time study may be employed to serve the managerial needs of an organization.

Motion study. The techniques of motion study (and other appropriate technologies) are employed in the cycle of managerial control in designing or selecting outputs, in planning ways of producing them, and in determining the workload required to produce the desired quantity of outputs.

Time study. The techniques of time study are employed in the cycle of managerial control in developing numerical coefficients for converting quantities of workload to quantities of manpower resources required. When the quantity of required manpower resources makes the outputs of the industrial enterprise unprofitable, the techniques of motion study (and other technologies) are employed to find cheaper (manpower or other resource-wise) ways of producing the outputs, or alternative outputs. When the quantity of required manpower resources makes the outputs of a government organization excessively expensive (manpower-wise) such that a reasonable achievement of the goals cannot be attained, the techniques of motion study (and other appropriate technologies) are employed to find more effective ways of producing the required outputs.

Motion study techniques for studying information flow. In any organization, when a production plan has been established, whether the outputs are substantive or service-type outputs, some means is needed to constrain the activity. Information flow systems must be designed to compare events to plans. Work measurement data will be the basis of manpower-use plans. The same data will be involved in the subsequent comparison of events with plans. Motion study (the procedure analysis aspects) will be basic to the design of such management information systems.

Time study and wage incentive systems. In order to provide additional motivation to meet the time standards embodied in the production plan, one of a variety of schemes that relate wages to output may be used. Such schemes are also referred to as *payment by results.* In the modern world they usually take the form of a guaranteed hourly wage with an incremental wage for production over a stated amount.

Personal requirements for getting results. The effective use of techniques, such as presented in this book, is limited only by the imagination and ability of the analyst who uses them. The ability must include not only the ability to use the techniques to obtain answers to problems but the ability to persuade people to use these answers. It would seem reasonable to suggest that basic to success in "selling" ideas is the need to carry on the work with openness, pleasantness, and honesty, working systematically and scientifically.

Institutional aids to effective work. Further, the work of a group, such as a motion and time study group, is usually aided when they have (1) policies and procedures for routine activities, (2) an adequate reporting system to higher levels of management, and (3) when they assist others in the organization by giving formal and informal training in their field of work. While these general aspects are equally applicable to both private industrial and governmental organizations, the details will differ greatly. Most private industry has a substantive product which serves as a natural framework, not only for the managerial control cycle, but for the whole of the industrial activity. Most government organizations do not have an adequate identification of outputs around which to structure not only their motion and time study activity but also their basic managerial controls. However, procedures for developing such an identification do exist and will be presented later in this book.

Policies and Procedures

Purposes and definitions. In the widespread use of motion and time study in an organization there are certain relatively routine activities that must be performed. The means of carrying on these routine activities should be formally stated. A group of policies is necessary to achieve consistency of actions. Formal procedures are necessary to routinize the carrying out of the policies.

Policies are statements of the aims to be sought in handling recurring types of situations. They are the rules by which the organization functions and are vital in all phases of plant activity. Procedures are the details of the methods to be employed to achieve these aims.

List of items needing policies and procedures. For the routine motion and time study work to be effective, the policies and procedures should cover at least the following seven items:

1. What does a standard time represent?
2. Who shall determine the standard method?
3. How will the standard method be made regular practice?
4. Who shall determine the standard time and how will it be determined?
5. Under what conditions may a standard time be changed?
6. How will production be reported?
7. If a wage incentive plan is in use, what are the rules covering its use?

The policies and procedures for items 2 through 7 can most readily be examined after the detailed techniques of motion and time study have been presented. However, the first question, "What does a standard time represent?" is basic to all techniques. A standard time is the final value resulting from both motion study and time study. Hence, its significance will be examined in detail in this chapter.

What a Standard Time Represents

Introduction. In Chapter 5 was shown that the term "standard time" could have many meanings. In much industrial practice however, a single specific meaning is attached to this term, and this definition represents an agreement between labor and management. Some fundamental aspects of potential definitions should be considered as a preliminary to achieving a satisfactory definition.

A standard time was defined as indicating how long a given rate of work input must be maintained to produce a unit of output. The factors or conditions concerned in a standard time were suggested to be:

1. *Unit of output* (*work unit*).
2. *Rate of work input:*
   a. Method.
   b. Conditions of work.
   c. Equipment.
   d. Type of individual:
      1. Skill.
      2. Aptitude.
   e. Degree of exertion of the specified type of individual.

General definition of a standard time. A *standard time* is a function of the amount of time necessary to accomplish a unit of work:

1. Using a given method and equipment.
2. Under given conditions of work.
3. By a worker possessing a specified amount of *skill* on the job and a specified *aptitude* for the job.
4. When working at a pace that will utilize, within a given period of time, the maximum physical exertion such a worker could expend on such a job without *harmful effects*.

Within this definition, the following word meanings are intended:

Skill—the ability to do a job in the proper manner; the ability to repeat a definite muscular pattern. Other worker characteristics being constant, the higher the skill, the faster the possible pace before the muscular coordinations fail.
Aptitude—physical fitness for the job.
Harmful effects—the results of excessive physical and mental activity, caused by the work, which are not dissipated during the typical usage of the interval between work days.

Physiological and sociological considerations in the definition. Consider the two heavy horizontal lines of Figure 6.1. The upper line, the physiological maximum, may be defined as the fastest an average man could work, day after day, using the time between work spells for rest, without physical deterioration. Such a pace is not attained in modern industry. This concept

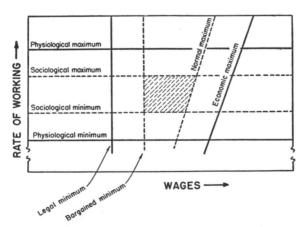

Figure 6.1. Wages versus rate of work and the area in which the actual work pace lies.

is not compatible with a democracy in which men work to live rather than live to work. The physiological maximum would connote a slave population with no energy to engage in family activity, leisure time, or civic pursuits. The lower, heavy line is the physiological minimum, the slowest pace at which work could be performed without increasing the rate of expending energy. This, too, is practically never encountered, and is certainly not compatible with a high standard of living based on high production.

For an organization to operate profitably within the economic environment, the work pace must be at least at or above the broken horizontal line labeled *sociological minimum*. For workers to accept this required rate of exertion as reasonable and proper (consistent with their concept of their society), the work pace cannot exceed the upper broken horizontal line labeled *sociological maximum*.

Economic considerations. The situation is complicated further by economic aspects. Leisure energy without sufficient income for comforts considered reasonable in the society is not satisfactory. Therefore, the two vertical heavy lines have been added to the chart to indicate, as do the first two heavy lines, conditions that are factually determinable but not always socially acceptable. The legal minimum wage is usually fixed by law. The economic maximum wage, the point where the organization would make zero per cent profit, a level economically determinable, is certainly not acceptable to the owners. Therefore, the two vertical broken lines have been added. The bargained minimum is the wage level above the legal minimum attained by virtue of the scarcity of skills, the desire of management to pay a good wage, or the pressure of collective bargaining. The line labeled *normal maximum* is the most the organization can pay, still making a reasonable profit and not pricing itself out of the market.

Wage versus rate-of-work limits. The four broken lines enclose a *wage versus rate-of-work area* within which the point denoting the actual conditions in an actual plant lies. Reference to Chapter 2 will suggest that Figure 6.1 is only a rough representation of two of the many variables affecting the situation. As will be seen in later chapters, a means of objectively documenting this point on the *rate-of-work scale* is feasible. However, the reader should recognize that the phrase *maximum physical exertion* (in the definition of standard time) refers to this sociologically determined point and not to a true, physiologically determinable condition.

Refining the definition of standard time. To make the general definition of standard time usable, it is necessary to replace the word *function* with a numerical value and to replace with a definite adjective each use of the word specified, thus defining a particular standard performance.

Many concepts of standard time are thus possible.

Industrial Examples

Some definition of standard time of this general type is usually found in collective bargaining contracts or in a statement of policy by management, although in many cases the detailing of the standard time is not carried out completely.[2]

From a steel foundry. The production standards will:

1. Be established for a specific set of conditions.
2. Reflect the production requirements as related to a fair day's work for a fair day's pay and will enable the normally skilled employee to increase his incentive earnings approximately 25 per cent above the occupational hourly rate for his occupation in effect (date) without unreasonable effort.[3]

From a manufacturing plant. Guaranteed piecework prices will be set so that a normal employee or group of employees, possessing normal skill and training, working under normal conditions, may, by normal incentive effort, after making an honest effort to attain incentive earnings over a reasonable trial period, have an opportunity to earn per pay period approximately 50 per cent above his piecework base rate or their piecework base rates.[4]

From a meat-packing plant. Standards will be set so that an average employee using the prescribed method, working under the prescribed conditions, and producing satisfactory work, will be able to earn 25 per cent above the evaluated rate after he has sufficient opportunity to become familiar with the job, has the required skill and ability to properly perform the job, and works at the incentive pace. It is understood, however, that there is no guarantee that every employee will achieve a performance under which he will earn not less than 25 per cent above the evaluated rate. Nothing herein contained shall be construed as imposing ceilings on earnings.[5]

---

[2]B. Gottlieb, "A Fair Day's Work Is Anything You Want It To Be," *Proceedings, 19th Annual Institute and Convention, American Institute of Industrial Engineers.* 1968, pp. 153–163. (This article also contains an extensive bibliography of related articles.)

[3]Agreement between American Steel Foundries and United Steelworkers of America, East St. Louis Works.

[4]Articles of Agreement Between Servel, Inc., and United Electrical, Radio & Machine Workers of America.

[5]Agreement between Cudahy Brothers Company and United Packinghouse Workers of America, C.I.O.

Effect of Uses of Standard Times on Definition

Summarizing the material in Chapter 5. A consideration of the uses of standard times enumerated in Chapter 5 and their requirements will show that it is desirable that the unit of measurement used in standard times should meet for widest use, if possible, the following three criteria, on the basis of which a convenient standard unit of measurement may be established:

1. Consistent with respect to difficulty of attainment. (See all uses listed.)
2. Representative, regardless of the job, of a performance that can be bettered by a constant amount by the average worker.[6] (See number 3 of the list of uses.)
3. If for incentive use, providing an attainable differential of adequate size. (See number 3d of the list of uses.)

Consistency. The first criterion may be fulfilled by setting up any definite set of specifications for the measurement or time study; adhering to these specifications in the determination of standard times; and, in addition, defining the *function of the amount of time* as that which would be taken by working at a pace requiring an expenditure of *some fixed proportion of the maximum physical exertion possible on the job.*

Attainability. The second criterion suggests two things: (1) wherever the word *specified* appears in the basic definition, as concerns the type of worker, it must be replaced with an adjective denoting the average worker, and (2) *the function of the amount of time* "that would be required with the maximum physical exertion such a worker could expend on the job without harmful effects," must be set as some definite proportion greater than unity,[7] so as to permit such a worker to exceed standard performance by a fixed amount.

---

[6] With some incentive plans, incentive payment begins at some fraction, less than unity, of standard performance. In such a case this criterion would need to be restated, but the implication would not change.

[7] If an incentive plan that starts the incentive below standard performance is used, the proportion of the time at maximum exertion that will be allowed may be set at unity and the point at which incentive begins placed at the point necessary to meet this second criterion and also the third, which follows. This is a less common situation, inasmuch as the method of expecting incentive to begin at standard allows wages to be in direct relationship to the percentage of standard performance, which appears to be more easily explained to working groups whose usual ability with mathematics is not high.

How much performance can exceed standard. The third criterion suggests that the *function of the amount of time* be set so that the proportion of time allowed for standard performance, as compared to the time at maximum exertion, is sufficiently more than unity to permit the worker to exceed standard performance by an adequate amount. What is "adequate" poses a real problem and is a function of the use to be made of the standards. If no incentives are involved, any amount is adequate inasmuch as the sole purpose in such a case is to avoid the frustration arising from an unattainable "standard." If the standard time is to be used also as a basis for wage incentives, or if this may eventually be contemplated, then the problem is more complex. An incentive in the general sense may be considered an incentive only when it is so regarded by the person to whom it is applied. Hence, in a particular plant the adequacy of size of the average attainable differential is a function of such factors as customs, and attitude toward standards and the incentive system. In the contract definitions cited earlier, the amounts were 25, 50, and 25 per cent. The common range is 20 to 50 per cent, with 30 per cent being a commonly effective value. Since most present-day incentive systems pay wages in direct proportion to output, the attainable differential in production should equal the adequate incentive increment.

Suggested Definition of Standard Time

On the basis of all the comments on the criteria involved, the following definition is suggested for general use.

The standard time for a job will be 130/100 of the amount of time that will be necessary to accomplish a unit of work:

1. Using a given method and equipment.
2. Under given conditions.
3. By a worker possessing sufficient skill to do the job properly.
4. As physically fit for the job, after adjustment to it, as the average person who can be expected to be put on the job.
5. Working at the maximum pace that can be maintained on such a job day after day without harmful effects.

Effect of changing the amount of performance expected in excess of standard. If the third criterion is interpreted differently, then the proportion of the time at maximum pace required for standard performance (doing the job in the standard time) can be altered accordingly; that is, in the sample definitions from the labor contracts cited earlier, the proportion would be 125/100, 150/100 and 125/100, respectively.

Other factors to be adjusted for. It should be noted that the use of standard times together with incentives, in shops where the work is not entirely manual, is accompanied by additional problems relative to the criteria given for standard times. Let us assume that the standard time for a manual job, job A, set according to the recommended definition, is 1.0 minute. An average incentive worker may be expected to perform this job in 0.77 minute. On another job, job B, where a man feeds an automatic machine whose fixed cycle time is 1.0 minute, the standard time for the manual work of feeding the machine while it is running is also 1.0 minute. The incentive worker who performs this work in 0.77 minute will still produce only one piece in 1.0 minute. (The range of jobs between A and B, with part of the cycle controlled, present the same problem to a lesser degree.) The standard time of job B does not violate any of the criteria established; it is the work situation that does not permit production to be solely a function of how hard the worker exerts himself. This may be thought of as an inequity which may be remedied with additives to be included in the job standard.

Importance of definition. There may be some who feel that the recommended definition is cumbersome and unnecessary. However, they should realize that time studies give time standards for human performance, on the basis of which equipment and manpower requirements, schedules, cost controls, supervisory objectives, prices, and even wages are often set. These areas of activity are vital to adequate managerial functioning. As was noted in the body of the chapter, the concept of standard time, like any standard of measurement, must eventually be placed in a usable, concrete form. This will be discussed in later chapters, but the determination of what concept is to be embodied is certainly a vital first step. It should be reasonable to suggest that this first step, on which all of these managerial activities are based, be performed in a manner that will make possible an approach to real measurement, adequate for the intended use.

PROBLEMS

6.1. Define standard time as it is used in a plant or an organization in which you have worked, or one with which you are familiar.

6.2. Obtain a copy of a union contract. Quote and criticize the clauses dealing with standard time.

6.3. A plant is going to use a wage incentive plan. Their labor contract stipulates that a typical worker should be able to earn a 30 per cent incentive over his base rate.

(a) If their incentive plan is

$$
\text{earnings} = \text{rate per hour} \times \left(\text{hours worked} + \frac{\substack{\text{standard hours earned} \\ - \text{ actual hours worked}}}{2} \right)
$$

what will be the necessary definition of standard time?
Note: This formula holds true only if the standard hours exceed the actual hours; otherwise, the worker is paid (actual hours × rate per hour).

(b) If the incentive plan is

$$
\text{earnings} = \text{rate per hour}
$$
$$
\times \left[\text{actual hours worked} + 2 \left(\substack{\text{standard hours earned} \\ - \text{ actual hours worked}} \right) \right]
$$

what will be the necessary definition of standard time? [See note under part (a).]

(c) If the incentive plan is

$$
\text{earnings} = \text{rate per hour} \times \text{standard hours earned}
$$

what will be the necessary definition of standard time? [See note under part (a).]

(d) What general relationship exists among the contract wage increment, the incentive plan, and the definition of standard time?

6.4. A large group of operators average 130 per cent of standard production. The range of performance is 125 to 135 per cent.

(a) Give five possible reasons for the existence of such a situation.

(b) Indicate what each of these reasons would suggest concerning the correctness of the standard if it was intended to represent the standard time as finally defined in this chapter.

part two

TECHNIQUES OF MOTION STUDY

7   POSSIBILITY GUIDES

Setting a Goal

How to start.   The first step in applying the scientific method to the solution of a motion and time study problem, where the outputs are adequately identified, is:

Aim—*determination of objective in terms of area of work to be changed, and establishment of criteria for evaluating the preferability or success of solutions.*

The performance of this step has three parts:

1. Selection of the criteria of success of solution.
2. Rough determination of the degree of change that is warranted or needed.

3. Contemplation of the apparently feasible areas of change and the selection of the most feasible area of change.

Part 1. Selection of the criteria of success of solution. To determine our degree of success we must know what it is we are seeking to achieve. Therefore, the person seeking a better method must determine his criterion first, using his knowledge of the activity to guide him.

Some possible criteria of success of solution are given in the list that follows. The list has been oriented toward manufacturing situations. For industries or activities other than manufacturing, these criteria would either need to be restated or special priorities be given to selected items. For instance, in department store work, as with other profit-oriented service activities, customer service is of extreme importance. In medical work, patient safety may have an overwhelming priority. In government activities, conformance with externally imposed limitations concerning money or manpower may be a primary objective.

1. *Maximum financial advantage through:*
   a. Less direct labor time.
   b. Less direct labor effort.
   c. Less indirect labor.
   d. Better balance of direct and indirect labor.
   e. Fewer or more steps (depending on product and volume).
   f. Less skill (lower-priced labor).
   g. More skill (more productive labor).
   h. Less equipment usage.
   i. Cheaper equipment.
   j. Less space.
   k. Less scrap.
   l. Higher yield of high-value product.[1]
   m. Less expensive material.
2. *Maximum conformance with externally imposed restrictions[2]:*
   a. Fewer people.
   b. Less time for critical skills.
   c. Less time on critical equipment.
   d. Less time in production.
   e. Less space.
   f. Less critical material.
   g. Less working capital.
   h. More output without more manpower.

---

[1] Particularly applicable to process industries.
[2] That is, availability, customer requirements, government restrictions, and so on.

3. *Better product in respect to function, salability, or acceptability:*
   This may be a long-range aspect of number 1, although it may involve greater cost of operation on a particular job. However, in many situations it may be accomplished with less cost and is then doubly desirable.

4. *Better material control:*
   This is also an economic objective, as it relates to inventory cost, scheduling and control functions, and customer service.

Part 2. Rough determination of the degree of change that is warranted.
The type of decision that must be reached in this part of the step can perhaps be best understood by considering a few illustrations. Let us assume that we are contemplating changing the method of doing one operation on a product. The operation requires 60 man-hours per week and the product will probably be made for 10 weeks. With a labor rate of $5.85 per hour, this means a labor expenditure of $3,510 on this operation. A 15 per cent cost reduction will return a gross of $526.50, against which we will have to charge all cost of analysis, physical changes, and possible training. In this case, our decision would be that changes would have to be quickly arrived at and be low in installation cost. This decision would temper our method of performing the next step. However, in contrast with the data cited, activities are encountered where more than 100 people are performing the same operation, day after day, and it is anticipated that the activity will continue indefinitely. Obviously, if in such a case there is a reasonable chance of obtaining an improvement, the amount of effort and money that may be expended is quite different and the techniques chosen will reflect these different facts.

For a second comparison, let us assume that a government agency is contemplating some change in an existing manner of distributing food to needy or hungry people. A sizable segment of the population and a large amount of money are involved. One of the existing complaints may be that the people, in general, are not satisfied (or feel denigrated) by the method of distribution. Some radical changes seem in order so that social acceptability may be obtained. Of course, in this situation some techniques other than motion and time study will also be involved, but this was also true in the first case. The economic analysis was not motion and time study; economic analysis and motion and time study both were used, even if the economic analysis used in the first problem was simple. In the case of the food distribution problem, the economic analysis aspect and the social analysis aspect will assume more importance and be more complex than in the first example cited, but the motion and time study aspect still exists. As mentioned in Chapter 1, the motion and time study techniques are merely an aid to managers; they are not a substitute for management or managers or for other disciplines or fields of knowledge.

For a third type of illustration, let us assume that we are attempting to increase the control of stock in order to cut down pilferage in a mail-order house or warehouse. In this case we may have only an estimate of present probable pilferage, an estimate of potential pilferage, and a realization that the less chance of pilferage taking place, the less chance of unpleasant consequences if and when a culprit is detected. This last consideration is hardly an economic one, yet all these data may assist us in arriving at a rough decision that an improved control of stock is desirable, provided that it is not more expensive than the present system (although there may be leeway even on this). Hence, our rough decision would be that if pilferage is a real danger, even extensive change is warranted, although we may merely break even on costs and savings.

For a fourth illustration, let us assume that we are studying the work done during an antiaircraft missile launch. Cutting a 10-second operation by 2 seconds may be of negligible economic value by itself (after all, the crew is available), but it may make all the difference in a war. If we were the analysts in such a case, we would probably decide that even extensive change is warranted and give minimum consideration to limiting the time and cost of creating the change.

To sum up this part: the analyst roughly considers the importance of the job on a monetary basis or otherwise, depending on which is suitable, and sets rough limits concerning the nature of the change and the expenditure of time and money.

Part 3. Contemplation of the apparently feasible areas of change and the selection of the most feasible area of change. As an aid in contemplating possible objectives, the method analyst may use a *possibility guide*. Even when one is not used, the thinking should follow its pattern.

The possibility guide is a device for systematically listing all possible changes that are suggested by the person familiar with the activity or output under scrutiny. The possibility guide also allows the showing of the consequences of each suggestion for the purpose of assisting in the selection of the most feasible type of change. Thus, it helps the analyst to select the appropriate analysis procedure for performing the next step of the motion and time study.

It is worth noting that a listing of possible changes often can be made by a person possessing only a slight familiarity with the work under scrutiny. In many cases, such a person can suggest more possibilities than a person so familiar with the job that he cannot see beyond it, although this relative blindness of the latter is only a state of mind and not an irremediable condition. The largest number of possibilities can usually be suggested by a person familiar with the work but still open-minded about it.

Making a Possibility Guide

The making of a possibility guide is most easily accomplished in two steps:

1. Listing the suggestions on a preliminary possibility guide form, sometimes called a possibility list.
2. Detailing the consequences of each suggestion on a detailed possibility guide form.

Typical forms for use during the two-step preparation of a possibility guide will be shown in illustrations later in this chapter.

Interrelationships with classes of change. The subject of a motion and time study scrutiny may be classified under one of the following five classes which, as was previously shown, are all interrelated:

1. A job.
2. The equipment on a job.
3. The process.
4. The product design.
5. The raw materials.

If the subject of the study is "a job," then the class 1 and 2 suggestions will relate to this task. The class 3, 4, and 5 possibilities relate to the job only as it is a part of the sequence in which it lies. However, these aspects should always be considered, since frequently one way to improve a job is to change the sequence so that the job is eliminated, combined with another job, placed more advantageously in the sequence, or, at least, made simpler.

If the subject of the scrutiny is "the equipment on a job," the remarks in the preceding paragraph again apply.

If the subject of the scrutiny is "the process," the class 4 and 5 possibilities concern the product design and the raw materials as they affect the process. With "the process" under scrutiny the class 1 and 2 possibilities may be exceedingly numerous, as they can be listed for each job in the process. One usual procedure is to find the most time-consuming job in the process and develop a better method for it, provided that it is found necessary to do this job at all. Although this is often most productive, still one should bear in mind that some of the less-time-consuming steps, if not eliminated by a process change, may yield a greater percentage of savings than can be gained from an improvement of a more important operation.

Possibility guides in respect to "the design" or "the raw materials" are similar in nature to those constructed in respect to "the process."

An Example

The procedure for using a possibility guide can perhaps best be illustrated by an actual example. The general procedure will be the same no matter in which of the previously listed five groups the subject of the scrutiny belongs.

Let us examine the example in detail so as to fully describe the procedure for "thinking through" a possibility guide. The example concerns a relay armature arm used in an automobile turn-signal flasher system as shown in Figure 7.1. Our objective or criterion of success is to *lower costs*.

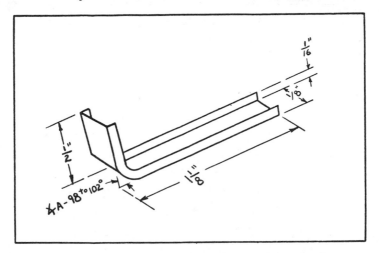

Figure 7.1. Details of relay armature arm.

Specifics of the problem. We would naturally have to have some familiarity with the process, a familiarity that could best be acquired by viewing the actual process, but the following description will have to suffice. The armature arm was used in a thermal relay designed to flash a light on and off. The raw material was in the form of U-shaped steel stock, 5 feet long, and had the dimensions of the U shown in Figure 7.1. The stock was issued as production orders required and was cut to length and formed in a punch press. The pieces were banked at the press, moved by tubs to an inspection station, and sorted by the angle A. Pieces outside the limits were rejected. If any sizable number of pieces were rejected, they were rebent to shape. The pieces were then sent in small tubs to the relay assembly-line station, where they were welded into the relay.

Using the checklist. The first thing we should do is to go completely through the checklist for possibility guide, jotting down all possible suggestions on a preliminary form, as is done in Figure 7.2. The body of typical form is divided into four columns, as shown. The first column is for a con-

BASIC CHART FORM

Prel. Poss. Guide ___ Type of chart ___ Mfg. ___ Department
Original ___ Original or proposed ___ L. Edmond ___ Chart by
Armature Arm ___ Subject charted ___ 2/16 ___ Date charted

Suggestion Number	Class of change	Description	other classes affected
1	5	Use sheet stock	4, 3, 2, 1
2	5	Purchase formed to size and shape	3, 2, 1
3	4	Redesign relay to allow more tolerance in arm eliminating inspection	3, 2, 1
4	3	Inspect at punch press as part of operation	2, 1
5	2	Dual inspection fixture, electric responce	1
6	1	Tip fixture to make easier, weigh to count at inspection	
7	1	Distribute work to both hands at inspection	

Figure 7.2. Preliminary possibility guide for manufacture of relay armature arm.

secutive numbering of the suggestions, so that others commenting on the list may easily identify what they are referring to, thus facilitating discussion. The second column identifies the class of change contained in the possibility or suggestion. The description column is the possibility itself. The last column lists the other areas that would probably be affected, and serves as a guide for future expansion of the possibilities for fuller examination.

A partial checklist begins below. The analyst may find it desirable to expand this list as he acquires more experience with this approach. In any case, the analyst will add to the list as he acquires experience in any particular field of work. To the experienced analyst each problem is no longer totally new; some previously acquired information may be applied to the new situation to assist in producing a solution. However, each field of activity has its own uniqueness. The checklist given here is primarily oriented around manufacturing. In any service activity, such as an area of government activity, a suitable checklist would be quite different; the questions in it would reflect a codification suitable for retrieving the specialized knowledge pertinent to the particular activity under scrutiny.

Organization of the checklist.   The questions in the checklist given here are numbered to correspond to the class of change they refer to, but are listed in the order used, the higher classes being taken first, as we usually start with the highest class of change and work down to class 1.

CHECKLIST FOR POSSIBILITY GUIDE

5. Can a slightly different raw material be ordered or can the same material be ordered in a form that would be more advantageous? Can we change:
   a. Shape?
   b. Size?
   c. Packaging?
   d. Quantity packed together?
   e. Material?
   f. Amount of processing done by supplier?
   g. Color?
   h. Finish?
   i. Any other specification?
   j. The product so as to make any material or auxiliary material unnecessary?
4. Can the product be made, sold, or sent out in a more advantageous form? Can we:
   a. Modify design?
   b. Pack differently?
   c. Change finish?

    d. Change weight?

    e. Change tolerances?

3. a. Can we do the different jobs along the route between receiving and shipping in a different order?

    b. Is any step unnecessary?

      1. What does it accomplish?

      2. Why is it done?

      3. What would happen if it were not done?

    c. Can we combine any steps?

    d. Can we advantageously break any job into two or more separate operations?

2. a. Can any new tools or equipment or a change in the workplace make any job in the sequence easier? (This is almost always possible.)

    b. Can any tool or equipment be eliminated advantageously?

    c. Can any two tools be combined?

1. Can a new motion pattern make any job in the sequence easier? This is almost invariably true. Specific suggestions are usually more easily made after the method analyst is more familiar with the man-analysis techniques. Hence, at this point, a mere list of the possible jobs that may be looked into or some rough suggestions will probably be all the student is capable of. (The analyst also tries to eliminate motions. In actual practice, as will be evident later, this part of the analysis may be done in more detailed fashion.)

    Application of the checklist. With the aid of the questions on the checklist, the preliminary possibility guide of Figure 7.2 would be made up as follows (it is suggested that the reader check these to follow the procedure):

Suggestion on Preliminary Possibility Guide	Checklist Question that Suggested Possibility
1	5a
2	5f
3	4e
4	3c
5	2a
6	1
7	1

    Following this, the analyst would expand the possibilities into the form shown in Figure 7.3.

POSSIBILITY GUIDE

Name of Operation Mfg. Process, Armature Arm – 124 R

Operation number _____

File Number 124 R

Analysis by L. Edmond

Date 2/16

Class of Change	Hand and Body Motions	Tools, Workplace and Equipment	Process	Product Design	Raw Material
1	Distribute work to both hands on inspection	Tip fixture Scale to weigh count (on inspection)			
2	Balanced hand pattern	Dual inspection fixture, electric response (See above also)			
3	Fit into pattern of Press operation	Fixture convenient for press operator	Inspect at punch press as part of operation		
4	Check assembly motions for affect of redesign	check assembly fixtures for affect of tolerance	Eliminate inspection from press send to assembly	Redesign relay to allow more tolerance in arm	
5	Eliminated No inspection See box above	Simple die to blank, no forming	Eliminate inspection	No change In shape, no channel	Purchase formed sheet stock instead of channel

Figure 7.3. Detailed possibility guide for manufacture of relay armature arm.

When the possibility guides are made. The possibility guides were made, as is usual, prior to any other formal analysis, by a person somewhat familiar with the manufacture of the armature arm.

Making Decisions with the Aid of the Possibility Guides

On the basis of the possibility guide, his experience, and an analysis of the economic and psychological factors concerned, the analyst must also determine which one of the five areas of possible change is the most promising one to start with as a tentative objective. This is particularly important because the specific analysis technique chosen for the second step of the logical procedure will be a function of the area in which change is sought.

Class of change and technique selection. Changes starting with a class 1 (hand and body motions) or class 2 (work station) change will affect only the way an individual does his job and the equipment used; the work of the persons who work with the product (or paperwork) before and after the workplace being studied is not affected. Hence, in such cases, the second main step of the logical procedure will involve *man analysis*, or a technique to determine how the man performs his job, or, in the case of a job that is not yet being performed, how it is proposed that the man perform it.

Changes starting with a class 3 (process or work sequence), class 4 (output design), or class 5 (inputs) change will involve a study of the flow of the output, or a *product analysis* technique as the main second step.

Factors affecting choice of class of change sought. The particular class of change chosen as an objective in any case is a function of many factors. The higher classes of change often take longer to install, affect more people, and usually require higher authority. Hence, in addition to the relative desirability or apparent feasibility of the individual suggestions, the following factors must be taken into consideration in detail when making a final decision.

Economic and psychological factors affecting change include:

1. How great is the actual or expected volume, and how often does the job occur?
2. How long will the job exist?
3. How much time, per unit, is spent on the job?
4. How much time is available for working up the change?
5. How much equipment is already invested in the job?

6. How much analysis time will be required?
7. How much loss of production, sales, or service will occur during a change?
8. How much retraining will be required?
9. What is the possible saving or gain in value by means of the improvement?
10. What is the position of the analyst in the organization?
11. What are the personalities involved?
12. What are the policies that affect the problem?
13. Is the product for internal use or for customers?
14. What do the other groups in the organization that will be affected think of the feasibility of the suggestion?

The importance of each factor will vary from case to case. Their particular order, as given here, does not necessarily indicate their relative importance. The analyst must weigh and evaluate them for each situation.

In all cases the class of change selected, together with the physical characteristics of the job or process in question, would determine which of the analysis techniques would next be used. This relationship will come into sharper focus when the analyst acquires greater familiarity with all the techniques listed.

Functions of the Possibility Guide

The possibility guide serves the purposes of:

1. Aiding in systematically listing possible changes and collecting material from which to determine an objective.
2. Aiding in determining a suitable analysis technique.
3. Indicating which divisions in the organization will be affected.

It should be noted that these are the prime functions of the possibility guide.

In addition to the list above, the supervisor should find the possibility guide of considerable assistance in:

1. Reviewing an analyst's initial thinking and approving of a clear-cut objective.
2. Delineating the scope of a project assigned to an analyst.

In any case the analyst should never lose sight of the desirability of eliminating a job. This is the ultimate improvement for any single job. Consequently, even when the objective is a class 1 or 2 change, it is usually good practice to make a rough product analysis after the possibility guide and before using any man analysis technique. This will allow the analyst to be reasonably certain that he is not analyzing an unnecessary job or an unnecessary phase of the job.

Summary—Possibility Guide

   1. Uses:
   a. Aiding in systematically listing possible changes and collecting material from which to determine an objective.
   b. Aiding in determining a suitable analysis technique.
   c. Indicating which divisions in the organization will be affected.
   2. How made:
   a. A form such as that shown in Figure 7.2 may be used or a blank sheet of paper may be substituted.
   b. The analyst states his criterion of success.
   c. The analyst makes a rough determination of the degree of change that is warranted.
   d. The analyst, with the aid of a checklist, lists the possibilities that occur to him, tempering his imagination with the decision reached in the previous step, identifies each possibility as to class of change and other areas affected, and gives each an identifying number.
   e. The analyst expands each possibility on a detailed possibility guide form such as Figure 7.3 or on a blank sheet of paper, so that he may examine the consequences of each item.
   3. How used:
   a. With the aid of business and manufacturing knowledge and with consideration of the economic and psychological factors involved, the analyst selects which class of change or which possibility appears most feasible. Several persons may well participate in this step.
   b. With this decision and the nature of the job in mind, the analyst selects an appropriate analysis technique.
   4. What then:
   The analyst is now ready for step 2 (analysis) of the logical approach. *Note:* The trained analyst may not make a formal possibility guide, yet the thinking should follow its pattern. Experience suggests, however, that actually following the steps is usually much more productive than the informal approach.

PROBLEMS

7.1. Make a preliminary and a detailed possibility guide of the problem area, "to augment the water supply of a seacoast town."

7.2. Make a preliminary and a detailed possibility guide with respect to the collection of cash fares on:
(a) An intracity bus system.
(b) A metropolitan subway.
Note: State any assumptions made with respect to current methods.

7.3. Make a preliminary and a detailed possibility guide with respect to using intermittently a large unabridged dictionary stored upright, with other books, on a shelf in an office.

7.4. Make a preliminary and a detailed possibility guide with respect to reducing the danger of blowing out your dual-voltage electric razor (adjustable to either 110 or 220) as you travel from place to place and the voltage varies from 110 to 220.

7.5. Prepare a preliminary and a detailed possibility guide for the following tasks which were described in Chapter 1.
(a) Enveloping plant magazine.
(b) Assembling roller wheels.
(c) Sorting day-old chicks.
(d) Handling materials to sandblast.
(e) Manufacturing transformer core.

7.6. (a) Make a preliminary and a detailed possibility guide for some job connected with a previous job or available to you for actual observation. (A household task may be used.)
(b) On the basis of the list of economic and psychological factors affecting change, select a suitable tentative objective for a motion study analysis of the job used for part (a). Justify your selection.

7.7. A professional society mails to its members, each month, a notice of the next meeting location and the subject to be presented. The notice is offset-printed on paper, $8\frac{1}{2}$ by 11 inches. It is folded into thirds and together with a business reply card is hand-stuffed into a No. 10 envelope and sent by first-class mail. The meetings are dinner meetings and meals are planned on the basis of the cards returned. The society has 100 members and about 80 come to each meeting.
(a) Prepare a preliminary and a detailed possibility guide for "notification of meeting." Select the most feasible possibility. Justify your choice.
(b) Assume that the society has 1,000 members with a typical attendance of 100. Prepare a preliminary and a detailed possibility guide. Select the most feasible possibility. Justify your choice.

7.8. A company manufacturing children's games has a platform in one of its boxes of games. The platform is made of gray chipboard and is stayed in the corners, similar to a box cover. It is then wrapped with colored paper on a box-cover wrapping machine. The covered platforms are moved by skid to the assembly line, where the game, box cover, and bottom are wrapped on wrapping machines. The platform and other items are hand-assembled into the box on a conveyor between the bottom-wrap and cover-wrap machines. Make a preliminary and a detailed possibility guide for the platform manufacture. If any additional information is needed, make reasonable assumptions and record these.

7.9. Examine the method used within your organization to collect and distribute mail. Make a preliminary and a detailed possibility guide for:

(a) In-plant mail.

(b) Regular mail.

8

WORK ACTIVITY ANALYSIS AND WORK SAMPLING

Introduction

Where the techniques are used.   In any office or plant there are usually jobs that appear to possess no cycle or repetitive pattern. Supervisory tasks tend to fall into this category. Many clerical, service, and maintenance jobs fit into a similar classification.

Purposes served by the techniques.   There are numerous reasons why one might wish to improve or facilitate such tasks: the backlog of work may be growing; the time lag between assignment and completion of work may be excessive; important but postponable aspects of the work may be neglected; the amount of manpower expended may be thought excessive; the effort required to perform the necessary work may be unreasonable. Indeed, the pressure for improvement may be intense because of the possible widespread influence, on other tasks, of a job of this general nature.

The reasons for choosing work activity analysis or work sampling. It should be obvious that a possibility guide for a task of this type is difficult to construct without some means of obtaining additional insight into the task. There is no clear-cut "present method" as with a repetitive or cyclic job. Likewise, a detailed motion study analysis of the manual work, made for a period of time, might yield helpful information but would probably include so much extraneous detail that it would be hard to find the real problem. Also, it would be very time consuming to prepare such an analysis.

An analysis of "products" that passed through such a job might also provide too great a bulk of data to deal with economically. Further, part of the solution to more productivity may be in possible changes with respect to the functions assigned such a worker, the sequence of the numerous tasks performed in discharging these functions, the organization, the physical method of performing, and the facilities used for each task. Only the effect of the physical method and the facilities would be revealed in the more common motion study analysis. However, such information might be deeply buried in the data and hard to discern.

List of available techniques. For jobs such as those being discussed, three techniques are available for performing the second step, analysis, of the scientific method: (1) work activity analysis, (2) work sampling, and (3) memomotion study. The first two are discussed in this chapter. Memomotion study, which is a multiuse technique, is presented in Chapter 19. It will be easier to persent the material later when it will be appropriate to examine all the uses.

Definition of work activity analysis. A *work activity analysis* is a chronological record, usually accompanied by a summary tabulation, of the nature of the activities performed, work-units produced, and the time spent at each activity by an individual performing a variety of tasks.

Definition of work sampling. "A *work sampling study* consists of a large number of observations taken at random intervals. In taking the observations, the state or condition of the object of study is noted, and this state is classified into predefined categories of activity pertinent to the particular work situation. From the proportions of observations in each category, inferences are drawn concerning the total work activity under study."[1]

Comparison of work activity analysis and work sampling. Work activity analysis observations preserve the sequence of work in the records, but they

---

[1]R. E. Heiland and W. J. Richardson, *Work Sampling*. New York: McGraw-Hill Book Company, 1957, p. 1.

may require a considerable expenditure of time, and are therefore seldom extended over a long period. Work sampling observations, on the other hand, do not record the sequence but can be made in an intermittent fashion and do not constitute a heavy workload. Hence, work sampling studies are often used to cover an extended period. In some cases the techniques are used to complement each other.

On jobs where the daily pattern is very different from day to day, work sampling yields, for a given amount of study effort, a more complete picture. On jobs where the daily pattern is similar from day to day, work activity analysis gives a more complete picture more rapidly. Both techniques are samples, in that only a portion of the total work is studied. Work activity analysis can be described as intensive sampling, while work sampling is usually an extensive sampling.

Either technique may reveal the following:

1. The need for a change in functions assigned an individual.
2. Which parts of the job occupy enough of the total time to merit being studied in detail with other techniques, such as process chart—man analysis, process chart—product analysis, operation chart, and so forth.

In addition, the work activity analysis may more readily reveal the need for changing the timing or sequence of the numerous tasks performed.

Because of the varied nature of the work to which such analyses are commonly applied, and because insight into the nature of the specific work is the first objective of such analyses, the basis for separating the work into pertinent steps is usually different from job to job.

Work Activity Analysis

Description of technique. With work activity analysis the division of the job into steps or categories of work is frequently on the basis of what might be called *outputs*. That is, the categories are based on the different results produced for each separate situation calling for a pattern of action or response. For example, with a clerk we might separate work into these categories: (1) letters typed, (2) phone calls answered, (3) persons located, (4) filed material located and delivered, and so forth. Of course, if the study is also going to delve into the work-generating mechanism in an attempt to reduce the amount of outputs needed, then these categories for activity must be compounded so that (1) *letters typed* is divided into subcategories. Each of the subcategories will represent letters generated from a different cause system. The same subcategorization might be made with the other categories: 2, 3, 4, and so forth.

How the technique is applied. A work activity analysis is frequently made by the person doing the work rather than by a separate observer, although this is not a necessary aspect.

A form such as the one shown in Figure 8.1 is commonly used, although an ordinary ruled sheet of paper easily may be divided into columns for this purpose. The essential features of work activity analysis are:

1. The chronological listing of work as performed.
2. The indication of the time spent at each task, as it occurs.
3. The noting of the work count for each occurrence of each task.

The *work count* is the quantity of actions performed or work-units produced of the type described, or the physical count of the product, such as pages typed, entries made, items filed, workers talked to, and so forth, as appropriate to the task being recorded. It is helpful if the work-unit selected for each separable activity is as closely related as possible to the amount of work called for (e.g., for a copy typist, "pages typed" rather than "documents copied"). As will be more obvious later, a work activity analysis shows the time for each separate action in the work count; a work sampling study does not.

When prepared by the person being studied. When the person doing the work is asked to maintain his own record, considerable care should be exercised to be certain that the supervisor and the worker both understand the purpose of the analysis. Their sincere cooperation must be obtained, or the final record may contain an excessive bias in any of several directions.

A common procedure is to:

1. Enlist the cooperation of the supervisor.
2. Explain to the worker the purpose of the record and how to keep it.
3. Help the worker until the procedure is fully understood.
4. Ask the worker to keep the record for a period adequate to sample that activity. The worker's and supervisor's advice may be sought on this aspect.
5. Analyze the summary of the data with the assistance of those who collected the data.

When a group studies itself. When a coordinated group is being studied, it is usually preferable to have the entire group maintain records simulta-

WORK ACTIVITY ANALYSIS 531216-01

Name, working title, grade, rank or classification. (Please print)
JOSEPHINE GALLAPAGOS, OFFICE MANAGER, R4

Organizational segment or unit, supervisor, date
CONF. CENTER, RLM, 16 DEC

TIME SPENT		HOW SPENT	UNITS	REMARKS
Began	Blank	Describe forms handled, work data assembled, errands run, idle periods, etc.	Quantity of actions completed	
8:45	5	Talked on phone to publicity section	1	
8:50	2	Answered phone call	1	
8:52	4	Entered part time help hours in ledger	4 entries	
8:56	4	Distributed checks to part time help	4	
9:00	4	Prepared material for publicity section	1 page	
9:04	1	Phone call for Mr. Jones	1 call	
9:05	1	Phone call for me from Mr. Alt wishing copy of letter he could not find	1	
9:06	6	Work conference with Clerk Typ. on what has transpired; what's to do	1	
9:12	5	Search for Mr. Alt's letter	1	
9:17	3	Organized work for day	1	
9:20	2	Phone call, Mr. Jones	1	
9:22	3	Called Mr. Smeet for Mr. Jones	1	
9:25	4	Called Mr. Toner for Mr. Jones	1	
9:29	1	Found Mr. Alt's letter and sent out	1	
9:30	2	Helped Mr. Gordo find scissors	1	
9:32	13	Conference with Mr. Gordo on layout for bulletin	1	
9:45	2	Began ditto's; material for Mr. Elm	0	
9:47	10	Located material in file for Mr. Jones	3	
9:57	4	Continued Elm's ditto's	0	
10:01	5	Work conference with Mr. Jones	1	
10:06	2	Explained and gave ditto work to part-time help (kept some)	1	
10:08	9	Ditto work, continued	1	
10:17	8	Mail opening and distribution	14 pcs.	
10:25	15	Ditto work, continued	2	
10:40	10	Helped locate file material for Jones	4	
10:50	11	Ran off dittos	2, 50 each	
11:01	2	Turned over ditto stapling and finishing to part time help	1	
11:03	9	Called M.G.S. Co. with information requested in morning mail; sent out bulletins	1	
11:12	2	Phone call for Mr. Jones	1	
11:14	3	Greeted visitors and sent to Jones	2	
11:17	2	Phone call requesting information	1	
				continued

Figure 8.1. Work activity analysis form for manager of small office.

neously. The explanation may be made simultaneously to all members of the group and all questions answered. Then each member of the group may be individually assisted to ensure that each individual starts and maintains an adequate record.

Case I—A Work Activity Analysis of the Office Manager of a Small Office

A section of a work activity analysis kept by the manager of a small office is shown in Figure 8.1. A summary of this section of the study follows. (To be meaningful, the summary used in practice would have to cover a longer period. This short summary is given here so that the reader can check against Figure 8.1 and understand the procedure of summarizing.)

Type of Activity	Minutes	Number of Occurrences
Type, run ditto, staple, etc.	41	5
Phone calls	31	10
Search for files or materials	28	5
Receive instructions	18	2
Instruct or plan	13	4
Handling incoming mail	8	1
Clerical duties related to part-time help	8	2
Originate material	4	1
Greet and direct visitors	3	1
Total	154	31

Why the study was made. The analysis was made because of an excessive time lag on high-priority work, difficulty in determining the status of various items of work in-process, and a need to increase the capacity of the office to handle an anticipated increase in work without an increase in office staff.

What the study showed. A study of the summaries of several days' records, together with the detailed records, indicated that the predominant activities were, in order of time spent:

FOR THE OFFICE MANAGER

1. Phone activity.
2. Typing and miscellaneous routine clerical activities.
3. Searching for filed material, supplies, or equipment.
4. Planning, supervision, and receiving and giving instructions.
5. Mail sorting.

FOR THE CLERK-TYPISTS

1. Various clerical activities.
2. Package-handling errands.
3. Receiving work assignments and instructions.

FOR THE PART-TIME HELP

1. Mailing activities.
2. Errands for supplies.
3. Delays for work assignments.
4. Duplicator operation.
5. Stapling materials.

A critique of the data. Considering these data together with the checklist, the following were discerned as defects in the office operation, as then staffed:

1. Inadequate time was given to planning, or planning was inadequate. Errands were not combined for part-time help. The clerk-typist was not excluded from this activity. Work to be done was not reorganized periodically and not "farmed out" effectively.
2. Inadequate gathering of similar material made it difficult to obtain extended activity of a given type by the clerk-typist and part-time help (i.e., duplicator operation, copy typing, and so forth). Further, work was not categorized nor were standard instructions prepared by category. Hence, each assignment consumed instruction time.
3. The office manager was not segregating and holding suitable work that could be performed between phone calls and supervisory activities. Probably doing an excessive amount of clerical work; neglecting planning.
4. Inadequate organization of the supply area and poor filing necessitated excessive time for locating items and discouraged the people served from getting their own materials.
5. The desk of the office manager was not arranged for the basic work of phone activity, note taking, and planning. Space for fill-in work was not available without usurping space from major functions.

Results obtained. A discussion of these items with the office manager and subsequent action (as implicit in each item of the critique) was responsible for a considerable increase in office productivity and effectiveness. The problems that had prompted the study disappeared.

CHECKLIST FOR WORK ACTIVITY ANALYSIS
AND FOR WORK SAMPLING

1. Are tasks appropriate to one job?
2. Are tasks occurring in too erratic a fashion?
3. Can like tasks be batched?
4. Can planning make continuous activity at each task possible for a longer period of time?
5. Can categories of assignments be identified and standard instructions prepared and communicated to the worker by an attached slip?
6. Are the workplace or workplaces appropriate for the distribution of tasks?
7. Can any work be eliminated by physical rearrangement or by use of different equipment?
8. Would a redistribution of work among the group increase overall productivity?
9. Would a different order of doing tasks eliminate or facilitate the work?
10. Is any recurring type of task of sufficient magnitude to justify more detailed study with other motion study techniques?

Work Sampling

Description of the technique. Work sampling, like work activity analysis, may be made by the worker or supervisor rather than by a staff analyst. In this respect, both of these techniques have certain inherent advantages, as one might deduce from the discussion in Chapter 2.

A work samping study differs from a work activity analysis in that the observations are made at intervals rather than continuously. This procedure is also described in the early literature as *ratio-delay study*; it was first described by L. H. C. Tippett.[2] This intermittancy requires some additional preplanning as compared to work activity analysis, although the subsequent carrying out of the study may follow a pattern very similar to that of a work activity analysis.

How the technique is applied. The first step is to define and prepare a list of the states or conditions in terms of which the object of the study can

---

[2]For a historical background, see L. H. C. Tippett, "Ratio-Delay Study," *Journal of Textile Institute Transactions*, Vol. 36, No. 2, 1935; R. L. Morrow, *Time Study and Motion Economy*. New York: The Ronald Press Company, 1946, pp. 176–199; C. L. Brisley, "How You Can Put Work Sampling to Work," *Factory*, Vol. 110, No. 7, 1952, pp. 84–89; J. S. Petro, "Using Ratio-Delay Studies to Set Allowance," *Factory*, Vol. 106, No. 10, 1948, p. 94.

be classified. If we are studying the activities of a person, these conditions may be such as to separate each *"where, doing what, with what."* The particular listing selected should be related to the object of the study. As a general principle, however, one should bear in mind that excess details obtained, as with work activity analysis, may be subsequently combined, but actions not separately identified when making the study cannot subsequently be separated. Therefore, too much detail is less likely to cause difficulty in interpreting the results than too little.

The second step in planning a work sampling study is to determine a suitable number of randomized observation times. The timing of the observations must be totally unrelated to the natural periods of the events being studied; randomization or stratified randomization is a means of obtaining this unrelatedness.

Why a number of observations are needed. The total number of observations made must be sufficient to obtain a reliable sample of recordings, and the recordings should be spread out over a sufficient number of days in order to include an adequate sample of days. The number of observations required may be computed after one has determined the objective of the study. If we are merely seeking insight into the task, loose limits of accuracy may be satisfactory. Studies where we may be seeking percentages to apply in work measurements affecting wages or task goals (described later in this book) require much more accuracy.

Computing the accuracy of the sample. The formulas that follow state the mathematical characteristics of the sample and may be used to compute the sample size for any desired limits of accuracy.[3]

$$\mathrm{SD}_p = \sqrt{\frac{p(1-p)}{N}}$$

where

SD_p = measure of the variability of the value of p obtained from N observations (called the standard deviation), possessing the following properties: 68 per cent of samples of N size would give a value for p within $p \pm \mathrm{SD}_p$; 95 per cent of the samples of N size would give a value for p within $p \pm 2\mathrm{SD}_p$

p = per cent of events of one type contained in N observations of the situation, expressed as a decimal

N = total number of observations

---

[3]For a complete treatment of this type of problem, see any text on statistical quality control. The usual Greek symbols have been replaced with acronyms using Roman letters in that I see no advantage in using Greek.

Hence, if we wish to determine a value for p with our sampling error reduced to the point where we may say the chances are 95 out of 100 that p is correct within ± 1 per cent, then

$$2SD_p = 0.01 = 2\sqrt{\frac{p(1-p)}{N}}$$

and

$$N = 40,000[p(1-p)]$$

If we wish to determine a value for p with our sampling error reduced to the point where we may say the chances are 95 out of 100 that p is correct to within ± 1 per cent of p, then

$$2SD_p = 0.01p = 2\sqrt{\frac{p(1-p)}{N}}$$

and

$$N = \frac{40,000}{p^2}[p(1-p)]$$

If we wish to determine a value for p with our sampling error reduced to the point where we may say the chances are 95 out of 100 that p is correct to within ± 5 per cent of p, then

$$2SD_p = 0.05p = 2\sqrt{\frac{p(1-p)}{N}}$$

and

$$N = \frac{1,600}{p^2}[p(1-p)]$$

A table for determining desired sample size. Table 8-1 gives the number of observations for 95/100 probability of p being within ± 5 per cent of the correct percentage, p being within ± 1 per cent of the correct percentage, p being within an error range equal to ± 1 per cent of p, and p being within an error range equal to ± 5 per cent of p. Values of p are given from 1 to 99 per cent.

Most of the values of N given in Table 8-I are larger than those required for qualitative studies. However, Table 8-I will be referred to later with respect to work measurement. It seemed desirable to give it but once, and in complete form. For the most part, in qualitative studies, the column headed "5 Per Cent of Total" will be a reasonable guide.

Planning the length of a study. When we have decided upon the total number of observations to be made, we may divide this by either the number

Table 8-I. VALUES OF N (NUMBER OF OBSERVATIONS) AT 95/100 PROBABILITY OF NOT EXCEEDING ERROR INDICATED, FOR VALUES OF p (PER CENT OF ACTIVITY)

		Error					Error		
p (%)	5% of Total	1% of Total	1% of p	5% of p	p (%)	5% of Total	1% of Total	1% of p	5% of p
1	16	396	3,960,000	158,400	51	400	9,996	38,431	1,537
2	32	784	1,960,000	78,400	52	400	9,984	36,923	1,477
3	47	1,164	1,293,000	51,720	53	399	9,964	35,472	1,419
4	62	1,536	960,000	38,400	54	398	9,936	34,074	1,363
5	76	1,900	760,000	30,400	55	397	9,900	32,727	1,309
6	92	2,256	626,667	25,067	56	395	9,856	31,429	1,257
7	102	2,604	531,429	21,257	57	392	9,804	30,175	1,207
8	118	2,944	460,000	18,400	58	390	9,744	28,966	1,159
9	131	3,276	404,444	16,178	59	387	9,676	27,797	1,112
10	144	3,600	360,000	14,400	60	384	9,600	26,667	1,067
11	157	3,916	323,636	12,945	61	381	9,516	25,574	1,023
12	169	4,224	293,333	11,733	62	377	9,424	24,516	981
13	181	4,524	267,692	10,708	63	373	9,324	23,492	940
14	193	4,816	245,714	9,829	64	369	9,216	22,500	900
15	205	5,100	226,667	9,067	65	365	9,100	21,538	862
16	216	5,376	210,000	8,400	66	360	8,976	20,606	824
17	226	5,644	195,294	7,812	67	354	8,844	19,701	788
18	236	5,904	182,222	7,289	68	349	8,704	18,824	753
19	246	6,156	170,526	6,821	69	343	8,556	17,971	719
20	256	6,400	160,000	6,400	70	337	8,400	17,143	686
21	266	6,636	150,476	6,019	71	330	8,236	16,338	654
22	275	6,864	141,818	5,673	72	323	8,064	15,556	622
23	284	7,084	133,913	5,357	73	316	7,884	14,795	592
24	292	7,296	126,667	5,067	74	308	7,696	14,054	562
25	300	7,500	120,000	4,800	75	300	7,500	13,333	533
26	308	7,696	113,846	4,554	76	292	7,296	12,632	505
27	316	7,884	108,148	4,326	77	284	7,084	11,948	478
28	323	8,064	102,857	4,114	78	275	6,864	11,282	451
29	330	8,236	97,931	3,917	79	266	6,636	10,633	425
30	337	8,400	93,333	3,733	80	256	6,400	10,000	400
31	343	8,556	89,032	3,561	81	246	6,156	9,383	375
32	349	8,704	85,000	3,400	82	236	5,904	8,780	351
33	354	8,844	81,212	3,249	83	226	5,644	8,193	328
34	360	8,976	77,647	3,106	84	216	5,376	7,619	305
35	365	9,100	74,286	2,971	85	208	5,100	7,059	282
36	369	9,216	71,111	2,844	86	193	4,816	6,512	261
37	373	9,324	68,108	2,724	87	181	4,524	5,977	239
38	377	9,424	65,263	2,611	88	169	4,224	5,455	218
39	381	9,516	62,564	2,503	89	157	3,916	4,944	198
40	384	9,600	60,000	2,400	90	144	3,600	4,444	178
41	387	9,676	57,561	2,302	91	131	3,276	3,956	158
42	390	9,744	55,238	2,210	92	118	2,944	3,478	139
43	392	9,804	53,023	2,121	93	102	2,604	3,011	120
44	395	9,856	50,909	2,036	94	92	2,256	2,553	102
45	397	9,900	48,889	1,956	95	76	1,900	2,105	84
46	398	9,936	46,957	1,878	96	62	1,536	1,667	67
47	399	9,964	45,106	1,804	97	47	1,164	1,237	50
48	400	9,984	43,333	1,733	98	32	784	816	33
49	400	9,996	41,633	1,665	99	16	396	404	16
50	400	10,000	40,000	1,600					

of observations that it is convenient to make per day, or we may divide by the number of days we wish to study. Dividing by the number of observations it is convenient to make per day gives us the number of days we must study. Dividing by the number of days we wish to study gives us the required number of observations per day. In either case, since the computation is only a guide, modifications may be made, provided they are compatible with the objective of the study.

Planning the observations. The actual observation times must be arranged in a random pattern. There are four general methods of doing this.

1. The first hour of the working day may be identified by the numeral 1, the second hour by the numeral 2, and so on. A table of random numbers is used to obtain a series of three-digit figures, the first digit representing the hour of the working day, and the next two digits, the minutes. Numbers representing hours not in the working day, or impossible minute values, are discarded. A sufficient number is obtained to give the required observation times for each day of the study. Each day should have a separate list. For example, let us say that our working day runs from 8 A.M. to 5 P.M. Let 8 A.M. be designated by the figure 1, 9 A.M. by 2, and so forth, down to 4 P.M. by 9. A short section of a random-number table and its interpretation are illustrated here[4]:

Random Numbers	Interpretation
907	4:07 P.M.
882	Impossible minute value; discard
544	12:44 P.M.
720	2:20 P.M.
838	3:38 P.M.
010	Impossible hour value; discard
413	11:13 A.M.

For actual use, the list of observation times would be arranged in time sequence.

2. A second method is to stratify the observations by hours. To do this, we divide the number of observations to be made per day by the number of working hours. Following this, we again use the random-number table to give us two-digit figures to represent the minutes for each hour.

---

[4]Table from Heiland and Richardson, op. cit., pp. 226–227.

Using the same working day as before, and the same values from the random-number table, the following would result, for one observation per hour:

90	Impossible minute; discard
78	Impossible minute; discard
82	Impossible minute; discard
54	8:54 A.M.
47	9:47 A.M.
20	10:20 A.M.
83	Impossible minute; discard
80	Impossible minute; discard
10	11:10 A.M.
41	12:41 P.M.

The list is in chronological order. If more than one observation per hour was required, the values for each hour would need rearrangement for actual use. Each day should have a different list of time values.

3. If a great number of observations is required, an observer may go continuously through the area where the observations are to be made, randomizing his route but constantly making observations.

4. A series of cards (i.e., 3- by 5-inch file cards) may be used. We may number these cards in order, one for each minute of the working day, say, 8:01, 8:02, 8:03, and so on, so as to have one card for each minute of the working day. The cards may then be shuffled and a list prepared for a day by taking the necessary number off the top of the pile. The times are then arranged in time sequence. The cards used are then returned to the deck and the procedure is repeated for each day to be studied.

In practice, the work of preparing categories, observation times, and lists may be done by an analyst, or by (or with) the supervisors of the working group. In the case of professional activities, the person on whom the observations are to be made may do this preparatory work himself.[5]

Making the actual observations. The actual observations may be taken by an analyst, by the supervisor, or by the worker. It is common practice when performing work sampling for methods study to have the supervisors make the observations. This not only reduces the workload of making such

---

[5]M. E. Mundel, "Motion and Time Study in Dentistry," *Journal of the American Dental Association*, Vol. 57, No. 10, 1958, pp. 520–524.

studies, but helps obtain participation, the desirability of which was indicated in Chapter 2.

The common procedure is similar to that given for a work activity analysis:

1. Enlist the cooperation of the supervisor and explain the purpose, nature, and method of making the study.
2. Explain to the workers the purpose of the study.
3. Help the supervisor prepare the necessary lists of categories and times.
4. Help the supervisor for a sufficient time to assure that he fully understands.
5. Have the supervisor obtain observations over the required period.
6. Analyze the summary of the data with the assistance of those who made the observations.

Case II—A Work Sampling Study of a Warehouse Crew

A large midwestern manufacturer had established a large warehouse to reduce seasonal manufacturing peaks. During peak activity the warehouse work-force was augmented by workers drawn from the production areas.

Why the study was made. The management considered the required amount of augmentation to be excessive. The industrial engineering department was requested to examine and improve the methods used by the 94-man warehouse crew.

To gain insight into the activities performed, the industrial engineering department decided to make a work sampling study prior to the use of other more intensive motion study analyses.

How the study was made. A series of meetings was held with the nine foremen of the warehouse. The objectives and methods of work sampling were explained. The foremen assisted in the compiling of a list of activity categories. This list was duplicated on $8\frac{1}{2}$ by 11-inch forms as shown in Figures 8.2a and b.

Each foreman was supplied with 10 sets of forms and a table of random numbers. Using the table of random numbers and method 2 previously described (drawn from random numbers and stratified by hours), each foreman inserted a specific reading time for each of the 8 hours of each of the 10 work days to be studied. (Inasmuch as the day started at 7:30 A.M., the interval between 7:30 A.M. and 8:30 A.M. was used as the first hour of the day and

WORK SAMPLING - MATERIALS HANDLING DATA SHEET _____DATE____

	7:30	8:30	9:30	10:30	12:00	1:00	2:00	3:00
PRODUCT								
Carry or lift								
Reach								
Roll drums								
Place or stack by hand								
Sample								
Package (to group)								
PACK (For Shipment)								
Make Wood boxes								
Staple boxes								
Fill								
Tape								
Band								
Stencil								
Mark or stencil								
Tie strings and bind								
Relieve jam on conveyor								
Weigh								
Check and inspect								
TRUCKING								
Loading								
Spotting								
Riding								
Empty								
Loaded								
Leading								
Empty								
Loaded								
Repair								
Inspect								
Walk - no load								
PALLETS								
Inspect								
Aside for repair								
Repair								
Stacking								
Unstacking								
CLERICAL								
Check orders								
Run press								
Write								
Telephone								
Check inventory								

Figure 8.2a. Work sampling observation sheet for warehouse study (*continues*).

so forth.) No samples were taken during the lunch period, 11:30 A.M. to 12:00 noon. Each foreman had a different series of times; each day was different.[6]

Each day, at each reading time on the sheet for that day, each foreman made a tour of the areas covered by his workers and inserted a tally mark

---

[6]Each foreman started at a different place on the random-number table so as to avoid duplication.

WORK SAMPLING - MATERIALS HANDLING DATA SHEET_____DATE_____

	7:30	8:30	9:30	10:30	12:00	1:00	2:00	3:00
DUNNAGE								
Prepare								
Place								
Remove								
CAR OR TRUCK								
Open								
Close								
Prepare								
Install load heaters								
Talk with supervisor or expeditor								
IDLE								
For material, truck elevator or stock								
For instructions								
For machine breakdown								
Avoidable								
Personal								
Break								
MISCELLANEOUS								

Figure 8.2b. (*Concluded*) Work sampling observation sheet for warehouse study.

in the column opposite the appropriate activity, one tally mark for each worker. The foremen avoided using a fixed route through their areas to avoid a possible bias of the tallies. (Actually, each foreman had 11 sets of sheets, but the first set was used for a practice run and was discarded.)

What the study showed. The summary of all the tallies and the percentage of time indicated for each activity are given in Figure 8.3.

A critique of the data. The data were examined with the checklist for work sampling previously given. A study of the data indicated that substantial opportunity lay in the *Idle* and *Clerical* categories and in the time for *Carrying or lifting* product.

Innovations made. Delays at the elevator were reduced by equipping the elevator with a truck to serve loading and unloading areas at each floor; avoidable delay was reduced by better supervision; clerical time was greatly reduced by supplying two full-time clerical personnel who were much more efficient at this work than the material handlers whose duties had previously included this activity; the areas where product was lifted or carried were subsequently studied in detail with other methods analysis techniques.

Summary—Work Sampling—Materials Handling Data Sheets

	Totals	Per cent
TRUCKING		26.1
Loading	61	.8
Spotting	105	1.4
Riding		
Empty	502	6.8
Loaded	515	6.9
Leading		
Empty	177	2.4
Loaded	171	2.3
Repair	18	.2
Inspect	16	.2
Walk—no load	381	5.1
IDLE		26.1
For material, truck elevator		
or stock	632	8.5
For instructions	37	.5
For machine breakdown	5	.1
Avoidable	344	4.6
Personal	176	2.4
Break	743	10.0
PRODUCT		17.1
Carry or lift	824	11.1
Reach	182	2.5
Roll drums	22	.3
Place or stack by hand	189	2.6
Sample	16	.2
Package (to group)	26	.4
PACK (for shipment)		8.0
Make wood boxes	8	.1
Staple boxes	23	.3
Fill	47	1.6
Tape	25	.3
Band	—	—
Stencil	11	.1
Mark or stencil	110	1.5
Tie strings and bind	51	.7
Relieve jam on conveyor	1	.0
Weigh	16	.2
Check and inspect	312	4.2
CLERICAL		7.7
Check orders	197	2.7
Run press	30	.4
Write	167	2.3
Telephone	86	1.2
Check inventory	312	4.2
CAR OR TRUCK		3.9
Open	5	.1
Close	22	.3
Prepare	78	1.1
Install load heaters	—	—
Talk with supervisor or		
expediter	179	2.4

	Totals	Per cent
DUNNAGE		1.7
Prepare	26	.4
Place	55	.7
Remove	44	.6
PALLETS		1.1
Inspect	2	0
Aside for repair	1	0
Repair	5	.1
Stacking	39	.5
Unstacking	37	.5
MISCELLANEOUS		8.1
Clean up area	25	.3
Deliver orders	33	.4
Replace tools	1	0
Making glue	3	0
Receiving instruments	3	0
Farmed out	2	0
Cutting stencil	15	.2
Moving equipment	6	.1
Assemble metal shelves	3	0
Hammer nail	3	0
Fill out time card	4	.1
Carry dunnage	2	0
Unload dunnage	4	.1
Deliver dunnage	2	0
Paint equipment	51	.7
Attend postal clinic	3	0
Vacation	4	.1
Check needs of blocking	2	0
Expediter talks to worker	81	1.1
First Aid	8	.1
Check loading	18	.2
Copy miscellaneous items	2	0
Seal cars	9	.1
Operate elevator	17	.2
Safety meeting	20	.3
Give instructions to truck		
driver	1	0
Sharpen pencil finished	1	0
Observing trains of goods	120	1.6
Pressing conveyor switch	22	.3
Miscellaneous	111	1.5
Lab delivery	3	0
Inventory meeting	12	.2
Patching bag	1	0
Pick up orders	2	0
Tank car sampling	2	0
Fill drums	1	0
Wiping drums	1	0
Paint and clean printing		
press	5	.1
Emptying cans	4	.1
Cut and scrape obsolete labels	4	.1
Charging conveyor truck	3	0
Intercommunications	4	.1
GRAND TOTAL	7423	99.7

Figure 8.3. Summary of data from work sampling study of warehouse crew.

Other savings. It is worth noting that previous to the work sampling, the foremen had felt that pallet repair was a major problem. This study indicated that this was a small problem and thus saved the time for a needless major investigation of this aspect of the work.

Summary—Work Activity Analysis

1. *Uses:*
   a. For analysis of a job that appears to possess no cycle or repetitive pattern.
   b. May lead to:
      1. Changes in functions assigned.
      2. Changes in organization or sequence of tasks.
      3. Determination of sections of total task worth analyzing with other motion study techniques.
2. *How made:*
   a. A form such as Figure 8.1 or a ruled, columned sheet of paper may be used.
   b. For a selected period, the worker usually keeps a chronological record of tasks performed, time spent on each task, and the work count for each occurrence of each type of task. This requires that the supervisor's cooperation be obtained and the worker be:
      1. Instructed as to the purpose of the study.
      2. Taught to keep the record.
      3. Checked as to method of recording.
3. *How used:*
   a. The data gathered are summarized in a fashion appropriate to the specific objectives of the study.
   b. The summaries and the details are scrutinized, usually by staff, supervisor, and worker, with respect to the checklist given and other appropriate questions.
4. *What then:*
   a. A revised general pattern may be evolved to achieve the objectives sought. A method of grouping activities for more expeditious handling or method of facilitating time-consuming ones may be evolved.
   b. Certain tasks may be designated for study with other motion study techniques.

Summary—Work Sampling (for Method Analysis)

1. *Uses:*
   a. For analysis of a job that appears to possess no cycle or repetitive pattern, particularly when a considerable period is required for the occurrence of all phases and activities within the job.

b. May lead to:
1. Changes in functions assigned.
2. Changes in organization of task.
3. Determination of sections of total task worth analyzing with other motion study techniques.

2. How made:
a. An objective is stated.
b. A set of categories is defined.
c. Using estimates of the magnitude of each category and a determination of the accuracy required, a sample size (number of total observations to be made) is determined. The number per work period is determined and observation times are selected so as to randomize these times.
1. The supervisors and workers are informed of the purpose and method of the study.
2. Their assistance is obtained as needed.
3. If the supervisor is to make the observations, he is so taught.
d. The observations are made.

3. How used:
a. The data gathered are summarized in a fashion appropriate to the specific objectives of the study.
b. The summaries and the details are scrutinized, usually by staff, supervisor, and worker, with respect to the checklist given and other appropriate questions.

4. What then:
a. A revised general pattern may be evolved to achieve the objectives sought. A method of grouping activities for more expeditious handling or a method of facilitating time-consuming ones may be evolved.
b. Certain tasks may be designated for study with other motion study techniques.

PROBLEMS

8.1. With respect to the procedure of having a worker prepare a work activity analysis, it was indicated that sincere cooperation must be obtained to avoid bias in the data.
(a) With poor cooperation, what kinds of bias might be obtained?
(b) What might motivate each separate kind of bias given in answer to part (a)?

8.2. With respect to work sampling, what might cause bias in the data obtained if observations were made at the same time every day?

8.3. Describe, in detail, how a work activity analysis and a work sampling study might complement each other.

8.4. Describe the nature and details of a job (or group of jobs) that you are familiar with which could be studied advantageously with work activity analysis. Explain why you feel that this is an appropriate technique for use with the work you have described.

8.5. Same as Problem 8.4, but with work sampling.

8.6. Prepare a work activity analysis:
    (a) For yourself, at work.
    (b) For yourself at home, for an evening.
    (c) For yourself, for a weekend.
    (d) For a gas station attendant.
    (e) For a store clerk.

8.7. After the necessary preparation, have a work activity analysis prepared by:
    (a) Your assistant.
    (b) Your secretary.

8.8. List the recommendations resulting from a study of the work activity analysis prepared in Problem 8.6 or 8.7.

8.9. (a) Treating minutes as strata, assign a random number of seconds to each minute so as to obtain 60 random reading times within the hour. Using these reading values, sample the "number of cars at the pumps in a gas station."
    (b) Decide what conclusions may be drawn from these data.
    (c) Evaluate the reliability of these data and results. (Several samplers may pool results.)
    (d) Design an adequate study to establish the tentative conclusions in part (b).

8.10. Same as Problem 8.9, for "status of students or personnel in library."

8.11. Same as Problem 8.9, for "status of personnel in lounge."

8.12. Same as Problem 8.9, for "status of personnel at some selected area of a cafeteria."

8.13. Same as Problem 8.9, for "some queue."

8.14. Same as Problem 8.9, for "personnel in an office."

8.15. Same as Problem 8.9, for:
    (a) Color showing on a selected traffic light.
    (b) Floor at which elevator is, in multistory building.
    (c) Status of a flashing sign.
    (d) Usage of a section of sidewalk, street, or stair.
    (e) Attention given an advertising display.

8.16. (a) If the largest category of activity is estimated to consume 41 per cent of the work time, how many observations are needed to make a rough work sampling study?
    (b) Why might more observations than the number obtained in part (a) be made?

9

WORK-UNIT ANALYSIS[1]

Introduction

Where the technique is used. As was noted in Chapter 4, a study of a service activity usually starts with the use of a technique designed to assist in the delineation of the outputs. It was further indicated that with respect to some service activities, a meaningful method for counting the outputs may not be readily apparent. In such cases work-unit analysis is usually the first motion study technique applied. —

Purposes served by this technique. The basic purpose of this technique is to provide a framework for identifying outputs and groups of outputs. Further, it assists with developing a method for counting outputs. It is

---

[1]The text material in this chapter is adapted from M. E. Mundel, *Measuring and Enhancing the Productivity of Service and Government Organizations.* Tokyo: The Asian Productivity Organization, 1975.

designed for use with service-oriented organizations rather than with organizations producing substantive products with intrinsic value.

With substantive outputs, such as home appliances, there is no particular difficulty in recognizing that there are different outputs, jobs, processes, and workloads connected with the production of each appliance, such as washing machines, dryers, and dishwashers. Further, each type of appliance can be represented by an engineering drawing of the total assembled unit.

The washing machine assembly drawing, for example, may be replaced by a set of part drawings, one for each part. Each part drawing and the information concerning how many of the parts are needed for each assembly will be supported by a material requirement statement and by a process description or other presentation of the production sequence for each part. Each step in the production sequence is readily identified; the requisite equipment may readily be specified. The subject of a possibility guide, or of one of the detailed motion study analysis techniques, may readily be identified. In addition, the jobs and their concomitant outputs are easily identified for time study.

With respect to service activities there is an information void created by the infeasibility of developing a set of drawings and a bill of materials describing the outputs of the organization. Outputs, to be counted in a meaningful fashion, must be those related to achieving the organization's objectives. Work-unit analysis is designed to assist in filling the information void caused by the impossibility of having drawings, bills of material, and so forth.

Reasons for choosing this technique. It is certainly foolish to improve productivity with respect to unwanted outputs. It is even more foolish to apply time study to produce unneeded measures. It is even worse to use time study data to support measures of productivity that relate to neither the organization's objective nor reflect management efforts.

In one country[2] the federal Budget Office accused the Weather Bureau of falling, in one year, from a "productivity" of 100 per cent to 65 per cent. Their computation was based on:

Base year, typhoons tracked $= 26$

Measured year, typhoons tracked $= 17$

Productivity $= \dfrac{17}{26} \times 100 = 65$ per cent

---

[2]Not the United States.

Of course, in this case it should be obvious that a Weather Bureau does not make typhoons; that no one wants them to; that their "gross output" is more properly identified as 365 days of "weather reported." However "weather reported" is a rather large "product" to study with either the usual motion study or time study techniques.

Hence, work-unit analysis, the only technique designed to delineate appropriate service outputs, is the first technique for use with service organizations. There is really not much choice; this use of this technique is a necessity if use is to be made subsequently of other motion and time study techniques.

Definitions

In the concept "work-unit analysis," the hyphenated term, "work-unit" has a special meaning. A variety of other special terms are also required to facilitate a meaningful discussion.

Work-unit defined. A *work-unit* is an amount of work, or the result of an amount of work, which is convenient to treat as an integer (an "each") when examining work from a quantitative point of view.

In this definition, the word "convenient" means that it provides a useful basis for:

1. Applying various motion and time study techniques.
2. Substantively supporting manpower aspects of budgets.
3. Developing unit labor costs.
4. Planning and assigning work.
5. Continually reviewing workload forecasts and on-going manpower utilization.
6. Continually comparing performance with plans.
7. Measuring the internal manpower productivity of an organization.

It should be obvious that, within the definition given, a wide variety of sizes of work-units could exist, covering a range from "one motion of the arm" to "the total annual work of an organization."

Orders of work-units defined. To facilitate an orderly discussion of work, in work-unit terms, a series of definitions identifying work-units of various sizes must be established. Each different size is referred to as a different *order of work-unit*.

Work-units of different orders defined. Table 9-I gives a series of definitions for work-units of various sizes suitable for talking, in a consistent fashion, about the quantitative aspects of any type of work. This list has been given starting from the 8th-order and going down to the 1st-order work-unit to conform to the sequence of analysis. Also, it seemed appropriate to use smaller numbers for smaller work-units. It should be noted that not all of these orders of work-units will necessarily be involved in a

Table 9-I. DEFINITIONS OF BASIC ORDERS OF WORK-UNITS

Numerical Designation	Name	Definition
8th-order work-unit	Results	What is achieved because of the outputs of the activity
7th-order work-unit	Gross output	A large total of end products or completed services of the working group.
6th-order work-unit	Program	A group of like outputs or completed services representing part of a 7th-order work-unit but which are a more homogeneous subgroup.
5th-order work-unit	End product	A unit of final output; the units in which a program is quantified.
4th-order work-unit	Intermediate product or component	A part of a unit of final output; the intermediate product may become part of the final output or merely be required to make it feasible to achieve the final output.
3rd-order work-unit	Task	Any part of the activity associated with, and all of the things associated with, the performance of a unit of assignment by either an individual or a crew, depending on the method of assigning.
2nd-order work-unit	Element	The activity associated with the performance of part of a task which it is convenient to separate to facilitate the designing of the method of performing the task or the time study of the task.
1st-order work-unit	Motion	The performance of a human motion. This is the smallest work-unit usually encountered in the study of work. It is used to facilitate job design or time study and never appears in control systems above this level of use.

work-unit analysis. On the other hand, in complex situations, there may be a need for orders of work-units between the ones given here. In such cases, decimals would be added (e.g., an order of work-unit between the 7th-order and the 6th-order would be identified as the 6.5th-order).

The list is offered not as a strict exercise in the taxonomy of work, but as a guide, as facilitating nomenclature. It is designed to provide a language for use when tracing the path from objectives to outputs and required resources when there is no obvious natural method. One must remember that a work-unit was defined as an amount of work *convenient* with respect to providing a basis for applying motion and time study techniques and assisting in managerial control. The concept of convenience suggests that different numbers of orders of work-units may, at times, be desirable.

Work-unit analysis defined. *Work-unit analysis* is the delineation of the outputs of an organization, and the subparts of these outputs, in work-unit terms. The analysis starts with the objective of the organization and proceeds through the delineation of smaller and smaller units of output (orders of work-units) until the following criteria are met:

1. There is a clearly visible relationship between objectives and outputs.
2. A suitable level of detail is reached such that meaningful forecasts can be made of outputs required for future periods.
3. A level of detail is reached such that other motion and time study techniques may be employed.
4. At each level of detail the list of outputs is all-inclusive.
5. At each level of detail the items in the list are mutually exclusive.

Work-unit structure defined. A *work-unit structure* is the hierarchical list of work-units resulting from a work-unit analysis. It is also called a *work-unit hierarchy*.

Work-Units and the Cycle of Control

The steps in the management control cycle (given in Chapter 1) may now be reexamined with the assistance of work-unit terms to discern more clearly the role of work-units. The different orders of work-units are shown in their place in the cycle of control in Figure 9.1. In this figure, the steps of the control cycle are merely numbered; the details of each step, which were given in Figure 1.1, are not repeated.

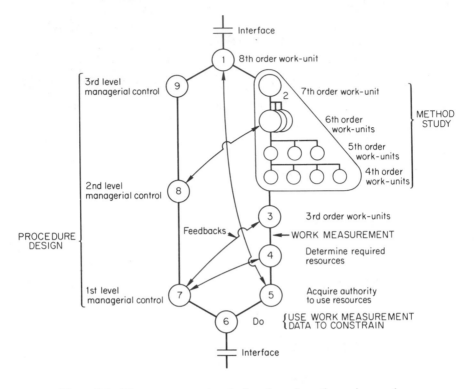

Figure 9.1. The management cycle showing where the various orders of work-units would appear. The major points of impact of method study and its subpart, procedure design, as well as the steps affected by time study or work measurement are also shown.

Procedure for Applying Work-Unit Analysis

Step 1. Stating the objective. To begin the delineation of the work-unit structure, a statement of the objective of the organization being studied should be set forth. In that higher levels of management will be confronted with many such statements of objectives, a specific format should be adhered to so as to assist in clear and rapid communication. This statement of the objective is referred to as the 8th-order work-unit. The suggested format has seven headings, listed below (and subsequently explained):

1. Type of service.
2. Mission area.
3. Purpose: Intent.

4. Purpose: Dimension.
5. Goal.
6. Limitations.
7. Freedoms.

1. TYPE OF SERVICE

In industrial service organizations there are six types of services:

a. Product output directly related
b. Output facilitative
c. Labor control
d. Finance control
e. Research and development
f. Customer service
g. Material acquisition

These categories are self-defining. Work-unit analysis is usually applied to categories b, c, d, e, f, and g. Category a is the classical area of application of motion and time study.

In government there are four basic types (or modes) of service, although an organization may use a combination of these. If a combination of these is used, the situation should be so identified. The four types and their definitions are:

a. *Constructive service*—do for the public that which it would otherwise have to do for itself.
b. *Social constraint*—restrain people, firms, or groups from undesirable or illegal activities.
c. *Grants, awards, and financial assistance*—achieving any effect by funding people or organizations, outside the federal government, to undertake the desired activity.
d. *Internally consumed*—the organization has no program relationship with the public; its efforts are totally consumed in a larger part of the total organization and are not a direct part of a final output (e.g., the internal budget section).

2. MISSION AREA

This refers to the sector or sectors of the enterprise, its market, its customers or suppliers where the results of the organization being studied are achieved; where the desired effects or "impacts" are achieved. This is not to be confused with the actual effects, which are the subject of subheads 3, 4, and 5, which follow. With government organizations this term refers to the sectors of the society, the economy or the geography "impacted."

3. PURPOSE: INTENT

Under this heading the type of effect to be achieved in each of the impact areas, listed under "Mission Area," should be described. Several intents may be listed, if appropriate. However, there should be at least one intent given for each mission area. If not, one can hardly think of it as an "impacted" area.

4. PURPOSE: DIMENSION

Under this heading describe the quantifiable attributes or characteristics of each intent, described under the subhead "*Intent,*" which are of value in quantifying the results achieved in the impact areas. At least one or more dimensions should be given for each intent. It is important to note that these dimensions need not be related, and are most usually not, to the manpower required by the organization. Many useless budget arguments are generated by a failure to note this usual lack of relationship.

5. GOAL

Under this heading indicate the amount, in terms of each of the previously listed dimension or dimensions of each intent, to be achieved. An organization usually has two goals: one long-range and one for each budget year. True managerial control requires the explicit stating of the goal for each forecast year as well as the current year. It is desirable, however, to assist in managerial control, to have a statement of the long-range goal with which each year's goal must conform.

6. LIMITATIONS

Under this heading list the unique (not typical of all organizations) restrictions on the operation of the organization being studied and its outputs. This statement should not include, "Manpower and budget limitations," in that this is hardly unique. More properly one may find, "Can only recommend change."

7. FREEDOMS

Under this heading list the areas of action where the organization being studied has some unique freedom of choice, such as if a contracts group is "Not limited to selecting the lowest bidder."

***Step 2. Delineating gross outputs** (7th-order work-units).* In order to achieve the purposes stated previously (with respect to a convenient work-unit structure) it is desirable first to divide the totality of outputs into a number of large categories. The separated categories should clarify the significance of the each outputs group with respect to:

a. Different mission areas served.
b. Different (radically) methods of producing the outputs.
c. Totally independent or different cost systems (e.g., in industry, volume of product related versus fixed funding base; in government, federally funded versus reimbursed work).
d. Different fund accounts separately appropriated.
e. Different benefits obtained.
f. Different mixes of intents.
g. Different types (or modes) of action.

These categories are called 7th-order work-units. To avoid confusing these outputs with tasks, particularly when the outputs are services, it is most convenient to describe them by means of a past-tense verb together with appropriate adjectives and nouns (e.g. "Product design made," "Economic evaluations reported," "Factories inspected and reported," "Audit services provided"). Further, a simple phrase such as "Product design made" must be thought of in the broadest sense. It is not merely the final making of a design. Rather, it is all the activities and services produced which finally culminated in "Product design made." It includes the sick and annual leave of all employees associated with the output, their payroll services, and so forth. The small list of 7th-order work-unit categories must be thought of as representing the net result of all the resources expended by the organization. The list must be devised to meet this concept.

Step 3. Delineating program outputs (6th-order work-units). Each 7th-order work-unit should be broken into two or more 6th-order work-units (of which the 7th is the sum), unless such a separation would serve neither the main criteria nor the additional criteria given here. The separation of additional categories at this point, while continuing to serve the main criteria of a work-unit structure, should make more clear the significance of the outputs with respect to:

a. Independent, separate cause systems which generate work.
b. Useful subaggregations for decisions with respect to a balance of benefits.
c. Subclasses of the separations initiated at the 7th-order, for the same reasons as with 7th-order work-units.
d. Outputs that appear alike but require different amounts of resource support.
e. Subgroups within a 7th-order work-unit which are more alike, within the subgroup, with respect to mission area served, or purpose, etc., than the remainder of the outputs in the 7th-order work-unit of which they are a part.

Step 4. Delineating units of output (5th-order work-units). In all cases 6th-order work-units are aggregations of outputs. That which it is convenient to identify as an "each," of the "all" contained in any 6th-order work-unit, is called a 5th-order work-unit. Separation at the 5th-order should, in addition to continuing to serve the main criteria, assist in:

a. Identifying a normal "each" of actual output with respect to utility.
b. Identifying an "each" in a manner related to the management decision making process with reference to "do/not do" types of decisions.
c. Provide a useful, meaningful basis for "pricing out" the outputs (unit costs).
d. Provide a basis for a work count for that which is contained in each 6th-order work-unit.
e. Provide a work-unit convenient for some type of workload forecasting, method improvement, and work measurement.

Additional Guidelines for Preparing Work-Unit Structures

Delineating additional levels of work units. In complex cases it may be convenient to proceed from one work-unit order to a lower order by way of an intermediate level such as 6.5 between 7 and 6 or 5.5 between 6 and 5. Such orders of work-units, between the whole-numbered levels, should be used to:

1. Assist in providing a clear relationship between objectives and the use of manpower resources.
2. Provide alternative ways of summarizing to higher orders of work-units.
3. Retain the use of whole numbers for work-units most readily identified with the definitions given for the integer-numbered work-units of Table 9-I.

Delineating component outputs (4th-order work-units). In many cases an "each," or a 5th-order work-unit, is such a complex service that work measurement cannot be performed directly at the 5th-order level. In such cases, a series of components, or 4th-order work-units, must be identified for each 5th-order work-unit. The criteria of separation are the main criteria given earlier.

Avoiding unduly restrictive terms. As far as possible, the description at each order of work-unit should not reduce the choice of alternatives at the next lower order. For instance, consider the 6th-order work-units that follow:

1. Departments maintained in conformance with laws and regulations.
2. Departments maintained in conformance with laws and regulations by means of four inspections per year.

The added material in the second version unduly restricts the freedom of choice of 5th-order work-units.

Additional statements related to main criteria

1. The complete list of any order of work-unit must represent the totality of work that results from the activity of the organization being studied, with the detail increasing as the lower orders are reached. The categories in each order must, therefore, as was stated, be "all-inclusive." In addition, they must not overlap; they must be "mutually exclusive."
2. The work-units that are forecast for budgeting or counted for performance measurement must be either:
(a). End-product outputs.
(b). Outputs having a known and fixed relationship with end products.

An *end-product output* is a unit of service or product on which no further work will be done within the organization being studied and which is assumed to contribute toward the achievement of the objective.

3. The work-unit structure must be acceptable to the program personnel and must be related to how they think of outputs, or how they should think of them when making program decisions. If the program personnel have participated in developing the work-unit structure, these criteria should be met readily.

Case I—Work-Unit Analysis Applied to an Industrial Engineering Department[3]

Introduction. Work-unit analysis can seldom be accomplished without a considerable involvement of program personnel in that they are usually the only ones with the necessary detailed knowledge of the programs. The concept

---

[3]This example is not to be taken as an ideal model of an industrial engineering department; it is merely an analysis of an actual department as it was constituted at the time of the study.

of "participation," in most cases, must be used fully to successfully accomplish this analysis.

Further, the organizational segment to be analyzed must be defined. The analysis may be applied to a section, a branch, an office, or a department. The most useful approach would entail application to the totality of an organization which has responsibility for programs that interface with the remainder of the organization or with the public. (In applying the analysis to government organizations, the interface is usually with a sector of the economy, society, or geography, where the effects achieved are evident.)

In the case examined here, an industrial engineering section was applying the analysis to itself in order to have a basis for applying more detailed techniques for self-improvement. They also wished to improve their own internal, managerial control.

The work-unit structure. Following the procedure given previously, the industrial engineering staff of the XXX Telephone Co., Equipment Works,[4] described its work-unit structure as shown in Figures 9.2 through and 9.5.

Each of the 7th-order work-units was assigned a two-digit number. Two additional digits were affixed at each successive order of work-unit so as to give each work-unit at each order a unique identifier. If these numbers are prefixed by another two digits, indicating the suborganization, a hierarchical chart of accounts results.

Results. The 8th-order work-unit clearly states the basis for measuring the organization's success. This cannot do other than help manage; it shows the way to go.

The 7th-order work-unit list serves as a basis for examining the allocation of staff resources (aggregated to this level) and for evaluating different allocation stratagies.

The 6th-order work-unit list serves as a basis for developing forecasts.

The 5th-order list serves as a basis for detailed managerial control. Further, the 5th-order list delineates the service outputs in a manner suitable for the application of other, more detailed motion and time study techniques.

Summary—Work-Unit Analysis

1. Uses:

For delineating the outputs of a service or government organization such that other motion and time study techniques (and other managerial tools) may usefully be applied.

---

[4]Name altered at company's request.

XXX TELEPHONE CO. EQUIPMENT WORKS,

INDUSTRIAL ENGINEERING SECTION

TYPE Output facilitative

MISSION AREAS

1. Transport
2. Building environment and services
3. Telephone equipment repair activities

PURPOSE: Intent

1.1 Attain economy of operation
1.2 Attain efficiency of operation
1.3 Provide basis for budgeting
1.4 Evaluation of alternative vehicles
2.1 Attain economy of operation
2.2 Attain efficiency of operation
2.3 Provide basis for budgeting
3.1 Attain economy of operation
3.2 Attain efficiency of operation
3.3 Provide basis for budgeting

PURPOSE: Dimension

1.1 Total cost per hour of operation
1.2 Percent down time of vehicles
1.3.1 Accuracy of forecasts of usage and cost
1.3.2 Timeliness of budgets
1.4 Relative update in areas of evaluations
2.1.1 Cost per square foot of operations building
2.1.2 Cost per square foot of warehouse
2.2 Percent of plant available for use when needed
2.3.1 Accuracy of forecast of usage and costs
2.3.2 Timeliness of budgets
3.1 Percent repair costs to replacement
3.2 Productivity
3.3.1 Accuracy of forecasts of activity and costs
3.3.2 Timeliness of budgets

GOAL:

1.1 Ten percent less costs or proof of infeasibility
1.2.1 Less than ten percent for light vehicles
1.2.2 Less than twenty percent for heavy lorries
1.3.1 Within five percent of actual
1.3.2 Available when due
1.4 Within six months of announcement of competing vehicles
2.1.1 Five percent reduction of variable costs or proof of infeasibility
2.1.2 Five percent reduction of variable costs or proof of infeasibility
2.2 100 percent
2.3.1 Within five percent of actual
2.3.2 Available when due
3.1 Less than, by at least twenty percent
3.2 Increase by five percent
3.3.1 Within five percent of actual
3.3.2 Available when due

LIMITATIONS:

1. Do not control use of transport, building, expenditures or actual operations
2. Must give "directed projects priority
3. Must comply with fiscal due dates

FREEDOMS:

1. Can select areas for study with manpower available above "directed" projects
2. Not restricted to specific techniques
3. Reports to plant manager rather than operating heads

Figure 9.2. The objective (8th-order work-unit) of XXX Telephone Co., Equipment Works, industrial engineering staff.

```
7th-order work-units
01   Transport performance recommendations made, including evaluations of
     alternative types of equipment
02   Building and service studies made
03   Tooling, equipment, method and time studies made
04   Economics of repair studies made
05   Budget support systems provided and serviced
06   Other special studies made
```

Figure 9.3. The output groups (7th-order work-units) of the XXX Telephone Co., Equipment Works, industrial engineering staff.

```
6th-order work-units
0101   Off-road equipment performance recommendations made
0102   Off-road, alternative vehicles evaluated
0103   On-road, equipment performance recommendations made
0104   On-road, alternative vehicles evaluated

0201   Building studies completed
0202   Services studies completed

0301   System economy studies made
0302   Individual workplace studies made

0401   Economics of repair studies made on vehicles
0402   Economics of repair studies made on works' equipment
0403   Economics of repair studies made on phone system equipment

0501   Transport budgets prepared
0502   Building budgets prepared
0503   Other services, budgets prepared

0601   Other studies made, directed
0602   Other studies made, selected
```

Figure 9.4. The programs (6th-order work-units) of the XXX Telephone Co., Equipment Works, industrial engineering staff.

2. *How made*:
   A four-step procedure for proceeding from objectives to unit outputs was given. It is to be noted that the analysis proceeds from the objective, through the gross aggregation of outputs, downward. Experience indicates that only in this manner can a definitive description of end-product-oriented outputs be obtained.

3. *How used*:
   The work-unit structure that results serves the same function as that served by the product drawings, parts list, and bills of material of the hardware-producing organization.

4. *What then*:
   Other motion and time study techniques (as well as other management improvement procedures) are more readily applied when a work-unit structure has been delineated.

```
5th-order work-units

010101  An off-road, equipment performance recommendation made
010201  An off-road, alternative vehicle evaluated
010301  An on-road, equipment performance recommendation made
010401  An on-road, alternative vehicle evaluated

020101  A building usage study completed
020102  A building repair study completed
020103  A building replacement study completed
020201  A warehouse service study completed
020202  A material handling study completed
020203  A maintenance service study completed

030101  An old system study updated with new economic data
030102  New economic study made of a system
030201  An individual workplace studied for economic improvement; limited
        to Class 1 changes.
030202  An individual workplace studied for economic improvement; in-depth
        changes
030203  An individual workplace time standard set; individual work.
030204  An individual workplace time standard set; group work.

040101  An economic study of vehicle repair made
040201  An economic study made of a works repair
040301  A study of the economies of a phone system component repair made

050101  A transport branch budget prepared
050201  A building budget prepared
050301  A service branch budget prepared

060101  A directed study completed
060201  An elected study, other than previously enumerated, completed
```

Figure 9.5. Units of output (5th-order work-units) of the XXX Telephone Co., Equipment Works, industrial engineering staff.

PROBLEMS

9.1. In the example cited in detail in Chapter 9 it was noted that the numbering system used would serve in a hierarchical chart of accounts. What uses would be served by such a chart of accounts?

9.2. In the discussion of "limitations" an example given was, "Can only recommend change." Why is this a limitation?

9.3. Construct the work-unit hierarchy for:
    (a) A school cafeteria.
    (b) A gasoline station.
    (c) A supermarket.
    (d) A national park.
    (e) A city taxing bureau.
    (f) A police force.

(g) A department store.

(h) A newspaper.

(i) A water-pollution-abatement project.

(j) A central-city-improvement project.

(k) A U.S. Navy ship-repair yard.

9.4. Evaluate the work-unit structure set forth in response to Problem 9.2 in terms of the criteria for a useful hierarchy of work-units.

9.5. Considering the work-unit structure set forth in response to Problem 9.2; at what level of work-unit does it appear that the work counts will be meaningful with respect to manpower resources required?

10 PROCESS CHART—PRODUCT ANALYSIS

Introduction

The three preceding chapters dealt with techniques designed primarily to assist in deciding on the most desirable and feasible class of change, or with the identification of outputs or time usage as a preliminary step in making such a decision.

This chapter is the first of a series of chapters dealing with techniques, with a primary focus on the collection of detailed facts to assist in designing a change of the anticipated class. This is not to say, however, that such detailed data collection is totally separate from the decision with respect to the preferred class of change. After all, with more facts in hand, one may frequently change his view with respect to the most feasible and desirable class of change.

This chapter deals with a technique designed to assist in fully performing the second step of the scientific procedure for determining better method—analysis.

Analysis: separating the whole of the work method into subdivisions or steps pertinent to the job, appropriate to its scope, possessing known characteristics, or concerning whose performance information is already available. The specific procedure selected for performing this step will be a function of the class of change finally sought and the characteristics of the subject of the scrutiny. The procedure selected will usually provide the framework for most of the succeeding steps as well. This chapter deals with only one device for performing these steps: the process chart—product analysis.

Process Chart—Product Analysis Defined

A process chart—product analysis is a graphic means of portraying (or a schematic model of) the separable steps of the procedure, applied to the product, required to modify the output from one stage of completion to another. A variety of graphic approaches may be used.

The Use of the Technique

Where the technique is used. This technique is primarily applicable when:

1. The timing of various batches moving through various steps or the time relationships among the steps are not primary aspects under consideration.
2. The relationship between successive steps is relatively simple.
3. The output is essentially substantive.

In most cases, if the problem does not meet the restrictions enumerated, other techniques, which will be discussed in subsequent chapters, will be more helpful in making an analysis.

Purposes served by the technique. A process chart—product analysis would be used under the following circumstances:

1. *As a last check prior to the application of a man analysis technique.* This would occur when the decision reached after step 1 of the scientific approach was to seek a class 1 or 2 change. The process chart—product analysis would be used to verify the necessity of doing a particular job before attempting to improve it.
2. *When the decision was to seek a class 3, 4, or 5 change.* When the volume is great and the job of long duration, and where the time for extended analysis and the authority for complex change are available,

it is usually desirable to exhaust all possible means of achieving a class 3, 4, or 5 change before modifying any single job. This may lead the analyst into a product-oriented study; into the use of an approach called *value analysis*. The term "value analysis" refers to subjecting each part of a product to a careful, searching scrutiny to determine what the particular part contributes to the functioning of the whole, and what alternatives of materials, configuration, and so forth, exist that might be advantageous from the viewpoint of the criterion of success. Of course, a study of the process by which a product and its components are produced is a useful part of such a study. Value analysis is concerned with an intensive consideration of class 4 and 5 changes; process chart—product analysis presents a broader picture. Of course, not all process chart—product analyses lead to class 4 or 5 changes; in some cases the change does not rise above class 3. This is outside the normal scope of value analysis. Hence, it would appear that the two approaches, rather than being separate, competitive, approaches, are parts of a total improvement procedure.

3. *When planning for a product or a layout.* Process chart—product analysis should be made whenever possible for all products, prior to the inception of production, to aid in designing the best possible process and layout before the beginning of work. This anticipatory planning is much cheaper, although less spectacular, than changes made subsequently. In all cases of plant layout and process planning, the process charts for the contemplated components or products should be made in connection with the plans for layout. These charts will be used as an aid in developing the best possible process and layout; that is, least handling, least number or cost of operations, least scrap, best control, smallest workable in-process inventory, or whatever are the appropriate criteria of preference.

Studying a multiproduct situation. When a variety of products are contemplated, the various products may have conflicting requirements. In such cases, a weighted evaluation of the various products' requirements must be made.[1] Adequate use of process chart—product analysis for such uses is facilitated by an intimate knowledge of product and process requirements and peculiarities. In competent hands, or with proper cooperative assistance from the various plant groups, the charts are valuable as a tool for the

---

[1]This may lead to a problem in linear programming or other mathematical model of the situation, as contrasted with the process chart, which is only a schematic model. If a mathematical model may be created, the best solution may be determined in a much more mechanistic fashion, but the process chart is frequently a vital aid in fashioning the mathematical model.

attainment of efficient plant layout and effective process planning. Further, they are a source of suggestions for new equipment and tool design. Process chart—product analyses are also a guide to the overall analysis of plant operation.

The steps separated with process chart—product analysis. The steps commonly used, and the symbols used to represent them, are given in Table 10-I. The process involved in manufacturing, processing, or handling parts, components, or product is usually portrayed graphically, in terms of these steps, by means of these symbols, to aid in ease of understanding and to break the sequence into relatively homogeneous steps. The use of these or some standardized set of symbols[2] is desirable because the classification of the steps into categories possessing known characteristics is of prime importance in this work, and the breakdown should be visually apparent. The symbols act as flags identifying the various types of steps. Also, the use of standardized steps and symbols permits the chart to be easily understood by others. The last is especially important with a technique leading to class 3, 4, or 5 changes, inasmuch as such changes frequently require a large number of cooperating individuals and groups. However, standardization should not be considered a rigid requirement if additional or different symbols will clarify the schematic model. The problem being charted, rather than the chart, should be the focus of attention.

Standardized charts. A standard analysis should be used when possible if we hope to develop standard procedures for improving processes, rather than remain content to approach each problem as something entirely new. However, the standard analysis should be thought of as the procedure of breaking the sequence into steps, rather than the rigid and blind use of a given set of steps. The standardization should be of the general approach rather than of the details of charting; these may change as dictated by the requirements of the problem.

---

[2]The A.S.M.E. proposed a standard set of symbols; see *Operation and Flow Process Charts.* New York: American Society of Mechanical Engineers, 1947. These differ from those used for some of the charts in this book. The symbols that appear in these illustrations most closely resemble those used in the past and better separate distinct activities. The symbols are to serve the function of *analysis* as defined, and not vice versa. In many cases, additional symbols may be of great help inasmuch as these charts are schematic rather than exact models of situations, and changes in schematic presentation are desirable if they contribute to clarification. However, a new variety of symbol should be introduced only if it offers distinct advantages. For a useful and novel set of symbols, see W. W. Phillips and S. W. Earle, "Cycle Charting—with 'Handlings,'" *Factory Management and Maintenance,* Vol. 3, No. 8, 1953, pp. 105–107.

Table 10-I. SYMBOLS FOR PROCESS CHART–PRODUCT ANALYSIS

Symbol	A.S.M.E. Symbol	Name	Used to Represent
○	○	Operation	Something done to the product at essentially one location.
□	□	Quantity inspection	A special form of operation involving the verification of the quantity of a product present against a record of the quantity that is supposed to be there.
◇		Quality inspection	A special form of operation involving the verification of an attribute or quality of a product against a standard.
○	⇨	Movement	A change in the location of a product which does not change it in any other way.
▽	D	Temporary storage	The storage of a product under conditions such that it may be moved or withdrawn from storage without a requisition.
▽	▽	Controlled storage	Storage of a product under controls such that a requisition or receipt is needed to withdraw it.

The steps given in Table 10-I have been chosen as appropriate to the scope of much industrial activity while separating different types of activity and different areas of responsibility, as follows.[3]

Operation—a step that usually requires labor and/or equipment, is the only type that may add value to the product, and is usually a responsibility of the production group aided by the tooling or "processing" staff.[4]

---

[3] At times, it may be desirable to use more or fewer steps, or a different set of steps. A scrutiny of those listed here and the areas of responsibility thus separated would suggest that this set approaches a minimum number. It should also be apparent that the selection of the steps is of much more importance than the selection of the particular symbols used to represent them. However, if symbols are used, a given set of symbols should be used consistently, to avoid confusion.

[4] It may assist analysis to separate *do* operations (those that add value to the product), from *get ready, put away*, or facilitative operations.

Quantity inspection—a step that usually requires labor, may require equipment, provides control, and is usually a responsibility of the production control group.

Quality inspection—a step that usually requires labor and equipment, provides control, and is usually a responsibility of the inspection or quality control group.

Movement—a step that usually requires labor and equipment, and is usually a responsibility of the materials handling group.

Temporary storage—a step that requires space, adds to inventory, and is usually caused by the process design or plant layout, hence is a responsibility of the group or groups having jurisdiction over these designs.

Controlled storage—a step that requires space and labor, represents inventory, and is usually a responsibility of the inventory, material, or stock control group.

Additional conventions. When the flow of a number of related items is made on one process chart—product analysis, the additional conventions shown in Figure 10.1 may be required.

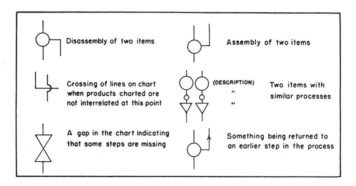

Figure 10.1. Conventions for process chart—product analysis.

Scope of the study. The length of the part of a process analyzed will be a function of:

1. *The scope of activity of the analyst.* A department foreman is usually not in a position to carry the work outside the department. The same is true of the worker who may use motion and time study to develop better methods for his own job.

2. *The certainty that class 3, 4, and 5 changes are out of the question.* Some analysts advise a complete process chart—product analysis prior to any man analysis, but this is often not expedient when immediate improvement is possible with class 1 and 2 changes.

Making the chart. In any plant, discrepancies between what is supposed to be happening, what the analyst may think happens, and what is actually happening are not at all uncommon. Hence, a process chart—product analysis is best constructed from an actual observation of the production process.

The analyst should trace the product step by step and make suitable entries for each step. An adequate description for each step may be found by answering questions such as those which follow. The analyst need not follow the outline rigorously but should approach the general pattern. With a small amount of practice, this becomes almost second nature.

Name of Step	*Answer These Questions in the "Description"*
Operation	Who does what where?
Quantity inspection	What units against what record by whom?
Quality inspection	What attribute of the product by what device to what limits?
Movement	From where to where by what means?
Temporary storage	Where?
Controlled storage	Where, under what control?

It is usually desirable to place a summary of the chart at the bottom, as will be shown in the example given later.

Flow diagram. In most cases it is desirable also to make a sketch of the part of the plant concerned and trace the path of the product on this sketch. This sketch is called a *flow diagram*.

Critiquing the process chart—product analysis. Our *criticism* of a process portrayed in process chart form will generally follow a very simple pattern. For instance, the first four symbols given in Table 10-I sort out and identify places where work is done with or on the product. We should learn not only to question the desirability of any occurrence of the first symbol, but also to seek its optimum location in the sequence. Optimum location is often a function of the peculiarities of the material being worked with. The next two symbols refer to special forms of the same type of step shown in the first symbol except that the purpose is control. We may well also question these symbols and seek their best place in the sequence. Optimum location of these is also often a function of the material worked with or the process.

Similarly, the fourth category, movement, is usually a labor consumer. A standard way of reducing an excessive labor consumption in any move represented by this fourth symbol is to change the method to a more economical means of transportation. What is more economical will depend on the exact nature of the case.[5]

The last two symbols classify steps that are primarily nothing other than space occupancy and represent inventory. One is also a control point. We may well question the desirability of such inventory occurring at the point where it does, its size, and the desirability of controlling it.

Further, the assignment of any step to one of the six categories suggested indicates the staff group from which cooperation, advice, and assistance must be sought.

Forms used. A form such as shown in Figure 10.2 or 10.4 or a blank sheet of paper may be used. However, too common dependence on a form, particularly the form of Figure 10.2, may tend to deter or preclude the analysis of processes of wider scope than the process shown in Figure 10.2. (A quick look at Figure 10.2 will reveal that a single column of symbols is sufficient to portray the process; the flow of only one item is charted. All processes are not necessarily this simple.)

Whatever form is used, the chart, for each item whose flow is traced, should usually have at least four columns: the first, for the quantity usually worked on or typical of that step (with storages this column usually shows the number that may accumulate at that point); the second, a distance column to be used with each movement symbol to show the distance covered by the move; the third, a symbol column to identify the general characteristics of each step; and the fourth, a description column to summarize the work done, the detailed nature of the step or its location. The unit time or cost for each step may sometimes be usefully included in additional columns.

Case I—A Simple Materials Handling Problem

Materials handling costs in a large home appliance plant were considered excessive. A plant-wide study was initiated to gather data from which cost-reduction changes could be planned. Process chart—product analyses were made for all procucts.

Analysis. Figure 10.2 is one of the process charts made. It gives the details of the handling of foodshelves for a refrigerator, from bulk storage in the plant to the plating department. This chart was made following the procedure described earlier in the chapter. Figure 10.3 is a flow diagram of the work shown on the chart.

---

[5]See any reference on materials handling or transportation.

PROCESS CHART FORM

Process chart, prod **Type**   _136-54_ **Department**

Original **Original or proposed**   _J.O._ **Chart by**

Refrig. shelves **Subject charted**   _3/20_ **Date charted**

Quantity	Distance	Oper.	Move	Qual.	Quan.	Temp. storage	Cont. storage	Explanation
X crates		○	○	◇	□	▽	▽	Bulk storage, Foundry
4 crates	100'	○	○	◇	□	▽	▽	By gasoline truck, Dept. 136 trucker
80 crates		○	○	◇	□	▽	▽	Daily bank, Dept. 45
1 crate	100'	○	○	◇	□	▽	▽	By hand truck, Dept. 54 trucker
10 crates		○	○	◇	□	▽	▽	Hourly bank, Dept. 54
1 crate		○	○	◇	□	▽	▽	Open crate, Dept. 54 trucker
100 shelves		○	○	◇	□	▽	▽	In crate
"	15'	○	○	◇	□	▽	▽	By hand truck, Dept. 54 trucker
"		○	○	◇	□	▽	▽	Automatic plater loading area
		○	○	◇	□	▽	▽	
		○	○	◇	□	▽	▽	
		○	○	◇	□	▽	▽	
		○	○	◇	□	▽	▽	Summary
		○	○	◇	□	▽	▽	○ 1
		○	○	◇	□	▽	▽	○ 3
		○	○	◇	□	▽	▽	▽ 5
		○	○	◇	□	▽	▽	Dist. 215'
		○	○	◇	□	▽	▽	
		○	○	◇	□	▽	▽	
		○	○	◇	□	▽	▽	
		○	○	◇	□	▽	▽	

Figure 10.2. Process chart—product analysis for original method of handling refrigerator food shelves from bulk storage to plating department.

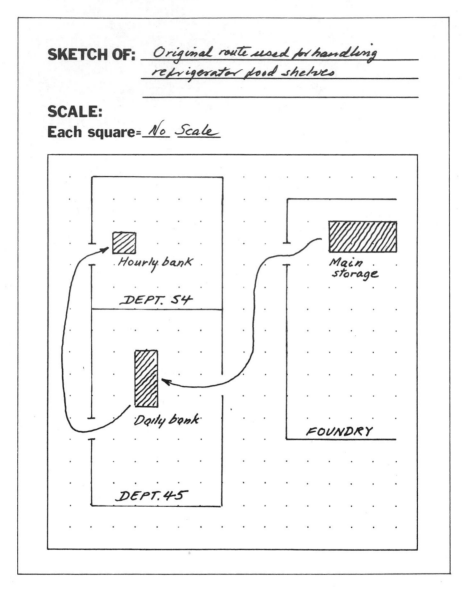

Figure 10.3. Flow diagram for original method of handling refrigerator food shelves from bulk storage to plating department.

Criticism: the application to the analysis of basic data or checklists of desirable arrangements of the steps into a preferable work pattern, and of information concerning desirable ways of performing each of the steps. This is step 3 of the logical procedure. The adequate performance of this step usually presupposes, as with the possibility guide, some familiarity with the type of process or the nature of the goods, or some general knowledge concerning the type of business.

The usual method of performing this third step consists of studying the chart with a checklist of questions, such as the list that follows, and with the flowchart (or floor plan) if necessary. Assistance may be sought from other staff groups concerned, as indicated by the steps shown on the chart.

If the analyst is familiar with the product and applies each of these checklist questions with an open mind rather than trying to find why each suggestion can be rejected, he will find, in most cases, a variety of suggestions for improvement.

CHECKLIST FOR PROCESS CHART—PRODUCT ANALYSIS

Basic Principles

A. Reduce number of steps.
B. Arrange steps in best order.
C. Make steps as economical as possible.
D. Reduce handling.
E. Combine steps if economical.
F. Shorten moves.
G. Provide most economical means for moving.
H. Cut in-process inventory to workable minimum.
I. Use minimum number of control points at most advantageous places.

1. Can any step be eliminated?
   a. As unnecessary. (Ask: Why is it done?)
   b. By new equipment. (Ask: Why is present equipment used?)
   c. By changing the place where it is done or kept. (Ask: Why is it done there?)
   d. By changing the order of work. (Ask: Why is it done in its present order?)
   e. By changing the product design. (Ask: Why is it made as it is?)
   f. By changing the specifications of the incoming supply. (Ask: Why is it ordered in its present form or used at all?)
2. Can any step be combined with another?
   Are there any possible changes that would make this feasible in:
   a. Workplace?
   b. Equipment?
   c. Order of steps?
   d. Product design?
   e. Specification of supply or any raw material?

3. Can the steps be rearranged so as to make any shorter or easier?
4. Can any step be made easier?
   (If this looks like a possibility, make further detailed analysis of this step. Analyses for this purpose will be discussed in later chapters.)

Detailed application of the checklist. For instance, in the example given in Figures 10.2 and 10.3, the analyst asked the questions listed later. The checklist that generated each question is given so that the reader may follow the thought process. The answers are also shown.

1a. Why do we have a daily bank?

Answer: Because the distance from the plating department to main storage is too great for the Department 54 tracker to be able to supply the plating machine crew.

1b. What if we cut a new door into the plating room?

Answer: Yes, the whole handling job would be shorter.

Innovation: the formulation of a new suggested procedure for performing the work. This is step 4 of the logical method. This step was performed when the process chart—product analysis shown in Figure 10.4 was constructed as well as the proposed flow diagram of Figure 10.5. (A different form is used in Figure 10.4, compared with Figure 10.2, to provide the reader with a variety of forms within a single example; the use of the forms is not associated with whether the original or proposed method is being charted.) The proposed flowchart is given in Figure 10.5.

Test: the testing, by means of the data previously used in step 3 (application of the checklist), of the desirability of the method formulated in step 4 (innovation), with respect to the objectives set up in step 1. This is step 5 of the logical method. In the case discussed here, as will be seen, the final suggested solution quickly meets this test. In actual practice, it is desirable to discuss each step, especially this one, with all those who will be affected by the change. The process chart—product analysis greatly facilitates these discussions.

The *test* of the proposal indicated:

1. In Department 45, 900 square feet of floor space, sorely needed for other use, would be made available.

2. The trucking distance for the Department 54 trucker would be reduced by 80 feet per trip, 80 trips per day, with the consequent saving of labor, time, and cost and damage due to rehandling.

3. The use of the gasoline lift truck was eliminated. This would save 20 trips per day, each a round-trip distance of 200 feet from the foundry to the daily bank. This would save $1,740 per year.

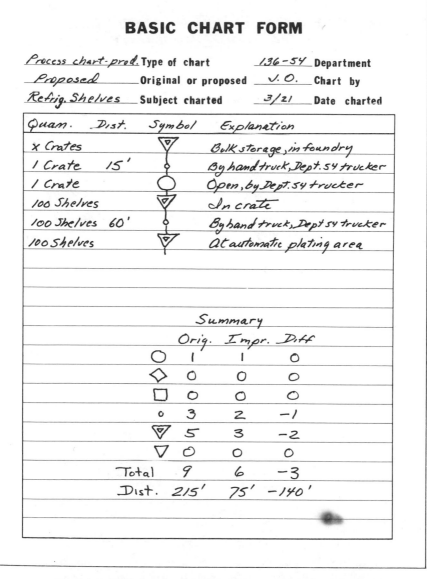

BASIC CHART FORM

Process chart-prod. Type of chart   *1.36-54* Department

Proposed   Original or proposed   *J. O.* Chart by

Refrig. Shelves Subject charted   *3/21* Date charted

Quam.	Dist.	Symbol	Explanation
x Crates		▽	Bulk storage, in foundry
1 Crate	15'	○	By hand truck, Dept. 54 trucker
1 Crate		◯	Open, by Dept. 54 trucker
100 Shelves		▽	In crate
100 Shelves	60'	○	By hand truck, Dept 54 trucker
100 Shelves		▽	At automatic plating area

Summary

Symbol	Orig.	Impr.	Diff
○	1	1	0
◇	0	0	0
▢	0	0	0
○	3	2	-1
▽	5	3	-2
▽	0	0	0
Total	9	6	-3
Dist.	215'	75'	-140'

Figure 10.4. Process chart—product analysis for proposed method of handling refrigerator food shelves from bulk storage to plating department.

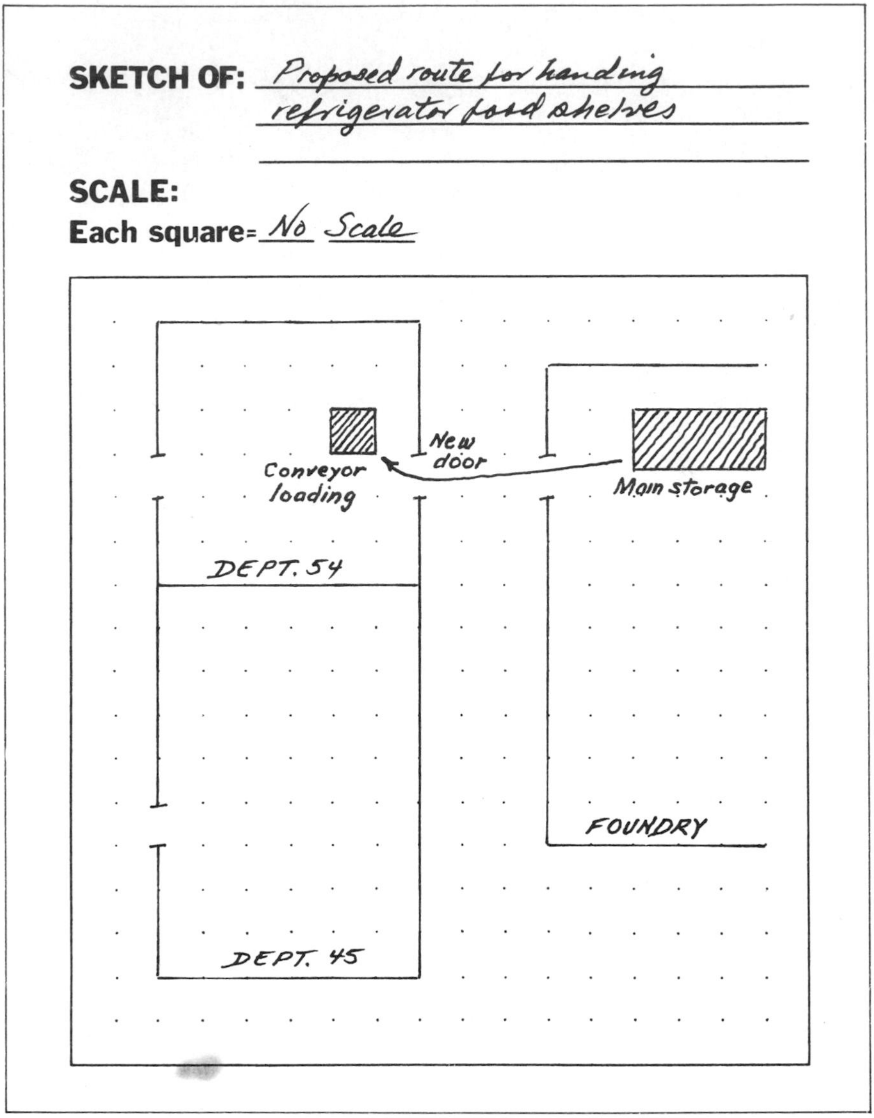

SKETCH OF: *Proposed route for handling*
refrigerator food shelves

SCALE:
Each square= *No Scale*

Figure 10.5. Flow diagram for proposed method of handling refrigerator food shelves from bulk storage to plating department.

4. The uncrating area in Department 54 would be eliminated, relieving congestion.

5. These advantages would be made possible by the installation of the new door shown in Figure 10.5, which, together with the movement of some equipment blocking access to it, would cost $580.

6. The savings from the reduced use of the gasoline truck would pay for all changes in only 4 months.

Trial: the sample application of the method tested in step 5 to ascertain the completeness with which all the variables have been taken into account. This is step 6 of the logical method. In the case being discussed, trial consisted of having plant maintenance verify the cost of the change.

Application: the final standardization, installation, evaluation, and maintenance of the improved work method. This is the seventh and last step of the logical method. The change was approved. The new door was installed. The assignments of the material handlers were altered to conform to the new method. The reassignment of the gasoline truck assured the savings.

Subsequent to Improvement

The analyst will facilitate his future activity if he will accumulate information on ways, peculiar to his industry or type of work, of improving the various types of step—that is, specific ways of achieving the general suggestions implied by the checklist questions. The analyst might well make a file sheet for each checklist question and subquestion and, as he changes processes, note the principles on appropriate sheets.

In the simple example shown it should be obvious that the analyst, in the future, would be very conscious of the possibilities of *new doors*.

Applying the procedure to other problems. The general procedure involving the seven steps of the scientific method is the same no matter what the class of change sought and no matter what the product is. Naturally, the details change from case to case.

Case II—Manufacture of the Plunger of a Diesel-Engine Fuel-Injection Pump

Introduction. This case was selected to illustrate (1) a different format of a process chart—product analysis as compared with the two forms shown in Figures 10.2 and 10.4, and (2) the depth of knowledge that may be required to apply the technique to a manufacturing problem.

Aim. The study was initiated because of excessive costs of production and the need for more output. Further, quality was low. Many rejects were discarded during inspection. A possibility guide had indicated that changes could not exceed class 3, process changes. A drawing of the injection-pump plunger is given in Figure 10.6. The critical dimensions are indicated with letters; the critical angles are indicated with numbers. All angles must be concentric around the center line.

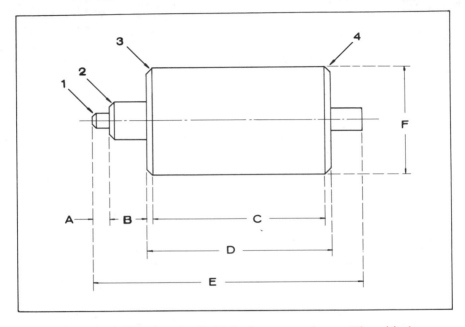

Figure 10.6. Diesel-engine fuel-injection-pump plunger. The critical dimensions are indicated with letters; the critical angles, with numbers.

Although in the actual study the machining of the body of the pump was also studied and improved, the method of study and improvement was the same as with the plunger. Hence, it seemed redundant to show both parts in this example.

Analysis. The process chart of Figure 10.7 was made. This indicates the sequence of machining, step by step. The symbols, because of the nature of the problem, are quite different from those used in the previous charts. This form was chosen because it was felt that the details of the process were more important in examining possible alternatives than the gross sequence of steps, moves, and so forth.

Criticism. The original method, as shown in Figure 10.7, had the following defects:

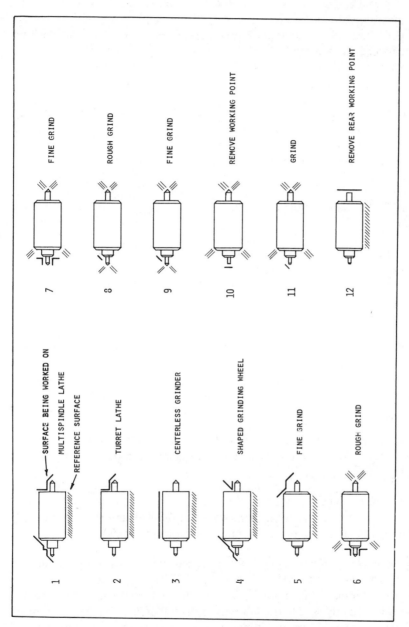

Figure 10.7. Process chart—product analysis of original machining sequence on diesel-engine fuel-injection-pump plunger.

1. All machining was not made with reference to one surface. For instance, examine step 6 in contrast with step 5; step 6 in contrast with step 8. This suggests a possible loss of control of relationships between surfaces and of concentricity.
2. Rough and fine grinds were used rather than accomplishing all grinding at one pass. This suggests excess metal left after machining.

The reason given for changing reference surfaces was: "The chucking is simpler." The reason for the rough and fine grinds was: "Some of the grinders are old and vibrate too much to hold good dimensions. These are used for rough grinding to reduce the fine grinding time."

Innovation. A new machining sequence was designed, as shown in Figure 10.8. An examination of the proposed process chart—product analysis will show that:

1. Only one reference surface is used throughout, after being thoroughly established with great accuracy.
2. Not all of the surfaces that were ground in the original method are ground. Only those that needed to be ground (critical dimensions) were worked on.

In addition to the changes above it was proposed that the grinders with excessive vibration be given new bearings and 100-pound blocks of cast iron be brazed to the static grinding extensions to provide sufficient mass to eliminate any residual vibration. In this way all grinders could be used for final grinding.

Test. The new method was tested over a weekend. The seven steps took less than half of the time of the old 12 steps. Rejects were practically eliminated.

Installation. The old grinders were rebuilt. The new process was fully installed. With the new process, without additional machines or personnel, production was raised from 20,000 units per month to 60,000.

Input of the analyst. Obviously, to alter the process studied, in the case just examined, the analyst needed to be familiar with the technical details of machining. This requirement does not reduce the utility of process chart—product analysis. Charts themselves do not create change. They must be thought of as an aid in obtaining facts and examining the consequences of changes suggested by the knowledge of the analyst; the analyst needs an open mind rather than an empty mind.

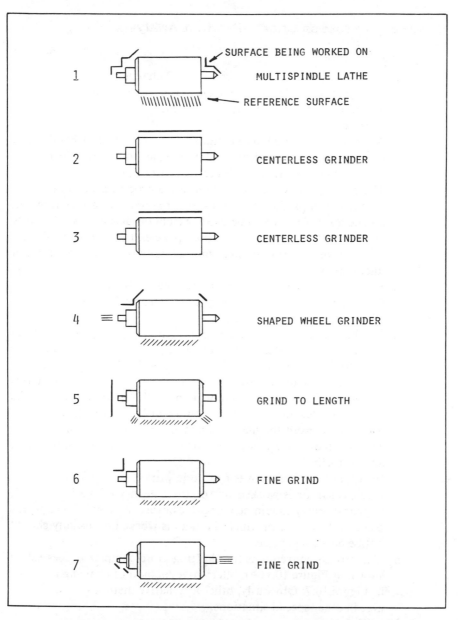

Figure 10.8. Process chart—product analysis of improved machining sequence on diesel-engine fuel-injection-pump plunger.

Summary—Process Chart—Product Analysis

1. Uses:
    a. For a last check prior to a class 1 or 2 change.
    b. For information leading to a class 3, 4, or 5 change.
    c. When planning for a product or layout.

2. How made:
    a. A form may be used, as in Figure 10.2 or 10.4, or a blank sheet of paper may be substituted. For simple processes a form with preprinted symbols, as in Figure 10.2, may be convenient.
    b. The analyst should actually observe the process if possible. If the item is not in production, or if direct observation is inconvenient, a scale floor plan should be used. Actual observation is frequently more desirable, inasmuch as discrepancies between what is supposed to be the process and what actually is the process are not uncommon.
    c. Pick a convenient starting place for the analysis.
    d. Classify the first step into the proper category by means of Table 10-I or other appropriate criteria.
    e. On the first line of the chart, place the proper symbol and description. If the step is a movement, pace off the distance or measure it by some means. In most cases, in the quantity column show the amount handled as a unit on that step, or if the step is a storage, the usual maximum quantity. If the time used for the step is desired, time the step with an ordinary watch or a stopwatch, or consult the file of standard times if it is available.
    f. On the second line place the proper symbol for the second step, and so forth.
    g. Every time the item moves from one workplace to another, waits, is inspected, or is worked on (or the category of work changes), a separate entry should be made. However, movements on a workplace in the course of which the item is worked on usually should not be noted separately.
    h. The various conventions for charting more complex processes are shown in Figure 10.1. An alternative method of charting is shown in Figure 10.7. Obviously, other alternative methods will be desirable in other special situations.

3. How used:
Each step of the process is questioned with the checklist given.

4. What then:
A process chart—product analysis is drawn for a resulting suggested improved method to permit a final check and to provide a means of describing the proposed new method.

PROBLEMS

10.1. A factory uses a number of screw machine parts in its finished product. The bar stock for these is kept in the company warehouse in stacks with wood slats between each layer of 10 bars. The warehouse employees consult their shipping requisition for type and size only, since the standard order is 50 bars, locate the correct stack, slip a crane sling under a layer at a time, and aid the craneman is guiding each slingful into a stake truck. An order usually constitutes a truckload. The truck carries the bars 1,000 feet to the screw machine department, where a monorail hoist capable of lifting five bars in a sling at a time carries the bars 50 feet to the proper storeroom rack. The storeroom is adjacent to the screw machine department. When an operator needs another bar of stock, he goes to the stockroom, where it is issued against his job order. The two stockroom employees then lift a bar from the proper rack and carry it to the machine, where the worker helps them load it into his spindle ready rack. When the stock in the machine is used up, the operator rolls the piece from the ready rack forward into the spindle and restarts the machine, which then works automatically. As the parts are completed, they are ejected to a chute, which drops them onto the steel-plate floor. When enough parts have accumulated, the machine operator calls for the craneman to bring a dump pan. When the crane operator complies, the machine operator shovels the dump pan full, while the crane operator waits. The crane then carries the filled pan and sets it down alongside a cleaning tank. When enough work of any one kind has accumulated in pans on the floor next to the tank, the crane comes back, hooks each pan of that kind in turn, without aid, and dumps the contents into the tank. The tank washes the parts in steam-heated and agitated water, and they are automatically lifted out by an open mesh conveyor, dried on a short section of the conveyor, and dumped onto the floor. The tank tender shovels the parts into carts and pushes these carts 60 feet to a storeroom, where they are weighed in the cart on a platform scale, moved under a good light, and sample-inspected for number of threads per inch, size, and so forth. They are then moved to the proper bin, dumped on the floor, shoveled up into the bin, and the net weight is entered on the bin control card. All incoming requisitions are charged off this card, so the balance of stores is known at all times.

(a) Make a process chart—product analysis of the manufacture of screw machine parts in this plant.

(b) Prepare a preliminary and a detailed possibility guide for the manufacture of screw machine parts in this plant.

(c) With the aid of the checklist, prepare a process chart–product analysis for an improved method not exceeding a class 3 change, assuming one of the following as assigned:

    1) A variety of parts are made.

    2) Only one type and size of part is made.

(d) List the information obtained from part (c) that you would add to your checklist file and indicate the question number and letter under which each would be filed.

10.2. The following is a running description of the preparation of a piece of direct-mail advertising in an advertising agency. The advertisement consists of four $8\frac{1}{2}$ by 11-inch sheets folded and stapled, as shown in Figure P10.1. Thousands

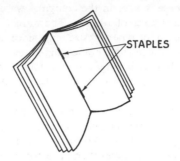

Figure P10.1

of sheets come in from the printer packed in ream (500-sheet) lots and are placed in a stockroom. They are withdrawn as called for by order and moved 100 feet to the collating room, where they are placed in a rack with four compartments. The collating operator takes a sheet at a time by hand from each of the four bins, jobs them, lays them in a cross-pile to keep each set of four separate, and builds a large stack. From time to time, a cross-pile stack is moved 20 feet to the folding room, where each set is jogged, folded, hand-creased, and piled into a stack. From time to time, a stack is moved 20 feet to an angle stapler. An angle stapler has an inverted V rest for the paper, so that the crease, when on the V, is correctly positioned for the stapler. Two staples are driven in. The stapled sheets are restacked. From time to time a pile of stapled sets is taken 200 feet to the mailing room, where the stack is eventually fed, one by one, into an addressing machine, addressed, and placed in a mailbag. All carrying of the stacks is done by hand.

(a) Make a process chart—product analysis for the whole process of preparation of the advertisements, as described, from the time they are in ream lots in the stockroom to the time they are in the mailbag.

(b) Prepare a preliminary and a detailed possibility guide for the process.

(c) With the aid of the checklist, prepare a process chart—product analysis for a better method involving a class 3 change.

(d) List the information gained from part (c) that you would add to your checklist and indicate the question number and letter under which each item would be filed.

10.3. The following is the process used in a large hotel to manufacture waiters' table towels. Bolts of material stored in the sixth-floor storeroom are issued by the head housekeeper to a handyman with a requisition for a bolt or more. He carries them to an elevator 30 feet away. After waiting for the elevator, he takes them down to the third floor, carries them 100 feet to the sewing room, and places them on a long cutting bench. Later, all but one bolt are removed from the cutting table by the cutter, and that bolt is spread out in a long single layer and cut to length, 120 pieces to a bolt. Each bolt is cut separately. The pieces are moved by the cutter, after each bolt is cut, 5 feet to the sewing bench and stacked. The sewer hems the cut ends and eventually counts and compares the number made to the requested quantity. Any excess is sent back to the sixth-floor stockroom. The finished towels are then moved 100 feet by the sewer to the elevator, where the sewer and the towels are carried to the fifth floor. The sewer leaves the towels in the laundry marking room, 60 feet from the elevator. The towels are then machine-marked, carried 60 feet to the washing machines, laundered, and wait in wheeled laundry hampers. A washroom assistant wheels these 30 feet to an elevator that carries them to the fourth floor, where an assistant wheels them 50 feet to the mangle room, where they are left. The entire batch waits in the mangle room until one of the mangle crews is ready to take them. Each hamper is then wheeled 25 feet to the designated mangle, where it is processed by a crew of four workers on a six-roll mangle. Coming off the back end of the mangle, the towels are stacked in piles on a table, then placed in a truck, moved 100 feet to the housekeeping department storage room on the fourth floor, where they are first checked against the requested amount, and then put into storage for future issue.

(a) Prepare a process chart—product analysis for the original method.

(b) Prepare a preliminary and a detailed possibility guide.

(c) With the aid of the checklist, prepare an improved process chart— product analysis, limiting yourself to the class of change designated by the instructor.

(d) List the information gained from part (c) that you would add to your checklist and indicate the question number and letter under which each item would be filed.

10.4. The following is the routine involved in handling a $3\frac{1}{4}$-lb jar of hard candy in a large department store. Twenty-four cases (two gross of jars) of hard candy arrive by truck and are unloaded by hand and stored on the receiving platform. Twelve cases are loaded onto a four-wheeled truck, and each 12, in turn, are transported 85 feet to the freight elevator in building A, wait, are loaded on and taken to the ninth floor of building A, and then are taken 50 feet across the floor to the checking area, where the dock helper leaves them and returns to the dock. Each case is eventually checked against the proper invoice and shaken to determine if any of the jars are broken. After the 24 cases (two trucks) have been checked and replaced on the trucks, each truck, in turn, is moved back to the freight elevator and left there. The elevator operator, when he stops at that floor, takes them on and down to the ground floor, where he pushes them off the elevator. A ground floor handler eventually moves them 400 feet through a tunnel under the street to the B building elevator and leaves them there. The B building freight elevator operator, when he stops for a load, takes them on, carries them to the fifth floor, and pushes them off. A fifth floor storeroom employee eventually takes them 85 feet from the elevator to the candy storeroom. The whole 24 cases are first stacked in the open space in the storeroom and then, in turn, the cases are carried one at a time 30 feet, and placed on the shelves until needed. When an order comes in for two cases (the usual quantity) from the floor sales department, the cases are removed from the shelves, loaded onto a truck, wheeled back to the elevator, down to the basement, across to building A, and to elevator 9, which is 525 feet away from the B building elevator. After waiting for the elevator, the truck with two cases is loaded on and brought up to the first floor selling department, 80 feet from the elevator, where the two cases are placed on a table. The storeroom employee returns with his truck. All but one of the jars are placed by sales personnel in undercounter storage. One jar is placed on display. When that jar has been sold (send sale), it is moved 12 feet to the sales counter, where a sales check is tied on, and then the jar is placed in a hamper kept there for delivery to elevator 9. Eventually, the hamper is moved by a floor handler to the elevator. The elevator operator ultimately takes it on and transports the jar plus other goods to the ninth floor, where he shoves the hamper off. A floor handler eventually wheels the hamper 270 feet to the packing room. In the packing room, the jar is removed from the hamper and checked against the sales check for correctness of merchandise and examined to see if it is in good condition. It is then placed on a shelf 2 feet away, from which it is taken by the packer when previously arrived goods have been packed. After packing, the jar is placed in a hamper alongside the packer with other packages that have been packed by the same packer. Later the hamper is wheeled 16 feet to elevator 19 or 30 and is left. The elevator operator takes this and other hampers down to the fifth floor, where it has to be transferred to elevator 23, 54 feet away. Elevator 23 then takes its load to the load level, where the hampers are unloaded and wheeled by handlers 40 feet to the sorting area, where the loads are sorted and marked and placed in a relay truck, which is parked 30 feet away. The relay truck then takes its load 5,000 feet to the shipping-center loading platform, where the goods inside the truck are unloaded and sorted into hampers for wheeling to driver's route bins, the average of which is 150 feet away.

(a) Draw a process chart—product analysis for the process described.

(b) Prepare a preliminary and a detailed possibility guide for the process.

(c) With the aid of the checklist, prepare an improved process chart for a send sale, not exceeding a class 3 change.

(d) List the information gained from part (c) that you would add to your checklist and indicate the question number and letter under which each item would be filed.

10.5. An acetylene generator separator plate (see B, Figure P10.2) is manufactured from strip stock, which is kept in large quantities in the plant stores warehouse. Fifty-eight strips are removed from the stack, one at a time, and loaded on a trailer 5 feet from the pile. A tractor is later hitched to the trailer and the whole load is transported 500 feet to Department 56, where it is unloaded and fed one piece at a time into a machine that pierces and blanks the strip. As the pieces come out of the machine, they are checked for size and number of holes and dumped into a pan that will hold 500 pieces. The pan is carried, by lift truck, to the next machine, 25 feet away, and again the parts are fed into

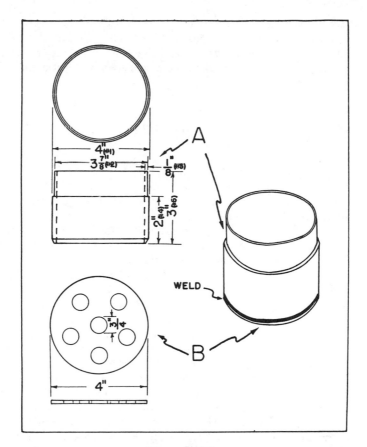

Figure P10.2

the machine one at a time for restriking and flattening. At this point, a production checker ascertains whether the quantity (about 3,000) in the tub under
the second machine corresponds to the production counter reading on the
machine. One tub (3,000 pieces) is picked up by an electric hoist and dumped
into an automatic washing machine 12 feet away. As the pieces come out of
the washer, they dump into tubs. Each tub is weighed to verify the count, and
is then taken by gasoline lift truck to Department 41 (1,265 feet), where the
tubs are deposited beside a sandblasting machine. The pieces are loaded into
the sandblaster one shovelful at a time until the full tub load has been accom-
modated. The pieces come out of the bottom of the sandblaster into a per-
forated container so that the grits can be extracted. They are shoveled from
these containers into tubs and loaded on an electric lift truck that takes them
to the inspection area 157 feet away, where the count is again verified, and
then by gasoline lift truck back to Department 43 stores, which is located
behind the washing area. This move is about 1,125 feet. At the Department 43
stores, the tubs are set down by the gasoline lift truck and the pieces in the

tubs put in boxes by the handful until each box contains exactly 3,000 pieces. These boxes are moved by gasoline lift truck to Department 42, where the boxes are emptied into bins by hand. The bins in this case are adjacent to the welding assembly fixture, where the pieces are positioned one at a time to be welded into the generator.

The tubes used for the generator body (see A, Figure P10.2) are processed as follows. Fifty-five tubes, one crane sling load from a tube cradle, are loaded on a trailer 20 feet from the cradle. A gaoline lift truck pulls the trailer 1,500 feet to Department 97 stores, where they are put on a rack with a crane capable of handling 10 tubes in a sling. These racks are later emptied one tube at a time by two men into a trailer that takes 11 tubes, and is hand-pushed 1,000 feet to Department 91, where the 11 tubes are put one at a time, by hand, into a wheeled rack that is then moved by hand 100 feet to the chamfering machine. At the chamfering machine, one piece at a time is chamfered, turned, cut off, the tube chamfered at the cutoff end, and so forth. The cut pieces are placed in a tub, 726 pieces to the tub. The tubs are taken by gasoline lift truck to the inspection area 100 feet away, where the count is verified and a visual quality inspection is made. After the inspection, the pieces are loaded back into the tubs by handfuls, and the tubs eventually lifted by the gasoline lift truck to be taken to Department 41, where they are loaded by the handful into a washer in which they are washed and automatically ejected into a container at the end of the washer. The pieces are loaded from the container, by the handful, into tubs that are taken, in turn, by an electric lift truck 85 feet to the sandblast room, and deposited beside the sandblast machine. At the sandblast machine, the pieces are taken out individually, sandblasted, dumped back into tubs, removed from the sandblast room and loaded, by the handful, into other tubs (dumping out sand in the operation) for transportation by electric lift truck to the inspection area, which is 157 feet away. At the inspection area the pieces are counted and retubbed, 725 to a tub. Eventually they are lifted by a gasoline lift truck and moved 587 feet to rest beside the fixture where part A is to be incorporated into the final assembly.

The welders clamp one separator and one generator body into the fixture and gas-weld them into a single unit, as indicated in Figure P10.2.

(a) Draw a process chart—product analysis for the manufacture of a generator unit from raw component stock to welded unit.

(b) Prepare a preliminary and a detailed possibility guide for the process of part (a).

(c) With the aid of the checklist, prepare a process chart—product analysis for a better process. You may go as high as a class 5 change but may not buy a completed unit for anything other than reasonable standard stock. Dimensions 1, 2, 3, 4, and 5 on B of Figure P10.2 may not be appreciably altered.

(d) List the information gained from part c that you would add to your checklist and indicate the question number and letter under which each item would be filed.

11   HORIZONTAL TIME BAR CHARTS

Introduction

In the preceding chapter it was indicated that with certain problems of process study, various techniques other than the process chart—product analysis possessed certain advantages. One set of conditions under which other techniques were more advantageous was given as: *the timing of various batches moving through the various steps, or the time relationship among the steps are the primary aspects under consideration.*

Where the technique is used. Under conditions such as those listed above, provided that the interrelationships among the steps are relatively simple, a horizontal bar time chart is of great assistance in making a useful analysis. This chapter will deal with such charts. When the relationships among the steps are complex, a network diagram, treated in Chapter 12, is usually a more advantageous analysis tool.

Horizontal time bar charts defined. A horizontal time bar chart is a graphic means of protraying the time relationships among the separable steps of the procedure involved in performing the work required to modify an output from one stage of completion to another.[1] The definition, one will note, is similar to the definition of process chart—product analysis, except for the emphasis on *time*. The process chart—product analysis, with the usual variety of conventions for portraying steps, emphasizes the sequence and nature of steps; the horizontal time bar chart is primarily an aid in examining the time relationships among a given set of steps and for considering alternatives wherein the primary change is with respect to timing.

Purposes served by the technique. A horizontal time bar chart is of assistance with studies of the sequence of the steps of production with both substantive and service outputs. With substantive outputs it is usually used to assist in determining class 3 type changes; with service outputs it usually assists in achieving class 4 or 5 type changes.

An Example

Making the chart. A horizontal time bar chart is most easily drawn on an ordinary sheet of graph or cross-section paper as shown in Figure 11.1.

Figure 11.1 is the horizontal time bar chart showing the time relationships among the various production steps of a component of a larger product. The component itself consisted of three parts. The component was produced on a batch-lot basis. In Figure 11.1 a separate line and a separate bar is used to indicate the processing of each part. The length of each bar is proportional to the time required to process the batch through the particular step. In the preparation of such a chart the analyst must select a scale appropriate to the time values encountered in the problem. The scale chosen for the problem illustrated in Figure 11.1 was one day per vertical line. The distance from the start of the first bar to the completion of the last bar is proportional to the amount of time required to produce the entire batch of the component.

---

[1] In many ways the horizontal time bar chart resembles the preliminary phase of a Gantt chart. I have avoided calling the horizontal bar time chart a Gantt chart in that it is not used in all the ways in which a Gantt chart is used; merely a few of the Gantt chart conventions are used; this chapter deals only with this one type of chart and is not at all a definitive treatment of Gantt charts. For a complete treatment of Gantt charts, one should consult any edition of the following two books: W. G. Ireson and E. L. Grant (eds.), *Handbook of Industrial Engineering and Management*, 2nd ed. Englewood Cliffs, N.J.: Prentice-Hall, Inc., 1971; H. B. Maynard (ed.), *Industrial Engineering Handbook*. New York: McGraw-Hill Book Company, 1971. The classic book on the subject is Wallace Clark, *The Gantt Chart*. New York: The Ronald Press Company, 1922. [The Clark book has also been printed in Britain (1934) and translated and published in 13 other countries.]

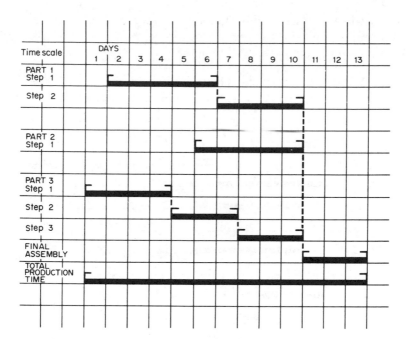

Figure 11.1. Horizontal time bar chart showing the time relationships among the various production steps of a component of a larger product.

Purpose of using horizontal time bar charts. The study was undertaken because of a problem whose basic nature concerned the use of time. The larger product, of which this component was a part, was manufactured to order. The lead time between the receipt of the typical order and the required delivery date did not leave enough time to permit orderly procurement of the raw material and the production of the components, even with overtime and night-shift work when materials were finally procured. Hence, a more rapid procurement procedure, a longer delivery time, or a shortened production time appeared as a necessary improvement. However, industry practices precluded a lengthening of the delivery time; procurement had been reduced to the apparent limit; changes were needed in the actual production cycle.

Critiquing the data. The problem, as portrayed in Figure 11.1, reveals some characteristics which frequently are found in problems of this kind. First, some change in the total time to produce the component could be achieved by changing only the time needed for the steps used to produce part 3. Parts 1 and 2, which are produced independently of the production of part 3, could readily be started earlier if part 3 took less time to produce. Under such conditions part 3 is referred to as the *critical* part. However, if the time for the steps on part 3 is reduced beyond a certain point, part 1 (whose production time exceeds that of part 2) will become the critical part. With most problems of this type the location of *critical* moves in a similar fashion as various improvements are made.

Innovation. The improved method shown in Figure 11.2 was finally devised and later successfully adopted. It should be noted that the improved

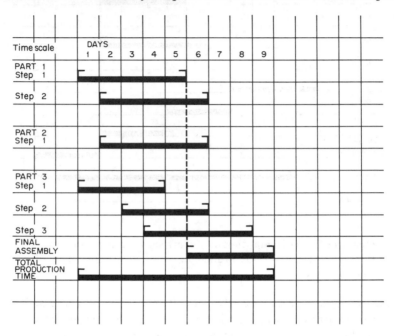

Figure 11.2. Horizontal time bar chart of improved method for product shown in Figure 11.1.

procedure meets the criterion of success, shorter production time, but that it differs from the original method in two important ways. First, the entire batch of a part is not processed completely through a step before the subsequent work on the batch begins. Second, the total time of machine usage is increased. The solution may be described as a trade-off; one advantage was gained at the cost of one or more disadvantages. In this particular case, the advantage of being able to meet customer delivery dates as well as avoid the use of overtime and night work (which previously only reduced rather than eliminated the lateness) more than offset the additional cost of both the more frequent handling of material from step to step (because of the overlapped steps) and the additional cost of increased total machine time. (The increased total machine time was caused by production stoppages affecting several steps simultaneously rather than merely the step on which the broken tool, and so forth, occurred.)

Criticism and innovation. In the development of the improved method the analyst considered various basic ways of evolving alternative time relationships, such as are given in the following checklist:

CHECKLIST FOR HORIZONTAL TIME BAR CHARTS

1. Can the time be shortened by reducing the time for a critical step or steps?
   a. By devising a new procedure for a step or steps
   b. By assigning more resources to work simultaneously on a step or steps
2. Can the time be shortened by overlapping steps?
3. Can troublesome steps be undertaken earlier in the procedure to allow defects to be detected sooner so that an increased amount of time is available for corrective action?
4. Can better service be given if some of the service outputs, or partial service outputs, are generated earlier in the process?
5. Can some of the required outputs be avoided if some of the outputs are generated earlier in the process?

Alternative Methods of Charting

In cases other than the one given as an example it may be desirable to differentiate different types of steps with different types of shading or different colors. It seems undesirable to lay down hard and fast rules, in that the purpose of the chart is to aid the analyst. The chart is not an end in itself; the analyst should introduce changes as he needs them.

Summary—Horizontal Time Bar Charts

1. Uses:
   For aiding in the development of a class 3, 4, or 5 change when:
   a. The timing of the various batches moving through the various steps of a process is the primary aspect under consideration and the interrelationships among the steps are relatively simple.
   b. The time relationship among the steps of a process is the primary aspect under consideration and the interrelationships among the steps are relatively simple.
2. How made:
   a. An ordinary sheet of graph or cross-section paper is usually adequate.

b. A separate line is used for each separable part of the output and separable step on each part.

c. A bar, whose length is proportional to the time required for the step, is drawn on each line in such a way that it shows the place of that time interval on the total time continuum.

3. *How used*:

The time relationship is examined with respect to both the criterion of success and questions such as given in the checklist.

4. *What then*:

A horizontal time bar chart is drawn for the resulting suggested improved method to permit a final check of the feasibility and to assist in evaluating the improvement with respect to its effect upon the time aspects of the problem. The new chart also provides a means of describing the proposed new method.

PROBLEMS

11.1. Make a horizontal time bar chart of the pattern of doffer team usage for serving the pattern described in Problem 3.3(b).

11.2. Make a horizontal time bar chart of:
(a) A session with the dentist.
(b) An efficient pattern for doing a large marketing at a supermarket.
(c) A typical evening's study.
(d) Completing registration.

11.3. Propose and chart an improved method for the situation charted for Problem 11.2. What was your criterion of improvement? What class of change is involved?

11.4. Make a horizontal time bar chart of:
(a) A national election.
(b) The administration of the judicial process with felons (in your area).
(c) The processing of a legislative bill by your local government.
(d) The processing of a new regulatory bill by:
  1) The House of Representatives of the United States.
  2) The Senate of the United States.
  3) The Congress of the United States.

11.5. Propose and chart an improved method for the situation charted for Problem 11.4. What was your criterion of success? What class of change was involved?

11.6. Make a horizontal time bar chart for the settlement of a damage claim in:
(a) A small claims suit.
(b) An automobile accident when no personal injury is involved.

12

NETWORK DIAGRAMS

Introduction

In Chapters 10 and 11, various ways of analyzing processes were discussed with a process defined as *all the steps required to modify an output from one stage of completion to another*. This chapter will deal with another method of analyzing processes.

Where the technique is used.   There are process design problems which are far too complex to handle with either of the two techniques discussed to this point. Such problems are encountered in designing the total process for such projects as the construction of a new ship, a new metropolitan subway, or similar complex object; the design and construction of dams or new aircraft; the planning of a new government service; the planning for and holding of a national convention for a scientific society; and so forth.

Dependencies, a new factor. Each of these problems differs from the type of problem previously discussed in that the process has within it a complex system of *dependencies*. The term "dependencies" refers to the requirement of completing some activities before being able to start other activities, regardless of the amount of resources (men or equipment) employed. In process chart—product analysis and in horizontal time bar charts the dependencies were essentially simple and the sequence of steps flowed, merged, or separated in a simple fashion.

Characteristics of problems for which network diagrams are needed. When the dependencies are tangled and do not progress in a uniform fashion, some other system of charting is required to assist in analysis. Such process design problems are also frequently characterized by one or more of the following additional differentiating characteristics (as compared with the type of problem for which the process chart–product analysis or the horizontal time bar chart is of assistance):

1. The output under scrutiny has a large number of components.

2. Many of the component outputs are service-type outputs, such as designs completed,[1] arrangements with other groups completed, information gathered, and so forth, and these service outputs are vital to the process.

3. The relationships among the steps of the process with respect to time are of vital importance to the design of a satisfactory process.

4. The complexity of the process and the scope of the work are such that an initial analysis of the whole precludes the feasibility of examining the work to the degree of detail typical of the usual process chart—product analysis.

Network diagram defined. With complex projects of the type that have been described, an analysis technique employing a *network diagram* has been shown to be of great value in developing an effective process design. A *network diagram*[2] is a graphic method of displaying the relationships between and among the steps in a process.

---

[1] In order to differentiate a service output from the activity of producing it, the description of the output has been given with the verb in the past tense, a procedure described in detail in Chapter 9.

[2] Diagrams of this type were originally given great publicity in connection with their use in the development of the U.S. Navy missile, Polaris. Such diagrams were employed in the design of the process and in the operation of a control system called the performance evaluation and review technique, referred to by its acronym, PERT. Hence, in the literature

A Simple Illustration

A simple network diagram[3] is shown in Figure 12.1. Each circle (or square) of Figure 12.1 represents a status achieved by the performance of activities. A *status* is a partial or complete service or substantive output completed. A square may be used in place of a circle to represent an important status, to make the chart easier to read; otherwise, the meaning of the square is similar to that of the circle.

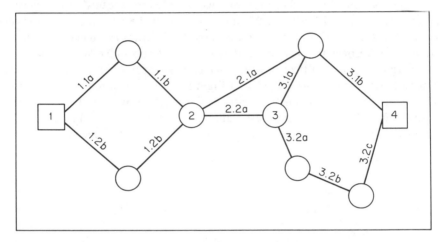

Figure 12.1. Simple network diagram.

Network diagrams and work-unit analysis. It should be recalled that in Chapter 9 we described a method for identifying outputs with respect to service organizations. "Outputs completed," as used in that chapter, should be identified as synonomous with "status achieved" in this chapter. A status of a network diagram represents the completion of a 4th-order or 5th-order work-unit. The activities are 3rd-order work-units. The utility of the work-unit-analysis approach as an input to a network diagram should be obvious.

---

such diagrams are commonly called PERT charts. They are also called *link diagrams*, *critical path planning*, and so forth. *Network diagrams* has been selected as the more generally descriptive term, not to be identified with a particular technique but to refer generically to all versions of the general approach.

[3]A typical diagram of a complex process may describe a process that will range from a hundred to several thousand steps. Obviously, such examples are beyond the scope of this book. However, as the discussion progresses it should become increasingly obvious that the more complex the problem of process design, the more useful the network diagram becomes.

Basis of separating status steps. The chart also may be drawn without squares or a variety of symbols may be employed, although differentiation of the status steps will be based on different criteria than used for the separation of the steps of a common process chart—product analysis. The common criteria for differentiation will be the importance of reaching the status, some aspect of the consequence of achieving the status, or some aspect of the resources needed to achieve the status, as contrasted with the process chart-product analysis where the basis of separation was the nature of the activities.

Charting activities. The performance of activities required to achieve a status is represented by the lines. In contrast with the varieties of status on network diagrams, the activities on a network diagram are usually differentiated with respect to only one criterion, their importance with respect to affecting the minimum time in which one can expect to complete the project. For instance, the network diagram of Figure 12.1 has been redrawn in Figure 12.2 with the expected time[4] for performing each activity added under the activity identification. From these values the critical path, indicated by the heavy lines used in Figure 12.2 for some of the activities, has been determined.

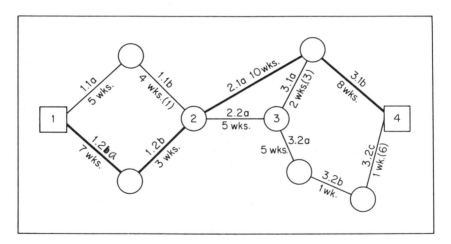

Figure 12.2. Network diagram of Figure 12.1 redrawn with the expected time for performing each activity added under the activity identification.

The critical path defined. The *critical path* is the system of dependencies that determines the minimum time in which the project can be expected to be

---

[4]The development of these time values, representing the time study aspect of this type of problem, will be discussed in Chapter 24.

completed. It can also, of course, be described as the maximum time pathway through the network. In the preparation of Figure 12.2 the slack times were also computed and are indicated by the numbers in parentheses under some of the activity lines.

Slack time defined. The *slack time* is the additional time that could be taken in performing the activities represented by a noncritical pathway through the network without causing any delay on the total project. Naturally, the slack times may not be totaled through the network but may be totaled only to the point where the noncritical path joins a status that is either dependent upon an activity associated with the critical path or is a dependency of an activity that is in the critical path. Hence, the slack times of Figure 12.2 do not appear under each noncritical activity. They are only under the final activity of the noncritical paths.

Normal procedure in making a network diagram. In actual practice, the preparation of a network diagram is usually a team project; a group of individuals knowledgeable in the various disciplines involved in the project will participate. The first step will be the setting forth of the network in general form such as shown in Figure 12.1. The nature of each activity and each status will be described and the dependencies indicated. In contrast with the horizontal time bar chart the network diagram may be drawn before the time values for each activity are known. The configuration (the sequence of activities, status points, and dependencies) can be checked for feasibility and altered to achieve apparent feasibility (if necessary). The configuration can also be subjected to simplification to reduce it to the apparently most simple process. Subsequently, the group is usually enlarged to include representatives of those who will undertake the various activities, in order to add the time values for each of the activities and to determine the critical path and the slack times.

Improving the process. Subsequently, the work of improving the method begins. Let us assume that the objective is to reduce the project time. The critical path, therefore, will be closely examined. For instance, in Figure 12.2 the critical path from status 1 to status 2 is the pathway 1.2a, 1.2b. The slack time of the adjoining noncritical pathway 1.1a, 1.1b is 1 week. Hence, the minimum required time to proceed from status 1 to status 2 may be reduced by not more than 1 week if the changes are confined to pathway 1.2a, 1.2b. There is no apparent advantage to expending more effort and money on this pathway to reduce the time by more than 1 week unless additional reductions can be achieved in pathway 1.1a, 1.1b. Of course, the reduction of the critical pathway 1.2a, 1.2b may not be worth the cost. It remains, then, in the critical path.

An Example

The status of the project. Let us examine an example that resembles the preceding general illustration. A city may be planning a subway. They are at status 1; the route and general design have been agreed upon. The first section of the critical path[5] may consist of the following two activities:

1. Determine who are the owners of the property under which the subway must pass or upon which the stations must emerge.
2. Acquire either an easement or a title to the property, as may be appropriate.

The noncritical pathway joining the pathway above may consist of the following two activities:

1. Locate which utilities (e.g. sewer lines, water lines, and so forth) will be interfered with by the subway.
2. Relocate these utilities.

Possibilities of improvement. The critical path time, perhaps, could be reduced by a change in the strategy of acquiring easements or titles, such as increasing the amount of money offered. However, any expenditure of effort or money beyond that required to cause the time for the critical path to be equal to the noncritical path would not decrease the total project time. If additional resources were available, to have any useful effect they would have to be expended thereafter on both pathways. If, in the illustration above, a way to shorten the critical path could not be found, it could be that the steps have not been divided into small enough activities. If the planning time is sufficient, an analysis may be made with smaller activity steps.

Basic values of a network diagram. It is still possible that even with smaller activity steps, no way can be found to shorten the critical path. The network diagram is still of great value. First, a complete and feasible plan exists. Second, during the actual activity managerial attention can be directed with greatest attention at the activities associated with the critical path so that delays beyond this controlling value are minimized. The noncritical paths permit some delays without delaying the whole project. There are, however, certain dangers if this kind of thinking is carried too far. For instance, in the illustration used, the utility crews who have the

---

[5]Obviously these steps in actual practice would be broken up into much smaller steps. I have avoided going into details in order to keep the illustration simple.

responsibility of moving the interfering lines have, without doubt, other work to do. If utilized properly by their management, they are fully employed. Hence, from their point of view they have no slack time; if they are late on the noncritical path associated with the subway route, they will be late on other work. Their management must think of the slack time only as a limit on their freedom of corrective action in rescheduling work if delays occur within their sphere of activity; the slack time is not really an indication of a totally permissive delay. The slack is only slack with respect to the subway schedule.

Usual Criteria of Improvement

Obviously, in most cases the object of improvement is to achieve one of the following criteria of success:

1. Reduce the overall time for the project.
2. Reduce the overall cost of the project.
3. Usefully employ the slack time to improve the project.
4. Control the impact of the project on the organization, society,or the economy with respect to employment, rate of expenditure, utilization of scarce resoures, and so forth.

Complex projects. With complex projects, such as are usually subjected to network diagram analysis, means of achieving such improvement cannot be generalized; they will be specific to the various fields of work. However, certain routine techniques can be and have been generalized. Procedures have been devised for transcribing networks into a data form suitable for entering into computers. Programs (computer routines) have been devised for examining networks and determining the critical paths and the slack times; and for examining potential reductions in activity times for the effect upon the critical path and the network. Routines have also been devised for examining the effect of changes in variables such as suggested by the fourth criterion of success given above. This computer aid, however, is routine. People must still be employed to generate either the change or the limitation (to conform to some specific statement of a criterion of success) to give to the computer to recalculate the network.

The preceding discussion is not meant to suggest that a computer is an inevitable concomitant of a network diagram. Simple network diagrams are handled much more economically by manual means. Even complex projects supported by a network diagram without a computer have a far better chance of having an effective process design than those planned by informal methods. However, when computers are available, there is a definite point at which they are more economical than manual methods.

Network diagrams and managerial control. The final network diagram forms a basis for managerial control. The time values for each activity are used to develop a schedule in real time. These data are the inputs to a production control system. Such systems are beyond the scope of discussion of this book.

Summary—Network Diagrams

1. Uses:
   As an aid in designing a complex process when the process has within it a complex system of dependencies. The output of the process may be a service or substantive output, or a combination.

2. How made:
   A group of individuals, knowledgeable in the disciplines involved in the process, set forth the configuration of activities, status points, and dependencies which represent an apparently most simple feasible process.

3. How used:
   Time values are determined for each activity and the critical paths and slack times are determined. Ways are examined to alter the network and its characteristics in order to conform to one or more criteria of success.

4. What then:
   The final network diagram forms a basis for subsequent managerial control of the process.

PROBLEMS

12.1. If in Figure 12.2, the time for the steps is altered to the following: 2.1a = 2 weeks; 3.1b = 2 weeks:
   (a) Define the new critical path.
   (b) Indicate the single activity that has the greatest potential for reducing network time.

12.2. Draw a network diagram of the activities associated with a curriculum change in:
   (a) A grade school.
   (b) A state university.
   (c) A private university.

12.3. Draw a network diagram, for your home area, of:
    (a) An air-pollution-abatement program.
    (b) A water-pollution-abatement program.
    (c) A noise-abatement program.

12.4. Draw a network diagram for a program of economic aid to assist a new nation toward the mainstream of the modern world.

12.5. Draw a network diagram for the development of a program to halt the population explosion.

12.6. Draw a network diagram of a program designed to rid a city of rats.

12.7. Draw a network diagram of a campaign by an individual for a city government office.

12.8. Draw a network diagram for the construction of a new 5-acre city park with a small lake, theater, bandstand, ballfield, and picnic area. (Provide a sketch of the completed park.)

13

MOTION ECONOMY

Introduction

The preceding three chapters have been devoted to a discussion of details of formal techniques for improving processes. A logical procedure was suggested for employing these techniques. The results obtained by applying these techniques have been illustrated with a variety of products. The making of an analysis, fundamental in the use of a logical procedure, has entailed learning ways of dividing tasks into steps or subdivisions. However, the emphasis has been on outputs; on products.

In the six chapters that follow the focus will be on the person doing the job; on the activities. A variety of techniques will be examined. If one examines the details of the illustrations that will be given, works on the problems that follow each chapter, and also applies these techniques to actual jobs, using the appropriate checklists to assist in developing improved methods, his way of looking at jobs will have undergone changes. These changes will include the following:

1. It will no longer be possible to look at tasks without seeing them as (a) falling into a category most readily studied with the aid of a particular technique, and (b) a series of steps related to this technique.
2. Certain basic improvements will be suggested from even a casual observation of a job; the checklists will be automatically applied.

It is the purpose of this chapter to aid in preparing the reader for the checklists of the various techniques, to lay a foundation for the second item from the preceding list, and to assist in eventually making possible a more complete approach to improving work. It is not the intent of this chapter to replace the formal analysis techniques with a quick visual method, but to provide a checklist of general principles for the analyst as a preview to learning to see jobs in terms of these analyses.

This chapter also traces the history of the development of these general principles. Hence, the references cited are old rather than modern.

Principles of Improvement, Historical Background

The general principles of improvement have a long history. Frank B. Gilbreth, a pioneer in motion study, first listed the variables in the work situation under several headings, two of which follow[1]:

A. *Variables of the surroundings, equipment, and tools:*
   1. Appliances.
   2. Clothes.
   3. Colors.
   4. Entertainment, music, reading, etc.
   5. Heating, cooling, ventilating.
   6. Lighting.
   7. Quality of material.
   8. Reward and punishment.
   9. Size of unit moved.
   10. Special fatigue-eliminating devices.
   11. Surroundings.
   12. Tools.
   13. Union rules.
   14. Weight of unit moved.
B. *Variables of the motion:*
   1. Acceleration.

---

[1]F. B. Gilbreth, *Motion Study*. New York: D. Van Nostrand, 1911; reprinted in W. R. Spriegel and C. E. Myers. *The Writings of the Gilbreths*. Homewood, Ill.: Richard D. Irwin, 1953, pp. 152–153.

  2. Automaticity.
  3. Combination with other motions and sequence.
  4. Cost.
  5. Direction.
  6. Effectiveness.
  7. Foot-pounds of work accomplished.
  8. Inertia and momentum overcome.
  9. Length.
 10. Necessity.
 11. Path.
 12. "Play for position."
 13. Speed.

Gilbreth went on to discuss each of these, to give specific examples and to generalize from these examples, thus creating general principles of improvement. He elaborated on these in *Fatigue Study* (written with his wife, Lillian M. Gilbreth),[2] and finally developed a set of "Rules of Human Motions."[3] These rules were rearranged and amplified by Barnes[4] as "Principles of Motion Economy," in an early text, and also given as "Laws of Motion Economy and Their Corollaries" by Lowry, Maynard, and Stegemerten,[5] in the first edition of their text.

Critique of General Principles of Improvement

However, it has been generally recognized, as was well stated by R. N. Blair, that

> These various listings do not represent the principles of motion economy in any fundamental sense, but simply provide some useful rules, of varying significance and of unequal importance. These range from basic points of general applicability, such as the preferability of smoothly continuous motions over motions involving sudden and sharp changes in direction, to very specific suggestions of the type describing a good kind of handle for large screwdrivers.[6]

---

[2]F. B. Gilbreth and L. M. Gilbreth, *Fatigue Study*. New York: Sturgis and Walton, 1916.

[3]F. B. Gilbreth and L. M. Gilbreth, "A Fourth Dimension for Measuring Skill for Obtaining the One Best Way," *Bulletin of the Society of Industrial Engineers*, Vol. 5, No. 11, 1923, pp. 6–7.

[4]R. M. Barnes, *Motion and Time Study*. New York: John Wiley & Sons, 1937, pp. 110–178.

[5]S. M. Lowry, H. B. Maynard, and G. J. Stegemerten, *Time and Motion Study*. New York: McGraw-Hill Book Company, 1940, pp. 95–112.

[6]R. N. Blair, "A Fresh Look at the Principles of Motion Economy," *Journal of Industrial Engineering*, Vol. 9, No. 1, 1958, pp. 3–5.

Checklists in Chapters That Follow

The individual checklists given in the six chapters that follow were derived from the basic Gilbreth material amplified by experience with each analysis technique. They are an attempt to present these principles without the ambiguity of generalities, avoiding the mixing of specifics phrased for different scopes of work. The general principles of improvement are presented now as a means of giving a preview summary of the various checklists that will follow. The material is phrased primarily with respect to the work of an individual. It should be noted, however, that, in many cases, references to "the hand" could also refer to crew members in group work. These general principles should be of great value in helping to discern potential job improvements and to assist in understanding the basis for selecting an appropriate analysis technique. Further, the list of principles may be used as a summary checklist after the job analyzed has been questioned with the aid of one of the detailed checklists. It may also be used as a summary checklist on new jobs once the initial work pattern has been "roughed out;" it can be used with all six of the technique chapters that follow.

The Human Factor

A final word of caution might be appropriate at this point. The list of general principles is directed primarily at the mechanical aspects of jobs. This is not meant to imply that the reaction of the worker is mechanistic. Indeed, as was indicated much earlier in the book, the worker's reaction is anything but mechanistic.

General Suggestions for Improving Jobs

A. *Elimination*
 1. Eliminate all possible jobs, steps, or motions. (This applies to body, leg, arm, hand, or eye.)
 2. Eliminate irregularities in a job so as to facilitate automaticity. Provide fixed places for things.
 3. Eliminate the use of the hand as a holding device.
 4. Eliminate awkward or abnormal motions.
 5. Eliminate the use of muscles to maintain a fixed posture.
 6. Eliminate muscular force by using power tools, power feeds, etc.
 7. Eliminate the overcoming of momentum.

   8. Eliminate danger.
   9. Eliminate idle time unless needed for rest.
B. *Combination*
   1. Replace with one continuous curved motion short motions which are connected with sudden changes in direction.
   2. With fixed machine cycles, make a maximum of work internal to the machine cycle.
   3. Combine tools.
   4. Combine controls.
   5. Combine motions.
C. *Rearrangement*
   1. Distribute the work evenly between the two hands. A simultaneous symmetrical motion pattern is most effective. (This frequently involves working on two parts at the same time.) With crew work, distribute the work evenly among members of the crew.
   2. Shift work from the hands to the eyes.
   3. Arrange for a straightforward order of work.
D. *Simplification*
   1. Use the smallest muscle group capable of doing the work, providing for intermittent use of muscle groups as needed.
   2. Reduce eye travel and the number of fixations.
   3. Keep work in the normal work area, the area reached without moving the body.
   4. Shorten motions.
   5. Adapt handles, levers, pedals, buttons, and so on, to human dimensions and musculatures.
   6. Use momentum to build up energy in place of the intense application of muscular force.
   7. Use the simplest possible combination of motions.
   8. Reduce the complexity of each motion, particularly the motions performed at one location (as contrasted with motions that change the location of things).

PROBLEMS

13.1. What "general suggestions for improving jobs" were embodied in the improved method of:
   (a) Figure 1.2?
   (b) Figure 1.5?
   (c) Figure 1.7?
   (d) Figure 1.9?
   (e) Figure 1.11?

13.2. Examine any task. Describe the present manner of performing this task and list "the general suggestions for improving jobs" that would be applicable. Describe your proposed method of embodying these suggestions. Suggested jobs:

(a) Checking out a library book.

(b) Bagging at the supermarket.

(c) Gas station service on a car.

(d) Putting paper into a typewriter:
      1) Paper only.
      2) With carbons.

(e) Serving at a cafeteria.

(f) Shoe sales activity at a shoe store.

(g) Yard goods sale at a fabric store.

(h) Any home task.

14   PROCESS CHART—MAN ANALYSIS

Introduction

After the possibility guide or other preliminary analysis is made, it may be decided to seek a class 1 or 2 change. This decision implies that the change will affect but a single job or work station and calls for the use of a man analysis technique for the main second step of the scientific method—analysis. A man analysis technique is also required if a single job or operation is to be designed.

The technique chosen to perform this step will usually and primarily be a function of the physical characteristics of the job studied. However, in some cases the choice will also be influenced by factors similar to those given in the list of the psychological and economic factors affecting class of change, such as the position of the analyst in the organization (and the techniques at his command). All this, of course, may be preceded by a preliminary study with a process chart—product analysis to make sure that the job is necessary.

Process Chart—Man Analysis Defined

A process chart—man analysis is a graphic means of portraying the separable steps that a person performs when doing a task that requires the worker to move from place to place in the course of the work.

The Use of the Technique

Where the technique is used. Some jobs or work stations are of such a nature that a considerable area is covered by the work of an individual, like many of the jobs of maintenance and service employees, machine tenders, materials handlers, warehouse employees such as stockpickers, mail clerks, roving inspectors, and certain clerical employees. In addition, many tasks in fields of activity other than manufacturing are similar in nature, such as much of the work of hotel maids and bellmen, restaurant busboys, department store salesclerks, hospital nurses and orderlies, librarians, farmers, and many members of the armed forces. In such cases the process chart—man analysis is frequently the proper technique for the main second step.

Purposes served by the technique. A process chart—man analysis is an analysis of what the *person does* and not of the steps performed in sequence on the product or material. The chart is an aid to clear understanding of the activities of persons performing work that requires them to move from place to place. Care must be exercised not to confuse this analysis with a process chart—product analysis. The end result of a process chart—man analysis is usually a class 1 or 2 improvement, although useful changes of greater complexity often become apparent.

Categorizing the types of jobs which may be encountered. Four general types of jobs may be encountered when applying process chart—man analysis techniques.

1. *The work has a single repeated cycle.* In such cases a single cycle will be charted with a cycle defined as all the steps necessary to bring a unit of output to the state of completion typical of the operation, or all the steps typical of a single performance of the task.

2. *The work is cyclic but there are several subcycles performed with different frequency.* For instance, the worker may perform subcycle A on

each part and then subcycle B for 10 parts together, subcycle B occurring one-tenth as often as A. In such a case a chart will be drawn showing not only one performance of each subcycle but also indicating the frequency of subcycles.

3. *The work varies from cycle to cycle*

(a) In some cases the variation may be primarily due to operator habit and not inherent in the work; consequently, the analyst may plot several cycles, to give him more material from which to develop a preferable work pattern.

(b) The variation may be inherent in the job, and each subsequent performance may differ in detail but not in general pattern. In such a case sample cycles are drawn up for study with the general pattern indicated and the details that may change so noted. Attention is paid on a weighted basis to the factors controlling the variation.

In either 3(a) or 3(b), in many instances, a more complex type of study, *memomotion study*, described in Chapter 19, may be more useful.

4. *The task may be such that there is no cycle or pattern.* This is usually true of supervisory and similar activities. In such cases the study of a process chart—man analysis may lead only to general suggestions, on the basis of which the individual who must constantly plan his task can in the future do a better job of planning. In such cases other techniques, such as *work sampling* or *work activity analysis*, described in Chapter 8, may be much more productive of useful changes.

The steps separated with process chart—man analysis. No matter where they occur, the jobs are usually broken down into the same types of steps. The steps given here appear appropriate for most active (physically) as opposed to cognitive activities. Experience has shown that with the breakdown into steps such as those given in Table 14-I, a considerable number of possibilities of elimination, rearrangement, combination, and facilitation are usually discernible. Hence, a breakdown of this type is often highly productive.

Because a limited number of easily drawn and differentiated symbols are available, the symbols for process chart—man analysis are the same as those used with process chart—product analysis, although they are fewer in number and are used to denote different steps to adapt them better to the problems peculiar to man activity.

The first three types of steps, it should be noted, refer to work done at one place but indicate different degrees or types of responsibility. Further, only the first type may indicate a step that adds value to the output.[1]

---

[1] It may assist analysis to separate *do* operations (those which add value to the product) from *get ready*, *put away*, or facilitative operations.

Table 14-I. SYMBOLS FOR PROCESS CHART–MAN ANALYSIS

Symbol	*A.S.M.E. Symbol*	*Name*	*Used to Represent*
○	○	Operation	The doing of something at one place.
□ ◇	□	Quantity determination	A special form of operation involving a person determining the quantity of an item present.
		Inspection	A special form of operation involving a person comparing an attribute of a product with a standard, or verifying the quantity present.
○	⇨	Movement	A change in location; moving from one place to another.
▽	D	Delay	Idleness. Waiting or moving, provided that the movement was not part of the job and the time could have been spent waiting.

The fourth type of step is affected primarily by the layout of the work area but is also influenced by the sequence.

The fifth type of step is usually undesirable unless it provides needed rest, in which case its position in the sequence of work is important.

Standardized charts. The graphic presentation achieved with a standardized chart is an aid to understanding, and it should be remembered that this is the main purpose of these analysis procedures. Hence, the standard steps and symbols should not be followed slavishly. If circumstances arise wherein the use of other steps or other symbols appears to be of more assistance in performing analysis, the analyst should not hesitate to use them. The format of the chart will also vary with the type of job. Hence, to keep down the size of this chapter, all examples will be of jobs with single repeated cycles. Jobs with more complex cycles will be treated in a later chapter.

Use in designing jobs. The presentation of the *original method and improved method* in the illustrations that follow should not cause the reader to lose sight of the use of the technique as a design tool for new jobs.

Case I—Wire Inspector (Simple Cyclic Work)

A possibility guide had suggested a class 1 or 2 change as the most feasible, and a process chart–product analysis of the production of wire had indicated that the job was necessary at its present place in the sequence. Because the task required the inspector, who checked the quality of wire from six machines drawing copper wire, to move from machine to machine as he worked, and because the work essentially consisted of a single cycle continually repeated, a simple process chart—man analysis was constructed as the main second step, analysis, of the scientific approach.

Analysis. As is preferred, the analyst actually observed the job being performed (if the job had been in the planning stage only, he would have had to work from the tentatively suggested procedure) and picked a suitable starting place for a cycle. Any place in the task would be satisfactory, but the first step attributable to some definite product or accomplishment is usually better; it makes the chart easier to understand. The first step of the cycle was classified into the proper category from Table 14-I and a suitable entry made on the first line of Figure 14.1, which is the original process chart—man analysis for the wire inspector. (A form with preprinted symbols may also be used; see Figure 10.4.) In contrast to a process chart—product analysis, this chart has only three basic columns: distance, symbol, and explanation or description. The time for each step may often be usefully included as a fourth column. The analyst proceeded to record each subsequent step on a new line of the chart until a whole cycle was charted. Note that the analyst chose in places to break the work done between moves into several steps, apparently feeling that some of them might later be eliminated, combined, or rearranged. This is quite satisfactory, but the analyst should be careful to use the same degree of detail when drawing proposed or revised charts, so that the charts will be comparable, rather than falsely indicating a saving by a mere change in amount of detail on the chart.

Man flow diagram. A flow diagram for the man, as shown in Figure 14.2, is usually a useful adjunct.

Criticism. The third step of the scientific method, *criticism*, is performed with the aid of the checklist for process chart—man analysis given. Each step in the original process chart—man analysis should be checked with the applicable principles and questions.

In the case of the wire inspector, the analyst found questions 1a, 2b, and 2c of value.

BASIC CHART FORM

Process Chart- man ____ Type of chart ____ *D Group* Department

Original ____ Original or proposed ____ *C. W. Mc.* Chart by

Inspect all drawn copper wire ____ Subject charted ____ *6/4* ____ Date charted

DIST.	SYMBOL	DESCRIPTION
24´	○	To next finished spool
	◯	Take spool
24´	○	To inspection bench
	◯	Place spool on bench
	◯	Strip and cut outer layer which is
		always damaged in drawing
15´	○	To scrap container
	◯	Dispose of scrap
15´	○	To inspection table
	◯	Cut 2´ from end, then 2" sample from 2´ piece
	◇	O. D. with box micrometer
	◯	Pick up spool
24´	○	To machine's "inspected" rack
	◯	Place spool

SUMMARY	
◯	7
▢	0
◇	1
○	5
▽	0
Dist.	102´ avg.

Figure 14.1. Process chart—man analysis for original method of inspecting production of copper wire.

181

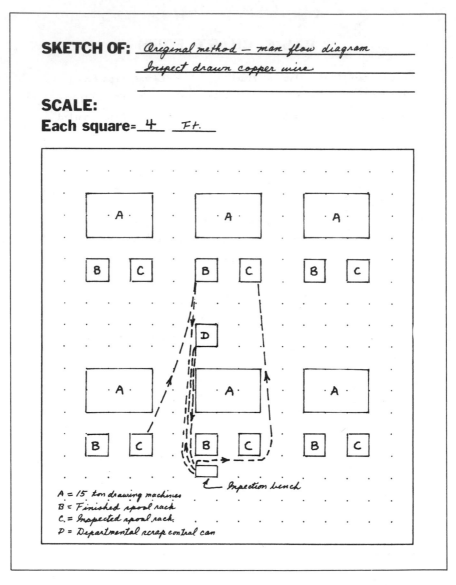

SKETCH OF: *Original method — man flow diagram*
Inspect drawn copper wire

SCALE:
Each square= 4 Ft.

A = 15 ton drawing machines
B = Finished spool rack
C = Inspected spool rack
P = Departmental scrap control can

Inspection bench

Figure 14.2. Man flow diagram for original method of inspecting production of copper wire.

182

CHECKLIST FOR PROCESS CHART—MAN ANALYSIS

Basic Principles

A. Eliminate all possible steps.
B. Combine steps.
C. Shorten steps.
D. Place in best sequence.
E. Make each step as economical as possible.

1. Can any operation be eliminated, combined, shortened, or made easier?
   a. As unnecessary?
   b. By new or different equipment?
   c. By changes in the layout; by grouping equipment better?
   d. By changing the form of the product sent out?
   e. By more knowledge on the part of the worker?
2. Can any movement be eliminated, combined, shortened, or made easier?
   a. By leaving out operations?
   b. By changing the places where things are kept?
   c. By shifting some operations to another job into which they fit more conveniently?
   d. By changing the layout?
   e. By changing equipment?
   f. By changing the order of work?
   g. By conveyors (make sure they are economical)?
3. Can delays be eliminated, combined, or shortened?
   a. By changing the order of work?
   b. By changing the layout?
   c. By new or different equipment?
4. Can countings or inspections be eliminated, combined, shortened, or made easier?
   a. Are they really necessary; what happens after they are done and the information obtained?
   b. Do they provide unnecessary duplication?
   c. Are they done at the best point in the sequence?
   d. Can sample inspection or statistical control be used?
5. Can any step be made safer?
   a. By changing the order of work?
   b. By new or different equipment?
   c. By changing the layout?

Innovation. The fourth step, innovation, is performed by preparing a proposed method, based on the suggestions obtained in step 3. A proposed method with only a class 1 change for the procedure used by the wire inspector is shown in Figures 14.3 and 14.4, which are the proposed process chart—man analysis and new man flow diagram. As the summary at the end of the chart indicates, the inspector's duty was considerably reduced. The savings could be used to give him more machines to check without increased effort, or to enable him to make more checks per machine, or to do a more thorough inspection job, or to inspect the product sooner after production, any one of which would be a worthwhile improvement.

Test. The fifth step, test, is performed by rechecking this proposed method with the checklist to make sure the best obtainable method is being proposed.

Trial. The sixth step, trial, would be aided by using the proposed process chart—man analysis or a suitably modified version, depending on the personality of the inspector, as a basis for training him to try this method. After a suitable acquaintance period with this new method, a time study was made to evaluate the new method, which was found to take only 52 per cent of the time of the original method.

Application. The seventh step, application, could consist of recording the method in such a form that the practice proposed in Figures 14.3 and 14.4 can be maintained until the next improvement. This type of recording will be shown in later chapters. The good analyst, it should be realized, recognizes subsequent improvement as almost inevitable.

Subsequent to the improvement. It is worth noting that with this technique, as with process chart—product analysis, the analyst may well build himself a file of inestimable value by preparing a file sheet for each question and subquestion, and recording, after each job studied with this technique, the specific way the inherent suggestion was applied, to facilitate future work with this technique. No complete list of such suggestions is given here, as they are usually peculiar to each industry or type of product. However, in the case being discussed, the analyst might well have made notations under questions 1a, 2b, and 2e, as follows:

1a. *Avoid unnecessary cutting of wire which does not aid task.*
2b. *Place scrap can close to inspector, even making it integral with inspection table to avoid extra travel with each piece of waste.*
2e. *Make inspection fixtures portable so inspector can take them to job. If workbench is needed, provide casters on workbench.*

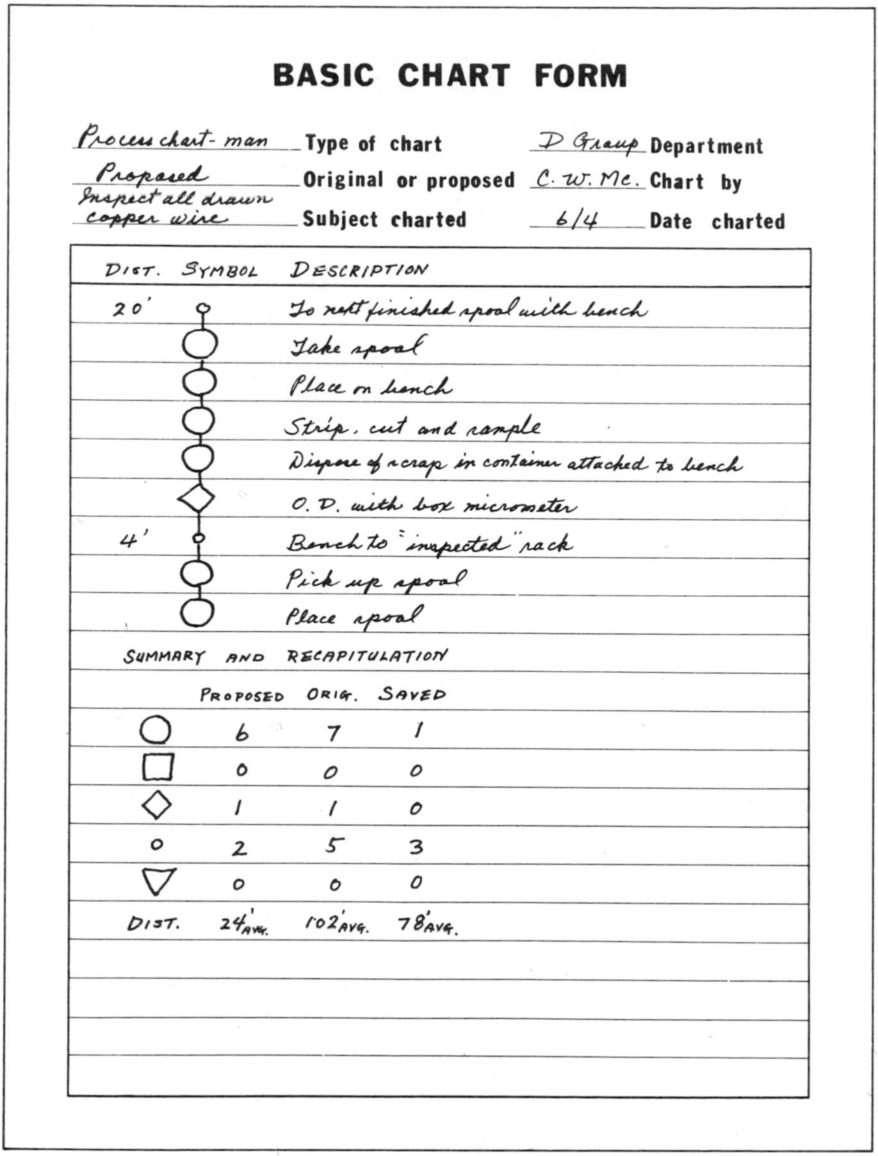

BASIC CHART FORM

Process chart-man ___ Type of chart   D Group ___ Department

Proposed ___ Original or proposed   C. W. Mc. ___ Chart by

Inspect all drawn
copper wire ___ Subject charted   6/4 ___ Date charted

DIST.	SYMBOL	DESCRIPTION
20'	◯	To next finished spool with bench
	◯	Take spool
	◯	Place on bench
	◯	Strip, cut and sample
	◯	Dispose of scrap in container attached to bench
	◇	O.D. with box micrometer
4'	◯	Bench to "inspected" rack
	◯	Pick up spool
	◯	Place spool

SUMMARY AND RECAPITULATION

	PROPOSED	ORIG.	SAVED
◯	6	7	1
▢	0	0	0
◇	1	1	0
○	2	5	3
▽	0	0	0
DIST.	24'$_{AVG.}$	102'$_{AVG.}$	78'$_{AVG.}$

Figure 14.3. Process chart—man analysis for proposed method of inspecting production of copper wire.

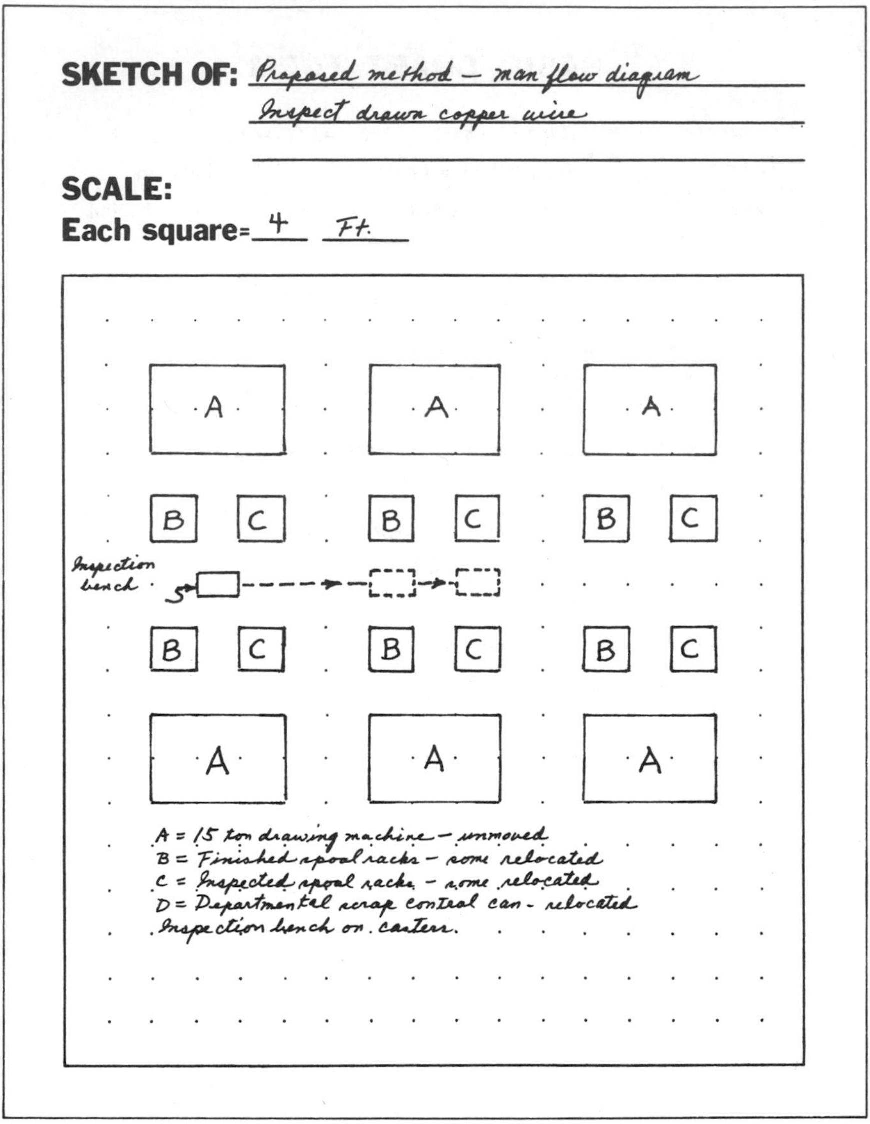

SKETCH OF: *Proposed method — man flow diagram*
Inspect drawn copper wire

SCALE:
Each square= 4 Ft.

Inspection bench

A = 15 ton drawing machine — unmoved
B = Finished spool racks — some relocated
C = Inspected spool racks — some relocated
D = Departmental scrap control can — relocated
Inspection bench on casters.

Figure 14.4. Man flow diagram for proposed method of inspecting production of copper wire.

Case II—Machine Setup (Simple Cyclic Work)

Introduction. This case was selected to illustrate the use of the technique in a situation wherein, in contrast to the first case, considerable specialized technological knowledge was required.

This application of the process chart—man analysis was made in a large steel-casting plant by cooperative action on the part of the general foreman of the core room, the shop superintendent, the general foreman of the pattern shop, the master mechanic, and the industrial engineer, who, at their request, supplied these supervisors with the time values shown on the charts. It would not be an overstatement to say that the ability of all of these individuals to understand and work with suitable motion and time study techniques was a fundamental prerequisite in the development of the improved method.

The core-blowing machine shown in Figure 14.5 was one of several in use. Many different cores had to be blown on each machine, and a possibility guide indicated that class 1 and 2 changes were desirable. The operation of the core blower naturally divides into two phases, "setup" and "do," with each core box. While a considerable number of changes were in the method of "do," this discussion is confined to the development of class 1 and 2

Figure 14.5. Core blower.

changes in the setup. The setup consists of switching core boxes, which is done by the materials handler, and changing blow plates, which is done by the blower operator. Each core box usually requires a different blow plate, which covers different holes in the master blow plate attached to the machine. The master blow plate and a specific core box blow plate are both shown in Figure 14.5. They have been removed from the machine for purposes of illustration. To show the parts involved in this job, the clamps, screw holes, and bolts are similarly identified on this photograph and in Figure 14.6.

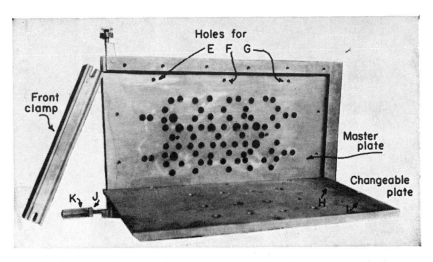

Figure 14.6. Master blow plate and a core-box blow plate, both disassembled from core blower of Figure 14.5.

Aim. The steel-casting plant was a job shop. A variety of products were made each day. It was felt that an excessive amount of time was spent on the setup work on the core blowers; if this time could be reduced, productivity could be raised.

Analysis. The process chart—man analysis for the original method is shown in Figures 14.7a and b. A time column was found to be a useful addition in this case, and the industrial engineer supplied existing time study values on request.

Criticism and innovation. Application of the checklist aided in replacing the original method with the improved method shown in Figures 14.8a and b. The method of holding the blow plate in place was changed to that indicated in Figure 14.9.

BASIC CHART FORM

PROCESS CHART - MAN **Type of chart**   COREROOM **Department**

ORIGINAL **Original or proposed** W.MOSBEY **Chart by**

CHANGE BLOW PLATE **Subject charted**   12/4 **Date charted**

DIST.	SYMBOL	DESCRIPTION	TIME
20'	○	TO DOLLY	.10 MIN
	○	GET DOLLY	.05
20'	○	CARRY DOLLY TO TABLE	.10
	○	PLACE DOLLY ON BLOWER TABLE	.05
3'	○	TO BLOWER BENCH	.02
	○	GET WRENCH	.05
3'	○	WRENCH TO BLOWER	.02
	○	PLACE WRENCH ON TOP OF BLOWER HEAD	.02
	○	WIND UP BLOWER TABLE	.40
	○	REMOVE 3 BOLTS (ABC) AND 1 SIDE CLAMP (D)	.73
6'	○	TO OTHER SIDE OF BLOWER	.05
	○	REMOVE 3 BOLTS (EFG) AND L SIDE CLAMP (H)	.73
3'	○	TO BENCH	.02
	○	ASIDE WRENCH AND GET SCREW DRIVER	.06
6'	○	TO REAR OF BLOWER	
	○	REMOVE 2 REAR SCREWS (HI) FROM BLOW PLATE	.35
6'	○	TO BENCH WITH CLAMPS, BOLTS AND SCREWS	.15
	○	PLACE PARTS, GET WRENCH	.19
3'	○	TO BLOWER	.02
	○	LOOSEN NUTS (JKLM) AND TAKE OFF FRONT CLAMP	.88
	○	PLACE WRENCH ON BLOWER	.02
	○	WIND DOWN TABLE	.13
15'	○	PULL BENCH TO FRONT OF BLOWER	.40
	○	PULL DOLLY OUT WITH PLATE	.13
15'	○	PLATE TO STORAGE	.20
	○	STORE PLATE AND GET NEXT ONE	.16

Figure 14.7a. Process chart—man analysis for original method of changing blow plates (*continues*).

DIST.	SYMBOL	DESCRIPTION	TIME
15'	○	NEW PLATE TO BLOWER	.20
	○	PLACE PLATE ON DOLLY	.15
	○	AIR CLEAN BLOWER AND PLATE	1.15
	○	PUSH DOLLY AND PLATE INTO PLACE	.79
	○	WIND UP TABLE	.15
	○	PUSH BENCH AWAY FROM BLOWER	.20
	○	AIR CLEAN CLAMPS AND 6 BOLT HOLES	1.08
	○	PICK UP WRENCH, FRONT CLAMP & TIGHTEN 4 NUTS	.18
6'	○	TO REAR OF BLOWER WITH DRIVER AND SCREWS	.10
	○	PLACE SCREWS	1.09
6'	○	TO BENCH	.10
	○	GET CLAMPS, BOLTS, ASIDE DRIVER	.11
3'	○	TO SIDE OF BLOWER	.05
	○	ATTACH CLAMP ON ONE SIDE; 3 BOLTS	1.59
6'	○	TO OTHER SIDE	.05
	○	ATTACH CLAMP; 3 BOLTS	1.59
3'	○	TO FRONT OF BLOWER	.02
	○	PLACE WRENCH	.03
	○	WIND DOWN TABLE	.22
	○	PULL OUT DOLLY	.05
20'	○	TO DOLLY STORAGE	.10
	○	STORE DOLLY	.05
20'	○	RETURN TO BLOWER	.10

SUMMARY

DISTANCE	164'	
TIME	14.83 MIN.	
○	31	
○	18	

Figure 14.7b. (*Concluded*) Process chart—man analysis for original method of changing blow plates.

BASIC CHART FORM

PROCESS CHART - MAN **Type of chart** COREROOM **Department**

PROPOSED **Original or proposed** W. MOSBEY **Chart by**

CHANGE BLOW PLATE **Subject charted** 12/10 **Date charted**

DIST.	SYMBOL	DESCRIPTION	TIME
	○	GET DOLLY FROM SIDE OF BLOWER	.10 MIN
	○	PLACE ON BLOWER TABLE	.05
	○	WIND UP TABLE	.40
3′	∘	TO BENCH FOR WEDGE (LOOSENING TOOL)	.02
	○	GET TOOL	.02
6′	∘	TO REAR OF BLOWER	.04
	○	STRIKE WEDGE	.10
3′	∘	TO FRONT OF BLOWER	.04
	○	PLACE MALLET, GET WRENCH	.05
3′	∘	TO BLOWER	.02
	○	OFF 4 NUTS AND FRONT CLAMP; SET ON HEAD	.60
	○	PULL OUT WEDGE, PLACE ON HEAD	.16
	○	WIND DOWN TABLE	.13
	○	PULL BENCH IN FRONT OF BLOWER	.40
	○	PULL OUT DOLLY	.13
15′	∘	TAKE PLATE TO STORAGE	.20
	○	STORE AND GET NEW PLATE	.16
15′	∘	TAKE PLATE TO DOLLY	.20
	○	PLACE PLATE ON DOLLY	.15
	○	AIR CLEAN BLOWER AND PLATE	1.15
	○	POSITION PLATE AND DOLLY	.79

Figure 14.8a. Process chart—man analysis for improved method of changing blow plates (*continues*).

BASIC CHART FORM

PROCESS CHART - MAN **Type of chart**   COREROOM **Department**

PROPOSED (CONT'D.) **Original or proposed** W. MOSEBY **Chart by**

CHANGE BLOW PLATE **Subject charted**   12/10 **Date charted**

DIST.	SYMBOL	DESCRIPTION	TIME
		WIND UP TABLE	.15
		PUSH BENCH AWAY	.20
		OIL WEDGE	.20
		PLACE AND DRIVE WEDGE	.23
		AIR CLEAN CLAMP	.26
		PLACE CLAMP AND TIGHTEN	.60
		WIND DOWN TABLE	.22
		PULL OUT DOLLY	.05
		HANG DOLLY ON SIDE OF BLOWER	.10

SUMMARY AND RECAPITULATION

	IMPROVED	ORIGINAL	SAVED
◯	24	31	7
○	6	18	12
DIST.	45'	164'	119'
TIME	6.92 MIN	14.83 MIN	7.91 MIN

Figure 14.8b. (*Concluded*) Process chart—man analysis for improved method of changing blow plates.

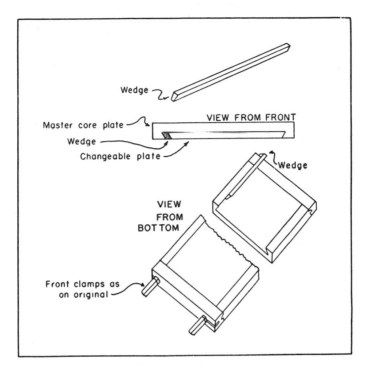

Figure 14.9. Details of new method of holding blow plates in place.

Evaluation of the results. The savings from this change were considerable. Analysis of the steps required by the proposed method indicates that the time required for changing a blow plate will be approximately 7.00 minutes. The present type of plate requires 15.00 minutes to change. A saving of 8.00 minutes per plate change is indicated.

An average of 15 plate changes per day are made. A saving of two blower hours per day would result from the change. Two blower hours cost $13.00 in direct wages. Since the hours saved can be used for needed production, we may include overhead in the saving, which increases it to $39.00 per day, or $9,672 per year.

In addition to measurable savings in blower time, it appears that a saving will also be made as a result of decreasing the incidence of blow plates sagging in the center, which has become a problem with the present method. The reverse tapers and support on all sides of the blow plate should help.

The cost of changing over to the new method would be $560.00 per blower and $24.00 for each plate. New plates would cost $18.00 more than present plates but may repay this in an extended life. An expenditure of $3,000 is anticipated to obtain the $9,672 increase in production, which should return the investment in less than 4 months from increased production alone.

Case III—Planting Corn (Simple Cyclic Work)

The corn belt of the United States raises a great deal of what is called hybrid corn. The seed for this is raised in special fields and results from the controlled cross-pollination of two corns—a seed parent, from which the ears are used but which is detasseled as it grows, and a pollinator, which is allowed to keep its tassels and supplies the male element for fertile kernels. The planted rows must not have any stray seeds of the wrong variety.

The original process chart—man analysis shown in Figure 14.10 was made from a study of a farmer planting corn for seed as described, using a four-row corn planter to produce a field as sketched in Figure 14.11.

Introduction. This case was selected for three reasons: first, to show how motion study techniques can be applied to agricultural jobs in the same manner as to industrial jobs; second, to reinforce the suggestion made, in Case II, that considerable technological knowledge may be required to employ the technique; and third, to show how the technique, designed essentially to lead to class 1 or 2 changes, may lead to changes of higher classes.

Innovation. These charts were critiqued with the checklist in a search for an easier and quicker method. A quicker method would be highly useful since more advantage could be taken of a break in the weather. The method that was proposed is shown in the process chart—man analysis of Figure 14.12 and in the sketch in Figure 14.13.

Class of change exceeds aim. Examination of these will show, however, that the analyst's intention of producing a class 1 or 2 change was over-reached; a class 4 change is suggested. The design of the product, *the planted field*, has been changed. The different arrangement of seed parent and pollinator rows could possibly affect pollination. When cases like this occur in industry, the suggestions must be discussed with more individuals than may have been originally contemplated—the design or sales engineers, in addition to the original list of tool designer, foreman, and worker. In the case of the corn, the equivalent additional consultant is the farm manager, hybrid corn seed specialist, or horticultural extension worker, who could check on the suitability of the proposed method. The reduction in work in the proposed method is obvious, and the writer, who is no agriculturalist, has been given to understand that the proposed method has been put to use.

BASIC CHART FORM

Proc. cht. - man _____ **Type of chart**   Field _____ **Department**

Original _____ **Original or proposed**   J.D. _____ **Chart by**

Planting seed corn _____ **Subject charted**   6/4 _____ **Date charted**

Distance	Symbol	Explanation
	○	Drive across field planting
	○	Turn around at end
3'	○	Climb down from tractor
6'	○	To back of planter
	○	Remove pollenator seed box
	○	Clean foot and planter plate
6'	○	To other side of planter with pollenator box
	○	Remove end seed parent box
	○	Clean foot and planter plate
	○	Attach pollenator seed box
6'	○	To other end
	○	Attach seed parent box
6'	○	To tractor
3'	○	Climb up
Summary		
○	8	
○	6	
Dist. walked	30'	

Figure 14.10. Process chart—man analysis for original method of planting corn for hybrid seed.

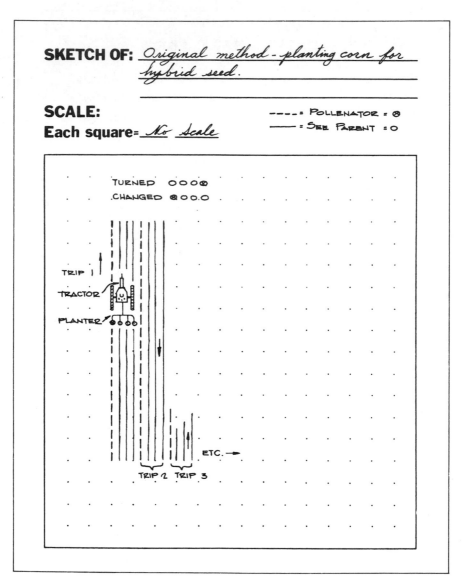

Figure 14.11. Movement of farmer and planter and layout of field with original method of planting corn for hybrid seed.

BASIC CHART FORM

Proc.cht. - man _____ Type of chart   Field _____ Department

Proposed _____ Original or proposed   J.D. _____ Chart by

Planting seed corn Subject charted   6/4 _____ Date charted

Distance	Symbol	Explanation
		Drive across field planting
		Turn around
Summary and recapitulation		
	Proposed	Orig	Saved (Per trip)
	2	8	6
	o	6	6
Dist. walked	0'	30'	30'

Figure 14.12. Process chart—man analysis of proposed method for planting corn for hybrid seed.

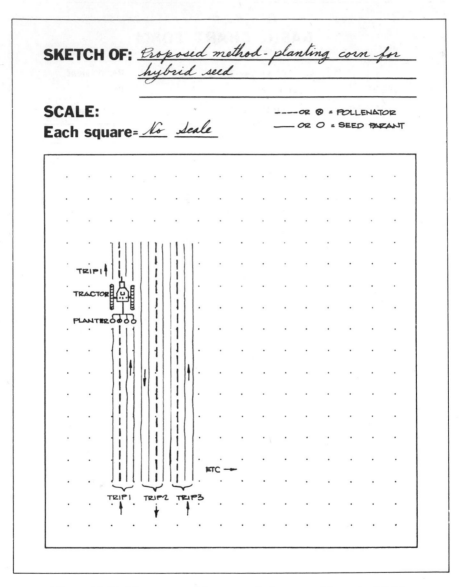

Figure 14.13. Movement of farmer and planter and layout of field with proposed method for planting corn for hybrid seed.

Summary—Process Chart—Man Analysis

1. *Uses:*
   a. For information leading to a class 1 or 2 change when the person moves from place to place while performing his job. *Note:* Higher classes of change are occasionally suggested from an analysis of this type and should not be overlooked.
   b. For designing a job that requires the workers to move from place to place while performing it.
2. *How made:*
   a. A form like that in Figures 14.1 or 10.4 or a blank sheet of paper may be used.
   b. The chart may begin at any point in a cycle of work, considering a cycle as the complete set of steps necessary to bring a unit of the output to the degree of completion typical of the work. However, it is usually most convenient to begin with the first step connected with a particular unit and end on the last step before the next similar unit is worked on or a new routine is started.
   c. The first work step should be carefully classified according to the categories of Table 14-I and the symbol and explanation (and, if the step is a movement, the distance) entered on the first line of the chart. The distance may be paced, estimated, measured on the actual floor plan, or scaled from a drawing. The explanation should be as succinct as possible.
   d. If the time for the step is desired, time it with an ordinary watch or stopwatch.
   e. Subsequent steps should be entered on subsequent lines, with the information placed in such a way that the symbols, explanations, and so forth, form separate columns for easy reading.
   f. Care should be exercised to make one entry for every separable phase of the work. Each time the worker moves from one place to another, an entry should be made. Each time the worker works at a workplace, an entry should be made. In some cases two distinctly separate activities may follow each other at a workplace without an intervening movement, and two operation symbols, one after the other, may be convenient.
   g. The steps should represent the activities of the worker, what he does to the product, and where he goes, rather than what happens to the product.
   h. If the job has several subcycles with different frequencies, each subcycle should be charted, separated from the others, and its frequency of occurrence noted.

    i. If the job has a variable cycle, classify as below and handle as indicated.

      (1) Due to operator habit: Chart several versions, to have more to work from.

      (2) Inherent and each different: Plot general pattern, marking parts that are constant and parts that vary from cycle to cycle.

      (3) No pattern or cycle: Plot a selected period of work as it occurs. (In many cases a work sampling or a work activity analysis, as described in Chapter 8, is more useful. Memomotion analysis, covered in Chapter 19, may also be more applicable.)

    j. In any case, a man flow diagram, which is a plan view of the area covered by the worker, with his path indicated thereon, is often a useful adjunct to the process chart—man analysis.

3. How used:

    a. Each step of the process is questioned with the checklist given.

4. What then:

    a. A process chart—man analysis is drawn for a resulting suggested improved method in order to permit a final check and to provide a means of describing the proposed new method.

PROBLEMS

14.1. Make a process chart—man analysis and a flow diagram of your activities in the morning from when you awake until you leave the house:

(a) For a normal morning.

(b) Same as part (a) except that you have seriously overslept and want to keep lateness to a minimum.

(c) Same as part (a) except that you wish to keep tension as low as possible and are willing to make a trade-off; less tension—more time.

(d) With the aid of the checklist, prepare a process chart—man analysis and a flow diagram for an improved method for performing the work charted in part (a).

14.2. Prepare a process chart—man analysis and a flow diagram from direct observation of a person:

(a) Doing the dishes.

(b) Baking a cake.

(c) Preparing a meal.

(d) Cleaning a room.

14.3. With the aid of the checklist, prepare a process chart—man analysis and flow diagram for a better method for performing the work charted in Problem 14.2.

14.4. (a) Make a process chart—man analysis of the best method for changing a tire on your car or your family's car.

(b) On what basis can you assert that the method shown in part (a) is the best method?

15   INFORMATION FLOW ANALYSIS

Introduction

When one designs some physical object consisting of a number of parts, one provides some mechanical means of keeping the parts together and some physical means for the parts to act upon each other. With a design for an integrated human group activity, the holding together of the various parts, and the acting of the parts upon one another in order to accomplish the desired goals, are attained by a phenomenon called communication.

Communication. *Communication* is the transmission of thoughts, opinions, information, or attitudes by speech, writing, or signs. This chapter is concerned with communication among people with or without the intervention of some mechanical means of transmitting, collating, interpreting, manipulating, or arraying the information. This definition of the scope of the chapter eliminates the details of the subjects of instructing machine tools, computational or communication equipment with tape or card programs, and so forth, but still leaves a very large area for scrutiny. No organization exists without a continuous flow of information among people. Continuous methods improvement in the area of communication is a necessity in the tempo of the modern world. An organization must avoid being overwhelmed by the problems of reacting in a timely manner to the enormous number of pertinent facts internal and external to its activity.

Improving a single step in the flow of information. The simplest type of information flow analysis is the analysis and improvement of a single step in a communication system, such as the preparation of a data input to a machine-aided system, the manual preparation of a data summary, the preparation of a dispatch schedule, a time ticket, a requisition, or any other formal communication. If the objective is to achieve a class 1 or 2 change on the particular step in the communication system, then either the process chart—man analysis or one of the other man analysis techniques described in later chapters may be satisfactory for the subsequent analysis and development of an improvement. The special notes on form design that appear later in this chapter may also be of value.

Improving the information flow system. However, as with any complex output, a larger problem almost always exists. The total procedure, or any sizable segment of it, for producing an output is either a design problem or an area for change that holds a much larger potential for improvement than any single step in the process. This is as true of an information flow system as it is of the processing of a substantive output, but the study of the flow of information presents a special type of problem as compared to the usual "product" analysis.

Special aspects of information flow problems. The communication means (card, tape, form, or voice channel) is of practically no value before information is inserted or entered upon it. Subsequently, the means of conveying the information is still, of itself, of little value, but the information it carries may have considerable utility. The card, tape, voice, or form is merely a means of retaining or conveying information as a part of a communication system. The means may be important to the system of flow, but the item of value is the information conveyed. The value of the information is in its utility to the individuals who either make decisions or take action based on the information. Hence, the timeliness, accuracy, and understandability of the information are vital criteria of the effectiveness of a communication system as well as the cost of operating such a system.

Timeliness and accuracy are aspects of a communication describing its relationship with the real world or information source; understandability describes the necessary characteristic of the communication at its point of impact. These are factors somewhat different from the factors that are considered in the analysis and improvement of the processing of a substantive product. It should therefore be obvious that analysis, step 2 of the scientific method, will be somewhat distinctive in a situation involving communication if we are to divide the work into units pertinent to such a job, appropriate to its scope, and possessing known characteristics.

Techniques that will be examined. This chapter will describe two techniques for aiding in the performance of *analysis* when the problem concerns an information system. The two techniques are:

1. Functional forms analysis chart.
2. Process chart—combined analysis.

Functional Forms Analysis Chart Defined

A functional forms analysis chart is a tabular presentation of the activities of an organization and the forms the organization uses for communication in performing each function in each activity in order to indicate areas with too many different or overlapping forms, or too few. It is used to analyze and improve an existing situation.

Process Chart—Combined Analysis Defined

A process chart—combined analysis is a graphic means of portraying the step-by-step procedure used with an information flow system, when the work done involves more than one work station. It may also show the interrelationships between any two or more items, such as several data collections, information and material, information and workers, and so forth. Such charts are particularly useful for analyzing the effectiveness of most control or information reporting procedures, or for designing a control or information flow procedure.

The Use of the Techniques

Where the techniques are used. These techniques are used in examining the flow of information, as a prelude to improvement, in such situations as:

1. The total communication system of an organization.
2. A job wherein the human physical activity and the flow of information are interrelated.
3. The processing and the use of information for decision making.

Forms and formats, a basic source of improvement. In almost all cases of information flow we find information being typed on forms. In even the simplest instance we find information being typed on blank sheets in a given format (letters). Hence, the following basic principles of improvement are of general applicability.

1. *Design the form and format for the user and his equipment.* This applies to billing, purchasing, posting machines, and so forth. Merely because semiautomatic machines will "chase" all over a form does not mean that we should use the extra time it takes to do this. However, in the application of this principle one should also consider the use made later of the typed form so that "origination" is not benefited at an undue cost to the subsequent user.

2. *On forms to be typed, provide spacing on the form to correspond to the typewriter.* This is quite obvious but often neglected, particularly with the lines on forms that are to be reproduced from photo-reduced printing plates.

3. *With typed forms, place as many items as possible against the left-hand stop to which the carriage is returned when the line is shifted.* This principle applied to ordinary business letters reduces the time by from 15 to 25 per cent.

4. *Eliminate all excess material from typing.* When similar items are typed on the form each time, use preprinted items with check boxes. Make sure the check boxes are placed in accordance with rule 3.

5. *Provide an automatic and consistent sequence of spaces on the forms wherever possible.* Many common forms have boxes for information scattered in such a way that setting typewriter tabulator stops is not feasible. The filling in of such forms by hand is also a laborious operation. In addition, the scatter of information makes the task of the reviewer or user of the information more difficult than it needs to be. (The U.S. Government Travel Request form has been, for decades, notoriously violative of this principle.)

Functional Forms Analysis

Form used. The functional forms analysis chart shown in Figure 15.1 was developed by a government agency, but it is applicable to any activity. Although it may be necessary to use different categories for the column headings, the principles will be the same wherever it is used. Figure 15.2 shows the definitions used for the column headings of Figure 15.1 as well as part of a filled-in chart, covering 514 forms, for a divisional office. The instructions for use of the form are contained in Figure 15.2.

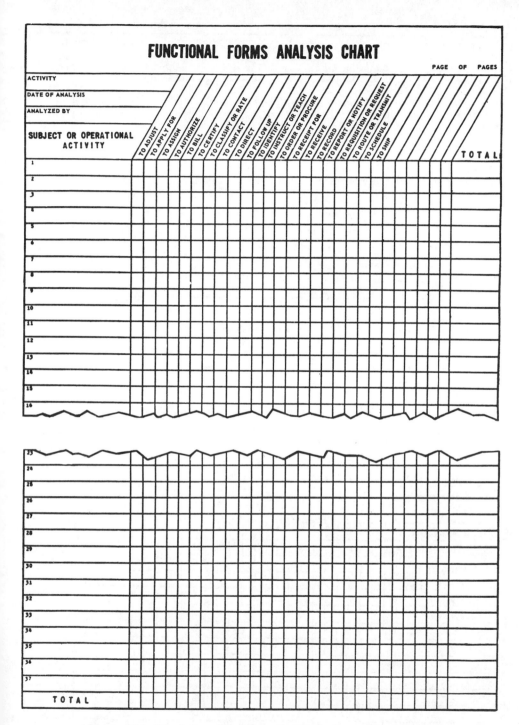

Figure 15.1. Blank functional forms analysis chart, Department of Army Form 553, Government Printing Office.

DEFINITIONS OF BASIC ACTIONS OR FUNCTIONS AS USED ON DA FORM 553

1.	TO ADJUST	to cause a revision to be made (to correct errors or discrepancies, obtain data or material omitted, increase or decrease quantities or amounts; to alter data, to amend, change, modify, cancel, reject)
2.	TO APPLY FOR	to request something which may or may not be granted
3.	TO ASSIGN	to appoint, allot, designate, apportion, or set apart
4.	TO AUTHORIZE	to empower, permit, sanction or to approve action (also passes and permits)
5.	TO BILL	to invoice or charge for
6.	TO CERTIFY	to attest to the truth of (affidavit, oath)
7.	TO CLASSIFY OR RATE	to relegate according to quantity, condition, grade, rank or class
8.	TO CONTRACT	to offer and accept in writing (an agreement)
9.	TO DIRECT	to give an order to (command) travel, duty, etc.
10.	TO FOLLOW UP	to remind of uncompleted action (tickler files, suspense files and forms, and follow up letters)
11.	TO IDENTIFY	to aid in proof or recognition of, to name (also personnel identification cards)
12.	TO INSTRUCT OR TEACH	to inform, advise, or counsel ("how to do it", not "you are instructed to, etc.")
13.	TO ORDER OR PROCURE	to secure or purchase, to give an order for labor, work, material, equipment or supplies
14.	TO RECEIPT FOR	to acknowledge delivery or payments; given by receipient in exchange for material delivered or payment made
15.	TO RECEIVE	to accept or admit as credible
16.	TO RECORD	to set down in permanent form for the sole purpose of registering or recording
17.	TO REPORT OR NOTIFY	to transmit a required notification, or account of status, progress, or action, usually in detail; to transmit an unsolicited statement
18.	TO REQUISITION OR REQUEST	to formally request by authority, to ask for
19.	TO ROUTE OR TRANSMIT	to send an attachment along a circuit of persons or places – (transmit – to send an attachment to another person or place) (messages)
20.	TO SCHEDULE	to outline regularly recurring events, or write a plan of future events; not merely a listing frequently entitled "Schedule of"
21.	TO SHIP	to send through any regular channel of transportation, as by rail

INSTRUCTIONS FOR USE

1. Basic actions or functions are listed across the top.
2. In the left hand column you will list the subjects and operational functions that pertain to your operation.
3. Enter in the proper columns the number of forms that apply to each subject and function.
4. This analysis will reveal the number of forms performing duplicate functions. Subject all such form to the detailed item analysis provided by DD FORM 85.

EXAMPLE

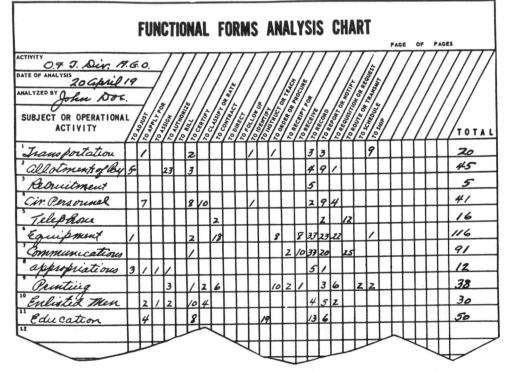

FUNCTIONAL FORMS ANALYSIS CHART

PAGE   OF   PAGES

ACTIVITY O.F. T. Div. A.G.O.
DATE OF ANALYSIS 20 April 19
ANALYZED BY John Doe.

SUBJECT OR OPERATIONAL ACTIVITY	TO ADJUST	TO APPLY FOR	TO ASSIGN	TO AUTHORIZE	TO BILL	TO CERTIFY	TO CLASSIFY OR RATE	TO CONTRACT	TO DIRECT	TO FOLLOW UP	TO IDENTIFY	TO INSTRUCT OR TEACH	TO ORDER OR PROCURE	TO RECEIPT FOR	TO RECEIVE	TO RECORD	TO REPORT OR NOTIFY	TO REQUISITION OR REQUEST	TO ROUTE OR TRANSMIT	TO SCHEDULE	TO SHIP			TOTAL
1 Transportation	1			2				1	1				3	3				9						20
2 Allotment of By	5		23	3									4	9	1									45
3 Recruitment													5											5
4 Civ. Personnel	7			8	10			1					2	9	4									41
5 Telephone					2							2	12											16
6 Equipment	1			2	18			8		8	33	23	22			1								116
7 Communications			1					2	10	33	20		25											91
8 Appropriations	3	1	1	1							5	1												12
9 Printing			3		1	2	6		10	2	1		3	6		2	2							38
10 Enlisted Men	2	1	2	10	4						4	5	2											30
11 Education	4			8				19			13	6												50
12																								

Figure 15.2. Sample functional forms analysis chart from reverse of Department of Army Form 553.

In some ways the *functional forms analysis chart* may be likened to the *work activity analysis* of Chapter 8: it helps find problem areas, it indicates where the greatest variety of forms exists within the communication system, and it suggests areas for further scrutiny. It does not, however, indicate where the greatest amount of time is expended.

Analysis; making the chart. All forms used for communicating within an organization are identified, classified, and the results entered on a functional forms analysis chart.

Criticism. All forms from any "box" on the chart showing a number greater than one are gathered and examined. A checklist is used as a guide. All boxes with a zero count indicate areas where new forms may be of help.

CHECKLIST FOR FUNCTIONAL FORMS ANALYSIS CHART

Basic principles

A. Eliminate unnecessary forms.
B. Combine forms.
C. Replace periodic freestyle narrative reports with forms.

1. Can any form be eliminated?
    a. Does it really serve a purpose?
    b. Could another existing form serve the purpose?
    c. Do we need all the copies of multicopy forms?
2. Can any form be combined with another?
    a. Do we really need separate forms?
    b. Could a single multicopy form be used in place of several separate forms?
3. Can any periodic reporting be benefited by the use of an appropriate form?

Discussion of the checklist. The checklist, one will note, appears to contain only simple suggestions. The functional forms analysis chart is, indeed, only a gross approach. Much additional improvement can usually be found later, when the flow of individual (or related groups of) forms is studied with the more detailed approach of the process chart—combined analysis.

Purposes served. However, there is little use in studying in detail the flow of unneeded forms. Such forms should be identified and eliminated. Also, prior to detailed analysis the total scope of the problem area should be assessed. The functional forms analysis chart serves both these purposes.

Case I—A Study of a Total Internal Communication System

Introduction. In order to avoid overwhelming the reader with excess detail, only the highlights of the case will be presented. The study concerned the improvement of the total internal communication system of a large, synthetic-textile-fiber plant.

In order to obtain a "population" of forms, all intraplant correspondence during a 1-week period was intercepted and a copy made of each communication. This was necessitated because the plant had no centralized forms control function that could be used as a starting point. In one way, the absence of such a control had triggered the study. As a rule of thumb, with the lack of such a control, one may presume an uncontrolled proliferation of forms. The existence of such a control function does not assure the absence of form proliferation, however. In addition, the existence of a control function does not assure the absence of unofficial or temporary forms, the latter often of long-standing use.

Analysis. The forms were classified and the totals entered on a functional forms analysis chart. The headings used on the chart were as given in Figure 15.1. The rows were such as: quality of pulp; quantity of pulp; temperature of retorting; and so forth.

Criticism. In this case, some criticism took place simultaneously with analysis; some of the forms defied classification; their purpose could not be ascertained. It could only be presumed that in the past they had served some purpose for some person; and that both were gone.

In addition, many forms that could be classified could not meet the test of checklist question 1a: "Can any form be eliminated?; Does it really serve a purpose?"

Further, in the absence of forms control, a large number of forms showed:

1. Overlap and repetition of information.
2. Excessive frequency of reporting.

Innovation. A suitable number of forms (and formatted reports) were retained to provide an information flow system. The organization was realigned to better utilize the new information flow.

Results. The reorganization, made feasible by the tremendously reduced information flow, made it possible to free a sufficient number of managerial personnel for a new mill that the company was building.[1]

---

[1] This was a real case. The company requested that its name not be disclosed.

Process Chart—Combined Analysis

Forms used. In simple cases a form like that used for process chart—product analysis may be used. In more complex cases a much larger blank sheet of paper, cross-ruled paper, or a special form may be used. This chapter will consider only simple problems.

Making the chart. In the preparation of a process chart—combined analysis, each item shown on the chart is usually displayed so that a horizontal line drawn across the chart at any point passes through symbols that indicate events occurring at the same time.

Relationship to organization. Inasmuch as a system of communication is being studied, the route of the information is a reflection of the formal and informal organization concerned. Alterations of a procedure may have far-reaching implications concerning the structure of an organization. An analyst seeking merely to improve a procedure must take proper cognizance of this possibility, and seek adequate participation in the analysis by all levels of management. On the other hand, it should be understood that widespread analysis of the procedures in an organization is an excellent approach to a critical study of the organization structure. The reader is reminded of the drastic effect upon the organization of the preceding illustration of the functional forms analysis chart. The process chart—combined analysis may also be used to assist in designing an effective organization. The reader is reminded also, in this respect, of a parallel aspect of process charts—product analysis, which may be used to design an effective manufacturing procedure and plant layout.

Steps separated with process chart—combined analysis. If the flow of information involves the flow or processing of forms, the part of the chart dealing with the flow of the forms is increased in clarity by the separation of the steps given in Table 15-I, together with the symbols used for charting them. Many of these steps are different from those used to chart a "product" and are designed to separate unique acts, pertinent to the type of work studied, as is the basic function of an *analysis* technique. The last two items given in Table 15-I may also be advantageously used on the *material* or *man* columns on the process chart—combined analysis if such items are also charted. It should be understood that in some cases there may be a need to differentiate the steps of the work to a greater degree, and subvarieties of these steps and symbols or additional ones may be used. For instance, it might be desirable to differentiate various types of operations by putting a "?" in the symbol if a decision characterizes the step, a "C" for a computation, and so forth.

Table 15-I. PROCESS CHART—COMBINED ANALYSIS SYMBOLS FOR USE WITH FORMS

Symbol	Name	Used to Represent
	Origin of form	Form first being made out.
	Origin of form	Form first being made out in duplicate.
	Origin of form	Form first being made out in triplicate, etc.
	Operation	Work being done on form, computations or additional information added, etc.
	Inspection	Correctness of information on form checked by comparison with other source of information. (Use broken line drawn to other source if other source appears on chart and line is aid to clarity.)
	Information take-off	Information being taken off form for entry onto another or for use by someone. Point of line indicates symbol on other parallel chart where information is going. (Use broken line to indicate destination if destination appears on chart and line is aid to clarity.)
	Disposal	Form or copy destroyed.
	Movement	A change in location of form, not changing it.
	Delay	Forms waiting to be worked on, such as in a desk basket.
	File	Forms in a file, organized in a formal fashion.
	Item change	Change in item charted.
	Gap	Activities not pertinent to study and hence not charted in detail.

Other sets of analysis categories. If the situation involves an analysis of the processing of information through a computer system, a totally different set of symbols may be necessary. For instance, the IBM system flowchart manual (C20-8152) recognizes, separates, and provides symbols for the following items:

Processing	Input or output
Punched card	Perforated tape
Document	Transmittal tape
Magnetic tape	Disk, drum, random access
Off-line storage	Display
On-line keyboard	Sorting, collating
Clerical operation	Auxiliary operation
Keying operation	Communication link

If the problem concerns a mixed communication system with voice, mechanical, and visual communication, such as at a missile launch site or in an airplane cockpit, a different set of separable steps, such as the following, will be more suitable than the two mentioned to this point:

Read	Change posture
Cognize	Transmit mechanically
Speak	Compute
Listen	Compare
Manipulate	Wait for data output
Change position	Feed data

Case II—C.O.D. Procedure in Department Store

Introduction. This application of the process chart—combined analysis was made in a large midwestern department store. With the original method, the truck drivers in the evening often had to wait their turn for the C.O.D. cashier, thus causing a considerable overtime expense for both drivers and C.O.D. cashiers at the "delivery center." It was this overtime that the analyst particularly sought to eliminate without unduly slackening the controls over the considerable amounts of money and goods that were handled.

The analyst, who had decided that a class 3, 4, or 5 change was desirable, had also decided that a process chart—combined analysis was the only feasible way of setting down the interrelationships among the various persons and forms involved in the task.

Analysis. To construct the process chart—combined analysis, the analyst first made separate notes, in the form of process chart—product or man analysis as necessary, using the special symbols from Table 15-I for the notes dealing with any form or written material. These notes were then used to construct the process chart—combined analysis of the original method shown in Figure 15.3. The section of the chart shown describes the procedure used to check in the C.O.D. receipts and the undelivered C.O.D. packages of the delivery truck drivers.

Criticism. To perform step 3, criticism, the original process chart—combined analysis was reviewed, using the checklist for process chart—product analysis and the checklist for process chart—man analysis, as well as the special checklist for process chart—combined analysis given in this chapter.

CHECKLIST FOR PROCESS CHART—COMBINED ANALYSIS

1. Each step should be necessary. If not, eliminate it.
2. Each step should have an ideal place in the sequence. Where should it be?
3. Each step should have a reason for being by itself. Can it be combined?
4. Each step should be as easy as possible.
5. Each form should have a real purpose. Verify it. Is the form necessary? Can it be eliminated, combined with another form, or replaced by a copy of another form?
6. Each file should have a unique purpose. Does it? Avoid duplication. Avoid excess files. File by subject used to enter files. Check on manner of use.
7. If form is finally destroyed, perhaps it should never have been originated. What purpose did it serve?
8. Information going from one form to another suggests more copies in the first place. Are all information take-offs and readings necessary? If so, which are going to be given priority in the design of the form? Which were given priority?
9. Are all copies of a form getting equal use? Sharing the load may speed up the procedure.
10. Does someone sign all copies? How can this be avoided? Signers are often busy people.
11. Is there excess checking?
12. Where is the best place to check? Calculate the risk.
13. What would happen if the form were lost?
14. What equipment might help the job? (See commercial catalogs.)
15. Does one person handle too much of the procedure?

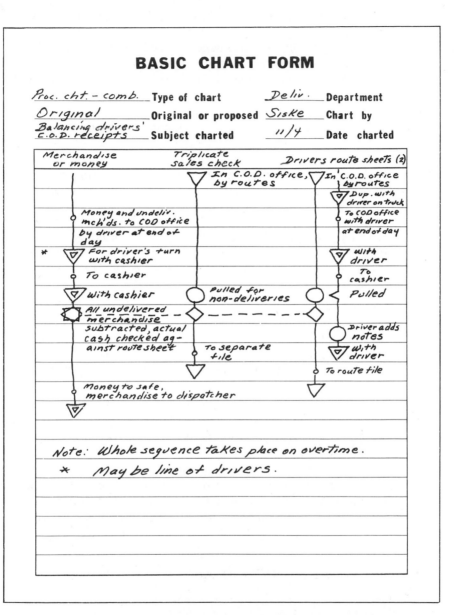

Figure 15.3. Original process chart-combined analysis for checking in C.O.D. receipts and undelivered C.O.D. packages from delivery truck drivers in department store.

16. Are as many steps as possible given to the lowest classification of personnel?
17. Can travel of forms be advantageously reduced?
18. Can the form be kept in action, out of file baskets?
19. Does the information arrive in a timely fashion so that it may be acted upon?
20. Has the information display been reduced to understandable form? (There is a limit to the amount of detail that can be comprehended).
21. Is the information accurate and reliable?

Innovation. Step 4, innovation, was carried out by constructing the process chart—combined analysis shown in Figure 15.4. It entailed the following innovations:

1. Form 490, the route sheet for the new method, was the same as the original, but one copy was printed on a manila envelope and spot-pasted to the original. The duplicate envelope-copy served as a report form and money container.

2. A night-depository type of money safe was installed to receive the envelope-form 490, so that the contents could be left safely after the departure of the C.O.D. cashier.

3. A package chute arrangement was installed to enable the driver to deposit the undelivered C.O.D. packages in the C.O.D. office for redispatch the following morning.

4. Tables were placed near the C.O.D. office in the delivery center to enable the drivers to balance their route sheets.

5. Several inexpensive listing-type calculators were placed on these tables to facilitate balancing route sheets.

The improved method reduced the overtime of the two C.O.D. cashiers by approximately $1,200 per year, with the accompanying reduction of comparable overtime for a much larger number of delivery truck drivers.

Summary—Functional Forms Analysis Chart

1. Uses:
  a. To study the total communication system of an organization so as to:
    (1) Assess the scope of the form-communication problem.
    (2) Determine activity-function areas with excess forms or without forms.

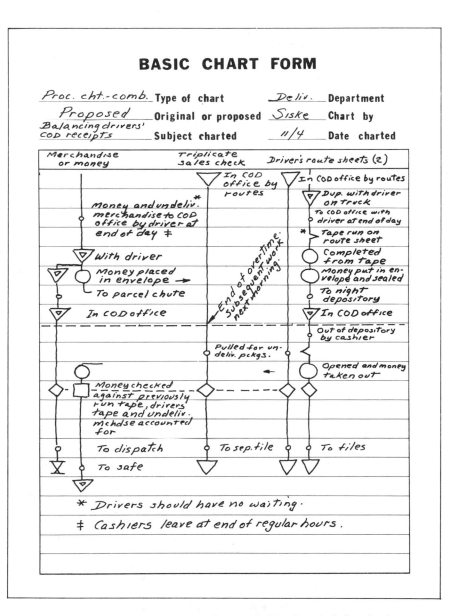

BASIC CHART FORM

Proc. cht.-comb. **Type of chart**   *Deliv.* **Department**

Proposed **Original or proposed**   *Siske* **Chart by**

Balancing drivers'
COD receipts **Subject charted**   *11/4* **Date charted**

Merchandise or money	Triplicate sales check	Driver's route sheets (2)
	▽ In COD office by routes	▽ In COD office by routes
		▽ Dup. with driver on truck
Money and undeliv.* merchandise to COD office by driver at end of day ‡		To COD office with driver at end of day
		* Tape run on route sheet
▽ With driver		○ Completed from tape
▽ ○ Money placed in envelope →		○ Money put in envelope and sealed
— To parcel chute		To night depository
▽ In COD office		▽ In COD office
		Out of depository by cashier
	Pulled for undeliv. pckgs.	
○ Money checked	←	○ Opened and money taken out
◇ □ against previously run tape, drivers tape and undeliv. mchdse accounted for	◇	◇ ◇
○ To dispatch	○ To sep. file	○ To files
▽ ○ To safe	▽	▽ ▽
▽		

End of overtime.
Subsequent work
next morning.

* Drivers should have no waiting.

‡ Cashiers leave at end of regular hours.

Figure 15.4. Improved process chart—combined analysis for checking in C.O.D. receipts and undelivered C.O.D. packages from delivery truck drivers in department store.

2. *How made*:
   Using a form such as that in Figure 15.1, the population of forms and formatted reports are classified by area of use and function.
3. *How used*:
   a. Each occurrence of more than one form for each activity-function combination is questioned with the aid of a checklist and excess forms eliminated.
   b. Each activity-function combination without a "dedicated" form is examined to determine whether a form would replace free, narrative reports.
4. *What then*:
   An approved forms list of only necessary forms is created. Remaining forms may be studied individually in detail, by other analytical techniques, for additional improvements.

Summary—Process Chart—Combined Analysis

1. *Uses*:
   a. For information leading to a class 3, 4, or 5 change in information procedures when the aspect under study concerns:
      (1) The flow of control information in relation to output flow.
      (2) The flow of information where its relationship to man work is important.
      (3) The processing and use of information for decision making, and so forth.
   b. For the design of a paperwork procedure.
   c. As an aid to the study of an organization.
   d. To assist in designing an effective organization structure.
2. *How made*:
   a. Following the instructions in Chapters 10, 14, and 15, as required, process charts are made of each person, product, form, or information item concerned. The steps with the product follow the breakdown of Table 10-I; use Table 14-I for the man activity and Table 15-I for the forms. In many cases it may be desirable to employ special steps and symbols more pertinent to the peculiarities of the process under study.
   b. A new chart, combining all pertinent information, is drawn with all the items on it so placed that a horizontal line drawn across the chart at any point passes through items that are happening at the same time. The chart may be drawn in several ways. Care should be exercised to select the format that displays information most clearly.

3. *How used*:
   a. Each step of the process is questioned with the applicable check-lists given in this chapter and in Chapters 10 and 14. With many charts of this type, it is extremely desirable to consult the people concerned as early in the analysis as possible. It may be desirable to note their remarks directly on the chart at the place on the chart that is the subject of their comments.
   b. The design of each form involved may be reviewed.
4. *What then*:
   a. A process chart—combined analysis is drawn for a resulting suggested improved method in order to permit a final check and to provide a means of describing the proposed new procedure.
   b. A new form (or forms) may be designed to facilitate the work, although this is not always necessary.

PROBLEMS

15.1. Redesign the wage analysis form shown in Figure P15.1 for use on a typewriter. Draw the revised form full-scale.

15.2. Redesign the intershop movement and identification ticket shown in Figure P15.2 for handwritten use. Draw the revised form full-scale.

15.3. Obtain and redesign any of your company or school forms.
(a) Prepare a full-scale drawing of the new design.
(b) Evaluate the effect of the changes made.

15.4. Obtain a complete set of the forms of your company, school, or of any organization whose forms are available to you.
(a) Prepare a functional forms analysis chart.
(b) Indicate any unnecessary duplication or any areas for which forms are missing.
(c) Comment on the general status of the forms available.
(d) Give your recommendations and justify them.

15.5. The following is the sequence of events that took place in a certain department store when a woman who expected a broadloom rug delivered on Wednesday failed to receive it. This is typical of several hundred calls per day concerning altered clothes, furniture, knocked-down items assembled by the store, drapes, shades, and other goods worked on before delivery. Some of these calls are caused by rash promises made to close a sale and some by customer anxiousness. A promise for "Tuesday" means Friday in store parlance, "Tuesday sure" means Wednesday, and only "Tuesday positively" means Tuesday. The workrooms overestimate time in self-defense, and the floor staff knows it; consequently, they promise more than the workrooms can do, in order to make sales. Mrs. A, who didn't get her rug, calls the store

DEPARTMENT LABOR PERFORMANCE ——————— ————

Department __3 Buffing__ Per cent incentive ___90.8__ Week ending 6/24/

Total hours 469.5 Hours on standard 426.5 Hours earned 506.5 Day work 43.0

Department standard performance 118.8 Department actual performance 117.0

No	Name	Hours Worked T.	Hours on Standard	Hours Earned	%Performance
2	S. Green	31.0	27.0	30.0	111.1
4	S. Doyle	37.0	35.0	34.0	97.1
5	J. Small	35.5	30.5	33.0	108.2
7	H. Suluski	27.5	25.0	30.0	120.0
8	B. Kirk	42.0	40.0	52.0	130.0
9	H. Auxford	45.0	39.0	48.0	123.1
11	J. Justin	44.5	42.0	48.0	114.3
14	R. Allen	39.0	35.0	42.0	120.0
16	K. Kall	40.0	35.0	44.0	125.7
17	M. Nedloeb	42.0	40.0	49.0	122.5
18	S. Kozoff	42.0	38.0	45.5	119.7
20	J. Marks	44.0	40.0	51.0	127.5

SCALE
inches

Figure P15.1

and asks for the rug department. The PBX operator rings the rug department telephone; whoever is near it answers, and then gets the section manager, Mr. B, who leaves what he is doing and goes, on the average 50 feet, to the telephone. He picks up the telephone, listens, and makes notes concerning the inquiry, while listening. He tells Mrs. A that he will call back and hangs up. After a time sufficient for the connection to break, he picks up the telephone, dials the rug workroom, usually gets a busy signal, and hangs up. He places the notes in his pocket and returns to what he was doing. In the usual course of events, Mr. B makes two more tries, returning to the telephone at intervals, taking out his notes, dialing the rug workroom before he manages to get the

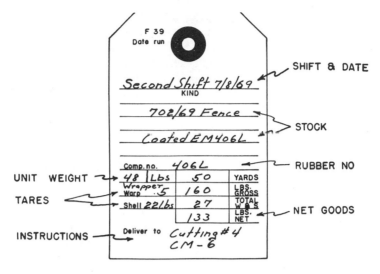

Figure P15.2

line. While he talks to the workroom clerk, the clerk makes notes concerning the inquiry. The workroom clerk tells Mr. B that she will call back, and they both hang up—Mr. B putting his notes back into his pocket and the rug workroom clerk, Miss C, going to the file to locate the order. She locates the order, finds the workroom number, hunts, and locates the merchandise in the workroom, determines its status and probable delivery date, adds this to her notes, returns to the telephone and calls Mr. B. He is called to the telephone and takes out his notes and adds Miss C's information to them. Miss C now destroys her notes. Mr. B finishes what he was doing, returns to the telephone, calls Mrs. A, tells her what he found out, soothes her, hangs up, and destroys his notes. (If Mrs. A calls Mr. B or Miss C later that day or the next, they must rely on memory to reinform her or go through the whole routine again.)

(a) Prepare a process chart—combined analysis for Mr. B, Miss C, Mr. B's notes, and Miss C's notes for the original procedure.

(b) Suggest steps to take to eliminate most of such calls.

(c) What class of change does each of your suggestions involve?

(d) Plan an improved procedure for handling the remaining calls that will inevitably take place and present it as a process chart—combined analysis. Do not exceed what would be a class 3 change if the product is the information Mrs. A receives.

15.6. A group of chain stores use a group of similar procedures in obtaining various types of nonperishable merchandise from the company's district warehouse that supplies these stores. Following is the procedure used to obtain their supply of all items of a type of nonperishable merchandise that we shall refer to as class B.

Once a week, using form 845, prepadded in duplicate, a complete list giving kind and quantity of needed class B merchandise is made out in longhand by the store manager. Both copies are sent by regular mail to the warehouse to the attention of the class B merchandise clerk. The manager makes out and seals his own envelope. The mail is received at the warehouse in a central mailroom, sorted, and delivered by house mail to the class B merchandise clerk. Using the form 845 and her balance-of-stores cards, the class B merchandise clerk eliminates the out-of-stock items on the list, reduces or increases the quantities according to general weekly instructions from the warehouse manager, adds items as also instructed, reduces her balance of stores by the quantities and kind of merchandise finally listed, and then types form 505, prepadded in triplicate, of the merchandise that will be shipped. She destroys both copies of form 845 and the triplicate of form 505. The first and second copy of the form 505 are sent to the warehouse class B stock room by house mail and there placed in the workbasket. When the shipping clerk comes to the form 505 we are following, he first uses it to partially fill out the two prepadded copies of the shipping ticket form 86. He enters all full cases only, longhand, on the ticket. The two copies of form 86 and the two copies of form 505, clipped together, are laid on a picking table, and all the items listed on the form 505 are pulled from stock and placed on this wheeled table. All items requiring a full original case lot or several full original cases are pulled first and checked off both sets of forms, which still have inserted carbons. Less-than-case lots are pulled next, checked off the form 505 as pulled and entered on the form 86 as they are packed into shipping cases with the contents of each case grouped. Any out-of-stock item, which is a rare occurrence, is crossed off both sets of forms. When the order is finished, the two copies of the form 505 are placed in the out-mail basket, and the order and its accompanying two copies of form 86 are wheeled to the shipping dock by the same clerk who filled the order. He then makes out a bill of lading in duplicate, again using prepadded forms. The duplicate copy of the bill of lading attached to the duplicate form 86 is sent to the warehouse file by house mail. They are filed by stores, chronologically for each store, with all bills of lading and forms 86 for all classes of merchandise. The original shipping ticket and the original bill of lading are sent by mail from the mailing room to the store and the merchandise is shipped by company truck. Whichever arrives first at the store is held until the other arrives. The two copies of the form 505 are sent back to the class B merchandise clerk who, using her unit price file, enters the unit price for all items and places both copies in her desk basket. The computer, who works next to her, takes these as she needs work, extends and adds them twice (to ensure correctness), separates the two copies, and puts one in her basket for the mail desk and one in her basket for the bookkeeper. These are moved periodically by the house mail boy. The duplicate form 505 is mailed to the store from the mail desk, and the original is used by the bookkeeper to post the store ledger and is then sent to the file room, where it is filed indefinitely.

The original shipping ticket is signed by the store manager as soon as feasible (when both it and the merchandise are on hand) and sent back to the warehouse. The store manager files the original bill of lading.

When the warehouse file clerk receives the signed copy of the form 86, the file clerk pulls the duplicate form 86 and bill of lading, matches them, then destroys the duplicates and files the original signed copy by stores by months.

The merchandise is also held intact at the store until the arrival of the duplicate form 505, or vice versa. When both are on hand, the merchandise is unpacked, checked against the form 505, and sent to storage or display in the store, as required. The form 505 is then used to post to the store's Monthly Buy Sheet, shortages or damages are posted to the store's Monthly Claim Sheet, and the form 505 is then filed at the store in a chronological file for 1 year.

Class C merchandise, also nonperishable, is obtained in the following manner. Form 491, prepadded in quadruplicate and set-numbered, is made out once each week, in longhand, by the store manager, giving kind and quantity of needed class C merchandise. Copies 1 and 2 are sent to the warehouse, attention of the class C clerk, copy 3 is destroyed, and copy 4 is left in the form book. The manager makes out and seals his own envelope. It is sent to the warehouse by regular mail. The mail is received at the warehouse in a central mailroom, sorted, and delivered by house mail to the class C merchandise clerk.

At the warehouse, the set number from form 491 is first entered on the Control List by the class C merchandise clerk, then the two copies are sent to the warehouse class C stock room by house mail and placed in the workbasket. When the shipping clerk comes to the form 491 we are following, he first uses it to fill out the two prepadded copies of the shipping ticket, form 86. He enters all items calling for full manufacturer's cases longhand on the tickets. The two copies of forms 86 and 491, clipped together, are laid on a wheeled picking table, and all the items called for on the form 491 are picked from stock and placed on this table and additional tables as necessary. Items that are out of stock are crossed off forms 491 and 86 as necessary, and marked OOS (out of stock). As the less-than-case lots of merchandise are pulled, they are checked off the form 491 and entered on the form 86 as they are packed into shipping cases with the contents of each case grouped. When the order is finished, the two copies of the form 491 are placed in the out-mail basket, and the order and its accompanying two copies of form 86 are wheeled to the shipping dock by the same clerk who filled the order, who now makes out a bill of lading in duplicate, again using a prepadded form. The duplicate copy of the bill of lading attached to the duplicate copy of form 86 is sent to the warehouse file clerk by house mail. They are filed by stores, chronologically for each store, with all bills of lading and copies of form 86 for all classes of merchandise.

The original shipping ticket and the original bill of lading are sent by mail from the mailing room to the store, and the merchandise is shipped by company truck. Whichever arrives first at the store is held until the other arrives.

The two copies of form 491 are sent back from the warehouse clerk to the class C merchandise clerk, who, using her unit price file, enters the unit prices for all items and places both copies in her desk basket. The computer, who works next to her, takes these as she needs work, extends and adds them twice (to ensure accuracy), and returns them to the class C merchandise clerk. The class C clerk then checks the set number off the Control List, thus making sure she has charge sheets for all class C merchandise orders sent by her to the stock room; she then places the duplicate form 491 in the out-mail basket for the mail desk to mail to the store, and the clerk also sends the original form 491 to the bookkeeping file. The original is filed without sorting until the end of the month, at which time all the originals are pulled by the bookkeeping department clerk. The bookkeeping department then posts them to the store ledgers and sends them back to the file room, where they are filed, chronologically by stores, for a 5-year period.

The original shipping ticket is signed by the store manager as soon as feasible (when both merchandise and shipping ticket are on hand), and sent back to the warehouse. The store manager files the original bill of lading.

When the warehouse file clerk receives the signed copy of form 86, she pulls the duplicate copy of form 86 and bill of lading, matches them, then destroys the duplicates and files the original signed copy, by stores, by months.

The merchandise is also held intact in the store until the arrival of the duplicate of form 491, or vice versa. When both are on hand, the merchandise is unpacked and is checked against the duplicate of form 491. The quadruplicate copy of form 491 and the duplicate of form 491 are then checked against each other. The quadruplicate copy is then destroyed, and the merchandise is sent to storage or display in the store as required. The form 491 is then used to post to the store's Monthly Buy Sheet, shortages or damages are posted to the store's Monthly Claim Sheet, and the form 491 is then filed at the store in a chronological file for 1 year.

(a) Prepare a functional forms analysis chart for the communication procedure described.
(b) Prepare and justify a set of recommendations concerning the elimination or retention of each of the forms.

16

OPERATION CHARTS

Introduction

In many of the jobs in plants or offices, the persons doing them remain at one work station. Indeed, each large circle on any of the graphic presentations so far discussed (except network diagrams) usually indicates such a situation. In many cases, the jobs are important enough to warrant spending time on their design or improvement. In most cases, improvement is the result of a class 1 or 2 change, although the higher classes may also be involved. Class 1 or 2 changes have the advantage, however, that they affect nothing other than the job under scrutiny. They may be accomplished without disturbing other aspects of the work, and thus often may be achieved more quickly. Frequently, changes may be installed on the same day on which the analysis is made.

Operation Chart Defined

An *operation chart* is a graphic means of portraying separable steps of a person's body members when he is performing a job that takes place essentially at one location. It is a schematic model of the method.

The Use of the Technique

Where the technique is used. When the objective of a work method problem involving but a single work station has been stated as a class 1 or 2 change, the second step of the logical approach is still:

> *Analysis—the separation of the work method into subdivisions or steps, pertinent to the job, appropriate to its scope, possessing known characteristics, or concerning whose performance information is already available.*

Under the conditions given, operation charts are a common technique for the performance of this step.

An operation chart is an analysis of the manual work performed by a person on any one operation on a process chart, either man, product, or combined analysis. It is a description of what the person does. It is used for the analysis of the work of individuals, not crews, and where the work of the machine, if any, is not the main controlling factor.

Purposes served by the technique. Operation charts are easy to use. All manual work performed at a workplace can be studied with the aim of making class 1 or 2 changes. No equipment is needed, and significant improvements are usually possible. Not much time is necessary for the making of this type of analysis, and the results are often immediately installable. The more experience one has with methods improvement, the more useful this technique becomes. Almost anyone can use operation charts profitably. They are also an ideal tool for designing the details of jobs in contemplated processes.

The steps separated with operation charts. The operation chart usually involves a breakdown, into steps, of the work performed by each of the body members engaged in the task. This usually means the right and left hands. Where the feet or eyes are important factors, they may also be charted. The operation chart shows not only the sequence performed by each body member charted, but also their relationship to each other while working. It does not indicate the time or relative time for the steps.

The steps into which the work is commonly divided as well as the symbols used to represent these steps are given in Table 16-I. The symbols are similar to those used for process charts, but because of the difference in the scope of an operation, as opposed to a process, the symbols are used to represent different steps.

Table 16-I. SYMBOLS FOR OPERATION CHART

Symbol	A.S.M.E. Symbol	Name	Used to Represent
◯	◯	Suboperation	Body member doing something at one place, such as taking hold, lining up, assembling, etc.
○	⇨	Movement	A movement of a body member toward an object or changing the location of an object.*
▽	◖	Hold	Body member maintains an object in a fixed position so that work may be done with or on it at that location.
▽	▽	Delay	Body member is idle or delaying for another body member.

*On very long operations the analyst may combine some of these steps into larger steps, using "get" in place of *reach for, take hold of,* and *bring object to work area;* "aside," meaning *move* object from work area, *let go* of object, and *return.* In such a case the chart is described as being made with a "gross breakdown" and is considerably shorter than when made with the usual steps.

How the operation chart is made. An operation chart is easily constructed from direct observation of the job. When the job is not yet being performed, such a chart may readily be used to set forth in detail the contemplated method. In either case, the analyst should first familiarize himself with the job cycle. The *job cycle* is all the movements required to bring a unit of the output to the stage of completion typical of the operation. It is most convenient to consider the cycle as starting with the first movement attributable to a unit of the output and ending with the last movement on that unit. This gives a more easily understood chart than one beginning and ending in the middle of the work on a part.

Categorizing the types of jobs that may be encountered. As with process charts—man analysis, four types of jobs, with the following characteristics, may be encountered:

1. The work has a single repeated cycle.
2. The work is cyclic, but there are several subcycles performed with different frequency.
3. The work varies from cycle to cycle.
4. The task has no regular cycle.

The method of handling these different types of jobs is essentially similar to the procedure used with process charts—man analysis, except that the work of the individual body members is studied rather than the worker as a whole.

Jobs of type 4 may be studied, more appropriately, with the aid of a work activity analysis or a work sampling study (Chapter 8) or memomotion study (Chapter 19).

Example that follows. Only the analysis of a type 1 job will be treated in this chapter. The more complex types will be examined in a later chapter.

Case I—Hand-Hole Cover Assembly for Diesel-Engine Crankcase

Aim. A class 1 or 2 change was indicated as desirable; the work was confined to one work station; hence, the analyst selected an operation chart as the most suitable man-analysis technique. The parts of the hand-hole cover are shown in Figure 16.1.

Analysis. The analyst, after first familiarizing himself with the job cycle, selected the busiest hand, in this case, the left, and classified the first step of that hand into the correct category from Table 16-I and made an entry on a form such as used in Figure 16.2. The analyst continued to analyze the complete cycle for that hand, showing the second step on the second line, and so on.

The charting of one hand completed, the analyst similarly analyzed the other hand. Finally, the chart was checked for correct simultaneity—that is, to see that items shown on one line happened at the same time. (In some cases, corrections for this are necessary, and sometimes the charts must be redrawn before they are clear.)

A summary indicating the total number of lines of each type of step was placed at the end to roughly indicate the effectiveness of the method and for comparison with alternative methods. Note that a hold by one hand extending during several steps on the other is counted as if the symbol were

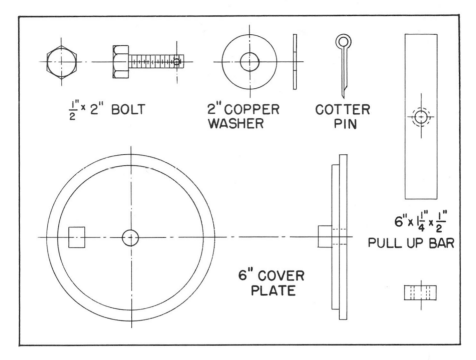

Figure 16.1. Parts of hand-hole cover.

repeated on each line, so that its importance is more adequately indicated than if it were only counted once. These quantitative figures should be used with care inasmuch as the steps are seldom equal in time value. Numerical aspects of schematic models are only a guide to their understanding.

The analyst also made a sketch of the workplace layout as shown in Figure 16.3; this is usually a useful adjunct to the operation chart.

Criticism. The general method of improving a job so analyzed is similar to that of improving a job charted with a process chart—man analysis. The checklist that is given is useful when carefully applied to each step on the operation chart.

The operation chart may be of an existing job or of a job that has not yet started. Both would still be subject to improvement.

Owing to the smallness of the separately identified steps, which in many cases are not unique to any one job or small group of jobs, the checklist is more detailed than are those previously given; furthermore, many of the items in it are experimentally verifiable in principle, apart from any job. Reference should be made to Chapter 13. Operation charts provide a level of detail suitable for contemplating from "a motion-economy point of view."

BASIC CHART FORM

Operation _____ Type of chart _____ 66 _____ Department

Original _____ Original or proposed _____ W. H. _____ Chart by

Assem. Hand-hole Cover Subject charted _____ 3/4 _____ Date charted

Left hand description	Symbols	Right hand description
To washer		To bolt
Pick up washer		Pick up bolt
To work area		To work area
Assemble		Assemble
To plate		Bolt and washer
Pick up plate		
To assembly		
Assemble		
To bar		
Pick up bar		
To assembly		
Bar		Thread bolt into bar
To cotter		Assembly
Pick up cotter		
To assembly		
Assemble		
Assemble		To pliers
		Pick up pliers
		To assembly
		Bend cotter
To finished parts		To pliers area
Place assembly		Place pliers.

	Summary		
	L.H.	R.H.	Both
O	8	6	14
o	9	5	14
▽	5	11	16

Figure 16.2. Original operation chart for assembly of hand-hole cover. (A form with a double row of preprinted symbols down the center may be prepared and used if desired.)

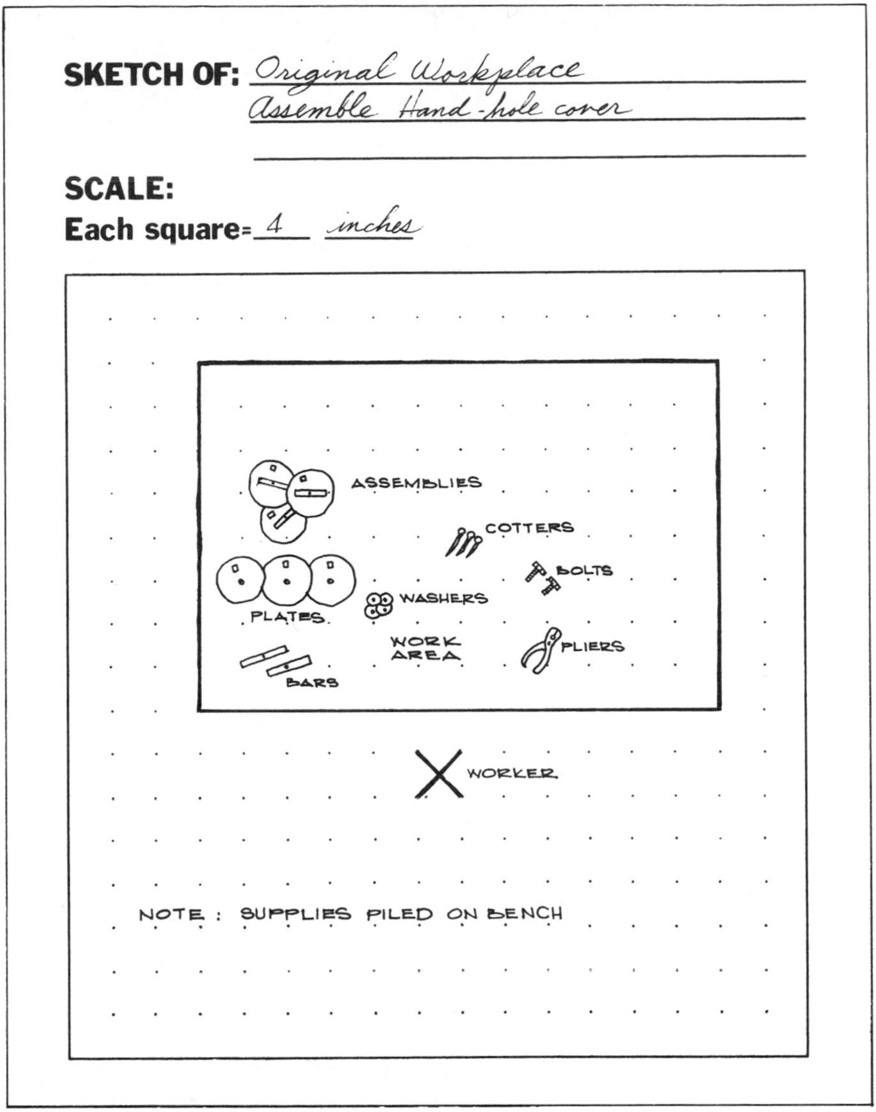

SKETCH OF: <u>*Original Workplace*</u>
<u>*Assemble Hand-hole cover*</u>

SCALE:
Each square=<u>4</u> *inches*

Figure 16.3. Workplace for original method of assembly of hand-hole cover.

Readings in the periodical literature also will assist the reader in obtaining a more complete understanding of the principles of improvement. The reader should review not only articles dealing with his industry but should read others as available, because many of the features of better methods are universally applicable to all types of work. The principles are sometimes more obvious when the work being done is not overly familiar.

CHECKLIST FOR OPERATION CHART

Basic principles

A. Reduce total steps to a minimum.
B. Arrange in best order.
C. Combine steps where feasible.
D. Make each step as easy as possible.
E. Balance the work of the hands.
F. Avoid the use of the hands for holding.
G. The workplace should fit human dimensions[1]:

1. Can a suboperation be eliminated?
   a. As unnecessary?
   b. By a change in the order of work?
   c. By a change of tools or equipment?
   d. By a change of layout of the workplace?
   e. By combining tools?
   f. By a slight change of material?
   g. By a slight change in product?
   h. By a quick-acting clamp on jig, if jigs are used?
2. Can a movement be eliminated?
   a. As unnecessary?
   b. By a change in the order of work?
   c. By combining tools?
   d. By a change of tools or equipment?
   e. By a drop disposal of finished material?
      (The less exact the release requirements, the faster the release.)
3. Can a hold be eliminated? (Holding is extremely fatiguing.)
   a. As unnecessary?
   b. By a simple holding device or fixture?
4. Can a delay be eliminated or shortened?
   a. As unnecessary?
   b. By a change in the work that each body member does?
   c. By balancing the work between the body members?

---

[1] For information on human dimensions, see C. T. Morgan, A. Chapanis, J. Cook, and M. Lund (eds.), *Human Engineering Guide to Equipment Design*. New York: McGraw-Hill Book Company, 1963; E. J. McCormick, *The Human Body in Equipment Design*. New York: McGraw-Hill Book Company, 1964; K. F. H. Murrell, *Human Performance in Industry*. New York: Van Nostrand Reinhold, 1965; A. Damon, H. W. Stoudt, and R. McFarland, *The Human Body in Equipment Design*. Cambridge, Mass.: Harvard University Press, 1966; and H. Dreyfuss, *The Measure of Man: Human Factors in Design*, 2nd ed., 1967. New York: Whitney Library of Design.

    d. By working simultaneously on two items?
      (Slightly less than double production is possible with the typical person.)
    e. By alternating the work, each hand doing the same job, but out of phase?

5. Can a suboperation be made easier?
    a. By better tools?
      (Handles should allow maximum flesh contact without sharp corners for power; easy spin, small diameter for speed on light work.)
    b. By changing leverages?
    c. By changing positions of controls or tools?
      (Put into normal work area; Figure 16.4.)
    d. By better material containers?
      (Bins that permit a slide grasp of small parts are preferable to bins that must be dipped into.)
    e. By using inertia where possible?
    f. By lessening visual requirements?
    g. By better workplace heights?
      (Keep workplace height below the elbow.)

6. Can a movement be made easier?
    a. By a change of layout, shortening distances?
      (Place tools and equipment as near place of use and as nearly in position of use as possible; see Figure 16.4.)
    b. By changing direction of movements?
      (Optimum angle of workplace for light knobs, key switches, and handwheels is probably 30° and certainly between 0° to 45° to plane perpendicular to plane of front of operator's body.)
    c. By using different muscles?
      (Use the first muscle group in this list that is strong enough for the task; see Figure 16.4 for visual items that may affect this order.)
      (1) Finger? (Not desirable for steady load or highly repetitive motions.)
      (2) Wrist?
      (3) Forearm?
      (4) Upper arm?
      (5) Trunk? (For heavy loads, shift to large leg muscles.)
    d. By making movements continuous rather than jerky?

7. Can a hold be made easier?
    a. By shortening its duration?
    b. By using stronger muscle groups, such as the legs, with foot-operated vises?

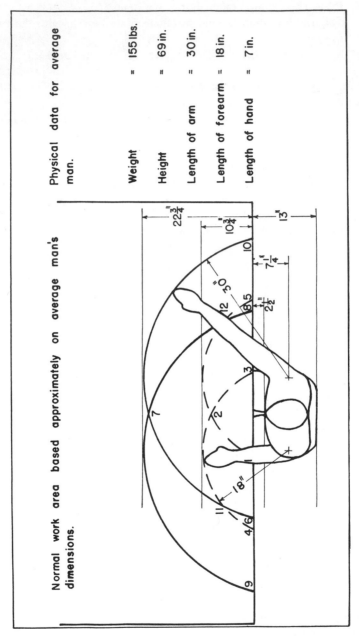

Figure 16.4. Normal work area.

232

Innovation. The application of the checklist to the job of assembling the hand-hole cover resulted in the improved method shown by the operation chart in Figures 16.5a and b, and in the new workplace layout shown in Figure 16.6. A chart form with preprinted symbols has been used in Figures 16.5a and b to show a variety of methods of charting within a single example. The form of the chart is not related to whether the original or proposed method is shown.

Test. This method was tested by rechecking with the checklist, and by discussion with the foreman.

Trial and application. It was given a trial in the shop, then standardized. Its application resulted in an increase in production of 82 per cent.

Subsequent to the improvement. As with process charts—man analysis, the analyst should keep a file of solutions that he develops so as to provide a source of more detailed suggestions for use with future studies.

The checklist questions that helped with this job and the detailed suggestions that would be noted are as follows:

3. *Can a hold be eliminated?*
   b. *By a simple holding device or fixture*
      Note: A hexagonal socket will hold a hex-head bolt firmly erect.
4. *Can delays be eliminated?* (Removing the hold created a delay.)
   d. *By working simultaneously on two items*
      Note: It is possible to perform most tasks simultaneously with both hands.
5. *Can a suboperation be made easier?*
   f. *By lessening visual requirements*
      Note: Small parts should be placed directly in front of the operator so that the eyes may direct both hands, and thus facilitate the work.
   d. *By better material containers*
      Note: Small material should be put in bins with lips, so that material may be picked up more easily and quickly.
   b. *By changing leverages*
      Note: A pull-up bar is easier to turn onto a bolt than a bolt into a bar because the bar gives more leverage.
   e. *By using inertia where possible*
      Note: A heavy pull-up bar will continue spinning onto a bolt when given a start. Bolts with low inertia will not spin.

RIGHT AND LEFT HAND OPERATION CHART

Improved **Method** _Bench_ **Mach. No.**

Assembly **Operation** _a-321_ **Opr. No.**

_____ Part _66_ **Part No.**

Hand-hole Cover **Part** _W. H._ **Chart by**

R. Allen **Operator** _3/4_ **Date**

SUMMARY /2/ PER PC.					
	Right Hand	Left Hand	Both Hands		
	ORIG	IMP	ORIG	IMP	ORIG	IMP	Saved

	ORIG	IMP	ORIG	IMP	ORIG	IMP	Saved
O	16		12	14	14		0
⚬	17		14	14	15.5		-1.5
▽	1		6	16	3.5		12.5
▽	0		2	0	1		-1
TOTAL	34		34	44	34		10

Left hand description	Oper.	Move	Hold	Delay	Delay	Hold	Move	Oper.	Right hand description
1. To bolt			▽	▽	▽	▽			To bolt
2. Pick up bolt			▽	▽	▽	▽			Pick up bolt
3. To work area			▽	▽	▽	▽			To work area
4. Assem. bolt to jig			▽	▽	▽	▽			Assem. bolt to jig
5. To washer			▽	▽	▽	▽			To washer
6. Pick up washer			▽	▽	▽	▽			Pick up washer
7. To assembly			▽	▽	▽	▽			To assembly
8. Assem. washer			▽	▽	▽	▽			Assem. washer
9. To cover plate			▽	▽	▽	▽			To cover plate
10. Pick up plate			▽	▽	▽	▽			Pick up plate
11. To assembly			▽	▽	▽	▽			To assembly
12. Assem. Plate			▽	▽	▽	▽			Assem. plate
13. To bar			▽	▽	▽	▽			To bar
14. Pick up bar			▽	▽	▽	▽			Pick up bar
15. To assembly			▽	▽	▽	▽			To assembly
16. Fit and spin on			▽	▽	▽	▽			Fit and spin on
17. To cotter pin			▽	▽	▽	▽			To cotter pin
18. Pick up cotter			▽	▽	▽	▽			Pick up cotter
19. To right side assem.			▽	▽	▽	▽			To right side assem.
20. Right s			▽	▽	▽	▽			Assem cotter

Figure 16.5a. Improved operation chart for assembly of hand-hole cover (*continues*).

RIGHT AND LEFT HAND OPERATION CHART

Improved **Method**   *Bench* **Mach. No.**

Assembly **Operation**   *a-321* **Opr. No.**

_____   *66* **Part No.**

Hand-hole Cover **Part**   *W.H.* **Chart by**

R. Allen **Operator**   *3/4* **Date**

SUMMARY / 2 / PER PC.

	Right Hand		Left Hand		Both Hands		
	ORIG	IMP	ORIG	IMP	ORIG	IMP	Saved
○	16	12	14	14			0
o	17	14	14	15.5			-1.5
▽	1	6	16	3.5			12.5
▽	0	2	0	1			-1
TOTAL	34	34	44	34			10

#	Left hand description	Oper. Move Hold Delay	Delay Hold Move Oper.	Right hand description
1	To left side assem.	○ ⟋ ▽ ▽	▽ ▽ ⟍ ○	To left side assem.
2	Assem. cotter	○ o ▽ ▽	▽ ▽ o ○	Left side assem.
3	Left side assem.	○ o ▽ ▽	▽ ▽ o ○	To pliers
4		○ o ▽ ▽	▽ ▽ o ○	Pick up pliers
5		○ o ▽ ▽	▽ ▽ o ○	To assembly
6		○ o ▽ ▽	▽ ▽ o ○	Bend left cotter
7	To right side assem.	○ o ▽ ▽	▽ ▽ o ○	To right side assem.
8	Right side assem.	○ o ▽ ▽	▽ ▽ o ○	Bend right cotter
9	For right hand	○ o ▽ ▽	▽ ▽ o ○	To aside pliers
10		○ o ▽ ▽	▽ ▽ o ○	Place pliers
11	To left assembly	○ o ▽ ▽	▽ ▽ o ○	To right assembly
12	Pick up assem.	○ o ▽ ▽	▽ ▽ o ○	Pick up assem.
13	To finished parts	○ o ▽ ▽	▽ ▽ o ○	To finised parts
14	Place assembly	○ o ▽ ▽	▽ ▽ o ○	Place assembly
15		○ o ▽ ▽	▽ ▽ o ○	
16		○ o ▽ ▽	▽ ▽ o ○	
17		○ o ▽ ▽	▽ ▽ o ○	
18		○ o ▽ ▽	▽ ▽ o ○	
19		○ o ▽ ▽	▽ ▽ o ○	
20		○ o ▽ ▽	▽ ▽ o ○	

Figure 16.5b. (*Concluded*) Improved operation chart for assembly of hand-hole cover.

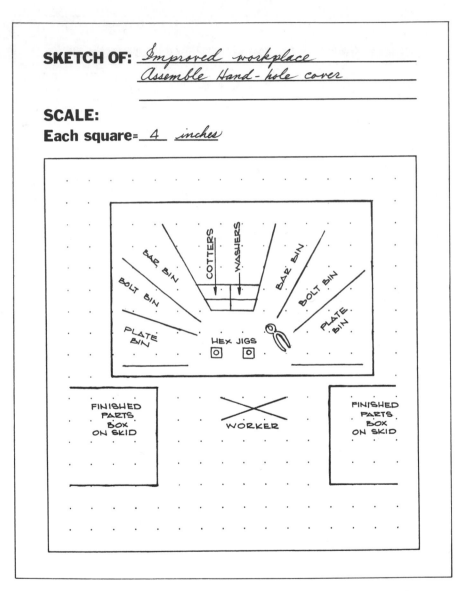

Figure 16.6. Workplace for improved method of assembly of hand-hole cover.

Summary—Operation Chart

1. Uses:
    a. For information leading to a class 1 or 2 change on a job taking place at one location when the operator rather than a machine controls the flow of work.
    b. For designing the details of a job of the type indicated in 1a.

2. How made:
    a. If the job is current, the chart is best prepared by actual observation of the worker. If the job has not yet started, it is prepared from a contemplation of the necessary work.
    b. The form illustrated in Figures 16.2 or 16.5a or a blank sheet of paper may be used.
    c. Although the chart may begin at any point in the work cycle, it is usually most convenient to begin with the hand that makes the first movement attributable to a unit of output.
    d. The first and last cycle of a work spell may be different from the other cycles. It is usually desirable to chart the most typical cycle.
    e. The first step of the hand that begins the cycle should be classified according to Table 16-I and an entry made, as on the first line of Figure 16.2. The symbols are usually drawn freehand, although forms with preprinted symbols are not uncommon.
    f. All of the steps of this hand should then be properly classified and plotted in order. It may be necessary to observe many cycles in order to accomplish this and check the completeness.
    g. The other hand should then be plotted, care being taken to place each step for this hand opposite the step of the other hand during which it occurs.
    h. Since step g often involves the squeezing in of lines and items to coordinate the chart properly, it is frequently necessary to redraw it so as to make it easily legible (Figure 16.2). However, since the chart is merely an aid to the understanding of the method, excess draftsmanship is a waste of time.
    i. A summary should be placed at the bottom of the chart.
    j. A sketch of the workplace is often a useful adjunct.

3. How used:
    a. Each step is questioned with the checklist for operation charts. The dimensions of the workplace and the movements required should be studied with the aid of data on human dimensions.
    b. The general principles listed at the top of the checklist are kept in mind as objectives.

4. *What then:*
   a. An operation chart is drawn for the resulting suggested method to permit a final check and to provide a means of describing the proposed new method.
   b. A revised sketch of the workplace is usually drawn, as well as drawings of the necessary tools, jigs, fixtures, bins, and the like.

PROBLEMS

16.1. Films of simple operations suitable for operation charting are easily made. Although operation charts are normally made from direct observation, these films may be used to bring the job before a group. They should be run steadily so as to replace a continuously working operator.

16.2. A flashlight assembly consists of a cylindrical case, a rear cap, a reflector, a lens, and a top ring that screws onto the case to hold the lens and reflector in position. The workplace is shown in Figure P16.1.

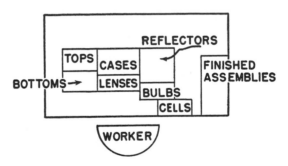

Figure P16.1

The operator picks up the case with the right hand and transfers it to the left hand. Then the right hand picks up the reflector from a pan about 15 inches away and drops it into the top end of the case. The right hand next picks up a bulb from a pan 10 inches away and screws it into the reflector. This part of the operation is slow because it is difficult to maintain a firm grasp on the bulb with the finger tips inside and near the bottom of the reflector. With the left hand still holding the case, the right hand places the lens, which was 8 inches away, on top of the reflector, where the first finger and thumb of the left hand hold it in position while the right hand reaches to a pan 12 inches away and picks up and assembles the top ring to the case. The right hand then places the cells into the case, getting them one at a time from an area 10 inches away, and then assembles the rear cap. The pile of rear caps is also 10 inches away. This completes the assembly. The flashlight is then transferred to the right hand, where the thumb operates the switch to make sure that the completed assembly is satisfactory. The rear cap is then removed with the left hand, the cells dumped out, and the rear cap replaced. The flashlight is then placed in a tray about 30 inches to the right of the operator. It is given a final inspection for appearance, by another operator, before packaging.

(a) Make a right- and left-hand operation chart of this operation.

(b) Prepare a class 1 or 2 change, and present the new method in the form of a right- and left-hand operation chart and make a sketch of the new workplace layout.

(c) List the information you would add to your checklist and for each item indicate the checklist question number and letter under which you would file it.

16.3. The workplace used to gage the thickness of piston rings is shown in Figure P16.2, and the rings and gage are shown in detail in Figure P16.3. The worker with her left hand picks up an ungaged ring from the uptilted box on the left and at the same time disposes of the gaged ring, either to the "passed" box on her right or the reject box in front of her. Using both hands she tries the ring against the slot under the no-go gage three times at points 120 degrees apart on the ring's circumference, each time sliding the ring on the gage flat. If the ring does not go under the no-go gage, she slides it with her left and up under the go gage and pulls it through with her right hand. This is the normal sequence. Rejects are disposed of as soon as detected.

(a) Make a right- and left-hand operation chart of the original method of gaging piston rings.

(b) Develop an improved method, with no greater than a class 2 change, and present an operation chart of this new method as well as a sketch of the new workplace or gage if required.

(c) List the information you would add to your checklist and for each item indicate the checklist question number and letter under which you would file it.

16.4. The following is a description of the original method used to finish-pack quart boxes of cultivated blueberries prior to crating. The workplace is shown in Figure P16.4. Field-picked boxes of blueberries are placed, by the packer, on the worktable, six at a time, on a flat from a stack of flats and boxes alongside the table. The operator reaches and with both hands gets a field-packed box, *PB*, from the left. (These are standard quart berry boxes.) She sets it down at *S* and adds enough berries from the box at *Ex* to flush-fill the box being packed. The extra berries are poured from the extra box into the box at *S*, and then the extra box is set back. A cellophane sheet is picked up with the right hand, and both hands set it over the box being packed. The left hand then picks up the frame at the left. The frame consists of a wood frame with heavy rubber flaps (marked *A*, *B*, *C*, and *D* in Figure P16.4). This is placed over the box at *S* by both hands and allowed to slip over the box, thus holding the

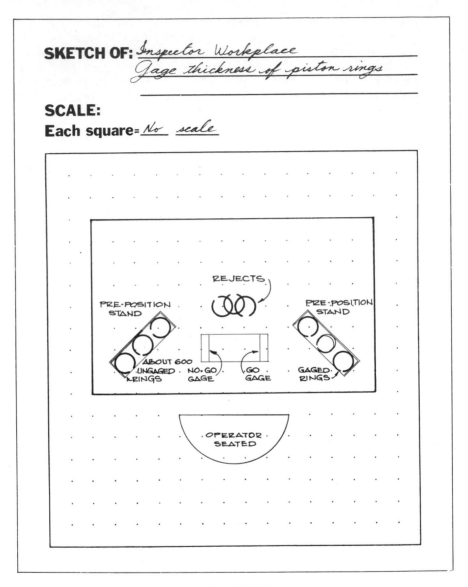

SKETCH OF: *Inspector Workplace*
Gage thickness of piston rings

SCALE:
Each square= *No scale*

REJECTS

PRE-POSITION
STAND

PRE-POSITION
STAND

ABOUT 600
UNGAGED
RINGS

NO-GO
GAGE

GO
GAGE

GAGED
RINGS

OPERATOR
SEATED

Figure P16.2

cellophane down on all four sides. The right hand then gets a rubber band from the supply and both hands place it over the box to hold the cellophane down. The two hands then place the finished box in the shipping box at the right rear. When six boxes are in the shipping boxes, the worker sets it on the

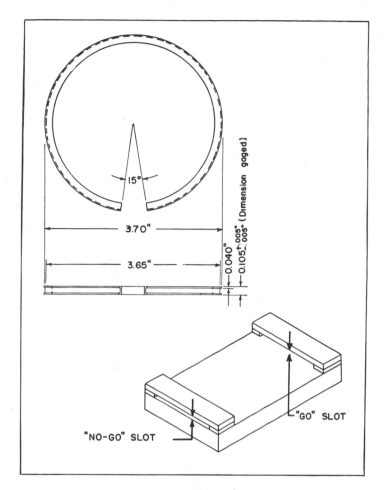

Figure P16.3

floor or stack at the right of the table, places the flat over it for a cover, gets a new flat, and finishes the boxes in it. When the box at *Ex* is used up, another box is taken from those on the flat on the table. One box will finish packing about 15 other boxes.

(a) What is the cyclic nature of the task described?

(b) Prepare a right- and left-hand operation chart for the method described.

(c) With the aid of the checklist, prepare an improved method and present a sketch of the revised workplace, if required, and a new operation chart.

(d) List the information you would add to your checklist and for each item indicate the checklist question number and letter under which you would file it.

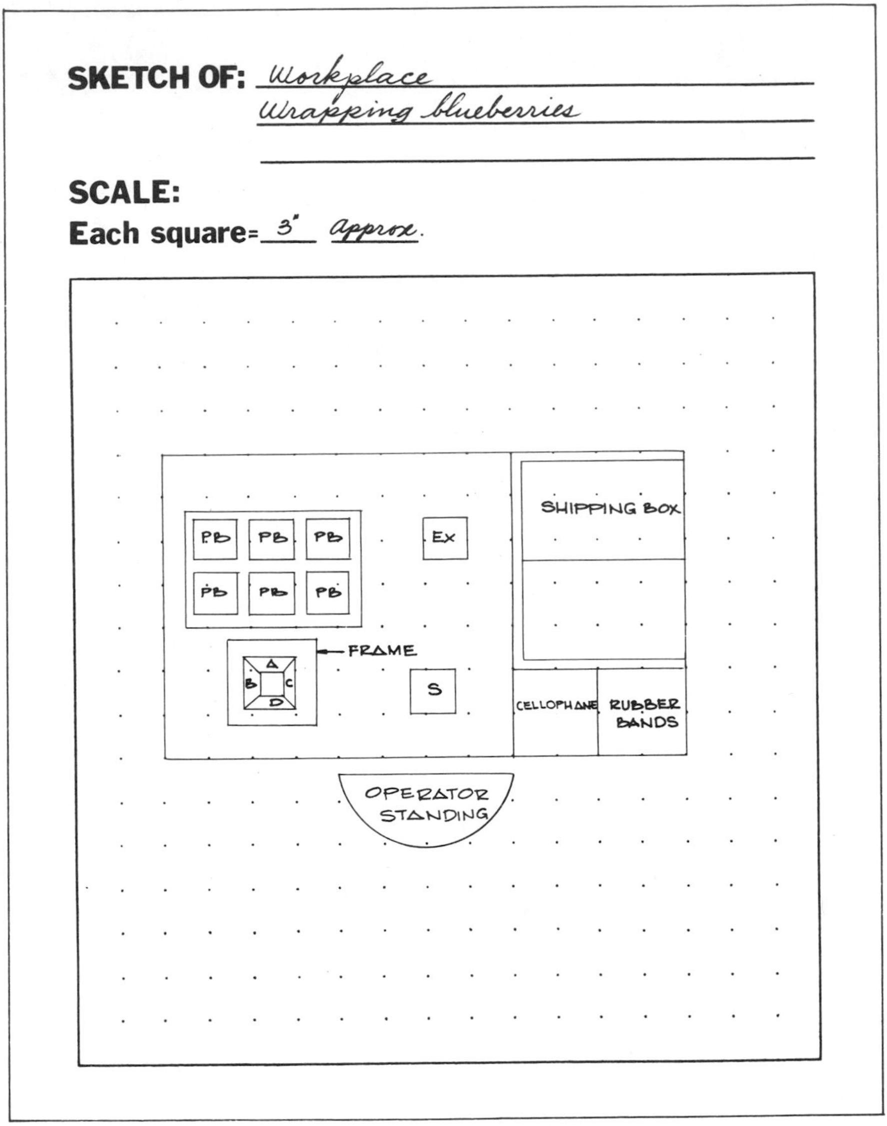

SKETCH OF: _Workplace_
Wrapping blueberries

SCALE:
Each square= _3"_ _approx._

Inside the sketch grid:

PB PB PB   Ex   SHIPPING BOX

PB PB PB

FRAME (A B C D)   S   CELLOPHANE   RUBBER BANDS

OPERATOR STANDING

Figure P16.4

17   MULTIPLE-ACTIVITY CHARTS

Introduction

Multiple activity refers to cases where a man works with one or more machines, or where a group of men work coordinately with or without machines. The essential element of similarity of cases to which multiple-activity analysis is applied is the requirement that any useful analysis show not only the sequence of steps of the two or more items charted but also the relative simultaneity of the steps of each. A variety of charting techniques can be employed.

Multiple-Activity Charts Defined

A multiple-activity chart is a graphic means of portraying the separable steps performed by a worker and a machine, or several workers with or without machines, in a manner such that the coordination of the subjects charted is also displayed.

Types of Multiple-Activity Charts

Man and machine operation chart. The symbols shown in Table 16-I, for use with an operation chart, are used to show the activities of the worker. An additional column is added for the machine, but only two symbols are used in this column: the operation symbol for the machine when working, and the idle symbol when it is not. Simultaneity is indicated by the horizontal alignment of symbols. A multiman version is conceivable but seldom employed.

Man and machine process chart. The symbols shown in Table 14-I, for use with a process chart—man analysis, are used to show the steps performed by the worker. An additional column of symbols is used for each machine, with the simultaneity of action indicated by the horizontal alignment of man and machine symbols as with man and machine operation charts. The time for each step may, if desirable, be shown next to each symbol.

Multiman and multiman and machine process chart. With this type of chart a separate column of symbols is used for each worker or machine. In every other respect it is exactly like a man and machine process chart.

Man and machine time chart. The steps separated on a man and machine time chart may be the same as those used with man and machine process or operation charts. The distinctive feature of *time* charts is that the charting symbols and conventions are such that the time for each step is indicated graphically and the exact simultaniety of the steps is shown.

Multiman and multiman and machine time charts. The steps separated are the same as with multiman and multiman and machine process charts. As with man and machine time charts, the charting symbols and conventions are such that the time for each step, as well as the simultaneity, are indicated graphically.

Where the techniques are used. Multiple-activity analyses are usually used when the objective of the study is either to design the operation or to make a class 1 or 2 change and where an analysis showing the relative simultaneity of the two or more items charted is required for a full understanding of the job. Like the techniques discussed in Chapters 10 through 16, they are devices for aiding in performing the steps of the logical approach.

Purposes served by the techniques. A multiple-activity chart is made usually to obtain better utilization of the machine or the worker (or workers). However, increased safety is also a common objective. The chart aids in determination of the most effective way of harmonizing the work of each individual with the group or with the machine. It may also be used to indicate how the machine might be altered to harmonize with the requirements of the individuals.[1] If the chart technique is employed prior to the finalization of a machine's design, the requirements of the operator may be properly taken into account so that the machine will be designed for human use as well as for mechanical function.

Categorizing the types of jobs which may be encountered. As with both process chart—man analysis and operation charts, four types of jobs may be encountered.

1. The work has a single repeated cycle.
2. The work is cyclic, but there are several subcycles performed with different frequency.
3. The work varies from cycle to cycle.
4. The task has no cycle or pattern.

As with the referenced techniques, jobs of type 4 are usually more appropriately studied with work activity analysis, work sampling or memo-motion study.

All the cases presented in this chapter will be type 1, simple cyclic work. More complex examples will be examined in later chapters. One example of each of the five types of chart will be shown.

Case I—An Analysis of the Operation of a Centerless Grinder Used on Engine Bearings (Simple Work)

Introduction. The production of engine bearings was being seriously hampered by the lack of centerless-grinding capacity in the plant. The cost of additional equipment was of sufficient magnitude to warrant careful study.

Form used. The same form is used as with a right- and left-hand operation chart, but an additional column is used for the machine.

---

[1] M. E. Mundel and R. C. Howell, "Applying Time and Motion Study Principles in Design," *Machine Design*, August 1948, pp. 121–126; E. D. Long, "Human Productivity and Tool Design," *Proceedings, 17th Annual Institute, American Institute of Industrial Engineers*, 1966, pp. 12–20.

Making the chart. In the three columns of the man and machine operation chart shown in Figure 17.1 for the original method of centerless-grind engine bearing, the activities of the right and left hand of the operator and of the machine are charted in such a manner that a horizontal line drawn across the chart at any point passes through simultaneous actions. This chart was constructed by first making an operation chart in the usual manner and then adding the machine column to it in the same fashion as the analysis of the second hand was added to the analysis of the first. The work shown on the chart as "finish grind" was a hand manipulation of the controls, so as to be certain that all play in the feeding linkage had been eliminated and the part ground to size. The plant "process engineer" insisted on this step. The length of the chart, with this type of analysis, is not proportional to the time required for the job. The original workplace layout is shown in Figure 17.2.

Analysis. The analyst, to obtain additional insight, used a stopwatch to obtain the time for each step shown on Figure 17.1. The time data were obtained in order to have sufficient information to continue the analysis by constructing a man and machine time chart.

Man and machine time chart. A man and machine chart that will indicate the time for each step graphically requires a new type of notation. A continuous vertical column is used instead of a column of symbols, and the column is scaled into time units. Since geometric symbols are no longer convenient, various shadings are used to indicate the nature of each of the steps of the operation. The length of each shading is used to indicate the amount of time spent on the step. The columns are kept in the same relationship to each other as with the first type of man and machine chart, except that the relationship is indicated much more accurately. A convenient system of symbols for man and machine operation time charts is given in Table 17-I. Man and machine time charts are still schematic models, but the charting of the steps on a scale of time makes them more complete models. The effect of shifting steps from one column to another may be more fully appraised and the consequences more carefully determined.

Advantages of time charts. A time chart facilitates rearrangement and balancing of the work of the two hands and the worker and the machine, and permits an evaluation of the magnitude of the probable effects of change. Together with labor rates and overhead figures, this makes possible the determination, in a reasonably accurate fashion, of the economic desirability of installing a change, prior to its actual application. Through the use of time values from predetermined time systems (commonly referred to as PTS) discussed in later chapters, alternative designs of machines and their operat-

Figure 17.1. Man and machine operation chart for original method of centerless grind engine bearing.

ing patterns may be compared, in terms of labor operating costs, while the machines are still in the drawing stage and the most preferable (depending on the criterion selected) design chosen. For ongoing operations, the required

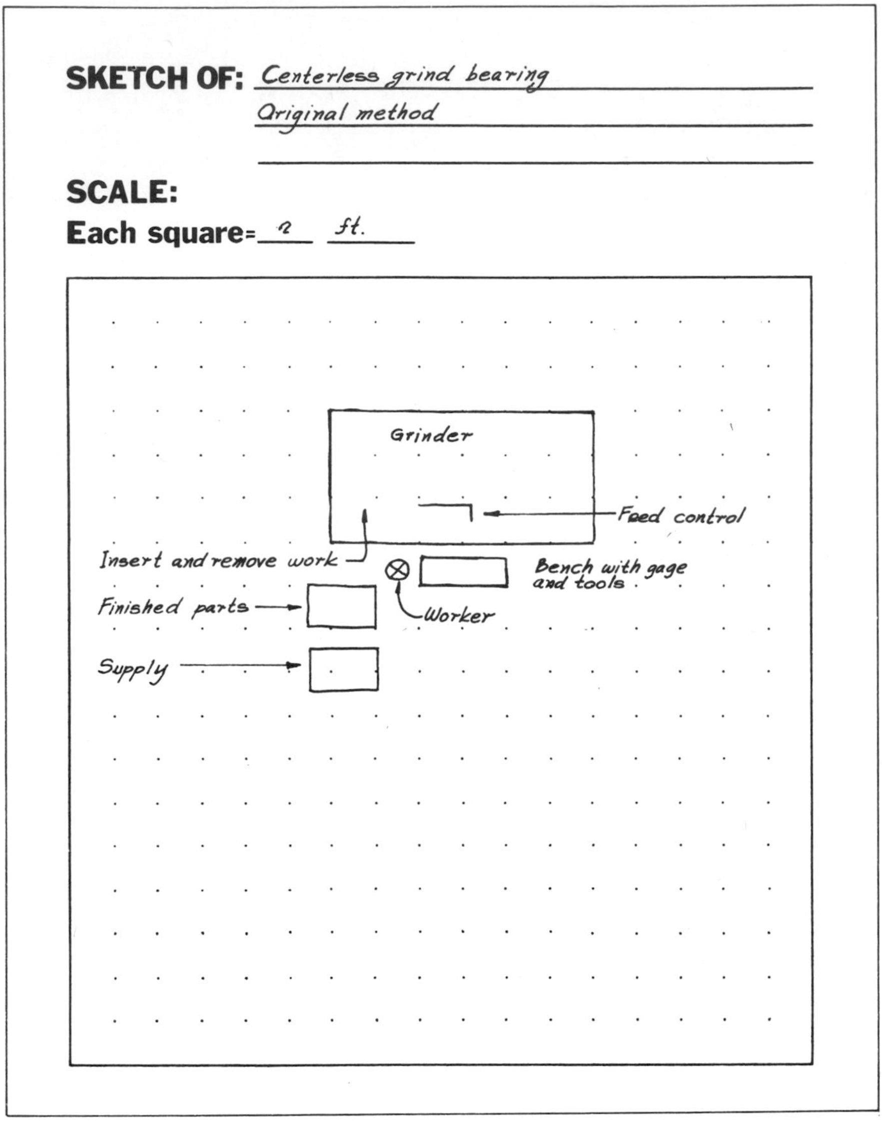

SKETCH OF: *Centerless grind bearing*

Original method

SCALE:
Each square= *2* *ft.*

Figure 17.2. Workplace for original method of centerless grind engine bearing.

time values may be obtained roughly with a watch, stopwatch, any of the motion-picture techniques, or with any of the methods that will be described in Chapter 21, which deals with predetermined times. Since the time values used on these charts are primarily for analytical purposes, they need not have the accuracy of time values recorded for wage determination.

Table 17-I. SYMBOLS FOR MAN AND MACHINE OPERATION TIME CHART

*Symbol**	*Name*	*With Man Activities Is Used to Represent*	*With Machine Activities Is Used to Represent*
	Suboperation	Body member or operator doing something at one place.	Machine working ("on" time), machine paced.
	Suboperation	Not used.	Machine working ("on" time), operator paced.
	Movement	Body member or operator moving toward or with an object.	Not used.
	Hold	Body member maintaining an object in a fixed position.	Not used.
	Delay	Body member or operator is idle.	Machine is idle ("down" time).

*The amount of shading is chosen to suggest automatically the general usefulness of the step. The less shading, the probable greater undesirability of the step.

Analysis, continued. The analyst, employing the conventions of Table 17-I, constructed the man and machine time chart shown in Figure 17.3.

To gain additional insight into the method, the whole of the man and machine activity may be classified into three additional categories, as follows: *independent man work*, when the operator is working and the work neither controls nor is controlled by the machine, or when the worker is idle and the idleness does not control the machine; *independent machine work*, when the machine is working and is not controlled by the worker; and *combined work*, when the worker is operating the machine, loading the machine, or when the machine is waiting for the worker.[2]

Consideration of the three categories suggested, which may be superimposed upon the usual time chart, should suggest in general that the com-

[2]Suggested by David B. Porter, New York University.

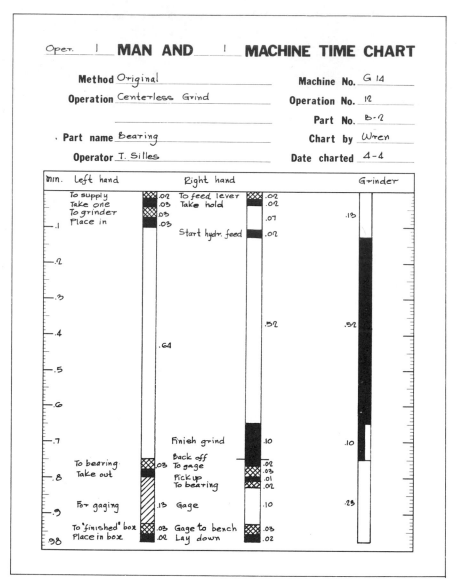

Figure 17.3. Man and machine operation time chart for original method of centerless grind engine bearing.

bined work be reduced to a minimum to approach the minimum cycle time, the independent man work plus the combined work be reduced to a minimum to reduce the man workload to a minimum, and so forth. Such an additional classification of the steps of the chart should assist in developing more effective work cycles.

Criticism. As with the preceding techniques, the third step of the logical procedure, *criticism*, is performed with the aid of a checklist. A suitable one follows.

CHECKLIST FOR MAN AND MACHINE CHARTS

Basic principles

A. Eliminate steps.
B. Combine steps.
C. Rearrange in best fashion.
D. Make each step as easy as possible.
E. Raise percentage of cycle of machine running time to maximum.
F. Reduce machine loading and unloading to minimum.
G. Raise machine speed to economic limit.

(The first seven questions that follow are similar to those used with operation charts, where more detail was given; hence, the reader is also referred to them. The bare questions are given here so as to provide, at one place, all the checklist items to be used.)

1. Can a suboperation be eliminated?
   a. As unnecessary?
   b. By a change in the order of work?
   c. By a change of tools or equipment?
   d. By a change in layout of the workplace?
   e. By combining tools?
   f. By a slight change of material?
   g. By a slight change in product?
   h. By a quick-acting clamp on the jigs or fixtures?
2. Can a movement be eliminated?
   a. As unnecessary?
   b. By a change in the order of work?
   c. By combining tools?
   d. By a change of tools or equipment?
   e. By a drop disposal of finished material?
3. Can a hold be eliminated? (Holding is extremely fatiguing.)
   a. As unnecessary?
   b. By a simple holding device or fixture?
4. Can a delay be eliminated or shortened?
   a. As unnecessary?
   b. By a change in the work each body member does?
   c. By balancing the work between the body members?
   d. By working simultaneously on two items?
   e. By alternating the work, each hand doing the same job, but out of phase?

5. Can a suboperation be made easier?
   a. By better tools?
   b. By changing leverages?
   c. By changing positions of controls or tools?
   d. By better material containers?
   e. By using inertia where possible?
   f. By lessening visual requirements?
   g. By better workplace heights?
6. Can a movement be made easier?
   a. By a change of layout, shortening distances?
   b. By changing the direction of movements?
   c. By using different muscles? (See Figure 16.4.)
      Use the first muscle group in this list that is strong enouth for the task:
      (1) Finger.
      (2) Wrist.
      (3) Forearm.
      (4) Upper arm.
      (5) Trunk.
   d. By making movements continuous rather than jerky?
7. Can a hold be made easier?
   a. By shortening its duration?
   b. By using stronger muscle groups, such as the legs, with foot-operated vises?
8. Can the cycle be rearranged so that more of the handwork can be done during running time?
   a. By automatic feed?
   b. By automatic supply of material?
   c. By change of man and machine phase relationship?
   d. By automatic power cutoff at completion of cut or in case of tool or material failure?
9. Can the machine time be shortened?
   a. By better tools?
   b. By combined tools?
   c. By higher feeds or speeds?

The first seven checklist questions are particularly useful for applying to the man activities during machine downtime, since any reduction in machine downtime increases the percentage of time the machine may be in operation. Because overhead costs often exceed direct labor costs, obtaining better utilization of the machine through reduction of downtime is often an extremely valuable type of improvement.

In some cases, the machine running time provides a needed rest for the operator. It is worth noting that increasing production by reducing the machine downtime through better hand motions may actually increase the amount of time per day that the operator may rest.

Innovation. By a thorough scrutiny of the original method, with the checklist questions as criteria, the new method shown in Figure 17.4 was devised. The improved workplace is shown in Figure 17.5.

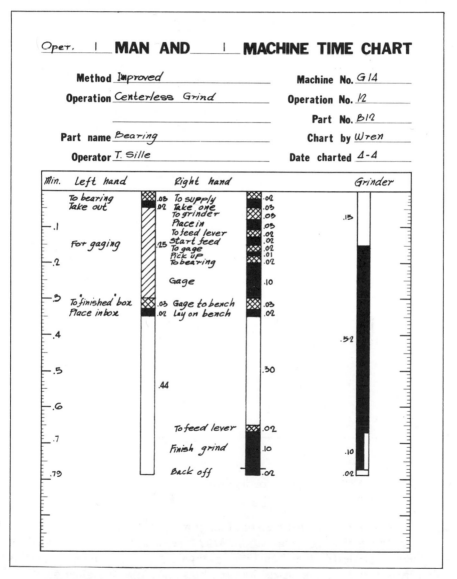

Figure 17.4. Man and machine operation time chart for improved method of centerless grind engine bearing.

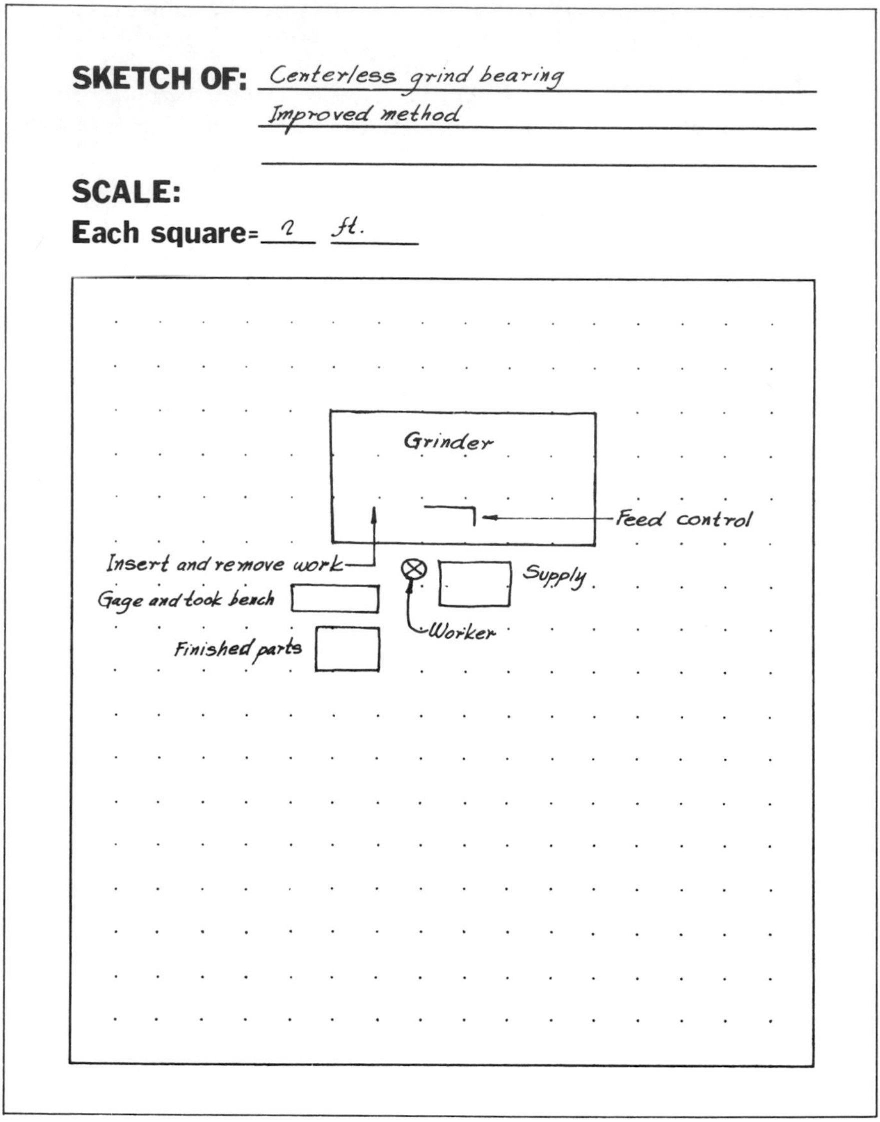

SKETCH OF: *Centerless grind bearing*

Improved method

SCALE:
Each square= *2   ft.*

Grinder

Feed control

Insert and remove work

Gage and tool bench

Finished parts

Supply

Worker

Figure 17.5. Workplace for improved method of centerless grind engine bearing.

Note, in the improved method, that some of the grinding is being performed during the gaging operation. This type of chart permits estimation of the magnitude of the expected improvement. A normal summary such as that used with operation charts does not make possible an evaluation of the effect of such changes as removing the sixth step of the original method, during which both hands were idle for a considerable time. The steps added to the improved method are performed in this interval, together with several of the steps originally performed while the machine was down. Close scrutiny of the chart will show that fewer steps are performed while the machine is idle, which is a better measure of the desirability of the improved method than the total number of steps, because with the machine set at optimum operating speed, the machine-contolled part of the cycle is fixed, and any improvement must affect idle or downtime. It is worth pointing out, however, that optimum operating speed is frequently a subject worthy of study, although the available information is much too extensive to include here.

The bearings could not be fed automatically into the grinder without considerable complications, because one end of the bearing had a shoulder. In many other cases, hopper feeding may make continuous machine usage feasible.

Results obtained. Output was increased by 23 per cent, the production bottleneck eliminated, and the amount of overhead and labor chargeable to each unit of the product was decreased.

Subsequent to the improvement. As with the preceding techniques, for future use, the analyst should keep a classified file of the details of improvement. The useful item, from questions 8 and 9 of the checklist, and the notes that were made to increase the ease of subsequent usage were as follows:

> 8. *Can the cycle be rearranged so that more of the handwork can be done during running time?*
>   c. *By change of man and machine phase relationship*
>   *Note:* If gaging of finished piece can be performed during machine running time, rather than during downtime, the cycle will
>   · be shortened by just that much.

Case II—The Use of the Man and Machine Process Time Chart

Introduction. A simpler man and machine chart, a *man and machine process time chart*, may be used as the preliminary step in determining the most economical number of machines an operator might run or the most desirable number of workers that might be used as a crew.

Analysis. To simplify the analysis, the worker may be treated as with a process chart—man analysis, and the coding given in Table 17-II is sug-

Table 17-II. SYMBOLS FOR MAN AND MACHINE PROCESS TIME CHART

Symbol	Name	With Man Activities Is Used to Represent	With Machine Activities Is Used to Represent
■	Operation	The doing of something at one place.	Machine working ("on" time), machine paced.
▯	Operation	Not used.	Machine working ("on" time), operator paced.
▦	Quantity determination	A special form of operation involving the person determining the quantity of an item present.	Not used.
▥	Inspection	A special form of operation involving the person comparing an attribute of a product with a standard, or verifying the quantity present.	Not used.
▨	Movement	A change in location; moving from one place to another.	Not used.
□	Delay	Idleness. Waiting or moving, provided the movement was not part of the job and the time could have been spent waiting.	Machine is idle ("down" time).

gested. This will permit the time values to be graphically represented on the chart and will permit an easier approach to the problems encountered than that possible with geometric symbols.

Figure 17.6 is the man and machine process time chart for an operator running an automatic pot-polishing machine (in a pot factory) and wiping and packing the pots. The operator has been treated as a single unit without detailing any single body member. Since this job does not require the operator to move from the workplace, no movements appear in the breakdown of this job.

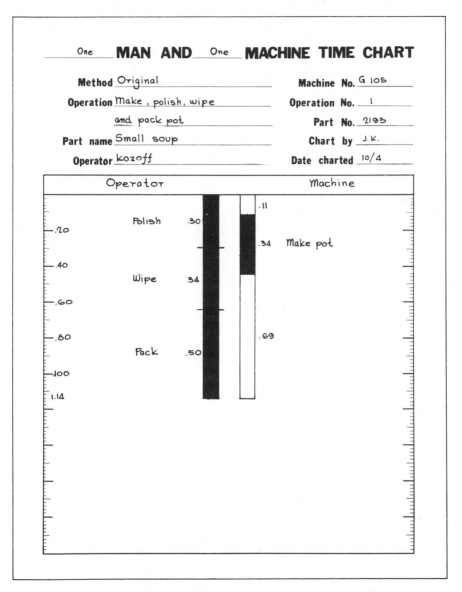

Figure 17.6. Man and machine process time chart for one-man crew on polish, wipe, and pack pots.

Note, with one operator, as charted in Figure 17.6, the machine time chargeable to each pot is 1.14 minutes, and the operator time chargeable to each pot is 1.14 minutes. This type of analysis makes possible the determination of the effects of using a crew of two or more, as will be examined in later chapters.

Multiman Charts

Introduction. A study of those jobs on which a group of workers work coordinately with or without machines requires a type of analysis that shows the relative simultaneity of the sequence of steps for all the workers and machines.

Multiman charts. Multiman charts are a graphic means of portraying the separable steps of the work of a crew, with or without machines, and indicating the relationships among them, when the work involves coordination of the crew.

Work in sequence. When the work is done in sequence, it is often best analyzed by considering the work of each crew member separately. The aim is to distribute evenly the amount of work for each member of the crew, by breaking the sequence at the proper places, or else to improve the methods for the work of the crew members having the longest part of the sequence. Techniques that are used for the improvement of individual jobs are commonly employed in this area as well. In complex cases, mathematical models may be valuable.[3] Such work in sequence is not usually classified as "multiple activity."

Making the multiman chart. A watch, stopwatch, or motion-picture or video-tape camera (as will be seen in Chapter 18) may be used for studying the work of the coordinate crew. Since details of the interrelationships of

---

[3] M. E. Salveson, "The Assembly-Line Balancing Problem," *Transactions of ASME*, Vol. 77, August 1955, pp. 939–948; M. D. Kilbridge and L. Wester, "A Review of Analytical Systems of Line Balancing," *Operations Research*, Vol. 10, September–October 1962, pp. 626–638; E. J. Ignall, "A Review of Assembly Line Balancing," *Journal of Industrial Engineering*, Vol. 16, July–August 1965, pp. 244–254; D. R. Freeman and J. V. Jucker, "The Line Balancing Problem," *Journal of Industrial Engineering*, Vol. 17, June 1967, pp. 361–364; D. Brennedee, "Two Parameter Assembly Line Balancing Model," *Models and Analysis for Production Management*, M. P. Hottenstein (ed.), International Textbook Co., Scranton, Pa., 1968; Heskia Heskiaoff, "A Heuristic Method for Balancing Assembly Lines," *Western Electric Engineer*, Vol. 12, October 1968, pp. 9–16; E. M. Mansoor, "Assembly Line Balancing—A Heuristic Algorithm for Variable Operator Performance Levels," *Journal of Industrial Engineering*, Vol. 19, December 1968; N. T. Thomopoulos, "Some Analytical Approaches to Assembly Line Problems," *The Production Engineer*, July 1968; R. Reiter, "On Assembly Line Balancing Problem," *Operations Research*, Vol. 17, July–August 1969, pp. 685–700; A. J. Nevins, "Assembly Line Balancing Using Best Bud Search," *Management Science*, Vol. 18, No. 9, 1972, pp. 529–539; C. L. Moodie, "A Systems Engineering View of Assembly Line Balancing by Computer," *Conference Proceeding of A.I.I.E.*, 1973; Ravinder Nanda and J. M. Scher, "Assembly Lines with Overlapping Work Stations," *A.I.I.E. Transactions*, Vol. 7, No. 3, 1975, pp. 311–318.

the time that is spent by each of the members of the crew are frequently the key to rearranging the task, a type of chart involving a time scale is of the greatest value in most situations, although it is not a necessity. If the time values are obtained, however, they are also an aid in determining the economic desirability of any innovation, since they facilitate a reasonably accurate evaluation prior to the application of the change. The time values are for analysis only, and not for wage determination, so extreme accuracy is not required. Also, as with man and machine analysis, predetermined time values, from tables, may be used to design crew jobs prior to the beginning of the actual work. Multiman charts are a desirable aid in developing such designs.

Steps separated in analysis. With coordinate crew work, the movements of the workers as units, particularly with heavy jobs, are of prime importance, and, therefore, a simple type of analysis, using the steps and conventions given in Table 17-II, is usually desirable. A coordinated right- and left-hand breakdown can be made of each worker, but for simplicity's sake it is better to make an analysis of the whole task, and left- and right-hand breakdowns only when parts of the job seem to warrant analysis in this detailed fashion.

Case III—Cut Studs in Prefabricated-House Plant (Simple Cyclic Work)

Aim. The objective chosen from the possibility guide was a class 1 or moderate class 2 change. The choice was influenced by the use for many other jobs of the equipment used on this job—a fact that tended to rule out any highly specialized equipment. A standard overhead-track cutoff saw was used.

Selecting the analysis technique. The decision to seek a class 1 or moderate class 2 change, the necessity (economic) of performing the job, the nature of the job (coordinate crew work), all indicated the desirability of a multiple-activity analysis—a multiman and machine time chart.

Analysis. Figure 17.7 is a plan view of part of the factory floor in a prefabricated-house plant, together with the paths of the three workers in the crew using the cutoff saw to cut 8-foot house studs from 14-foot mill-length 2- by 4-inch lumber. The work pattern of the crew and saw is shown in the multiman and machine process chart in Figure 17.8. On this chart, one column has been used for each man and one column for the saw.

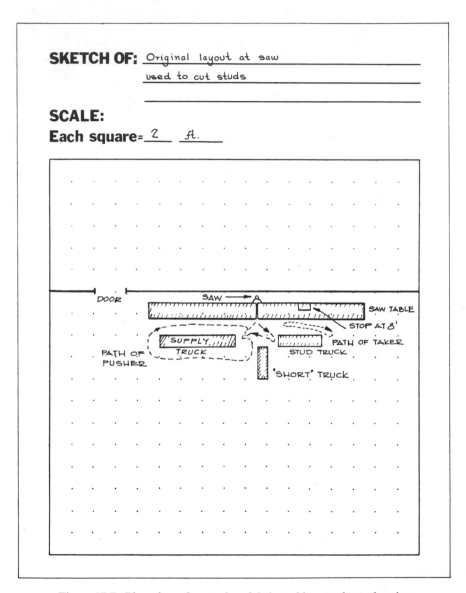

Figure 17.7. Plan view of part of prefabricated house plant, showing location of material and equipment and path of three-man crew cutting studs, using original method.

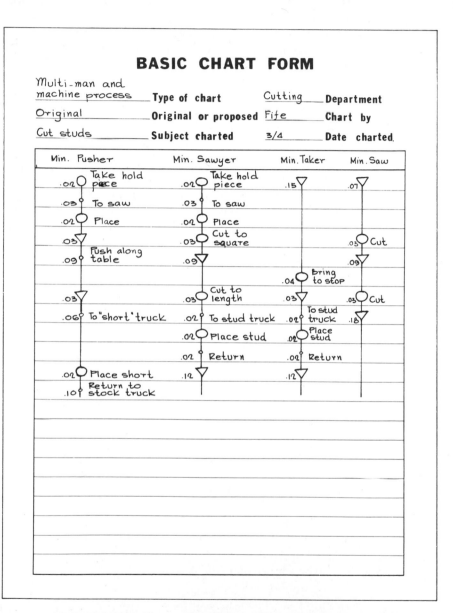

Figure 17.8. Multiman and machine process chart for original method of three-man crew cutting 8-foot house studs from mill length 2 by 4-inch stock.

The chart was constructed by making a process chart—man analysis for each worker, roughly keying them together with the operation of the machine, obtaining rough stopwatch time values for each step, and adjusting these as the actions of the workers indicated, so that they meshed properly, as shown in Figure 17.8. For a person with considerable experience with the type of work charted, this form of analysis may be sufficient, but the time chart form, even though it takes a little longer to prepare, is usually more productive. In the present case, the time chart shown in Figure 17.9 was drawn next. Note that all these steps and time values may be obtained simultaneously from a motion-picture film or video tape, thus greatly reducing the analytic work, particularly with large crews (see Chapter 18).

The coding of the columns follows the standards given in Table 17-II. The length of each section of each bar is proportional to the time spent on that step of the task. As with previous charts, the columns are arranged so that a horizontal line drawn through them at any place passes through simultaneous actions of the workers and machines involved.

Criticism. Each item on the chart was criticized with the aid of a suitable checklist such as the one shown here.

CHECKLIST FOR MULTIMAN AND MULTIMAN AND MACHINE PROCESS CHARTS

Basic principles

A. Balance the work of the crew.
B. If a machine is involved, consider increasing percentage of use.
C. Ease the job of the most-loaded man.
D. Eliminate steps.
E. Combine steps.
F. Make steps as easy as possible.

1. Can any operation be eliminated?
   a. As unnecessary?
   b. By changing the order of work?
   c. By new or different equipment?
   d. By changes in the layout?
2. Can any movement be eliminated?
   a. By leaving out operations?
   b. By shifting some operations to another job into which they fit more conveniently?
   c. By changing equipment?
   d. By changing the layout?
   e. By changing the order of work?
   f. By conveyors? (Make sure they are economical.)

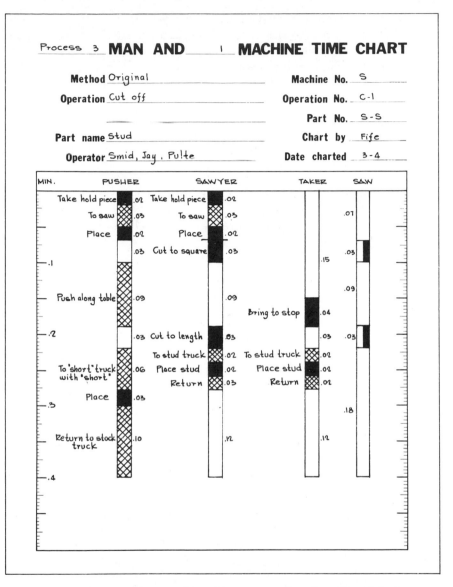

Figure 17.9. Multiman and machine process time chart for original method of three-man crew cutting 8-foot house studs from mill length 2- by 4-inch stock.

3. Can delays be eliminated?
   a. By changing the order of work?
   b. By changing the layout?
   c. By new or different equipment?

4. Can countings or inspections be eliminated?
   a. Are they really necessary? What happens after they are done and the information obtained?
   b. Do they give unnecessary duplication?
   c. Can they be performed more conveniently by another person?
   d. Are they done at the best point in the sequence?
5. Can operations be combined?
   a. By changing the order of work?
   b. With new or different equipment?
   c. By changing the layout?
6. Can movements be combined?
   a. By changing the order of work?
   b. By changing the layout?
   c. By changing the quantity handled at one time?
7. Can delays be combined?
   a. By changing the order of work?
   b. By changing the layout?
   c. If they provide rest, can they be grouped better?
8. Can countings or inspections be combined?
   a. By changing the order of work?
   b. By changing the layout?
9. Can any step be made safer?
   a. By changing the order of work?
   b. By new or different equipment?
   c. By changing the layout?
10. Can any operation be made easier?
    a. By a better tool?
    b. By changing positions of controls or tools?
    c. By using better material containers or racks, bins, or trucks?
    d. By using inertia where possible and avoiding it where worker must overcome it?
    e. By lessening visual requirements? (See Figure 16.4.)
    f. By better workplace heights?
    g. By using different muscles?
       Use the first muscle group in this list that is strong enough for the task:
       (1) Finger.
       (2) Wrist.
       (3) Elbow.
       (4) Shoulder.
       (5) Trunk.
    h. By jigs or fixtures?

11. Can any movement be made easier?
    a. By a change in layout, shortening distances?
    b. By a change in the direction of movements?
    c. By changing its place in the sequence to one where the distance that must be traveled is shorter?
12. Can any delay of one crew member, caused by another crew member, be eliminated?
    a. By changing the number of the crew?
    b. By changing the number of machines that the crew uses?
    Note: Different crew sizes and numbers of machines may create one of the following four conditions. Additional criteria of desirability may need to be stated in order to help select the "best" combination.
    (1) Reduction of operator delays to the minimum required for rest and personal time. There may be considerable machine delay.
    (2) Reduction of machine delays to the minimum required to provide the operator with rest and personal time, at which times the machine is unattended. There may be considerable other operator delay.
    (3) Reduction of machine and operator delays such that they will provide the most economical balance.
    (4) Reduction of both operator and machine delays to the minimum required to provide the operator with rest and personal time.
    c. By a redistribution of the work among the crew?
    d. By changing the order of work of the crew?

Innovation. As a result of the application of this checklist, the improved method shown in Figure 17.10 was devised and output was increased by 100 per cent. Saw usage was thus doubled. The changes in the workplace and the new pattern of movement of the crew are indicated in Figure 17.11.

Subsequent to the improvement. The changes that made this new method possible were suggested by the following questions from the checklist. The notes the analyst made for future reference are also given.

12. *Can any delay of one crew member, caused by another crew member, be eliminated?*
    c. *By a redistribution of the work among the crew*
       Note: Have the sawyer, instead of the pusher, place the shorts. Have the pusher place the stock on the table by himself. (The second suggestion would be impossible as the job was done, so further improvement was necessary.)

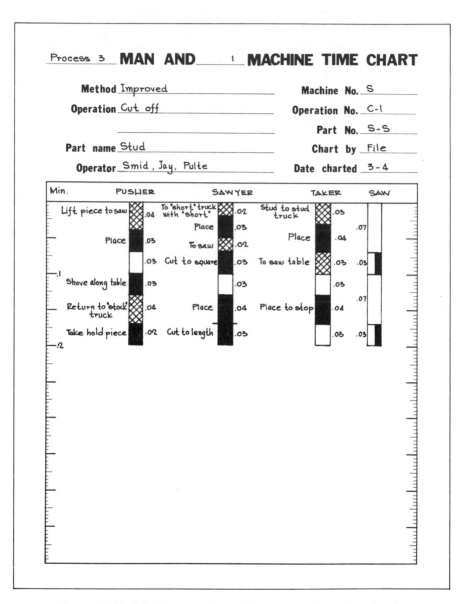

Figure 17.10. Multiman and machine process time chart for improved method of three-man crew cutting 8-foot house studs from mill length 2- by 4-inch stock.

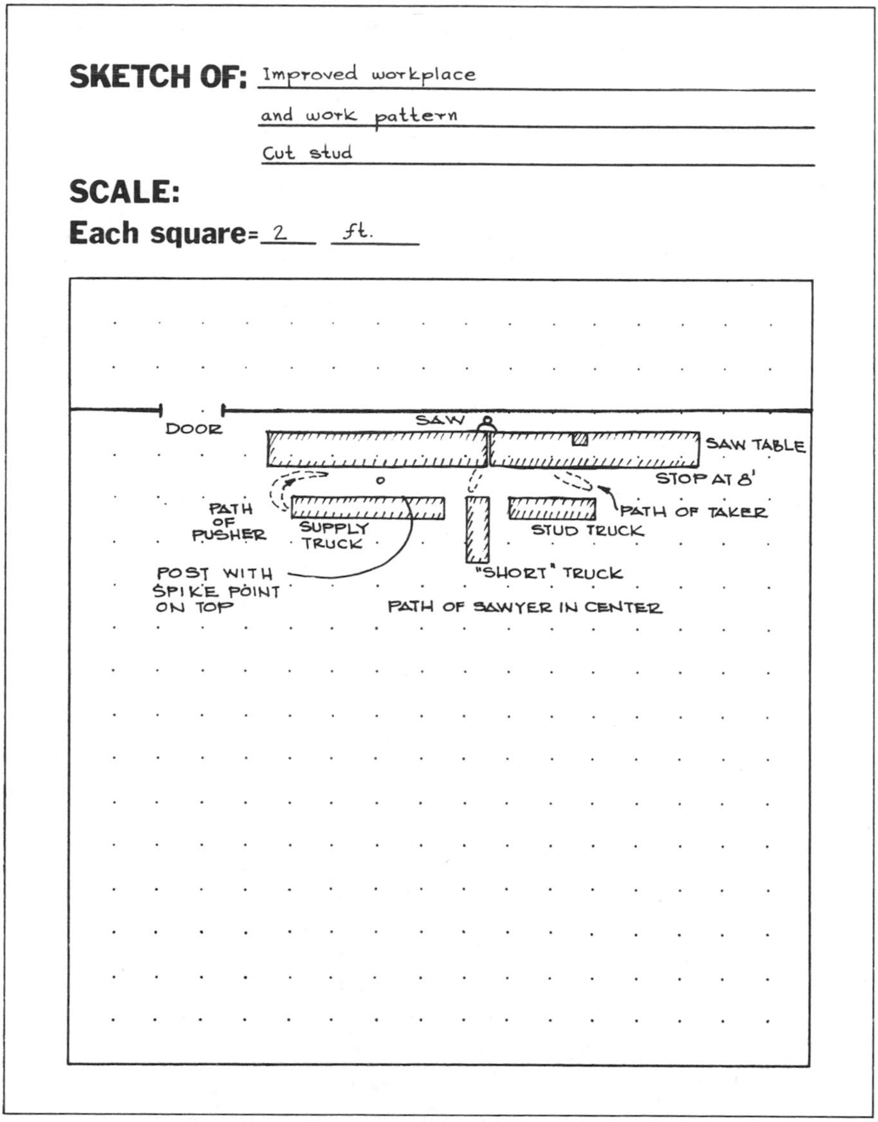

SKETCH OF: Improved workplace

and work pattern

Cut stud

SCALE:
Each square= 2 ft.

DOOR

SAW

SAW TABLE

STOP AT 8'

PATH
OF
PUSHER

SUPPLY
TRUCK

PATH OF TAKER

STUD TRUCK

POST WITH
SPIKE POINT
ON TOP

"SHORT" TRUCK

PATH OF SAWYER IN CENTER

Figure 17.11. Plan view of part of prefabricated-house plant, showing location of material and equipment and path of three-man crew cutting studs using improved method.

10. *Can any operation be made easier?*
    h. *By jigs or fixtures*
       Note: A post with a spike point on top (located as shown in Figure 17.11) makes it possible for a pusher to lift one end of the 14-foot stock, place its middle over the point, lift the opposite end onto the table by depressing and swinging his end of the board, and then lifting his end over onto the table, thus placing the stock on saw tables without help.
    d. *By using inertia where possible*
       Note: Instead of pushing material all the way down a saw table, it is possible to give it a thrust, and it will slide to its destination.

Summary—Man and Machine Charts

1. Uses:
   a. To obtain better utilization of worker or machine and better integration of the two on jobs where the person works with one or more machines and the interrelationship between the worker and machine affects productivity.
   b. To aid in the selection of a desirable person and machine work pattern in view of production requirements and cost conditions.
   c. To assist in setting up a mathematical model of the person and machine relationship for the eventual determination of the true optimum arrangement.
   d. To assist in designing a machine for operation (as contrasted with function).
   e. In the simpler forms, to aid in determining where more detailed analysis techniques may be usefully applied.
2. How made:
   a. Man and machine operation charts are constructed as follows:
      1. A right- and left-hand operation chart is constructed as detailed in Chapter 16.
      2. The machine is then charted by the use of a procedure similar to that used for placing the second hand on right- and left-hand operation charts, with the machine information being placed on the chart as shown in Figure 17.1. Usually, only two symbols are used for the activities of the machine—the symbol for suboperation when the machine is working and the delay symbol when it is idle.
      3. A summary is placed at the bottom of the chart.

b. Man and machine operation time charts are constructed as follows:
1. A man and machine operation chart is drawn.
2. By the use of the second hand of a watch, or a stopwatch, or from film or standard time tables, sample time values are obtained for each step.
3. A second chart is drawn on the form shown in Figure 17.3 or on a sheet of graph or cross-section paper, using the conventions established in Table 17-I. The length of each step on the chart is made proportional to the time involved. A convenient scale should be chosen by assigning to each line of the chart a time value such that the chart is of convenient size. The scale should be the same throughout the chart.
4. A summary may be placed at the bottom of the chart, although the length of the chart is, of itself, a rough summary.

c. Man and machine process time charts are constructed as follows:
1. A process chart—man analysis is constructed as described in Chapter 14, but the information is confined to half the chart, with the right half usually left blank.
2. The machines used are charted on the same sheet, as with man and machine operation charts, with the steps of the machines properly keyed into the process chart.
3. By the use of a watch, stopwatch, camera, or table of time values, the time is determined for each item on the chart.
4. By the use of a form similar to that used in Figure 17.6, or a sheet of graph or cross-section paper, the chart is redrawn, using the conventions of Table 17-II and making each item of a length proportional to the time involved. A convenient scale should be chosen and a time value assigned to each line of the chart so that the chart is of a convenient size. The scale should be the same throughout the chart.
5. A summary may be placed at the bottom of the chart, although the length of the chart is, of itself, a rough summary.

3. *How used:*

Each step of the method analyzed may be questioned with the aid of suitable checklists.

With any worker and machine task, the desired type of solution should be stated, so as to have a criterion on the basis of which to select a solution.

The time chart may be used to locate sections of the task that need more intensive study.

4. *What then:*

A proposed chart is either constructed or selected from the available solutions for the subsequent steps; or more detailed analyses, for further study, may be made of sections of the task.

Summary—Multiman Charts

1. *Uses:*
   a. Arranging crew work for the best balance.
   b. Estimating the effect of crews of different sizes.
   c. Designing a crew task.
   d. As a basis for instructing the crew.
   e. Locating sections of the task where a detailed right- and left-hand analysis is needed to develop an improvement.
2. *How made:*
   a. A process chart—man analysis is constructed for one of the crew members. The information is confined, however, to one side of the chart. The classification of steps, symbols, and procedure is the same as with process chart—man analysis.
   b. The other crew members are charted, one at a time, in a similar fashion on the same sheet as in step 1, care being exercised to place next to each other the symbols that indicate simultaneous activities.
   c. The machine or machines are analyzed next in a similar manner and are placed on the same chart.
   d. By the use of an ordinary watch, stopwatch, motion picture, video tape, or table of preapproximated performance times, the time is obtained for each item on the chart. If watches are used, the times will be collected from several cycles; hence, some adjusting may be necessary to obtain comparable values. If films or tapes are used, steps a through d may be carried out on the same cycle of work.
   e. By the use of a form similar to Figure 17.9 or a sheet of graph or cross-section paper, the information obtained in the previous steps is recharted, using a time scale such that the chart is of convenient length. The conventions of Table 17-II are used.
   f. A summary may be placed at the bottom of the chart, although the length of the chart is, in itself, a rough summary.
3. *How used:*
   Each step of the operation is questioned with a suitable checklist. The general principles listed at the top of the checklist are kept in mind as objectives.
4. *What then:*
   A new multiman chart is drawn for the resulting suggested method (there may be several alternative charts if effects of variations in crew size are being examined) in order to permit a final check and to provide a means of describing the proposed new method.

PROBLEMS

17.1. (a) Prepare a man and machine process time chart for centerless grind bearing (see Figure 17.3) for one operator and one machine.

    (b) Prepare a man and machine process time chart for the same job, for one operator and two machines.

    (c) If the machine-hour rate is $10 and the direct labor rate is $6 per hour, what is the cost per machine load with the two-machine method as compared with the one-machine method?

    (d) If the machine-hour rate is $5 and the direct labor rate is $8 per hour, what is the cost per machine load with the one-machine method as compared with the two-machine method?

    (e) Prepare a general equation for solving this problem with any dollar values of man and machine time.

17.2. What reactions might you expect from the operator in Problem 17.1 when his job is changed from a one-machine to a two-machine operation? What would you do if you were his supervisor?

17.3. Under what circumstance would a combination of a number of machines per operator that is not the most economical be a preferable method?

17.4. (a) Observe a task that involves some machine-controlled time and prepare a man and machine time chart of a suitable variety.

    (b) Review the operation with the proper checklist and prepare a proposed method in the form of a man and machine time chart.

    (c) List the information you would add to your checklist, and for each item indicate the checklist question number and letter under which you would file it.

17.5. (a) Prepare a multiman and machine process time chart for a two-man method of doing the operation shown in Figure 17.10. Do not exceed a class 1 change. Show the new workplace layout and flow diagram.

    (b) Compare the man-minutes per stud using this two-man method to the man-minutes per stud using the three-man method.

    (c) Under what conditions would this two-man method be preferable?

17.6. Prepare a multiman and machine process time chart for a class 2 change for the job shown in Figure 17.10. Evaluate your proposal.

17.7. (a) Observe some group work activity and prepare a suitable multiman time chart.

    (b) Evaluate the method with the aid of the checklist and suggest an improved method in the form of a comparable multiman time chart.

    (c) List the information you would add to your checklist and for each item indicate the checklist question number and letter under which you would file it.

18 PHOTOGRAPHIC AND ELECTRONIC DATA-RECORDING AIDS

Introduction

Human limitations. The human eye, ear, and hand, when used in a real-time data recording system, impose a severe restriction with respect to the number of bits of information that may be recorded per unit of time. Further, they limit the number of aspects of a situation which may be maintained under simultaneous observation. In addition, the eye–ear–hand data recording system, when extended to its limit, is subject to errors of both observation and recording.

Situations needing aids. There are many method study situations wherein some additional data-recording aid is needed in order to meet the requirements of effective analysis. Such situations are encountered when analysis requires the study in detail of all the motions of a repetitive job, the path of motions with work involving a high level of skill, the relationships among a

272

large and complex crew, or the simultaneous use of multiple communication channels in an information-flow procedure carried on in a restricted time frame.

The purpose of aids.   In situations such as have been listed, a variety of photographic and electronic data recording aids have been used. These aids assist by recording data from real-time events and holding the information in a way that permits its transcription into a form serving the purpose of analysis. These aids serve as a time buffer; they record within the time frame set by the events under scrutiny but may be transcribed, in a time frame imposed by human limitations, into a form suitable for analysis. The form of the analysis may be any one of the techniques discussed to this point or any of the approaches presented in the next chapter. The techniques of the next chapter follow this chapter because they are totally dependent upon the use of some aid; their use presumes data gathering at a level of detail beyond the capability of the unaided human data-recording system.

List of aids for data gathering.   The devices that have been found to be of assistance for gathering data are as follows:

1. Motion-picture camera with timing device:
   a. High-speed photography.
   b. Normal-speed photography.
   c. Slow-speed photography.
2. Video-tape recorder.
3. Sound-tape recorder.
4. Special-purpose still camera:
   a. Cyclegraphic photography.
   b. Chronocyclegraphic photography.

The Motion-Picture Camera and Video-Tape Recorder

Similarities.   With the advent of relatively low priced, portable video-tape recorders, the motion-picture camera and the video-tape recorder may be considered as interchangeable devices. They both employ a camera. The motion-picture camera records on photographic film; the video-tape recorder on magnetic tape. They both can record visual and auditory information simultaneously. The speed of the motion-picture camera and of the video-tape recorder can be held constant with respect to a time datum such as the frequency of the alternating current in use. The video record can be converted to a film record, and vice versa. There are, however, some small differences which affect the desirability of employing one or the other of these devices in particular situations. These differences will be noted in the few places where pertinent in the discussion that follows. In most cases, that which is given as a characteristic of motion-picture photography should be understood as applying equally to the use of the video-tape recorder.

An aid for many kinds of analysis. As has been briefly discussed, the motion-picture camera (or video-tape recorder) is merely an aid for performing the second step, analysis, of the scientific procedure. The motion-picture film provides the means of obtaining as much detail as desired and possesses greater accuracy and flexibility than the human data-recording system. However, a common misconception is to think of motion-picture-aided motion study (the use of a camera and timing device) as being only a means of studying, in great detail, the motions of a highly repetitive job performed at one location. Although it is extremely useful in such cases, camera aid should actually be thought of as an alternative means of observing and recording the work activities during any job, regardless of whether one worker, a worker and a machine, a crew, or several workers and machines are engaged, and whether the cycle is long or short. Indeed, it is usually the superior way to study long and complex cycles, particularly those with variable cycles and coordinated crew activities. The timing device may be a special time indicator included in the camera's field of view or a synchronous (constant-speed) motor for driving the camera; the video tape is taken at a known time rate.

Time and method recording. The use of a motion-picture camera and film, or video recorder and tape, provides a permanent, complete record of the job method. The timing device provides a permanent time record on the film or tape for each step in the job, and the film or tape itself provides a record of the interrelationships among the men or machines involved. It is important to realize at the outset that motion pictures, or the electronic images on the tape, are merely a successive series of still pictures and that the time lag between successive pictures can and should be varied to apply conveniently to the job being studied. No other data-gathering method is as flexible in this respect. The size of the time values involved, thich depends on the filming speed used, may be as large as or larger than those commonly associated with the breakdown on process chart—man analysis or smaller than those on the finest man and machine operation time chart. Even work sampling studies may be made with the aid of an intermittently operated motion-picture camera. Although small video-tape recorders of the type implicit in this discussion do not lend themselves to such radical changes in speed, the set used for analysis may be adjusted to present images on the screen of the image tube as if they had been taken at such slow speeds or random intervals; the set merely skips the information between the desired pictures. Video tape is not particularly applicable to work sampling.

How film or tape records are analyzed. With motion-picture- or tape-aided study, a selected portion of the data recorded (usually a typical cycle of work, although in some cases a series of cycles or a selected period of time is used) is transcribed into form for analysis and criticism, first by film or tape analysis, and second by presentation by a suitable tabular or graphic technique. Any of the job breakdowns and accompanying graphic techniques that have already been described may be used, as well as any one of the variety of techniques to be described in the next chapter.

Analysis at different levels of detail. The techniques described in the next chapter include one with a very detailed set of steps for analysis, but the very detailed breakdown is used only when the physical conditions of the job require it; it is *not* a necessary part of motion-picture or tape-aided study. With some jobs the breakdown will resemble the *work activity analysis*. Whatever type is used, the essential feature is the use of film or tape to record the data as the event or job being studied occurs, with the subsequent analysis of the recorded data being the basis of dividing the task under study into appropriate steps pertinent to the job and convenient to the objectives of the study.

Costs. Film for motion-picture camera study costs approximately $0.16 to $3.50 per minute of record taken, depending on the speed of picture taking. A video tape may cost about $2.50 per minute of recording, but the tape may be "wiped" and reused. Equipment for filming and film analysis may be assembled for approximately $1,000, although completely adequate and flexible equipment will run to about $3,000. The video-tape setup will run about $3,000 to $4,000; more if color capability is desired. However, if with the aid of this equipment costs can be reduced by only one-half of a man-year, the investment will be completely recovered. This should readily explain why the industrial use of the various forms of camera and video-tape study has been rapidly increasing.

Where filming helps most. If the work of a method study group is extensive enough to make considerable use of the equipment, the camera or video-recorder-aided technique is, in many instances, the most economical means of analyzing many jobs. For example, it would be much easier to determine the interrelationships of a three-man crew, such as those in the operation of cutting studs for prefabricated houses, described in Chapter 17, from a film or tape than from direct observation of the actual operation.

The film or tape can be stopped, run backward and forward, and repeated as often as is necessary. As the work increases in size and the task becomes more complex, the advantage of film or tape is even greater. A 40-man crew in a large foundry, which would normally have presented an almost insuperable problem, was studied by means of film with ease and accuracy.[1] The use of certain slow filming speeds has led to a much broader application of motion-picture films than may be generally realized and, as has been noted, tape may be used in a similar manner. The recording cost for the three-man stud-cutting job of Chapter 17 would have been about $0.20 (with pictures taken at 60 to 100 frames per minute as contrasted to the normal motion-picture speed of 960 frames per minute) for three cycles of work. It is worth comparing this with the probable cost of the analyst's time when obtaining the data by observation with paper, pencil, and watch. Such unusually slow picture-taking speeds, not used industrially until 1947, expanded the use of film and are referred to as "memomotion study."

Characteristics of the equipment. Modern 16-mm film equipment, the size most commonly used,[2] and video-tape equipment have both been brought to such a degree of perfection that the data recording usually requires no special preparation of the workplace and has little chance of failure. The equipment is reliable, sturdy, compact, extremely portable, and easy to use. The motion-picture-camera equipment is slightly smaller than the tape equipment.

Advantages over other data-gathering methods. Film recording and video-tape recording have the following advantages over other methods of gathering data:

1. *Permits greater detailing than eye observation.* Of course, the analyst should obtain from the record only that degree of detail best suited to his analysis requirements. However, with crew work or work with irregular cycles, the simultaneity of actions of the crew or the sequence, as it occurs, may be recorded with complete detail by the camera or tape far beyond the capacity of any observer to observe and record without aids. For instance, a film study record taken at one frame per

---

[1]Although the particular plant did not have enough work to make the purchase of filming equipment desirable, they could easily afford the analysis equipment. They hired an outside photographer to take their films, to obtain the use of superlative equipment.

[2]The introduction of Super-8 film and advances in the quality of film make it feasible to use this smaller size.

second (memomotion) of a pharmacist revealed that his prescription-filling time was consumed as follows:[3]

Description of Activity	Rank Importance	Per Cent of Time Used
Work on labels or prescription blanks	1	23.3
Work wrapping	2	10.5
Work putting material into prescription containers or with containers	3	10.1
Inspection of prescription blanks	4	7.4
Work applying labels	5	7.2
Travel to and from register	6	5.4
Work counting items	7	5.2
Travel to shelves or cupboards for material	8	5.0
Work with balance and accessories	9	4.3
Work getting down items	10	4.1
Work compounding	11	3.9
Work on drugs	12	3.3
Talking to customers	13	2.8
Work at cash register	14	2.3
Inspection of shelves	15	2.0
Inspection of drug containers or contents	16	1.3
Travel to shelves or cupboards to put away	17	1.1
Work putting up items	18	0.6
Work with liquid measures	19	0.2
Total		100.0

Such data could have been gathered without film only with difficulty, because the work pattern was highly irregular. This analysis almost automatically suggests steps that might be taken to facilitate the work. For instance, the time expended on the most time-consuming part of the work could be reduced with a special label attachment on the platen of the typewriter; the next-most-time-consuming task could be facilitated through the use of a prescription bag (introduced later), such as that shown in Figure 18.1.

2. *Provides greater accuracy than pencil, paper, and watch techniques.* The time from picture to picture is approximately 0.001 minute when

---

[3]Study made for "The Pharmaceutical Survey," of the American Council on Education.

Figure 18.1. Prescription bag used to reduce wrapping time.

an 8-mm or 16-mm camera is used at the normal speed of 16 frames per second. Smaller time intervals may be obtained by running the camera faster. Slower speeds of 100 frames per minute (0.01 minute per frame), 60 frames per minute (1 second per frame) and 50 frames per minute (0.02 minute per frame) are often extremely useful and economical.[4] Of course, even larger time units may be used if desired.

3. *Provides greater convenience.* The operation may be studied after a short run or after an experimental run. The film or tape can be stopped at will when being studied, so that each phase of the operation can be studied without bothering the operators. Even with such irregular jobs as building custom truck bodies, a single unit could be usefully studied

---

[4]Speeds of 60 frames per minute appear to be just about right for maximum film economy and adequate detail, despite the disadvantage of using seconds instead of decimal minutes. In some industrial applications, however, the advantages of decimal minutes may outweigh the film economics, making 100 frames per minute preferable. In any case, the time value used should be selected by considering all the requirements of the problem. Video tape may be used in similar fashion.

by means of one-per-second films (or a video tape used to produce equivalent data), developing data for better methods and better cost estimates on any other order that even partially resembled the one studied. Films or tapes also permit group observation without interrupting factory routine.

For instance, a large company acquired a license to manufacture, for domestic sale, a cooking-range unit developed by a foreign firm. Video tapes were made, by the foreign firm, of all operations on the cooking unit. These tapes were sent to the licensee. Using the detailed analysis method described in Chapter 19, the methods and the worker–equipment relationships were recorded and charted. The charts and video tapes were studied, in the plant conference room, by the methods analysts, tool designers, and process designers. This group developed more productive and less costly equipment and methods than were in use in the originating plant. No travel was involved; the original operations could be observed readily when needed; every detail could be examined.

4. *Provides a true record.* The film or tape is without errors of the kind that may enter into pencil-and-paper records. This is particularly true of complex jobs such as crew tasks, where the recording of simultaneity depends upon the ability of the observer. A pencil and paper record is, at best, partly subjective, whereas a film or tape is an almost true record of what happened and is available for reexamination at any time. With larger crews, it is almost a necessity. For instance, to study a five-man extrusion-press crew in an aluminum plant, the analyst has a choice of:

   a. Observing each of the crew members in turn and attempting to make the tasks key in properly by "adjusting" the figures.
   b. Getting four other observers to help him, so that all data come from the same cycles, or
   c. Making a film or video tape of the task, studying each crew member separately on the same recorded cycles, and combining these data afterward.

The ease of procedure c and the positive nature of the record it provides should be obvious. The slower (memomotion) speeds are commonly used in such applications.

5. *Aids in developing methods men.* Training the use of film or video-tape-aided method study, particularly with the detailed breakdown, tends to develop a "motion-mindedness" that almost invariably makes a person better able to develop more effective methods. It appears that greater familiarity with the detailed movements of which jobs are com-

posed helps one gain more facility for seeing possibilities of shifting them to achieve better methods. This is extremely important.

6. *Aids in preserving a record of the method.* Films or video tapes (or films made from tapes) are useful for preserving adequate records of job methods of intermittent jobs, for training, and for certain phases of time study work.

Time lag between data recording and use. The main disadvantage of film study is the time lag between the taking of the data and its transcription into usable form. However, with developing facilities available in many cities to fill the needs of television news films, films can usually be made available for study on either the same day or the day after taking. Video tapes, of course, are usable immediately after taking.

Relative cost compared with visual methods. Although the cost of film is listed as a disadvantage as compared with visual methods, there is in many cases an actual saving in the total cost of analysis, because of the reduced study time made possible by using film or tape for recording. It has been the experience of many plants that the recording equipment often more than pays for itself on the first job on which it is used.

The three steps in using film or video tape. Film- or video-tape-aided study involves three separate phases of activity for performing step 2, analysis, of the scientific method.

1. Recording.
2. Transcribing (called *film analysis* or *tape analysis*).
3. Graphic presentation of the method for study, criticism, and improvement.

Sound-Tape Recorder

Types of work studied. Reference was made in Chapter 15 to a special set of steps that might be recognized in an analysis of a mixed-means communciation system. Such systems arise in the operation of an aircraft, the aircraft–control tower relationship, a missile launch, and so forth. The recording of data in a form suitable for real analysis is beyond the limits of the human data-recording system. However, a multichannel tape recorder (or a number of tape recorders, one on each channel, cross-cued for a time datum base) may be employed. These tapes may later be transcribed in the same manner as dictation records, and the time obtained by one of two methods. For one method, the tape may be replayed, after being transcribed, and the sequential times for each sequential transmission may be determined by a stopwatch. As an alternative, for somewhat greater accuracy, the tape may be played to an oscillograph and the length of the oscillograph tape measured later to obtain both the time of the inception of the transmission and its duration. Various types of time charts may be used to present such data for analysis.

A routine use. The U.S. Federal Aviation Authority, as a routine procedure, tapes communications between aircraft and control towers so as to provide an aid in analyzing any undesirable events that may transpire. The aim is to assist in designing remedial measures.

Other uses. Sound tapes may also be used to record visually observed data; the use of tape decreases the limitations of the eye–ear–hand recording system. In many cases such data may be presented for analytical consideration by any one of the analysis techniques described in preceding chapters.

Special-Purpose Still Camera

Some tasks performed at one location are of such a nature that details of the motion path are of primary importance and concern. This is particularly true of short-cycle skilled operations for which the operator must be taught a motion path as well as a sequence of motions. It is also true of motions in space, such as folding flatwork in a laundry, or parts of jobs containing similar motions. For the study of such tasks, F.B. Gilbreth developed two techniques, cyclegraphic and chronocyclegraphic analysis. He was still perfecting these when he died in 1924.[5]

Gilbreth used a stereoscopic (three-dimensional) camera so as to record fully the true motion path; subsequent analysts have, for the most part, done likewise. In some cases, however, to expedite the availability of data, a Polaroid (instant-development) camera has been used with a simple two-dimensional presentation.[6] However, converting such a camera to three-dimensional recording is relatively easy.

---

[5]The techniques reported here have not been used extensively in the United States. However, they appear to have their place of use and are described here to make available this additional approach to the determination of the "one best way." The writer is indebted for much of this material to Anne G. Shaw of Cheshire, England. Shaw, who studied under Lillian Gilbreth, has taken a leading role in applying motion study in Great Britain.

[6]G. E. Clark, "A Chronocyclegraph That Will Help You Improve Methods," *Factory*, Vol. 112, No. 5, 1954, pp. 124, 125. This article describes a photographic wedge slightly different from Connolly and Wiltshire's.

Cyclegraph Defined

A *cyclegraph* is a photographic record on still film of the path of motion of a body member taken with a light attached to the body member. A long shutter opening is usually used, equal to the time for a motion cycle.

Chronocyclegraph Defined

A *chronocyclegraph* is a similar photographic record except that the light is caused to flicker with a nonsymmetrical peak of brilliance[7] so that both speed and direction of movement may be determined.

Use made of the techniques. In the past these techniques received but little use in the United States, probably because of the scant literature available concerning them, the lack of experience with them, and the lack of simple commercial equipment. However, in England, under the direction of Anne G. Shaw, these techniques have been used extensively and appear to represent another valuable approach to the selection of a preferable way of doing a job. The photographs illustrating this technique were supplied by Shaw. She wrote concerning Figure 18.2 as follows (the job used to illustrate the technique was the collating of a report of five pages):

> When the . . . chronocyclegraph [Figure 18.2] is examined as a whole, three points are immediately obvious:
> 1. The right hand is moving each sheet a considerable distance and the spacing and shape of the spots shows that the speed of movement varies along the path followed in picking up each sheet of paper. The round, closely spaced spots indicate how the movement begins slowly as the sheet is picked up. They then spread out and elongate as the hand gains speed, contracting as it slows down to allow the left hand to take the paper. As the hand travels down to pick up the next sheet they become long and thin, showing that this is the fastest part of the movement.
> 2. The left hand moves in jerks and the congestion of spots shows that it travels much more slowly than the right hand.
> 3. Both hands travel over a considerable distance from the moment when they pick up the first sheet to the point at which they shake the sheets together before the right hand puts them down and moves back to pick up the next sheet.[8]

---

[7]The same result may be obtained by rotating a photographic wedge in front of the film, a method proposed by Shaw and developed by R. Connolly and H. C. Wiltshire at the College of Aeronautics, Cranfield, England.

[8]A. G. Shaw, *The Purpose and Practice of Motion Study*, 2nd ed. Buxton, England: Columbine Press (Publishers) Ltd., 1960, p. 106.

Figure 18.2. Chronocyclegraph of original method of collating sheets of a duplicated five-page report. (Photograph courtesy of A. G. Shaw.)

In her discussion of this chronocyclegraph,[9] Shaw points out how the motions might be shortened and the task of the left hand, which merely accompanies the right hand and acts as a holding device, eliminated, thus freeing it also for picking up sheets. She goes on to indicate that a method such as that shown in Figure 18.3 permits both hands to move more freely and productively. Note the more regular and more widely spaced dots. (The finished reports, piled directly in front of the operator, are separated from one another by causing sheet 5 to project somewhat from the other four.) In addition, Shaw points out that this method is immediately available, although a fixture might further facilitate it. She goes on to note that the conclusions (quoted previously) were drawn from a general examination of the chronocyclegraph without counting the spots; further, that in a simple task such as this one, an experienced analyst would probably not need a record of this type. She adds, however, that in many cases a proposed method may be conveniently tested with this technique and additional sources of savings found or operator faults detected and corrected.

---

[9]Ibid., pp. 78, 79.

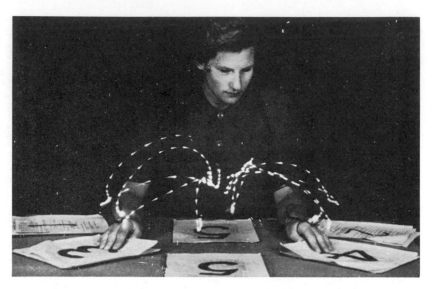

Figure 18.3. Chronocyclegraph of improved method for task of Figure 18.2. (Photograph courtesy of A. G. Shaw.)

Summary—Motion Pictures, Video Tapes, Sound Tapes, Special Still Cameras

1. Uses:
   a. To collect method and time data when the situation presents more information than the analyst can observe and record.
   b. To transmit, in almost primary form, methods and time data.

2. How made:
   A photographic or electronic record is made, as may be appropriate.

3. How used:
   a. The recording is made.
   b. The data are taken from the record by reviewing it, in an appropriate manner, such that it is feasible to transcribe all desired detail.
   c. The data are presented for study in the form of a suitable chart or table.

4. What then:
   The usual steps of the logical method, subsequent to analysis, are employed.

PROBLEMS

18.1. Examine the cameras and projectors available in the laboratory:

    (a) Prepare a checklist of desirable camera features for film study and using this as a basis compare the cameras.

    (b) Prepare a checklist of desirable projector features for film analysis and using this as a basis compare the projectors.

18.2. Assuming the use of Eastman Tri-X film, determine with the aid of an exposure calculator the lens speed required to take pictures without special or extra artificial illumination at 16 per second ($\frac{1}{30}$-second exposure), 8 per second ($\frac{1}{15}$-second exposure), and 1 per second ($\frac{1}{12}$-second exposure and $\frac{1}{8}$-second exposure) in the following places:

    (a) Classroom.

    (b) Laboratory.

    (c) Various parts of the shop.

18.3. Prepare a budget for equipping your industrial engineering department for film or video-tape study. Assuming a 10-year life for equipment, compute the annual cost. Find the average direct labor rate of pay, and compute the number of hours of direct labor that must be saved per year to pay for the equipment.

18.4. From the literature, report on the use of filming or video taping to study an actual problem.

19

MICROMOTION AND MEMOMOTION ANALYSIS

Introduction

The data recorded on a motion-picture film or video tape may be transcribed into a form suitable for presentation by any of the graphic techniques presented in previous chapters. However, it may be preferable to take advantage of the peculiar characteristics of the data record and transcribe a greater variety of facts, or greater detail, if doing so will provide additional assistance in the development of a better method. One of the advantages of a tape or film record is that a tentative transcription of the data may be made from the film or tape. If the details of the transcription fail to suggest alternatives leading to improvement, then other transcriptions at other levels of detail or with other types of details may be made; the raw data remain on the film or tape for reuse as often as necessary.

Two basic methods of analysis. Two basic methods of film or tape analysis have evolved. One is a detailed breakdown called micromotion analysis; the other is a "family" of grosser breakdowns called memomotion analysis.

These two basic methods of analysis have important similarities and important differences. To highlight these aspects in this chapter the basic material with respect to the both methods is presented first under appropriate general heads. The details, with respect to each technique, are separated under these headings. Subsequently, the details of use of each technique will be given independent treatment.

Definitions

Micromotion analysis defined. Micromotion analysis is a detailed recording of the motions involved in performing a job. The activities of the hands (and other body members as may be desirable) are usually recorded in terms of 17 separate categories. These categories are called *therbligs* (Gilbreth backward), after Frank B. Gilbreth, pioneer in motion and time study, who, together with his wife Lillian M. Gilbreth, first identified these categories and developed the film-analysis technique.[1] Originally, 17 were identified and another was added later, but one original category, *find*, has been dropped from the list given later in this chapter, since it never occurs by itself, has no duration, and is really the end point of *search*, which is included in the 17 given. These 17 therbligs are common to all human activity and provide a most convenient set of categories for the classification of all physical acts and for use as a framework for the classification of basic ways of improving these acts.

Memomotion analysis defined. Memomotion analysis is the name given to the analysis of the special forms of film or video-tape study in which pictures are taken at unusually slow speeds. (With tape, as was noted, the shift to the equivalent of slow speed is obtained by manipulation of the analysis set.) Sixty frames per minute (one per second) and 100 frames per minute are the speeds most commonly used. Like all film or tape study, it is primarily another means of performing the second step of the logical method, analysis, with man activity, and requires three phases: filming, film analysis, and graphic presentation.

---

[1]Edna Yost, *Frank and Lillian Gilbreth, Partners for Life.* New Brunswick, N.J.: Rutgers University Press, 1949.

Where the Techniques Are Used

Micromotion analysis. The detailed breakdown of micromotion analysis, supported by film or tape study, is designed principally for routine application to short-cycle jobs which involve mainly hand motions. Cases will arise in which the motions of other body members (the leg, trunk, or eyes) are also worth analyzing. However, an analysis of the activities of the two hands is the most common case. When other body members are also involved, the procedure is still similar.

Long-cycle jobs should be analyzed first with one of the grosser techniques until these procedures have provided all the improvements they can suggest, since a detailed micromotion analysis of the whole original long cycle often provides an overwhelming amount of detail. However, in competent hands a large number of valuable suggestions can be developed, and a detailed analysis is often exceedingly effective. Suggestions for a whole series of jobs may come from detailed analysis of one job. The selection of this analysis technique requires discretion and careful weighing of the time at the analyst's disposal for profitable application to the job. However, as will be explained later, familiarity with this type of analysis has several by-products of extreme importance that make familiarity with this technique almost a necessity for fully effective method and time study work.

Micromotion study with the detailed breakdown is also particularly useful in reorganizing jobs to fit handicapped workers. Here the aim is not only to improve the whole method, but also to locate and eliminate required movements of crippled or missing body members or joints. Conversely, for therapeutic work, the aim may be to increase the movements involving affected body members or joints. Since micromotion study with the therblig breakdown analyzes the task in terms of body movements, it is a highly effective tool for both of these special uses.[2]

Memomotion analysis. The primary field of use of memomotion analysis is with any of, or any combination of, work with long cycles, irregular cycles, or coordinated crews and for long-period studies. Memomotion study may also be used to examine the flow of material or the use of materials handling equipment in an area, or to study simultaneously the human work, equipment usage, and flow of material.[3] In such cases, if a time-

---

[2]The Gilbreths also pioneered in these applications. For a historical perspective, see F. B. Gilbreth and L. M. Gilbreth, *Motion Study for the Handicapped.* London: George Routledge and Sons Ltd., 1920.

[3]See W. J. Richardson, "Memomotion and Fork Truck Time Standards," *Modern Materials Handling*, Vol. 8, No. 4, 1953, pp. 67–70; B. F. Coggan, "Why Not Try Area-Wide Camera Studies?" *Modern Materials Handling*, Vol. 8, No. 12, 1953, pp. 74–76.

shortened visual presentation is desired, film must be used or a special film made from the tape record. In this way, 1 hour of activity can be viewed in about 4 minutes. The information contained on the film may be analyzed in numerous ways, and alternative presentations of the data in graphic form are possible, depending on the objectives of the study. The development of an adequate method of analysis will be discussed later in this chapter.

Advantages of Using the Techniques

Micromotion analysis.   Micromotion study is an analysis technique that assists in examining the small details of manual work. In addition to both this basic use and the basic advantages of film records (as described in Chapter 18), there are other, indirect advantages. Familiarity with therbligs and their characteristics facilitates three other aspects of motion and time study work:

1. Therbligs may be used to increase clarity in writing job descriptions for time studies or training material. They provide a descriptive terminology for jobs in terms of what the worker must do, rather than describing only what happens to the material. In addition, as will be seen in later chapters, the time for many tasks may be approximated prior to the inception of the task from tables of therblig times. Designing a job in terms of therbligs (when such detail is appropriate) facilitates dimensioning it in terms of time.

2. Since all manual jobs consist of various combinations and amounts of therbligs, they provide a convenient and extremely effective framework for the thinking of the method analyst or time study man. They are sufficiently small so that information gained in improving the therbligs on one job is frequently directly applicable to another job to a greater extent than with any of the breakdowns of the analysis techniques previously discussed.

3. After a moderate amount of training with film or tape analysis, the analyst is usually able to perceive a job (where such detail is appropriate) in terms of therbligs and can transfer his knowledge of ways of improving the performing of therbligs to the job being studied visually with significant results.

Any one of these three aspects would be sufficient reason for becoming familiar with the technique of micromotion study and therbligs.

Memomotion analysis. Memomotion, in addition to the basic advantages obtained with the use of film or video tape, has three unique advantages over other analysis techniques.

1. It will assist in recording interrelated events more accurately than visual techniques. In addition, it facilitates studying tasks that consist of irregular sequences of events that cannot be predicted in advance and for which records of both method and time are desired.

2. It reduces film cost to about 6 per cent of the cost with micromotion study film speeds. The amount of film to be analyzed also is reduced, without reducing the period covered.

3. When film is used, it permits rapid visual review of an extended period of performance. When a film taken at one frame per second is projected at the normal speed of 16 frames per second it permits viewing a film of an hour of operation in 4 minutes. In addition to saving time, viewing with a compressed time scale frequently brings to light novel aspects of the subject being studied which are often instrumental in developing new ideas for better methods.

The Use of Micromotion Study

Introduction. The details of use will be presented with the aid of an example. To overcome the difficulty usually encountered in perceiving small parts in figures in books, a job has been selected with readily visible parts. The operation is from a utility company's pole transformer rebuild shop. The job is rolling 6- by 14-inch sheets of insulating paper into a sleeve and glueing the final flap of the sleeve with a cement that is not affected by the oil with which the transformer container is later filled. The sleeves are used to insulate the leads from the transformer windings to the insulated terminals on the top of the transformer can.

A sketch of the workplace, the sleeve, and the location of the camera during filming are shown in Figure 19.1.

Gathering the data. A film or video-tape record is made of several properly performed cycles of the operation under scrutiny. Care should be taken to be certain that the camera angle is such that all pertinent motions are visible. Available light should be used; the purpose of filming is to gather data; the filming is not an artistic effort.

It is good practice to notify the worker of the filming plans a day before the filming; the worker may want to wear other than normal attire. Further, the worker may prefer that the picture not include his face. If the head and eye motions are not important, one should readily comply with such a request.

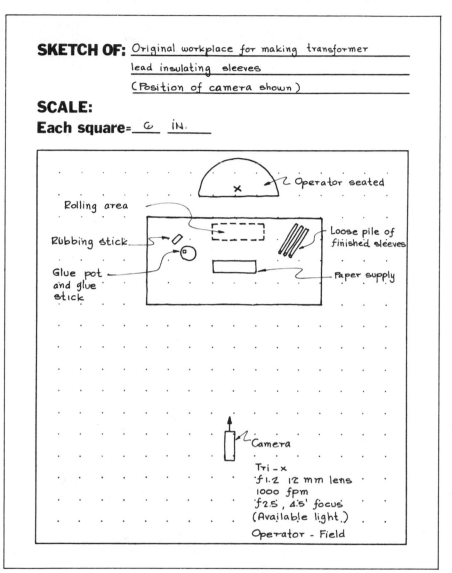

Figure 19.1. Workplace for making insulating sleeve in transformer rebuild shop. The location of the camera used to film the operation is also shown.

The steps separated with micromotion analysis. As was noted earlier, the usual analysis is in terms of 17 categories of motion called therbligs. The usual graphic means of presenting the data obtained from the film includes a graphic display of time, as with multiple-activity time charts. The therblig time chart is called a *simo-chart*.[4] With 17 categories of motion to be differentiated, black-and-white shadings of the sort used with the multiple-activity charts would be confusing. Hence, colors are used. Names, definitions, letter symbols, and colors used to represent the therbligs on the simo-chart are given in Table 19-I. For ease of recognition, each therblig is defined by its beginning point, its content, and its end. The therbligs are grouped by colors in such a manner that even a quick glance at a simo-chart will give an idea of the overall effectiveness of the method. There are five color groups, and the order of appearance in Table 19-I roughly indicates the order of desirability of the groups.

To facilitate an examination of this technique, part of the film of the task of making insulating sleeves is reproduced, frame by frame, in Figures 19.2a through d. Only enough of the film appears in the chapter to permit the reader to check a few steps of the analysis. Far too many pages would be required to present a whole cycle. In addition, the presentation would become repetitious.

A large special clock[5] is commonly included in the picture taken. The clock's large hand rotates 20 times per minute; the clock face has 100 divisions. In that the usual filming speeds are either 960 or 1,000 frames per minute, the faster clock scale (2,000 units per minute) assures that each frame will contain a readily identified, separate clock value. In the film shown in Figures 19.2a through d, the camera was driven at 1,000 frames per minute by a constant-speed electric motor. No clock was used. The film was analyzed using a projector with a frame counter. To assist in analyzing the section of film shown, the frame counter readings have been added to each of the pictures.

Transcribing the data from the film. The analysis of the part of the film shown appears in Figure 19.3. (A ruled columnar sheet of paper could have been used if a form was not readily available.) It is suggested that the reader check each step of the analysis given in Figure 19.3 against the pictures of the film, for the short section of film which is shown, using the procedure that follows.

In making a film analysis with the therblig breakdown, it is usually desirable to use the following steps:

1. *Identify the data.* The data at the top of the sheet are obtained from the film or tape data sheet, so that the information is properly identified for future reference.

---

[4]Gilbreth's short name for *simultaneous motion cycle charts*.
[5]Called a microchronometer.

2. *View the film*. The entire film or tape is viewed, and a typical whole cycle is selected from those available on the film.[6] A cycle is considered to be the complete series of motions required to bring a unit of output to the degree of completion characteristic of the operation. As with the previous techniques, it is usually most convenient to select a cycle starting with the first motion connected with the production of a unit and ending when the same motion is repeated with the next unit.

3. *Transcribe the data*. The actual recording of the therbligs usually begins with the busiest body member, following it entirely through the cycle. The analyst then returns the film to the starting place, repeating the procedure for each body member being analyzed. The film or tape is examined frame by frame.[7] The beginning time (clock or projector counter reading for the first frame or picture in which the therblig appears), the letter symbol, and the explanation for the first therblig of the cycle are noted in the proper columns of the first line of the analysis sheet. The second notation for the cycle is made for the first frame in which the next therblig appears. The explanation should assume that the therblig is the verb, and should tell what and where the action is. Actual therblig times may be obtained later by successive subtractions. For therbligs such as *RL*, which can take place in the interval between two pictures and thus not appear on the film at all, the time interval itself is usually arbitrarily assigned (e.g., at 16 frames per minute, approximately 0.001 minute). On the form given, space is provided for the two hands, and the column headed "Notes" is for the analysis of any other body member (eyes, feet using pedals, and so forth). This is the most common type of analysis. Special forms may be designed for more complex analyses.

Some analysts prefer to construct the simo-chart directly from the film or tape without recourse to the analysis sheet, but most people will find it more convenient to use this intermediate step. Some, after making the analysis, have a clerk construct the simo-chart.

Making a graphic presentation. It is regretted that printing economies do not allow the presentation of the simo-chart in color. The reader will find it worthwhile to color the chart while studying it, to see what it actually looks like.

---

[6]In some cases, a considerable number of cycles or all of the film covering a selected period of time may be analyzed, depending on the nature of the task and the objectives of the study.

[7]When using tape a convenient interval between "stop-motion and hold" points must be selected.

Table 19-I. THERBLIG DEFINITIONS AND SYMBOLS

Color Group and General Characteristics	Therblig	Symbol	Color	Eagle Pencil*	Dixon Thinex Pencil	Definition
	Grasp	G	Lake red	744	369	Begins when hand or body member touches an object. Consists of gaining control of an object. Ends when control is gained.
	Position	P	Blue	741	376	Begins when hand or body member causes part to begin to line up or locate. Consists of hand or body member causing part to line up, orient, or change position. Ends when body member has part lined up.
	Pre-position	PP	Sky blue	$740\frac{1}{2}$	418	Same as position except used when line up is previous to use of part or tool in another place.
Red-blue— terminal therbligs	Use	U	Purple	$742\frac{1}{2}$	396	Begins when hand or body member actually begins to manipulate tool or control. Consists of applying tool or manipulating control. Ends when hand or body member ceases manipulating tool or control.
	Assemble	A	Heavy violet	742	377	Begins when the hand or body member causes parts to begin to go together. Consists of actual assembly of parts. Ends when hand or body member has caused parts to go together.

Table 19-I. (CONTINUED)

Color Group and General Characteristics	Therblig	Symbol	Color	Eagle Pencil*	Dixon Thinex Pencil	Definition
	Disassemble	*DA*	Light violet	742	422	Begins when hand or body member causes parts that were integral to begin to separate. Consists of taking objects apart. Ends when hand or body member has caused complete separation.
	Release load	*RL*	Carmine red	745	383	Begins when hand or body member begins to relax control of object. Consists of letting go of an object. Ends when hand or body member has lost contact with object.
	Transport empty	*TE*	Olive green	739½	391	Begins when hand or body member begins to move without load. Consists of reaching for something. Ends when hand or body member touches part or stops moving.
Green— gross movement therbligs	Transport loaded	*TL*	Grass green	738	416	Begins when hand or body member begins to move with an object. Consists of hand or body member changing location of an object. Ends when hand or body member carrying object arrives at general destination or movement ceases.

Table 19-I. (CONTINUED)

Color Group and General Characteristics	Therblig	Symbol	Color	Eagle Pencil*	Dixon Thinex Pencil	Definition
	Search	*SH*	Black	747	379	Begins when hand or body member gropes or hunts for part. Consists of attempting to find an object.
Gray-black— hesitant movement therbligs						Ends when hand or body member has found location of object.
	Select	*ST*	Light gray	734½	399	Begins when hand or body member touches several objects. Consists of locating an individual object from a group. Ends when the hand or body member has located an individual object.
	Hold	*H*	Gold ochre	735	388	Begins when movement of part of object, which hand or body member has under control, ceases. Consists of holding an object in a fixed position and location. Ends with any movement.
Yellow- orange— delay therbligs	Unavoid- able delay	*UD*	Yellow ochre	736	412	Begins when hand or body member is idle. Consists of a delay for other body member or machine when delay is part of method. Ends when the hand or body member begins any work.

Table 19-I. (CONTINUED)

Color Group and General Characteristics	Therblig	Symbol	Color	Eagle Pencil*	Dixon Thinex Pencil	Definition
	Avoidable delay	*AD*	Lemon yellow	735½	374	Begins when hand or body member deviates from standard method. Consists of some move- ment or idleness not part of method. Ends when hand or body member returns to standard routine.
	Rest for over- coming fatigue	*R*	Orange	737	372	Begins when hand or body member is idle. Consists of idleness which is part of cycle and necessary to over- come fatigue from previous work. Ends when hand or body member is able to work again.
Brown— accom- panied by thinking	Plan	*PN*	Brown	746	378	Begins when hand or body members are idle or making ran- dom movements while worker decides on course of action. Consists of determining a course of action. Ends when course of action is determined.
	Inspect	*I*	Burnt ochre	745½	398	Begins when hand or body member begins to feel or view an object. Consists of determining a quality of an object. Ends when hand or body member has felt or seen an object.

*The colors of some of these pencils vary somewhat from the standard colors. They have been selected to match the standard as closely as commercial pencil colors allow.

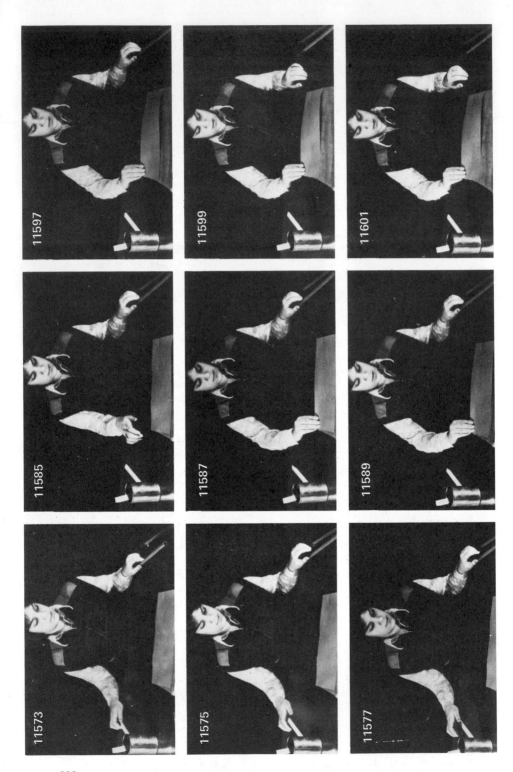

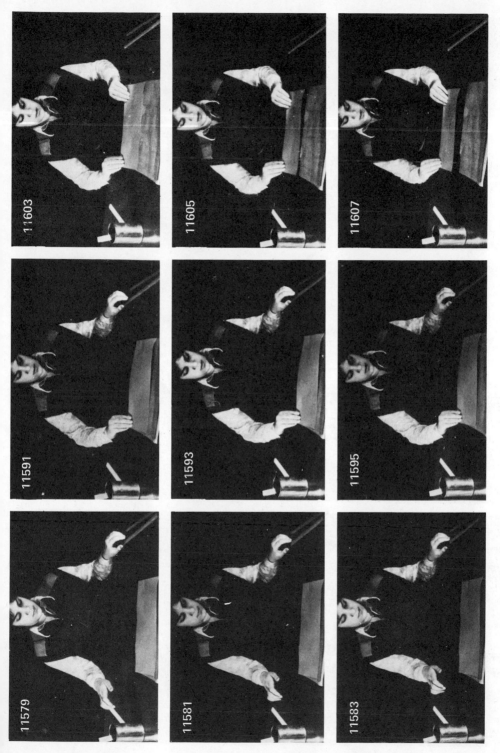

Figure 19.2a. Print of a movie film for original method of making insulating sleeve (*continues*).

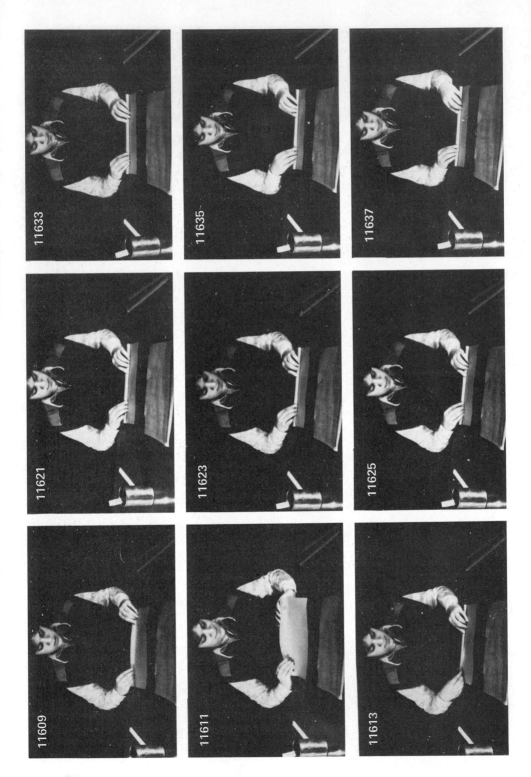

300

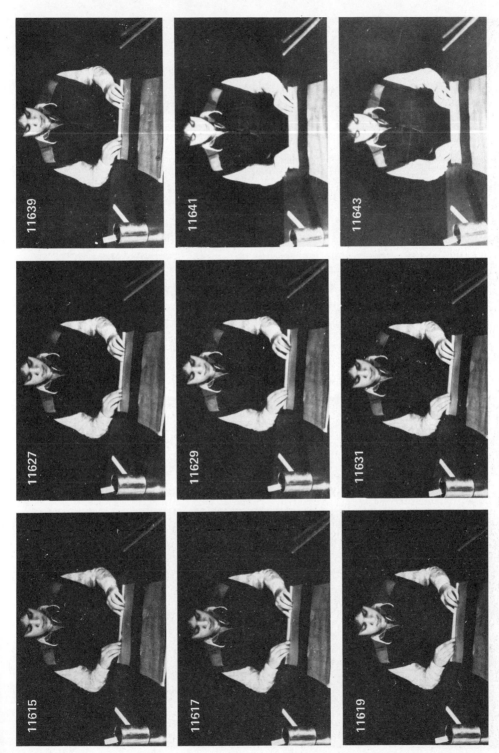

Figure 19.2b. Print of movie film for original method of making insulating sleeve (*continues*).

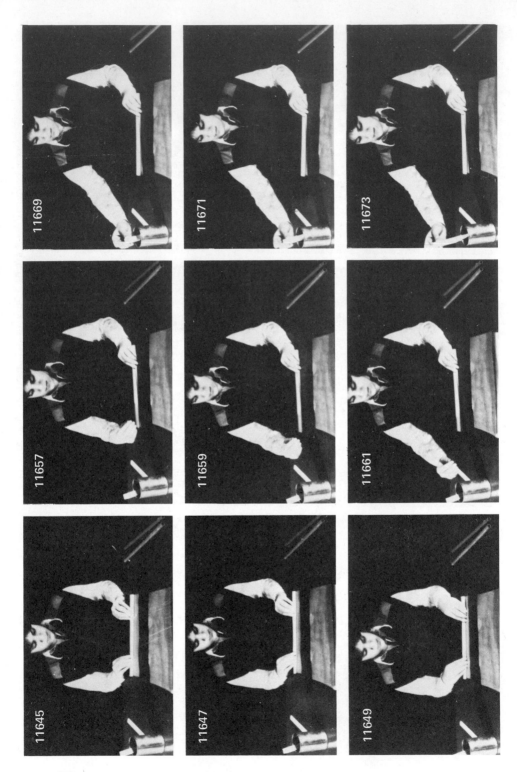

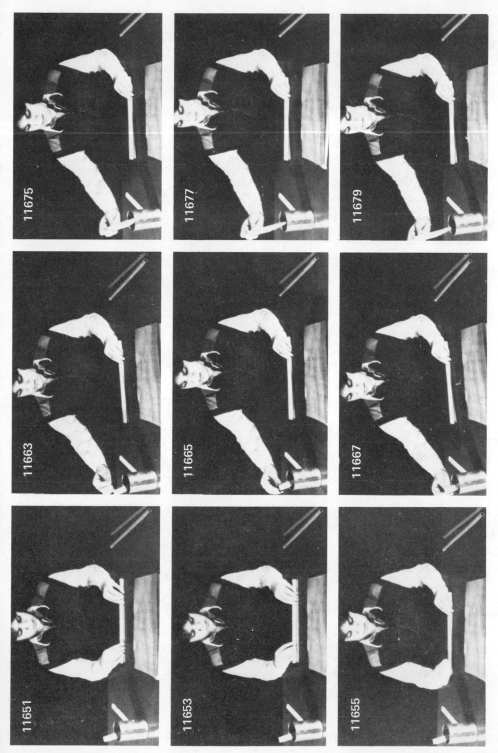

Figure 19.2c. Print of movie film for original method of making insulating sleeve (*continues*).

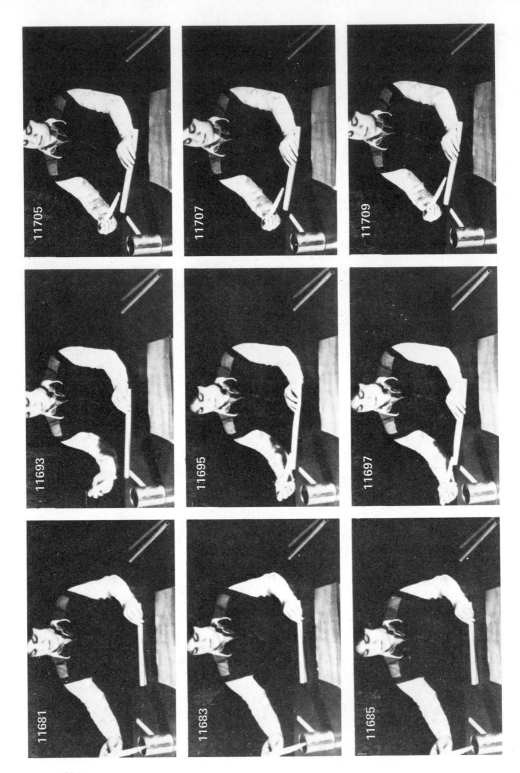

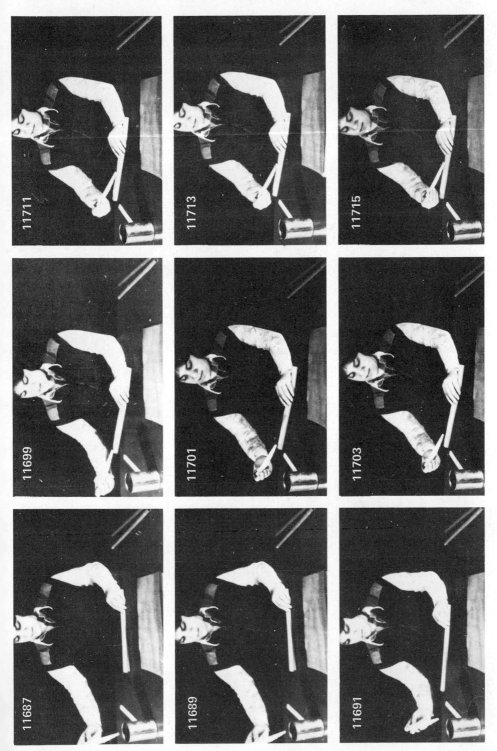

Figure 19.2d. (*Concluded*) Print of movie film for original method of making insulating sleeve.

RECORD OF FILM ANALYSIS

Film number __1217__
Date filmed __1/8__
Analysis by __JMC__
Date __1/9__

Operation __Roll__
Operator __Smith__
Part name __Tube__
Part No. __10-297__

Sheet of __1__
Dept __63__

Therblig symbol	Clock reading	Subtracted time .0005 min	LEFT HAND DESCRIPTION	RIGHT HAND DESCRIPTION	Therblig symbol	Clock reading	Subtracted time .0005 min	Body member	Therblig symbol	Clock reading	Subtracted time	NOTES
UD	11575	24	For R.H.	To paper supply	TE	11575	12					
TE	599	3	To paper in R.H.	Paper	G	587	8					
G	602	3	Paper	Paper to work area	TE	595	10					
P	605	4	For rolling	For rolling	P	605	4					
A	609	48	Roll into 3/8" tube	Into tube (roll)	A	609	45					
H	657	32	For RH	Tube in LH	RL	654	1					
TL	689	4	Toward glue	To glue stick	TE	655	12					
H	693		For glue application	Glue stick	G	667	6					
				On can to remove excess	U	673	14					
				To roll seam	TL	687	6					
				Stick to seam	P	693	2					
				Glue to 1st section	U	695	12					
				To 2nd location	TL	707	2					
				Stick to seam	P	709	2					
				Glue to 2nd section	U	711						
		118					136					

Figure 19.3. Film analysis using therblig breakdown for film in Figures 19.2a through d.

306

The colors are quickly and easily learned since they are in groups, and the colors and patterns representing undesirable and desirable job characteristics are soon recognized. In general, their desirability follows the order of their groups as given in Table 19-I. The understanding and improvement of the entire job pattern are aided considerably by the colors.

The simo-chart for the part of the film of making insulating sleeves is given in Figure 19.4. Each hand has been charted in a separate column. The two columns are so aligned that a horizontal line drawn through them at any place would indicate simultaneous action of both hands. A clock scale appears on both sides of the chart, to facilitate construction and comparison with the film, and a cumulative time scale appears in the center. The scale used on simo-charts is usually adjusted to give convenient length to the chart while avoiding undue compression of the data. The chart may, if appropriate, run to more than one page. Each space may represent any time unit, as long as the same scale is used throughout the chart. Since the film was taken at 1000 frames per minute, the basic time unit on this chart is 1/2,000 minute.[8]

In the case being discussed, the simo-chart represents the final phase of performing the *analysis*, the second step of the logical procedure.

Criticism. To perform step 3, criticism, each therblig in the job is questioned with the proper checklist questions from the checklist for therbligs, together with the eight important basic rules that appear at the head of the list, in Table 19-II. The checklist is much more detailed than any previously given, since detailed analysis permits the checking of many more possibilities of improvement.

Innovation. A suggested improved method, developed by trying to utilize as many checklist suggestions as possible for this operation, is shown in operation chart form in Figure 19.5. The chart was made to describe the method and to check the desirability of the improved method prior to its trial. With considerable micromotion experience, this may be done with a reasonable degree of reliability.

Test and trial. The proposed method was then rechecked against the checklist. Subsequently, a sample application was made.

Application. The new workplace layout that was used is shown in Figure 19.6. The new method made possible a 40 per cent increase in output.

Subsequent to improvement. The analyst will find it useful to add to the checklist from time to time as he finds suggestions that are peculiar to his industry.

---

[8]Gilbreth called the value 1/2,000 of a minute a *wink*.

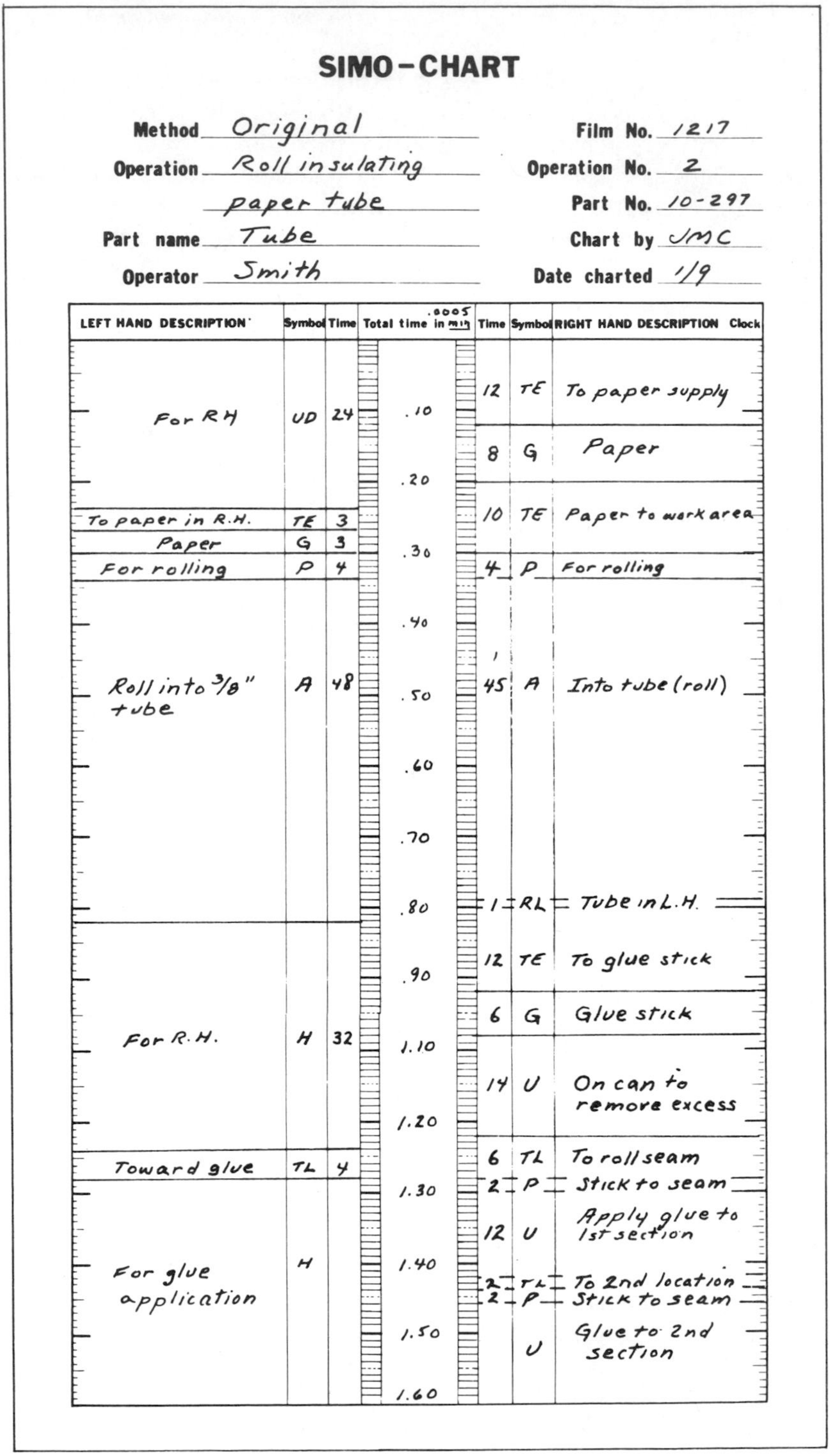

SIMO-CHART

Method _Original_ Film No. _1217_
Operation _Roll insulating_ Operation No. _2_
paper tube Part No. _10-297_
Part name _Tube_ Chart by _JMC_
Operator _Smith_ Date charted _1/9_

LEFT HAND DESCRIPTION	Symbol	Time	Total time in min (.0005)	Time	Symbol	RIGHT HAND DESCRIPTION	Clock
For RH	UD	24	.10	12	TE	To paper supply	
				8	G	Paper	
			.20	10	TE	Paper to work area	
To paper in R.H.	TE	3					
Paper	G	3	.30				
For rolling	P	4		4	P	For rolling	
			.40				
Roll into 3/8" tube	A	48	.50	45	A	Into tube (roll)	
			.60				
			.70				
			.80	1	RL	Tube in L.H.	
			.90	12	TE	To glue stick	
				6	G	Glue stick	
For R.H.	H	32	1.10	14	U	On can to remove excess	
			1.20				
Toward glue	TL	4		6	TL	To roll seam	
			1.30	2	P	Stick to seam	
				12	U	Apply glue to 1st section	
For glue application	H		1.40	2	TL	To 2nd location	
				2	P	Stick to seam	
			1.50		U	Glue to 2nd section	
			1.60				

Figure 19.4. Simo-chart for original method of making insulating sleeve.

Table 19-II. CHECKLIST FOR THERBLIGS

Basic principles

1. Try to have both hands doing the same thing at the same time or balance the work of the two hands.
2. Try to avoid the use of the hands for holding.
3. Keep the work in the normal work area. (See Figure 16.4.)
4. Relieve the hands of work whenever possible.
5. Eliminate as many therbligs or as much of a therblig as possible.
6. Arrange the therbligs in the most convenient order.
7. Combine therbligs when possible.
8. Standardize method and train worker.

			Examine		
Therblig	*Design of Product*	*Tools*	*Jigs*	*Workplace Layout and Equipment*	*Motion Pattern*
G	Easy to pick up	Combine	Easy to take parts from or self-ejecting	Ejecting bins	Avoid hand-to-hand grasp
	No hazard	Preposition	If portable, design for grasp	Lip bins	*PP* parts
		Assign place		Slide bins	Slide parts
		Design for grasp		*PP* boxes	Use bins to advantage
		In holder		No barriers to vision	Use best type of grasp
				Tool holders	
				Tweezers or tongs	

Table 19-II. (CONTINUED)

Therblig	Design of Product	Examine			Motion Pattern
		Tools	Jigs	Workplace Layout and Equipment	
P and PP	Less weight Maximum tolerances Bevel holes Round tops of pins Bevel screw ends Make parts for easy line-up Easy access Remove burrs	Self-guiding or locating Easy grip Good leverage Preposition in holders	Hold parts at convenient angles Receive parts from convenient TL path Stops, guides, funnels Maximum tolerance in jig Large locking motion Self-locating for parts	Paint for seeing Maximum PP of tools and material Arrange for easy TL to place of P and PP	Natural, free motions with accuracy supplied by stops Combine P with TL Combine several P's into one
$U, A,$ and DA	Minimum tool work Reduce screw lengths Easy to get at Combine parts Subassemble Remove burrs	Power Ratchets Combined tools PP tools Design for task Easy to use Best leverage Utilize momentum	Allow free action of tools Guide tools Bevel bushings Bullet top on locating pins Hold parts firmly Uniform type of fastening, preferably clamp levers At convenient height and angle Rotatable Few fastenings	Not in way of tools Tool holders Convenient height	Natural motions Lightest muscle group able to do job Proper leverage Proper posture Back brace on chair Combine U's and A's

Table 19-II. (CONTINUED)

				Examine		
Therblig	Design of Product	Tools	Jigs	Workplace Layout and Equipment	Motion Pattern
RL	Droppable Easy to let go of	Suspended or in PP holder at all times	Easy to fit parts into Will automatically locate parts Kick, blow, drop, slide, or spring parts out	Chutes for RL near work area or in TL path Self-counting trays	As soon as possible Foot ejector As part of TL Without P
TE and TL	Fewer parts Less weight	Within easy reach Light Easy to hold Balanced Counterbalanced Self-returning Foot control	Near parts Chutes and drops Make following P less exacting Attach levers, wheels, and wrenches	Arrange parts for natural sequence Get parts and tools close to point of use	Use smooth continuous motions, circular paths, avoid backtracking Coordinate with use of eyes Use both hands systematically Use smallest amount of body required Two or more parts at once, provided this does not interfere with subsequent P

Table 19-II. (CONTINUED)

			Examine		
Therblig	Design of Product	Tools	Jigs	Workplace Layout and Equipment	Motion Pattern
SH and ST	Standardize parts Make nontangling Color code	Not tangle with other tools Minimum number Special eyeglasses Combine Paint in contrasting color Preposition Definite location	Fixed in place Levers or wrenches attached Paint controls in contrasting color	Lip bins Definite places for tools and materials Label or color bins Bins contrast with parts Illuminate workplace Paint workplace for seeing	Use eyes to do work Use uniform motion pattern Use bins or trays of material systematically
H, UD, and AD	See basic rules. These therbligs are undesirable. Balance work with machine cycle if machine is used.				
R	If other therbligs are improved, this will be reduced to a minimum. Rest is preferably provided by a rest pause rather than as a regular element in the cycle. If it occurs as part of a machine operation, it should take place during machine running time.				
PN	See basic rules. This therblig is undesirable. Balance work with machine cycle if machine is used.				
I	Easy reference points Minimum requirements	Easy reading Go-no go Optical Rugged Combined gages	Minimum number of fastenings Uniform fastenings Light Built-in gages Easy reading	Good light, free of glare and flicker; of proper color, direction, and contrast	Fixed and definite pattern even for eyes Arrange so part is stationary when being viewed

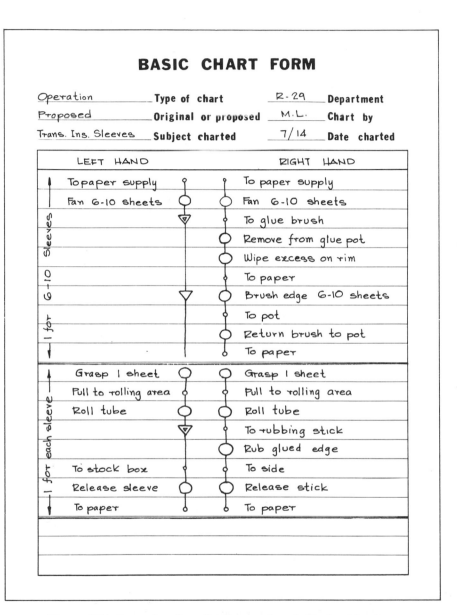

Figure 19.5. Operation chart for improved method of making insulating sleeves.

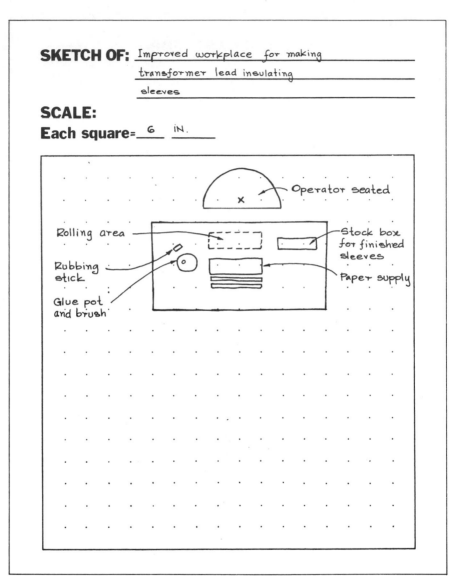

SKETCH OF: Improved workplace for making transformer lead insulating sleeves

SCALE:
Each square= 6 IN.

Operator seated

Rolling area

Rubbing stick

Glue pot and brush

Stock box for finished sleeves

Paper supply

Figure 19.6. Layout of workplace for improved method of making insulating sleeves.

Using the Checklist with Operation Charts

It is probably obvious that the checklist for the right- and left-hand operation charts may also be used with simo-charts. On the other hand, the body member activities involved in each step of a right- and left-hand operation chart are usually more clearly understood by the analyst with some micromotion training. The therblig checklist can then be profitably applied to right- and left-hand operation charts, without recourse to film analysis.

Micromotions Analysis Without Film or Tape

Right- and left-hand operation charts may also be drawn in terms of therbligs without using films by employing the symbols indicated in Table 19-III and placing the proper therblig symbol alongside each operation chart symbol on the chart.

Table 19-III. SYMBOLS FOR OPERATION CHARTS WITH THERBLIG BREAKDOWN

Symbol	A.S.M.E. Symbol	Used with Therblig
◯	◯	G, P, PP, U, A, DA, SH, ST, RL, I
○	⇨	TE, TL
▽	D	H
▽	▽	AD, UD, R, PN

However, the time values will not be available, and an accurate evaluation of the possibilities inherent in a proposed method requires much more experience on the part of the analyst than is needed with a simo-chart.[9] Making a right- and left-hand analysis in terms of therbligs from observation also requires a high degree of familiarity with micromotion and a discerning

---

[9]Time values may be supplied from one of the predetermined time systems described in a later chapter.

eye, but a well-trained analyst may readily make such an analysis and find it very useful.

The Use of Memomotion Analysis

Introduction. It should be noted that memomotion analysis takes a variety of forms, depending on the situation being studied. Hence, only a general description can be given. From study to study the details will change as compared with the relatively fixed procedure of micromotion analysis.

Memomotion analysis has been used with advantageous results to study such activities as the following[10]:

1. Gas company street work.
2. Twenty-four-man steel-casting mold line.
3. Prefabricated-house-section manufacture.
4. Railroad car humping in a classification yard.
5. Aircraft service on the ramp at a commercial airport.
6. Dry-salt meat packing line.
7. Stripping at the delivery end of a cutting press used to cut cartons from printed cardboard sheets.
8. Package handling at a packing-house sorting center prior to truck loading.
9. Two-man welding crew on water-heater assembly line.
10. Municipal garbage handling.
11. Dental activity.
12. Household activities.
13. Department store clerks.
14. Fifty-man paper-making machine repair crew.
15. Ice-house crew.
16. Railroad car loading crew.
17. Auto and passenger pattern at large airport passenger terminal entrance.

To demonstrate the similarity of the general approach employed as well as to highlight the differences from study to study, two illustrations will be used.

Case I—Replacing Household Gas Service Lines

Introduction. From its records a public utility found that the time to install a gas service renewal averaged 64 man-hours, although some crews

---

[10]The list cited is not exhaustive but merely indicates the wide range of activities to which the technique may be usefully applied. (A film showing samples of the actual

consistently took only 24 man-hours. (A nine-man crew was used on the job, although only three or four were employed on a single site at any one time.) Memomotion analysis was selected as the technique for determining the details of the 24-man-hour method; improving upon it if possible; and to serve as a basis for developing training materials to help in reducing the time taken by the average crew.

Gathering the data. A motion-picture camera operating at exactly one frame per second was used to film a whole day of the 24-man-hour method.[11] The crew completed a service renewal during this period. The field of view covered by the camera is shown in Figure 19.7. A clock was not required.

Figure 19.7. View through memomotion camera of gas service renewal crew beginning street excavations.

The steps separated in the analysis. The analysis categories were created by compounding the phases of the work (*preparing the site, opening the excavations, making the connections, closing the excavations*) with the basic man activities (*do, move, inspect, wait*) and with the tools and equipment used (*shovel, air drill, torch*, etc.). Typical categories were:

WPB   work preparing barricades for traffic
WOD   work opening excavation with air drill
MPA   move air compressor to location

---

films taken for the first nine applications in the list is available from the author.)

[11] To obtain a complete record, the workers were instructed to stop work at one blast of a whistle and resume work at two blasts. The interval in between, approximately 30 seconds, was used to change camera film magazines.

Transcribing the data from the film. Analysis proceeded in the same manner as with micromotion study, with the following differences:

1. In that no clock was used, an initial frame was marked so that the starting point was identified.

2. Columnar paper was used to record the work categories and projector frame counter readings.

3. The film was analyzed nine times. Three of the analyses followed a worker from his arrival to departure. One of the analyses was used to note the intermittent arrival and departure of the foreman, who supervised four crews. The remaining five analyses of the film were to study equipment usage at the site.

Making a graphic presentation. Multiple-activity charts, with a time scale, were made of the entire operation.

Criticism. Visual observations were made of other crews, who invariably took longer than the crew filmed and their performance was compared with the multiple-activity chart. In this way, the differences of the slower methods were detected.

Innovation. The comparative analysis revealed that the quick performance was due to the organization of the work and not to the speed of working. For instance, the foreman very carefully sequenced the three groups of men who worked on each job (excavators, connectors, hole fillers) so that no idle overlap time occurred. Also, the excavations were made as small as possible. Further, the equipment used was laid out at the site in a pattern facilitating use without undue relocation during the work cycle. In addition, numerous ways of making the task easier and less costly were found. In this respect, particular attention was paid to the most time-consuming activity categories. For instance, the cart carrying the connection tools was fitted out to hold the tools in sequence of use rather than as a jumble; a wider variety of bits were provided for the air hammers so as to get maximum efficiency with different pavings, and so forth.

A job instruction manual was produced.

Test and trial. The job instruction manual was used with the crews of a second foreman. The goal of 24 man-hours per job was reached. The instruction manual was revised after this initial test so as to remove difficulties encountered during the first use.

Results. Due to the unusually large number of service renewals scheduled for the year, the utility had been planning to contract for approximately $1,000,000 worth of work by crews from outside companies. These contracts were never let. Using the method developed from the memomotion analysis, all the work was done by the existing in-house crews.

Case II—A Study of a Dental Operatory

Introduction. A work sampling study had been made by a group of dentists of their own activity. The study indicated that the major portion of time was spent in the operatory; that the largest single category of work was restorative dentistry. A technique that preserved a record of the flow of work and gave the movements of the dentist as a whole, together with the activities of the assistant, seemed most appropriate for making a more detailed study. The variety and rapidity of the activity and the lack of a repetitive pattern made it impossible to expect to record the events directly from visual observations. Therefore, a memomotion study was made of a 1-hour period of restorative dentistry.

Gathering the data. The camera and motor drive were placed high in the corner of the operatory, as shown in Figure 19.8. The camera was started before the patient entered the operatory and allowed to run unattended.

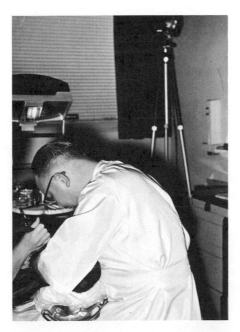

Figure 19.8. Memomotion camera unobtrusively placed in the corner of a dental operatory.

The steps separated with the analysis. A viewing of the film at normal speed compressed the 1 hour of work into 4 minutes. It was readily apparent that there was excessive movement around the operatory of both the dentist and the dental assistant. Hence, it was decided to analyze the film with respect to two categories:

1. Move from one place to another.
2. Do things at one place.

The details, of course, with respect to "moves," from where to where, were recorded. With respect to "do's," what was done was recorded in normal dental terms.

Transcribing the data from the film. The film was analyzed twice; once for the details of the dentist's activity, and a second time for the assistant.[12]

Making a graphic presentation. The path of the dentist is shown by the solid line and that of the assistant by a broken line on the plan view of the operatory in Figure 19.9.

Criticism. It should be obvious that most of the enormous amount of time and energy-consuming travel was caused by the poor placement of equipment. The operatory, laid out as was customary for one-man operation, was inefficient for a two-man team.

Innovation. The operatory was rearranged as shown in Figure 19.10. All the equipment that was worked with by the assistant or picked up and handed to the dentist was stored on the assistant's side of the operatory. The travel of the dentist and the assistant are again shown.

Test and trial. These new travel lines were taken from a second memomotion film of a case of restorative dentistry requiring almost the same amount of dental work as that in the first film.[13] The change is so great that quantification is not needed.

Summary

A summary for all photographic and electronic techniques appears at the end of Chapter 18, hence is not repeated here.

---

[12]As might be expected, in order to describe things properly, the dentist who was filmed participated in the film analysis.

[13]A film showing the "before" and "after" as memomotion films is available from the film library of the American Dental Association and is titled, "Motion Study in the Dental Operatory."

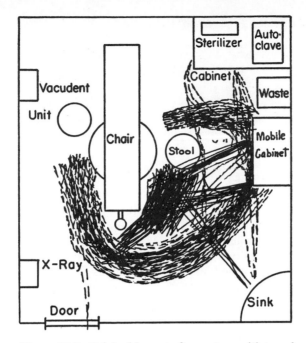

Figure 19.9. Original layout of operatory with travel paths of dentist and assistant.

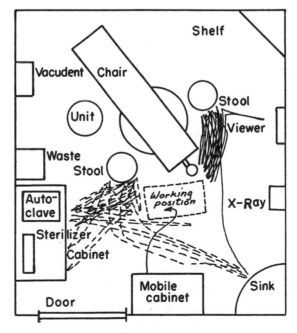

Figure 19.10. Improved layout of operatory with travel paths of dentist and assistant.

PROBLEMS

19.1. (a) Prepare a film analysis sheet in terms of therbligs, leaving out the clock readings and subtracted time, for the operation of Problem 16.2.

    (b) Prepare an operation chart in terms of these therbligs for this method of performing the operation of Problem 16.2.

    (c) With the use of the checklist for therbligs, prepare an operation chart in terms of therbligs for a better method of doing this operation as well as a sketch of the proposed workplace. Do not exceed a class 2 change.

19.2. Film for film analysis may easily be made as necessary.

    (a) Make an analysis, in terms of therbligs, of an assigned film.

    (b) Prepare a simo-chart from the data obtained in part (a).

    (c) With the aid of the checklist, prepare an improved method not exceeding a class 2 change and:

        1) Synthesize a simo-chart of this new method.

        2) Prepare a sketch of the revised workplace.

        3) Compute the estimated per cent of time saved.

        4) Compute the estimated per cent increase in production.

19.3. Devise a set of categories for analyzing films of housework with the eventual aim of making it possible for the work to be performed by:

    (a) A person with a poor heart.

    (b) A person without the use of legs.

    (c) A one-armed person.

    (d) A person of very short stature.

    (e) A person of extreme height.

19.4. Devise a set of categories for examining the work of:

    (a) A men's store clerk.

    (b) A women's store clerk.

    (c) A gas station attendant.

    (d) A house carpenter.

    (e) A plumber.

    (f) A repairman.

    (g) A tool setter.

    (h) A line crew.

    (i) A street-repair gang.

    (j) A surgeon.

    (k) A nurse.

19.5. Make a memomotion film and:

    (a) Make a suitable analysis and graphic presentation including a workplace sketch or flow diagram.

    (b) Apply the proper checklist and propose, in the form of a chart similar to that used in part (a), a better method.

    (c) Calculate the estimated per cent saving in time.

    (d) Calculate the estimated per cent increase in productivity.

part three

TECHNIQUES OF TIME STUDY

20

DIRECT TIME STUDY— INTENSIVE SAMPLING

Introduction

Definition.   *Direct time study—intensive samplings* (*DTSIS*) is a procedure in which the performance of a task is observed directly and continuously for a limited period of time. Data are recorded concerning the work time and the associated work count, together with an appraisal of the performance in comparison with the standard concept of performance. An allowance for nonwork time is usually added in conformance with policies that have been established by the organization. All these data are used to compute a standard time.

Other names for the technique.   In much of the literature the procedure is called either "stopwatch time study" or "direct time study." In that recording devices other than stopwatches are often used, the term "stopwatch time study" seems misleading. Also, there are other methods of time study in which direct observations are made, but the method of sampling is different; hence, a compound name is used here to fully identify the general technique. Direct refers to actual direct observations; time study refers to setting standard times; a small sample is observed intensively.

Alternative definition. As an alternative definition, using the terms of the equation for a standard time given in Chapter 5, direct time study—intensive sampling may be defined as a procedure wherein:

[handwritten: work count] *[handwritten: P.61]*

1. *W/C*. Obtained, usually with details concerning the method, by a direct and continuous observation of a period of performance. The period is limited and fully watched. It is an intensive sample.
2. *[handwritten: Work Time]* *W/T*. Obtained by direct observation with a stopwatch, motion-picture camera, video-tape recorder, or time study machine.
3. *[handwritten: Modifier]* *M*. An appraisal of the performance in comparison with a standard concept of performance; a *rating*; a mental evaluation made by the observer.
4. *[handwritten: Addition]* *A*. An increment added to allow for nonwork time; usually determined by policy; the value may vary with the type of work and with the purpose of the standard time; referred to as an *allowance*.

Engineered standards. The results of direct time study—intensive sampling are sometimes referred to as "engineered standards." The term "engineered standard" is used when the value for *W/T* contains only actual work time and when the values for *M* and *A* are obtained systematically and the basis of *M* is known. However, as will be seen in later chapters, there are other means of obtaining *engineered standards*. It will also be seen that there are situations that preclude the setting of engineered standards. However, in such situations, other techniques may be used to obtain standard times which satisfy all the requirements for such values. Hence, the term "engineered standard" does not in all cases denote a preferable type of value.

Types of work for which it is used. Direct time study—intensive sampling is primarily employed when the task for which a standard time is sought is repetitive. Repetitive work is work with a cyclic pattern which is repeated over an extended period of time much greater than the period required for the sample or observation period. Repetitive work is usually a third-order work-unit as defined in Chapter 9. The technique may be used when the work has a single repeated cycle, subcycles, or a limited variety of cycles.

Much industrial production work and routine office work will exhibit characteristics that allow this technique to be employed. Of course, as will be seen later, other techniques may also be used, but all techniques must be understood if the most preferable technique is to be chosen. This chapter will examine pertinent details of alternative methods of employing direct time study—intensive sampling.

Limitations on the use of the technique. The use of the technique is limited to manual, repetitive work which is actually being performed; the technique may not be used to set a standard prior to the start of work. However, a short experimental run may suffice to provide the necessary data.

If a large variety of cycles are employed, which are not repeated in a limited time, other techniques which are described later, such as *direct time study—extensive sampling*, may well be preferable.

Steps in use of technique. There are five distinct steps in the setting of a standard time by means of direct time study—intensive sampling. They are:

1. Defining the standard of measurement so as to provide a basis for determining the modifier M. This definition needs to be established only once for all studies in a plant.
2. Recording the standard practice; describing the work-units and their relationship to the next higher order of work-unit.
3. Observing and recording the work time taken by a particular operator, together with data concerning the associated work count.
4. Rating or relating performance to the standard; determining the modifier (or rating) M.
5. Application of allowances; the determination of the adjustment factor A.

Steps 2, 3, 4, and 5 must be performed for each job studied.

Step 1, the defining of the standard of measurement, is the basis (as was noted) for the determination of the modifier M. With direct time study—intensive sampling, this modifier is determined toward the end of the work measurement process. Therefore, it seems convenient to delay the discussion of step 1 until the discussion of M (the modifier or rating), at which point the significance of step 1 can more readily be seen. Hence, details of steps 2, 3, 4, 1, and 5 follow, in that order.

Recording the Standard Practice

The recording of the standard practice requires that the units in which the output is to be counted (usually the fourth-order work-unit), the method, the equipment, and the conditions of work be made a matter of record. The carrying out of this step results in what is commonly called a "written standard practice."

An implicit requirement. Any time study predicates either the previous existence of, or the creation of, a written standard practice.[1] This is one aspect of motion and time study where the motion study phase and the time study phase are hard to separate. Certainly, any plant where the production is such that formal motion study techniques would be used could hardly afford to be without formal time study.[2] Consequently, it would seem obvious that the final statement of a job method design should be put into a form suitable for use in conjunction with the subsequent time study of which it will be a necessary and integral part.

Prior to recording the standard practice. Before recording the standard practice, the time study analyst should first make the method as effective as can be conveniently accomplished. The experienced motion and time study analyst is often able to suggest several class 1 changes in a job after a simple visual inspection. A trained motion study analyst should be able to regard a job in terms of steps such as are used in constructing process chart—man analysis, operation charts, man and machine charts, and so forth. He should be sufficiently familiar with the basic principles of improvement so that desirable changes are obvious to him. Sufficient practice with the analysis techniques will develop this attitude, often referred to as "motion-mindedness." Such a cursory analysis during time study is sometimes the economical limit of analysis.

If more complex changes are possible, it may still be desirable to determine the standard time for the job, as it exists, and revise the standard when the changes are installed. In some plants, though, it may be desirable to hold off the determination of the standard time until the job is fully improved. The individual plant situation will determine which is preferable.

Developing methods at the same time as the written standard practice. It must be recognized that in many plants, because of the lack of trained personnel or a lack of realization of the possible benefits, there is no formal motion study activity, but only time study activity with informal or casual method design, which is performed by line supervisors, tool engineers, workers, and time study men. In such cases, the written standard practice must be prepared during the carrying out of the time study function. We must

---

[1] Written standard practices may exist for process or operation. Only the latter are considered in this chapter.

[2] This was the case, however, with U.S. Government Arsenals and Depots from 1917 to 1947 because of restrictions placed upon them in congressional appropriation bills. The statement in the text summarizes the testimony offered the Armed Forces Appropriations Committee of the U.S. Senate by a group of engineers interested in performing a public service. The restrictions on time study were removed in 1948. Since the removal of the restriction, much federal activity has been subjected to work measurement.

also realize that production exigencies will frequently necessitate starting a job (and perhaps even completing the run) with a method designed "on the spot," and even accompanying it with a standard time set long before any formal methods analysis for the determination of the optimum method is made. Indeed, many cases will exist where it would be uneconomical to use any but the most cursory procedures for method design, as in adapting past practice on somewhat similar operations to the requirements of the new job.

General criteria for an adequate written standard practice. Before any other phases of the measurements for standard time are made, the method, equipment, and conditions should be recorded in sufficient detail so that the job and the conditions surrounding it may be reproduced or examined at any time in the future exactly as they were at the time of study. This is the guiding principle for preparing an adequate written standard practice.

Other values of a written standard practice. An adequate record of standard practice also increases the value of the time study for the eventual determination of time values for synthesized standards based on "standard data." (This aspect will be covered in Chapter 22.) Experience has also shown that, if the standard practice is not adequately recorded, the standard time that is determined will eventually become "wrong." The actual practice is not static and, as time passes, almost invariably changes. Without an adequate description of what was timed, there is no means of evaluating a change. It should be obvious, therefore, that if the standard time is to have any real significance, it must be accompanied by a reliable record of what this time allows for or what actions are involved.

In addition, without a good description of the job, an undesirable plant attitude is developed. Standard times become associated with amounts of production rather than with amounts of physical work that are called for from the worker. This brings about resistance to changes that lead to higher productivity, even though they may call for the same amount of physical work. The cry of "speed-up" is often raised. We must separate "speed-up" into two categories: The speeding up of workers may or may not be justifiable, depending on the original pace; speeding up production without increased exertion, although it raises certain other personnel problems, is a vastly different proposition from increasing the exertion, and the two must not be confused.[3] An adequate method record helps differentiate these.

---

[3] Increased exertion may also take two forms: raising the exertion to the level called for in the definition of standard time, or exceeding this level. Although only the second variety is theoretically improper, the first also raises definite problems in the maintenance of satisfactory industrial relations.

A written standard practice is also vital in the maintenance of union relations. Clauses such as the following are frequently found in labor contracts:

> Once an incentive rate has been established and has been given a fair trial under normal operating conditions, no change in rates shall be made during the life of this Agreement, except as follows:
>
> (a) If a change of design, methods, equipment, tools, job duties, or material makes a difference in the time required to perform the job, a revision may be made in the piece rate. If the change clearly affects definite elements in the existing job, only the elements affected will be changed. However, if a job has been so completely changed that a comparison between the old job and the new job is not reasonably possible, a re-study of the entire job may be made to establish the rate.

The need for an adequate method record for proper operation within such an agreement should be readily apparent.

Photographic and electronic aids. Motion-picture film or video-recorder tape is sometimes used to make a record of the method. Either provides an extremely accurate record of that which may be seen and is often an economical aid, since it may also be used as an aid in future training of operators. It is particularly valuable on important but intermittent jobs or on jobs so complex that written descriptions would be almost impossible (i.e., boning out a veal carcass in a meat-packing plant). The photographic or tape record is frequently only a partial record. Speeds, feeds, tools, and the like must still be manually recorded.

Identifying the work-unit. The first step in recording the standard practice is to identify the work-unit or end-product of the job, usually a fourth-order work-unit.

Several alternatives are often possible, although in most cases, one is preferable. For example, in operating a screw machine, the preferable unit may be a *piece* (or some number of pieces); alternatives would be *per bar of stock* or *per foot of stock*. For the operation of washing the oil off the pieces produced by the screw machine, the preferable work-unit may be a 100-pound basket of stock; an alternative would be pieces.[4] It should be noted that in most cases the work-units are interconvertible. After taking a time study in terms of one, it may be desirable subsequently to convert to another. Whatever work-unit is selected, it should be adhered to in the preparation of the standard practice so as to produce a record that will relate to the

---

[4]Substantive outputs have been given as examples. Direct time study–intensive sampling is not usually applicable to service-type outputs.

standard time finally obtained. In general, the work-unit for direct time study should be selected so as to be:

1. Directly related to the work-input.
2. Suitable for forecasting the workload.
3. Convenient to time.
4. Easily identified.
5. Convenient for scheduling and recording production (if it is the final, reported work-unit).

Breaking the job into elements. In actually describing a job for time study, it is usually found desirable to break the job down into steps referred to as *time study elements*. These are second-order work-units. The use of an element breakdown, as will be seen subsequently, facilitates timing, comparison of the times with the times from other tasks, evaluation of the data, and later use of the times for the development of standard elemental data for synthesized standards.

The seven criteria for separating elements. The elements, which will be timed separately, should be chosen in accordance with the following requirements:

1. *Easily detected and definite end point.* This will facilitate timing because the well-defined end point is easier to note accurately. The most desirable type of end point is one which permits some means of anticipating its occurrence so that the time study observer may prepare to read his watch at the correct instant. For example, it is easier to determine the instant a lathe chuck key is laid upon a bench than the instant the worker is finished tightening the chuck preparatory to removing the key. However, a compromise is sometimes necessary when this requirement conflicts with one of the others which follow.

2. *As small as is convenient to time.* When stopwatches are used, the smallest practical unit is about 0.04 minute, or 3 seconds. Even this, however, requires skill on the part of the time study man. If motion-picture films or video tapes are taken and later analyzed, or if time study machines involved with a digital memory are used, a much smaller time unit is possible. Some hand operations involving highly repetitive performance with highly skilled operators can be much more adequately studied with these finer discriminating techniques.

3. *As unified as possible.* The element should consist of a well-unified group of motions such as reach for, take hold of an object, move it, and place it (in terms of therbligs *TE, G, TL, P, A,* and *RL*), rather than as part of a series of movements with one object and part of another series with another object. The worker usually tends to perform the

unified activity as a pattern rather than as a series of acts. Hence, the motions should be timed as a group and should not be divided at an odd place. Also, an operator may possess different amounts of the different skills required by the various steps of the job; thus, his performance on each step is best judged when separated. In addition, breaking the job down into elements involving only manipulative ability, visual perception, strength, and so forth, facilitates the subsequent job of rating.

4. *Hand work time should be separated from machine work time.* Hand time is subject to the control of the operator; machine time, with automatic feeds or fixed speeds, is not. Even where hand feeds are used, the proper time for the machine work is determinable with a long established, highly mechanistic form of experimentation. Machine time also permits a high degree of standardization and cross-checking from study to study. Hand time is much more variable and harder to determine precisely. Hence, these two types of times should always be separated.

5. *Internal work time should be separated from external work time.* Hand work done while the machine or process controls the total elapsed time (internal time) should be separated from hand work done while only the hand work controls the total elapsed time (external time).

6. *Constant elements should be separated from variable elements.* The act of starting a machine is usually independent of the piece being worked on. The act of setting aside a given type of material box is often independent of the nature of the pieces in it, within limits,[5] and is usually independent of the operation performed. Elements of this type should be kept separate from those involved in the actual handling of the piece, which will probably vary with the size, shape, and weight of the individual piece, its location, orientation, and how easily it may be handled. For instance, in small-assembly work, the getting and placing of a nut may be a function of the size, bin, destination, or nature of assembly—a variable from job to job. If the assemblies move progressively in standard boxes, the box handling may be a constant from job to job. The separation of these two types of times aids in the development of standard elemental data, and in cross-checking between studies. The standard-elemental-data aspect will be discussed more fully later in the chapter.

7. *Regular and irregular elements should be separated.* Elements that do not occur in every cycle should be kept separate to facilitate proper prorating. Often, considerable observation is required to find them, but they are a valid part of the job.

---

[5] Weight, roughness with which it may be handled, distance, and so on.

Detailed criteria for a written standard practice. The criteria of adequacy of a written standard practice are:

1. Does it contain everything the worker has to do?
2. Could the job be reproduced from it?

The following items must be included to meet these criteria. These are also the requirements of a written standard practice when it is made following the motion study phase, even if no time study is to be made. These items may also be called the requirements for recording a method design.

1. The department in which the job is.
2. Job number.
3. Work-unit identification.
4. Product, material specifications, and identification as related to the opration and work-unit.
5. Workplace layout[6] and dimensions.
6. Equipment description and its condition; if abnormal, so noted.
7. Tool descriptions.
8. Feeds and speeds of machines, welding currents used, and so forth.
9. Surrounding environmental conditions.
10. Services in the way of machine and tool maintenance and delivery and material handling rendered to, or required of, the worker.
11. A description of the actual manual details of the job.
(a) The seven requirements of good time study elements should be observed as far as possible in this description.
(b) The type of terminology used in the description of the elements varies with the nature of the job. For heavy work involving moving from place to place, a description of the activities of the person as a whole is most suitable. For heavy work with crews, the activities of the individual crew members, as well as their coordination, must be indicated. For heavy and moderately heavy work, done mainly at one place, the activity of the individual as a whole is sufficient, unless the coordination of the body members of the worker is a critical factor, in which case the activities of each body member should be detailed and the coordination indicated. The latter is the preferable procedure with light work. Where the time study observer is familiar with the therblig breakdown, he will find this a useful

---

[6] A Polaroid camera may be employed advantageously to assist in making this part of the record.

terminology to use in his descriptions, particularly with small handwork. Where the therblig time studies are shown to persons who are unfamiliar with the terminology, the use of the therbligs may be undesirable, as they will be an unintelligible shorthand and may be subject to mistrust, depending on plant attitude. However, more and more plants are wisely undertaking training programs in motion and time study for foremen and supervisors, and therbligs or equivalent first-order work-units are constantly coming into more common use. Also, they form a good shorthand to use on this record, which may be expanded for instructional material later.

Actually making the record. The written standard practice is usually checked and updated, if in existence, or made as part of the time study procedure. In that an actual worker or group of workers is being studied, the first data recorded are:

a. The name, employee number of the worker being studied, and the work location.
b. The date and time the study began. *and ended.*

Subsequently, the 11 items of the written standard practice are recorded.

Using a time study form. In some cases, a single time study form is sufficient for the recording of the standard practice. In other cases, it is necessary to attach drawings of the tools, and so forth. In still others, the written standard practice may run to several pages. No one form has yet been devised that will handle the problems of all plants satisfactorily.

The preparation of the description of the manual details of the job (item 11 of the requirements of an adequate written standard practice) may be done most easily in two steps:

1. Using a scratch pad, list a rough description of the elements and check these against the previously given requirements for time study elements, adjusting the elements as necessary.
2. Using the time study form, detail the descriptions of the elements, one by one.

As an example, let us examine a simple machine job wherein the operator inserts a piece in a machine and activates the machine control lever. This activation causes the piece to be clamped, the machine to advance to the work, to tap three holes in the piece, retreat, and release the piece clamp. The worker replaces the finished piece with another, and so forth. At the completion of every 50 pieces the worker gages the three tapped holes.

The scratch pad recording would be:

Element Number	Preliminary Description
1	Replace piece in tapping machine.
2	Automatic machine cycle.
3	Internal work; regular; put away finished piece; get new piece.
4	Internal work; irregular; gage 3 holes, 1 piece in 50. (May be too long to be all internal.)
5	Irregular; external; make adjustment to machine.

It should be noted that the separation of the elements conforms to the criteria given earlier for element separation but that the detail does not yet conform to the criteria of an adequate written standard practice. Hence, the next step is to write the details.

The detail on the first step would be as follows. [The terminal point of the element is underlined. This is a desirable way of indicating when the watch (or time study machine or film or tape) is to be read as the job is actually timed.]

Element Number	Left-Hand Description	Right-Hand Description
1	Hold new piece, insert move to control lever, and activate machine	Reach for finished piece, lift out of fixture, move to safety control lever (which makes activating lever operable) and hold
	In terms of a therblig breakdown, this would be:	
	H (new piece), TL, P, and A (in fixture), RL, TE, G (control lever), and U	TE (to finished piece) G, DA (from fixture), TL, G, and H (safety control lever which makes activating lever operable)

Examples. A complete sample time study sheet ready for time study, with the entire written standard practice included on it, is shown with an ordinary detailed breakdown in Figure 20.1, and with a detailed therblig breakdown in Figure 20.2. Note the columns for feeds and speeds as well as

TIME STUDY SHEET

SHEET __1__ OF __1__

DATE __10/30__

PART NO. _31-13642_

PART NAME _BODY FOR #4 SYLPHON ATTACHMENT_		DRG. NO. 38309

OPERATION _TAP (3) 5/16" #18 HOLES_	DEPT. 32	OPER. NO. 6

ORDER NO. 337295

MATERIAL _IRON CASTINGS_	QUANTITY 150

ATTACHMENTS

SKETCH ☐
PHOTO ☒
PHOTO IN FILE ☐
OBSERVER _R. DUCHIN_
FOREMAN _A. GAULKE_
APPROVED _L. Piel_

OPERATOR NAME _Charles Ukanov_	CLOCK NO. 2250	SEX M	TOOL NO. 5/16" #18 Tap H.S.S.	GAUGES 5/16" #18 thread plug gage
MACH. NAME _NATCO_	MACH. NO.			
STOCK ☒ TRUCK ☒ ORDER RECEIVED ☐ PAN ☐ RANDOM	STOCK ☒ TRUCK ☒ ORDER REMOVED ☐ PAN ☐ RANDOM			WORK UNIT _Body (1)_

NO.	FEED	SPEED	DESCRIPTION OF OPERATION BY ELEMENTS		NO. OF OBSERV.	BASE TIME
			L.H.	R.H.		
1			Idle, help position pc. and hold piece	Pick up piece from truck (at right), help position under spindle, pick up brush in oil can, oil taps, return brush to can while right foot moves to pedal and engages feed		
2		1196	Tap (3) 5/16" #18 holes with machine on automatic			
3			Remove piece, dispose of in truck at left	Help remove piece, idle		
4			Pick up piece from truck, hold piece, return piece to truck	Pick up plug gage from bench, gage (3) 5/16" #18 holes, return gage to bench		

REMARKS: _Element #4 too long to be internal_

TOTAL TIME ALLOWED _____

HOURS PER 100 PCS _____

PIECE PER HOUR _____

SET-UP _____

Figure 20.1. Front of time study sheet for tapping operation using ordinary terminology for describing the elements.

the special boxes for items commonly associated with the machining operation in this shop.

Figure 20.3 is part of a standard practice for a job in a steel mill. Two

TIME STUDY SHEET

PART NO. 31 -/36 42

SHEET __1__ OF __1__

DATE __10/30__

ATTACHMENTS

SKETCH ☐
PHOTO ☒
PHOTO IN FILE ☐
OBSERVER *R. DUCHIN*
FOREMAN *A. GAULKE*
APPROVED *LPiel*

PART NAME *BODY FOR #4 SYLPHON ATTACHMENT*			DRG. NO. *38309*
OPERATION *TAP (3) 5/16" #18 HOLES*			DEPT. *32* / OPER. NO. *6*
			ORDER NO. *337295*
MATERIAL *IRON CASTING*			QUANTITY *150*
OPERATOR NAME *Charles Ukanov*	CLOCK NO. *2250*	SEX *M* / TOOL NO. *5/16" #18*	GAUGES *5/16" #18*
MACH. NAME *NATCO*	MACH. NO. *363-364*	*H.S.S.*	*thrd. plug*
STOCK ☒ TRUCK ☒ ORDER / RECEIVED ☐ PAN ☐ RANDOM		STOCK ☒ TRUCK ☒ ORDER / REMOVED ☐ PAN ☐ RANDOM	WORK UNIT *Body -1*

NO.	FEED	SPEED	DESCRIPTION OF OPERATION BY ELEMENTS		NO. OF OBSERV.	BASE TIME
			L.H.	R.H.		
1			*UD, TE (to piece), G, P and H*	*TE (to piece), ST, G, P (under spindle), RL, TE (to oil brush), G, TL (to piece), U (lube holes) TL (to oil can), RL (brush) Foot TL to and engage feed*		
2		1196	*Tap 3 5/16" #18 holes with UD, TE (to piece)*	*machine on automatic UD, TE (to piece)*		
3			*G (piece) TL, P, RL in truck*	*G, TL, P, RL, UD*		
4			*TE (to pieces in truck) G, TL, H (for RH) TL, P, RL (piece in truck) (1 pc / 50)*	*TE (to plug gage) G, TL, (P, A) x 3, TL gage to bench), P, RL (on bench)*		

REMARKS: *Element 4 too long to be internal*

TOTAL TIME ALLOWED _____

HOURS PER 100 PCS _____

PIECE PER HOUR _____

SET-UP _____

Figure 20.2. Front of time study sheet for tapping operation using therbligs in the description of the elements.

separate studies are required here: one of the operator, and one of the helper. The descriptions are of such size that they are kept on sheets separate from the time study. The time study sheets merely refer to these elements by number.

| STANDARD METHOD FOR BURR ROLL – "SUPER STRIP", 2 MAN CREW | | | 1 sheet |
| 150 lb. COIL   3/4" WIDE   4 WIRE TIES | | | of 2 |

Element No.	Operator	Element No.	Helper
1	Stop machine and walk from control lever to coiler	1	Pick up 4 pieces of wire from holder, place 4 pieces of wire 90 degrees apart on top of finished coil on skid and step to coiler
2	Inspect last end of coil, pick up snips from machine framework, cut off last end of coil and lay end and snips on machine framework	2	Remove handwheel and front plate, pulls coil a little away from back plate and slip on coil clamp
3	Walk from coiler to spare reel		
4	Index swivel reel, pick up end of coil on spare reel and feed thru Burr Roller to feed rolls		
5	Walk from feed rolls to coiler		
5a	With helper lift coil off coiler and place on skid	3	With operator lift coil off coiler and place on skid
6	Walk from coil on skid to control lever	4	Remove core from coil on skid and place on coiler
7	Start machine and feed end of coil thru feed rolls and past coiler and stop machine	5	Delay
8	Walk to coiler		
9	Inspect first end of coil, pick up snips from machine framework, cut off first end of coil and lay end and snips on machine framework		
10	Walk from coiler to control lever	6	Rethread coiler, take up slack and replace front plate and handwheel
11	Delay		
12	Start machine and walk from control lever to spare reel	7	Step to coiled strip on skid
13	Unscrew reel handwheel from reel shaft and lay on floor out of way	8	Pick up one end of wire under finished coil on skid, make 1 wire tie around coil by hand, twists wire tight around coil with pliers, and bends ends flat on coil
14	Pull reel front plate from reel shaft, roll to support and lean front plate against support	9	Repeat No. 8
		10	Repeat No. 8
15	Pick up hook, pull coil about 8" off pile on skid and lay hook down	11	Repeat No. 8

Figure 20.3. Part of a method description for two-man burr roll (would be accompanied by machine drawing).

Observing and Recording the Work Time and Work Count

This section concerns the recording of time values from an actual operator and the reduction of these figures to values representative of the actual performance observed.

Equipment used for timing. Time values for a time study may be recorded in the following three ways:

1. *Stopwatch*—the most common type is the decimal minute watch shown, with the usual clipboard, in Figure 20.4. Electronic timers with digital readouts are also available to replace the watch.

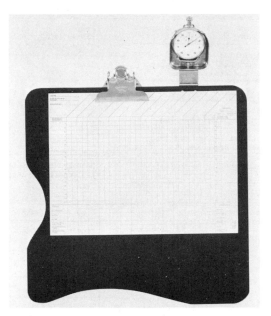

Figure 20.4. A typical time study board as used with continuous or repetitive timing. A left hand model is also available. (Courtesy Meylan Stop-Watch Company.)

2. *Motion-picture camera and film analysis*—until the introduction of available light photography and video-tape recording, the cost of using motion pictures was usually too high, but this aspect has been vastly altered as indicated in Chapter 18.

3. *Time study machine*—a small, portable solid-state memory device is available. The observations are keyed in and stored in computer language. These data can be transmitted by the device to a computer for all further processing. Such a device is shown in Figure 20.5.

Figure 20.5. Solid-state time study data collector. (Datamyte® courtesy of Electrol General Corp., Minnetonka, Minn.)

Of these three methods, the use of the stopwatch (or electronic timer) is the most common, but under certain conditions, as will be seen in a later chapter, the use of photographic or electronic recording is preferable. The basic steps are the same with all three approaches; it is merely the details that change. Hence, to reduce the size of this chapter, the details will refer only to the use of the stopwatch. The details of the other methods will appear in a later chapter.

How stopwatches are used. Three[7] commonly used methods of operating stopwatches for the taking of time studies are the following:

---

[7]A fourth method of stopwatch timing, called *cycle timing*, is frequently described in motion and time study and industrial engineering literature. This method consists of timing the elements in groups of one less than the number of elements in the cycle. It is followed by a mathematical procedure for the determination of the individual element times. This

1. Continuous timing.
2. Repetitive, or snapback, timing.
3. Accumulative timing.

1. CONTINUOUS TIMING

In continuous timing the watch runs continuously throughout the study. The watch is started at the beginning of the element of the first cycle being timed, and is not stopped until the study is completed. At the end of each element the time is recorded. The individual element times are obtained by successive subtractions after the study is completed. This is one of the most commonly used methods.

2. REPETITIVE TIMING

In repetitive, or snapback, timing, the watch is started at the beginning of the first element of the first cycle being timed, and is sumultaneously read and snapped back to zero at the completion of this, and each subsequent, element. This allows the element times to be entered directly on the time study sheet without the need for subtractions. Consistent over- and under-reading of the watch will cause cumulative errors with this method but would not affect the continuous method. Also, considerable manipulation of the watch is required. Many labor groups look upon the repetitive method as being highly liable to error. With extremely short elements, any errors that occur may represent large percentages of the elements. Some time study men use another watch to accumulate the total time so as to check these errors. However, in competent hands the repetitive method is successful enough to make it widely used.

3. ACCUMULATIVE TIMING

Accumulative timing is a method involving either two or three watches. In one method, two watches are mounted in a special holder with a mechanical linkage between the watches. For continuous timing, the linkage is manipulated so that at the end of each element one watch is stopped and

---

method involves the assumption that the average value for each element computed on the basis of a small sample of cycles will be the same as the average value for each element computed on the basis of a sample $(X - 1)$ times the size of the original sample, with X equal to the number of elements in the cycle. Chance alone, in each use of this method, would control the relative validity of this assumption and hence the accuracy of the results. The assumption would only be valid with an enormous number of readings obtained from a stable performance. In practice, such a large number of readings would not be from a stable performance, and hence, new random variations would be introduced because of variations in the work pace. An introduction into time study measurement of inaccuracies such as are caused by cycle timing is not desirable, and this method is not recommended.

the other restarted. The stopped watch is read, and element times are obtained later by subtracting alternate readings. For repetitive timing, the stopped watch is returned to zero after being read, and element times are read directly. An effective three-watch accumulative mechanism is also used and is shown in Figure 20.6. The watches used on this board function as follows. The

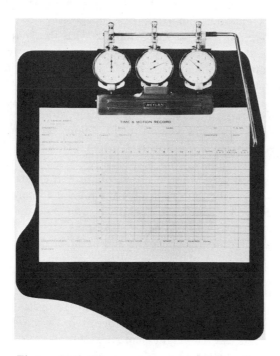

Figure 20.6. Three-watch, accumulative time study board called Quick-Click produced by the Meylan Stop-Watch Company. A left hand version is available.

first press on the crown returns the watch to zero, the second press starts it running, the third press stops it at the reading. On board in Figure 20.6 each watch is one step out of phase with the others. Each press on the lever, which presses the crowns of all three watches, stops one watch, returns one to zero, and starts another watch running. In this fashion, the time for each element may be read, without subtraction, directly from a stopped watch. For short elements the board is very useful. A fourth watch is sometimes added as a check on the total elapsed time. It should be noted that a single digital electronic timer is available which replaces all four watches while being used in the accumulative mode.

Stopwatch errors. A study of the errors in reading stopwatches, with both the continuous and the repetitive method, has been made by Irwin Lazarus.[8] He analyzed the readings made by 37 time study men familiar with the repetitive method and 13 time study men familiar with the continuous method. The standard deviation[9] of the errors, about the mean error, made with the continuous method was 0.0081 minute, and 0.0081 minute with the snapback method. The average error with the continuous method was +0.000097 minute, and −0.00082 minute with the snapback method. We may conclude that in competent hands either of these two methods is satisfactory, inasmuch as the error is not large enough to influence any of the subsequent calculations.[10]

Recording the values. For recording the time values with a stopwatch, the form shown in Figure 20.7 is convenient. Many variations of this form are used with equally good results. It is often printed on the back of a form such as was used to record the standard practice illustrated in Chapter 19. It should never be used without an adequate written standard practice. In the example shown, the time values for the machine operation, discussed earlier in this chapter, have been recorded. The continuous method of stopwatch operation was used. Note, in the small description column for each element that the observer wrote the end point of the element as an aid in timing. (This is the part of the element description that was underlined in the example given earlier in this chapter; this small reference to the element should not be confused with an adequate description.)

Space is provided on this form for recording 15 cycles of 15 elements. If more elements or cycles are to be recorded, two or more sheets may be used. If the study has less than eight elements, two sets of lines may be used, thus allowing space for 30 readings of each element. If the use of two sheets is a common occurrence, a larger form should be designed and printed. For each element of each cycle, two boxes are provided in columns labeled *R* and *T*. The *R* column is for "readings" when continuous timing is used, as

---

[8]I. P. Lazarus, "Nature of Stop-Watch Time Study Errors," *Advanced Management*, Vol. 15, No. 5, 1950, p. 15. Lazarus provided, by means of a modified telegraph code signal generator, a series of warning and reading stimuli, one series auditory and one series visual, so as to provide a controlled situation that approximated the actual time study situation. The data on which these statements are based were obtained with experienced time study men, using their own watches, in their own plants, using the method they were accustomed to, but reading visual signals from Lazarus's equipment.

[9]This is the error probably not exceeded 68 per cent of the time. Twice this value would probably not be exceeded 95 per cent of the time, and three times this value would probably not be exceeded 99.7 per cent of the time.

[10]While the difference in the average error was of statistical significance (not ascribable to chance), it was not large enough to be of practical significance. Differentiation between these two concepts of significance is vital in interpreting time study research data.

ELEMENTS — CYCLES

No.	TERMINAL POINT	1 R	1 T	2 R	2 T	3 R	3 T	4 R	4 T	5 R	5 T	6 R	6 T	7 R	7 T	8 R	8 T	9 R	9 T	10 R	10 T	11 R	11 T	12 R	12 T	13 R	13 T	14 R	14 T	15 R	15 T
1	Engage feed	20	20	62	27	98	23	92	27	85	27	13	23	62	22	2	24	42	25	79	22	28	23	54	21	92	24	30	23	69	23
2	Tap out of piece	30	10	70	8	°09	11	52	10	95	10	33	10	71	9	11	9	51	9	99	10	37	9	63	9	°1	9	40	10	79	10
3	Rl pc. in truck	35	5	75	5	15	6	58	6	°0	5	40	7	78	7	17	6	57	6	°5	6	43	6	68	5	7	6	46	6	85	6
4	Rl gage on bench																														
5																															
6	/	°8	23	49	25	89	25	29	24	67	22	X	F	°20	23	60	24	98	23	40	27	81	27	°2.0	25	97	24	38	27	73	20
7	2	18	10	58	9	98	9	39	10	76	9	92	X	29	9	69	9	°7	9	49	9	89	8	30	10	°6	9	48	10	82	9
8	3	24	6	64	6	°5	7	45	6	82	6	97	5	36	7	75	6	13	6	54	5	95	6			11	5	53	5	88	6
9	4																							°73	1:48						
10																															
11	/	°15	27	54	24	91	22	27	24	68	26	°07	23	41	20	75	18	°93	24	48	20	84	20	°18	20						
12	2	24	9	63	9	°0	9	36	9	77	9	16	9	51	10	84	9	22	9	57	9	93	9	27	9						
13	3	30	6	69	6	6	6	42	6	84	7	21	5	57	6	89	5	28	6	64	7	98	5	33	6						
14	4																														
15																															

o = New minute; began study at 2:30 P

RECAPITULATION

ELEMENTS	1	2	3	4	5	6	7	8	9	10	11	12	13	14	15
Amount of body															
Foot pedals															
Bimanualness															
Eye-hand coordination															
Handling requirements															
Weight or resistance															
TOTAL TIME IN Min.	9.58	3.81	2.41	1.43											
NUMBER OF OBS.	41	41	41	1											
PRO-RATE DIVISOR	1	1	1	50											
AVERAGE PER CYCLE	.233	.093	.059	.029											
RATING															
RATED TIME															
PER CENT OF CYCLE															
1+ADJUSTMENTS															
BASE TIME															
1+ALLOWANCES															
ALLOWED TIME															

SYMBOLS USED
C-Extra unnecessary motion
D-A.D. Dropped part
F-Unnecessary fumble
H-Unnecessary hesitation
P-Includes personal time
R-Made reject by improper work
X-Deviation from std. routine

ALLOWANCES
Personal _____ %

TOTAL LOOP VALUES		%
112	100	91
85	81	71
65	55	49
45	41	37

Figure 20.7. Continuous timing of tapping operation. (See Figures 20.1 and 20.2.)

343

in the example shown. The subtracted times, which are the time values for each element, are placed in the *T*, or "time," column. A colored pencil for this keeps the *R* and *T* values separate and facilitates correct calculations later. If repetitive timing is used, the values will be placed immediately in the *T* column, or a form with *T* columns only may be used. On some forms the letter *C* (for clock) is used instead of *R* and the letter *I* (for interval) instead of *T*. It is usual practice to time enough cycles to obtain a representative sample of performance.

Obtaining a reliable sample of time recordings. What constitutes a representative or reliable sample may be mathematically determined. It is well worth pointing out that this is not a substitute for rating but merely a means of determining if the value to which the rating is to be applied is reliable. We must realize that there is almost invariably some difference from reading to reading for any element, even if the worker is not attempting to vary his pace. This difference will be caused by the following random variations, among other causes:

1. In operator movements and pace.
2. In the positions of the parts worked with.
3. In the position of the tools used.
4. In the slight errors in watch reading.

For any observed pace of performance, timings of 10 cycles will tend to produce a more stable average than readings of five cycles; the average of 15 timings will tend to be better than that of 10, and so forth. It should also be understood that two sets of 15 readings on the same element will seldom, if ever, in practice, produce an identical average, even though the pace may be the same in both cases, and thus result in the same performance ratings. Hence, without a control on the number of readings, an error may creep into our standards. A study of this possible error shows that it frequently is of considerable magnitude.

However, too many readings would be unduly expensive. Therefore, it would be reasonable to require only enough to make the chances 95 out of 100 that we are within ±5 per cent of the true average for the element for the pace at which it was performed.[11] Some may prefer a looser criterion of 95 chances out of 100, ±10 per cent. In the latter case, the odds may be restated as 68 out of 100 being within ±5 per cent.

---

[11]The reliability of the sum of the elements will be better than the individual elements. If the time values are to be used only in a single standard, it will be less important that all elements meet this criterion. For combined tolerance see any book on statistical quality control. On the other hand, if the elemental values are to be used in constructing standard data, meeting the criterion given will simplify subsequent work with these data.

If the time studies are to be used for incentive wages, either of these criteria would seem to be reasonable minimum reliabilities, since ± 5 or 10 per cent usually approximates a bargainable increment in wages. Errors of more than this magnitude are to be avoided.

Measures have been developed by mathematical statisticians for determining the probable accuracy of a sample. These may be adapted to fit this time study problem conveniently.

Computing the sample accuracy. A relatively simple procedure, based on formula (A), which follows, is particularly suitable for application on the shop floor while the study is being made. Formula (A) gives a measure of the variability of the data about its average. The variability is represented by SD, the standard deviation, which is expressed as follows[12]:

$$\text{SD} = \frac{\bar{R}}{d_2} = \frac{\overline{H - L}}{d_2} \tag{A}$$

where

$\bar{R}$ = average range; average difference between the highest and lowest values in a series of samples of readings of correct performances of an element

d_2 = constant based on sample size (see any text on statistical quality control)

H = highest value in a sample of readings of correct performances of an element

L = lowest value in a sample of readings of correct performances of an element

$\overline{H - L}$ = average of differences between high and low values in samples of correct readings of an element

Note: $H - L = R$ for a single sample; R is an estimate of $\bar{R}$. Further, $\bar{X}$, the average, may be estimated as follows:

$$\bar{X} = \frac{\overline{H + L}}{2} \qquad \text{(for a series of samples)} \tag{B}$$

$$X = \frac{H + L}{2} \qquad \text{(for a single sample)} \tag{C}$$

If we assume that SD, as computed, may be used to represent the variability of a huge group of similar readings from the parent population (a commonly tenable assumption), another measure $\text{SD}_{\bar{x}}$, the standard error of the mean (or average) may be computed by (D). The value computed by

---

[12]The SD here is only an estimate of the sample SD, which is only an estimate of the population SD. However, as will be shown, a quick estimate has considerable utility.

(D) indicates the probable variability of the averages of groups of N values of X about the obtained $\bar{X}$.

$$\mathrm{SD}_{\bar{x}} = \frac{\bar{R}}{d_2\sqrt{N}} \tag{D}$$

Setting 5 per cent of $\bar{X}$ equal to 2 $\mathrm{SD}_{\bar{x}}$,

$$0.05\,\frac{\overline{H+L}}{2} = 2\,\frac{\overline{H-L}}{d_2\sqrt{N}}$$

Rearranging the terms so as to make N into N', a variable, so as to determine the number of required readings to meet the criterion:

$$\sqrt{N'} = \frac{4(\overline{H-L})}{0.05\,d_2(\overline{H+L})} \tag{E}$$

and N' is a function of $\overline{H-L}/\overline{H+L}$.

All expectable values of $\overline{H-L}/\overline{H+L}$ for samples of 5 and 10 readings of an element have been used with equation (E), and the answers for N' are presented in Table 20-1.

Table 20-I. NUMBER OF READINGS REQUIRED FOR ± 5 PER CENT; 95/100 PROBABILITY*

$\dfrac{H-L}{H+L}$	Data from Sample of:		$\dfrac{H-L}{H+L}$	Data from Sample of:		$\dfrac{H-L}{H+L}$	Data from Sample of:	
	5	10		5	10		5	10
0.05	3	1	0.21	52	30	0.36	154	88
0.06	4	2	0.22	57	33	0.37	162	93
0.07	6	3	0.23	63	36	0.38	171	98
0.08	8	4	0.24	68	39	0.39	180	103
0.09	10	5	0.25	74	42	0.40	190	108
0.10	12	7	0.26	80	46	0.41	200	114
0.11	14	8	0.27	86	49	0.42	210	120
0.12	17	10	0.28	93	53	0.43	220	126
0.13	20	11	0.29	100	57	0.44	230	132
0.14	23	13	0.30	107	61	0.45	240	138
0.15	27	15	0.31	114	65	0.46	250	144
0.16	30	17	0.32	121	69	0.47	262	150
0.17	34	20	0.33	129	74	0.48	273	156
0.18	38	22	0.34	137	78	0.49	285	163
0.19	43	24	0.35	145	83	0.50	296	170
0.20	47	27						

*For ± 10 per cent, 95/100 probability, divide answers by 4.

Using the table to determine needed sample size. To use Table 20-I, $H - L$ from one sample may be used as an estimate of $\overline{H - L}$ and $H + L$ from the sample used as an estimate of $\overline{H + L}$. Using this procedure, the time study observer, by inspection, takes the H and L values from either the first 5 or 10 readings of an element, as available, and computes the value $(H - L)/(H + L)$. (It is to be noted that 10 values form a good basis for using Table 20-I in order to determine the appropriate number of readings, although 10 readings are seldom an adequate sample on which to base the final average.) Entering the table with this value and going across the table to the proper column, corresponding to the sample size, the time study observer may find an estimate of the required number of readings without further computation.

An example of the use of the table. For example, let us assume the element readings were 6, 7, 6, 8, 7, 5, 6, 8, 7, 6, 7, 6, 6, and 7 (all in 0.01 minute). Taking the first 10 values, we find:

$$H = 8$$

$$L = 5$$

$$\frac{H - L}{H + L} = 0.23$$

For this value, based on a sample of 10, Table 20-I indicates 36 readings as required. (This is somewhat higher than the number that would be indicated as necessary by a more rigorous form of computation using a computed SD rather than an estimate based on the range. However, the value errs on the safe side and the number indicated is not an excessive number in most situations. The conclusion from the quick computation is that 14 readings are insufficient; they do not give a sufficiently reliable average. The study must be continued until at least 36 readings are obtained.)

As an alternative procedure, the readings may be divided into samples of 5, and $\overline{H - L}/\overline{H + L}$ used to enter the table. For the illustration given, the results would be as follows:

First 5:   6, 7, 6, 8, 7: $H - L = 2$, $H + L = 14$

Second 5:   5, 6, 8, 7, 6; $H - L = 3$, $H + L = 13$

$$\overline{H - L} = \frac{2 + 3}{2}$$

$$\overline{H + L} = \frac{14 + 13}{2}$$

$$\frac{\overline{H - L}}{\overline{H + L}} = \frac{5}{27} = 0.18$$

and $N' = 38$ (from Table 20-I samples of 5). This compares with the $N' = 36$ obtained from the use of the 10 readings as one sample of 10.

Checking the sample size later. Note that this method, based on R, is sensitive to extreme readings, particularly with short elements. Further, since the use of a single sample R as an estimate for $\bar{R}$ is only an approximation, the results have a limited validity. However, it may be used immediately, while the time study is being made, as a guide to subsequent action. In questionable cases, one may check the estimate, using a more rigorous procedure after the time study has been taken back to the office.

How to hold the clipboard and read the watch. The time study observer should have the time study form clipped to a clipboard and hold the watch in such a way that it is directly in his line of vision to the job. He should stand slightly to the side and to the rear of the operator, if possible, so that his motions do not disturb the operator (a proper position is shown in Figure 20.8). The worker should be aware that a time study is being taken. The main requirement is to hold the watch so that the hands of the watch can be kept in view while the job is being studied. The watch may then be

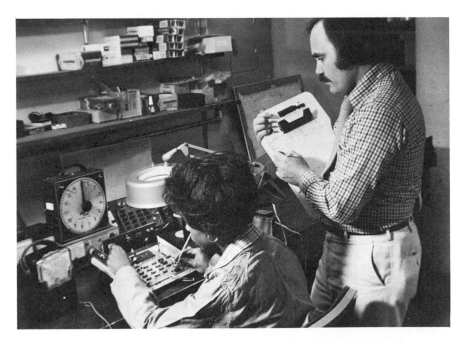

Figure 20.8. A correct position for taking a stopwatch time study. The timer on the board is a small, electronic, digital timer. Left-hand equipment is also available. (Courtesy Meylan Stop-Watch Company.)

read by refocusing the eyes without hunting for the watch and with little loss of time. It is important to watch constantly the performance of the operator to be sure that the time entries made represent valid performances of the elements as the method description defined them.

Recording unusual occurrences. Such occurrences as the following may be noted:

1. A fumble during an element.
   a. Due to lack of operator skill.
   b. Inherent in the job.
2. A false movement.
3. A personal movement, such as scratching. (Note that time will be allowed for this in the allowance for personal time, but that this is not part of the element being timed.)
4. A minor machine adjustment or repair.
5. Faulty work due to low skill or faulty movements.
6. Faulty work due to poor material.

Any element containing any such variation or any variation from the prescribed pattern of motions should be marked so as to identify it on the time study sheet. (See the notes given under the heading "Symbols Used" in Figure 20.7.)

Handling the unusual data. The unusual occurrences should be handled as follows:

1. If it is not a necessary part of the job, as the job is set up, or if it represents wrong movements or work improperly done, the recorded time value should be discarded and have no influence on the final result.

2. If it is inherent in the element (e.g., fumbles with tangled material), it should be allowed to remain in the study.

3. If it is an irregular occurrence (e.g., faulty material or a machine adjustment), it should be evaluated separately and added to the final time standard in proportion to its rate of occurrence. If, on a sewing machine operation, the thread tends to break once every 50 cycles, one fiftieth of the time required to rethread, after evaluation to standard performance level, will be added to the standard total cycle time.

Obtaining average times for elements per cycle, including pro rata effects. The actual element times for each element are computed by subtraction, if continuous timing was used, added, and the total entered in the proper total box, in the columns at the bottom of the form. The number of observations included in this total is entered in the box under the total. The relative

ADDITIONAL NUMBERS   TIME STUDY STARTED | 2:30 PM

		1	2	3	4	5	6	7	8	9	10	11	12	13	14	15	16	17	18	19	20
1 Engage feed	C																				
	I	20	27	23	27	27	23	22	24	25	22	23	21	24	23	23	23	25	25	24	22
2 Tap out of piece	C																				
	I	10	8	11	10	10	10	9	9	9	10	9	9	9	10	10	10	9	9	10	9
3 RL piece in truck	C																				
	I	5	5	6	6	5	7	7	6	6	6	6	5	6	6	6	6	6	7	6	6
4 RL gage on bench	C																				
	I																				
1	C																				
	I	F	23	24	23	27	27	25	24	27	20	21	24	22	21	26	23	20	18	24	20
2	C																				
	I	10	9	9	9	9	8	10	9	10	9	9	9	9	9	9	9	10	9	9	9
3	C																				
	I	5	7	6	6	5	6	—	5	5	6	6	6	6	6	7	5	6	5	6	7
4	C																				
	I					143															
1	C																				
	I	20	20																		
2	C																				
	I	9	9																		
3	C																				
	I	5	6																		
4	C																				
	I																				
	C																				
	I																				
	C																				
	I																				
	C																				
	I																				

RECAPITULATION

ELEMENT NO.	1	2	3	4	5	6	7	8	9	10	11	12	13	14	15	16
Amount of body																
Foot pedals																
Bimanualness																
Eye-hand coord.																
Handling Require.																
Weight																
Total time in min.	9.52	3.91	2.41	1.43												
Number of obs.	41	42	41	1												
Pro-rate divisor	1	1	1	50												
Average per cycle	.232	.093	.059	.029												
Rating																
Rated time																
Percent of cycle																
1 + Adjustments																
Base time																

SYMBOLS USED IN STUDY

Figure 20.9. Repetitive timing of tapping operation. (See Figure 20.7.)

frequency of the cycle to the occurrence of the element is entered in the box provided for this value. The arithmetic average per cycle is calculated and entered in the appropriate box. The average is the total divided by the product of the number of observations and the prorate divisor, so as to indicate the time to be charged to each cycle. The units of time used (seconds, decimal hours, or minutes) should be noted on this row. The statistical reliability should be calculated and used to decide whether to proceed further with the calculations. Figure 20.9 shows these computations completed for the time study shown, with time recordings only, on Figure 20.7.

Why the arithmetic average is used.   The arithmetic average is suggested (instead of other values such as median, mode, and so forth, used by some time study observers) because it is the only figure that is representative of the total sample of observations. It is simple to compute and includes both high and low values, which, if they represent valid performances of the elements, most certainly should be included in the representative measure. Also, the representativeness of this measure may be, as was shown earlier, evaluated.

Rating or Relating Performance to Standard (Determining the Modifier, *M*)

The manner of defining standard time was discussed in Chapter 6. It should be obvious that in practically all real situations, whatever specific definition of standard time is used, the observed operator is neither of the type specified by the definition of standard, nor working at the pace required for standard performance. Hence, two questions remain: (1) how to evaluate the performance observed as compared with the requirements given in the definition of standard used as the basis of the measurement, and (2) how to reduce this evaluation to a mathematical value (the modifier *M*) that will allow the adjusting (if necessary) of the average time values actually obtained, so as to determine a base for a standard time. These are the aims of rating.

Base times.   The result of the adjustment made by rating still will not include a pro rata proportion of the time during which the operator will be away from the workplace in order to attend to personal needs or for performing such activities as final, daily workplace cleanup, and the like (the additive *A*), as required for the *standard time*. Hence, the time values after being adjusted by ratings are commonly called *base times*.

Many rating procedures.   Many time-study rating procedures have been proposed. Indeed, this is still one of the major areas of controversy in present-day direct time study. Many of these rating procedures have been in wide use with varying degrees of success, depending on the inherent relative validity of the procedures, the ability of the time study observers who use them, and the plant attitude toward standards.

Why some poor rating methods appear to work. Plant attitude is particularly important. Many standard times in plants may appear to function well, even though they are incorrect in that they are either inconsistent with each other or with the concept of standard they purport to represent, because the workers have learned that it is advantageous to have them function as if they were correct. The workers may vary their exertion on the different jobs to produce at all times in some relatively fixed relationship to standard, thus covering up time study inconsistencies. They may complain sufficiently on the standards that are difficult to achieve so that the standards are revised, and restrict their output on the easy ones so that the inaccuracy there is undisclosed. The workers may even set up a system such that a consistent differential is maintained between individuals.

Indeed, the "game" may become even more complex, particularly when good labor relations exist. Most workers may feel that the standard times in use are fair, and in a time study situation may more or less honestly attempt to give their concept of a fair performance. The time study observer may be generally aware of this. While certain workers may not be in the "game," being inclined either to slow down or speed up abnormally when being studied, the time study observer may know who these workers are and tend to avoid studying them. Hence, the observed time represents almost by tacit agreement the correct rated time. The time study observer may, by rating, adjust this value a small amount, almost at random, to complete the study. This "game" often continues until (1) somebody not "in on the game" becomes part of a study, or (2) a new type of work enters the system and the workers' original concept of what is a fair output is grossly in error, or (3) until the standards, which are, in reality, a heterogeneous group of concepts of standard (representing essentially many workers' concepts), are subjected to real pressure, either for wages through incentives because of external economic changes, or for sociological reasons. The subsequent demonstration of the inconsistencies of the standards creates many problems.

Results of inconsistent standards. These "games," encouraged by past rate cutting, are sufficiently widespread so that many of the poorly conceived systems of rating or leveling often appear to be functioning properly. Almost inevitably in such cases costs become out of line with profitable operation, or labor unrest develops. Certainly, the relationship between management and labor is deleteriously affected and morale suffers. Good labor relations are hard to build when one group feels it must continually deceive the other. With poor labor relations productivity usually suffers.

Basic criterion for all measurement. It should be realized at the outset that time study measurement is not an absolute measurement, nor for that matter are most measurements. Measurement is usually characterized by a residual error or inherent variability, which can be demonstrated by repetition of the measurement. The important question is, as with all measuring techniques: "Is the size of the residual error, in the procedure to be used, such that the measurement is satisfactory for its purpose?"

Other criteria affecting the choice of rating procedures. The basic criterion of a measurement, however, is not the sole criterion with which either the student, the practicing industrial engineer, technician, or the industrial supervisor can select the time study rating procedure for study. In the first place, through the force of circumstances of industrial employment, an individual may be confronted with a procedure neither embodying the best of available techniques nor based on the most tenable theory.[13] Whether the person is in a position wherein there is no choice but to acquiesce or whether it is possible to suggest or introduce change, it is still necessary, if actions are to be planned on an intelligent basis, that the actual procedure in use be understood as well as its underlying theory, and its limitations. In the second place, most proposals have been made in an attempt to avoid specific shortcomings inherent in what was previous practice; hence, they require some knowledge of previous procedures for a complete understanding.

Practice is needed. Further, knowledge about a time study rating procedure does not automatically confer skill in its use. Practice is necessary in all cases.

Types of rating procedures. Commonly employed time study rating procedures can be divided into four main groups:

1. Mathematical.
2. Requiring judgment, unanchored; *subjective rating.*
3. Requiring judgment, unanchored, some guidelines given; *subjective rating.*
4. Requiring anchored judgment; *objective rating.*

Mathematical Procedures

The mathematical plans require a statistical sorting out, on the basis of the time recordings alone, of the effect of the operator's skill, aptitude, pace, relative rate of exertion, capriciousness, and so forth, from normal job variation. These must be separated in order to obtain a measure that

---

[13]This may not, of course, have been the situation when the procedure was instituted at the particular plant.

would be relatively the same, regardless of whatever conditions of the abovementioned variables were in existence at the time the data were recorded. It is not surprising that this has never been done successfully. Any mathematical method would require the existence of an outside reference point from which the variation of the particular observations obtained could be measured. The only reference point usable for such a purpose would be the standard time, but if this were known, there would be no reason for the rating.

Procedures Requiring Judgment, Unanchored; Subjective Rating[14]

Many rating systems involving judgment have been proposed and used. These, for the most part, involve a procedure that meets the following definition: *Rating* is that process during which the time study observer compares the performance of the operator under observation with the observer's own concept of normal performance.

For the most part, the subjective procedures that have been proposed involve the same basic routine because of the inference, inherent in the usual interpretation of this definition, that the time study observer must perform two separate steps in rating, which are as follows:

1. The observer must judge the difficulty of the job and form a mental concept of what the performance of the job under observation would look like if it met the requirements of standard performance as defined by the definition the observer is working with.

2. The observer must appraise the actual performance under observation as compared with the concept formed in step 1 and place a numerical value on this appraisal.

This type of rating procedure is referred to in the literature by various names, *effort rating* and *speed rating* being the most common.

In order to think realistically about "what the performance of the job would look like," and so forth, one must identify an observable characteristic (or observable characteristics) on the basis of which one will render

---

[14]The author wishes to point out that he makes no claim whatsoever to originality in this subsection. He has drawn on the works of Carroll and Presgrave, two of the best-known advocates of this procedure, and is deeply in their debt for the concepts and procedures described herein. Phil Carroll, Jr., *Timestudy for Cost Control*. New York: McGraw-Hill Book Company, 1954; and R. Presgrave, *Dynamics of Time Study*. New York: McGraw-Hill Book Company, 1945.

his judgments or form his concepts. "Effort," the characteristic used in "effort rating," is defined as *the apparent exertion and speed exhibited in doing the work*. Obviously, not too much attention should be given to the apparent exertion or it will also be necessary to appraise and take into account the type of operator; high exertion from an inferior operator would have to be discounted, and low exertion from a superior operator would have to be taken at more than face value. If, as would appear proper, we are to take speed as the primary characteristic, we need to have some concept of the proper speed (or rapidity of motion) for the job, in short, what speed we can typically expect from the typical operator.

Presgrave describes "normal effort" (or speed) as equivalent to either of the following two illustrations:

1. *Dealing 52 playing cards into four piles in 0.50 minute.* The dealing of cards is done with the cards in the left hand, the thumb advancing the top card each time, the right hand grasping the prepositioned corner of the top card between its thumb and first finger, carrying it to the proper pile before releasing it, and then reversing the motion back to the pack. Four piles are formed by the dealing, one in front of the dealer and the other three at the other three corners of a 1-foot square.

2. *Walking at 3 miles per hour, taking 27-inch steps.* One should bear in mind that Presgrave's concept of normal is such that the average worker, with incentive, can exceed standard by 30 per cent. If one maintains the same physical concept of what can be expected from an incentive worker, but defines typical incentive performance as 125 per cent, then the speeds at normal become 0.50 minute $\times$ 125/130, or 0.48 minute, for card dealing and 3 mph $\times$ 130/125, or 3.12 mph, for walking.

The Society for Advancement of Management has prepared an extensive series of films and circulated them among an enormous number of time study observers in order to obtain a general consensus on the effort or speed exhibited and has published these films and these values. (The late T. U. Matthew, formerly of the University of Birmingham, England, performed the same task, with these films, in the United Kingdom.) Such a group of films is designed to provide the time study observer with a means of becoming familiar with the appearance of normal or standard speed for a variety of jobs.

In practice, the time study observer is expected, while the time study is being made, to select from the concepts (gained by experience on similar jobs, knowledge of similar tasks, exposure to films available, and so forth) the concept that will serve for step 1 of the two steps previously given as necessary for rating. Then step 2 is performed and a value is placed on the

RECAPITULATION

ELEMENTS	1	2	3	4	5	6	7	8	9	10	11	12
TOTAL TIME IN _MIN_	3.20	1.29	A									
NUMBER OF OBS.	15	15	B									
PRO-RATE DIVISOR	1	1	C									
AVERAGE PER CYCLE	.213	.086	D = A/(B x C)									
RATING	110	110	E ◄ Rating of Effort									
RATED TIME												
PER CENT OF CYCLE												
1+ADJUSTMENTS												
BASE TIME	.224	.095	F = D x E									
1+ALLOWANCES	1.07	1.07	G									
ALLOWED TIME	.240	.101	H = F x G									

Figure 20.10. Section of time study sheet with effort rating and base time.

time study sheet as shown in Figure 20.10. This figure is used to multiply the average time taken for the element and thus produce a *base time*, as previously defined. A machine-controlled element would be assigned a rating of 100.[15]

Some find it advantageous to rate individual element readings which they feel certain of and to use these to check the rating of the average of all readings. Exact agreement is unlikely (due to the sampling error of individual readings), but some check is considered possible.

It is to be noted that the type of rating that has been described is essentially simple in concept, although the crux of the problem appears to be the necessary exposure to performances of a variety of tasks, the ratings of which are known, in order to develop a broad-enough basis for performing step 1 of the rating procedure, However, it should be obvious that the more similar the jobs in a shop are, the easier this procedure will be.

Procedures Requiring Judgment, Unanchored, Some Guidelines Given; Subjective Rating

These procedures vary from the pure subjective procedures with respect to the details given for performing step 2. The differences are (1) the number of subfactors into which the second step is divided, (2) the terms used to describe the basis of the comparison between the performance observed and the concept formed in step 1, (3) the numerical scales employed, and (4) whether benchmark jobs are suggested.

---

[15]The entire procedure may be based on 60 being defined as normal, in which case the numbers will be different, but the results will be the same.

The number of subfactors into which the performance of step 2 is divided varies from one to as many as six, depending on the procedure examined.

The terms used to describe the basis of the comparison between the performance observed and the concept formed in step 1 include *speed, effective speed, tempo, effort, pace, speed of movement, skill, conditions, consistency, attitude, application, mental–physical coordination*, and so forth. In some cases, several of these are given as synonymous.

Various descriptive terms such as *excellent, good,* and *fair*, or more elaborate phrases, are frequently provided as guides to various typical numerical values to be determined in step 2. These numerical values are intended for use as a rating or leveling factor by which the final representative values for the observed element times are to be multiplied so as to adjust them to *base times*. In some applications, one rating factor is used with all the elements of a time study. In other applications a separate rating is developed for each element (not each reading). A separate rating for each element is considered a better procedure than the use of the single rating.

The best known system of this type is frequently called the *Westinghouse Leveling System*. (It was originated at Westinghouse and the authors called it leveling rather than rating.) The categories for which judgments are made and the general guides are given in Table 20-II.

In the text referenced, under Table 20-II, additional, extensive, descriptive guides are given for each of the individual categories of the table.

For a quick example of the use of the table, let us say that the performance associated with all the reading for an element was classified as: Skill, B1; Effort, B2; Conditions, C; Consistency, B. The adjustments would be: $+0.11$; $+0.08$; $+0.02$; $+0.03$. The total of these is $+0.25$. This is the same as an "effort" rating of 125 per cent. It would be applied as was the rating in Figure 20.10.

In many applications only the skill and effort sections of Table 20-II are used.

Procedures Requiring Anchored Judgment; Objective Rating

Objective rating is the name given to a procedure wherein the two steps of the rating procedure are performed in the reverse order of the usual subjective procedure. As will be shown later, this permits anchoring, to external standards, of the judgments used for both steps. The procedure was first used in an industrial plant in 1939. The first publication describing

Table 20-II. WESTINGHOUSE LEVELING GUIDES*

Skill			Effort		
+0.15	A1	Superskill	+0.13	A1	Excessive
+0.13	A2	Superskill	+0.12	A2	Excessive
+0.11	B1	Excellent	+0.10	B1	Excellent
+0.08	B2	Excellent	+0.08	B2	Excellent
+0.06	C1	Good	+0.05	C1	Good
+0.03	C2	Good	+0.02	C2	Good
0.00	D	Average	0.00	D	Average
−0.05	E1	Fair	−0.04	E1	Fair
−0.10	E2	Fair	−0.08	E2	Fair
−0.16	F1	Poor	−0.12	F1	Poor
−0.22	F2	Poor	−0.17	F2	Poor
Conditions			Consistency		
+0.06	A	Ideal	+0.04	A	Perfect
+0.04	B	Excellent	+0.03	B	Excellent
+0.02	C	Good	+0.00	C	Good
0.00	D	Average	0.00	D	Average
−0.03	E	Fair	−0.02	E	Fair
−0.07	F	Poor	−0.04	F	Poor

*From S. M. Lowry, H. B. Maynard, and G. J. Stegemerten, *Time and Motion Study and Formulas for Wage Incentives*. New York: McGraw-Hill Book Company, 1940, p. 233; as reprinted in H. B. Maynard (ed.), *Industrial Engineering Handbook*. New York: McGraw-Hill Book Company, 1971, pp. 3–41.

it appeared in 1944,[16] and it first appeared in a text in 1946.[17] Subsequent research developed refinements to the procedure and made all the necessary

[16]M. E. Mundel, "An Analysis of Time Study Rating Systems and Suggestions for a Simplified Systematic System," *Industrial Engineering*, April 1944.

[17]M. E. Mundel, *Systematic Motion and Time Study*. Englewood Cliffs, N.J.: Prentice-Hall, Inc., 1946.

materials readily available. It is in current use in many plants throughout the world.

Criteria for a good rating procedure. We must recognize that rating is relating observed performance to standard by means of judgment. The judgment, to be of real significance, should be such that the judgment is explainable by other means than mere recourse to "experience." The basis must be concrete.

Advantages of objective rating. Such a basis would permit real agreement between labor and management on at least the measuring unit involved, or in other words, the concept of adequate performance. Whether unions are involved or not, it is likely that any group of workers will cooperate more fully if they understand the procedure of rating time studies, provided that the procedure is sound. Further, such a basis would make for more effective cooperation between all the parties involved in a time study, and, if a single objective standard of reference could be substituted for the usual multiplicity of mental concepts, we could expect to reduce the error of measurement.

The basic judgment in rating. The judgment made must involve the evaluation of the observed performance, as compared with that required by the definition of standard performance in use, so as to permit the computation of the standard for the job. For any given job, the primary objective phenomenon from which all inferences concerning performance are made is the pace or rapidity with which the parts of the job are performed. The actual pace of performance observed must be understood to be a function of the skill, aptitude, and exertion of the operator, but these variables are neither separately identifiable nor is a separate appraisal pertinent.

Effect of operator skill. The operator's *skill* determines how rapidly the person may do the job properly: less skill showing up in a slower maximum pace, more skill in a faster maximum pace, provided that the method remains constant. Hence, with a given method, skill may be reflected in pace. If the worker does not possess sufficient skill to perform the job in the proper manner, even at a slow pace, then the job cannot yet be time-studied. It should be obvious that no consistent set of units exists with which to measure deviations in method; consequently, even the first step of expressing method as a function of time (entailed in rating method) is not feasible. However, in most cases, even partially trained people can perform a job in a prescribed manner if allowed to work at a suitable pace.

Effect of operator's aptitude. The operator's *aptitude* determines how fast a pace can be maintained, or how long a period it takes to acquire the skill required for a rapid pace. Other things being equal, poor aptitude permits but a slow pace, and high aptitude permits fast pace, provided that the proper method is followed. Hence, both aspects of aptitude are also reflected in pace.

Effect of operator exertion. The *exertion* of a given operator is a function of two items: the difficulty of the job and pace. Hence, exertion, which determines the physical effect of the work upon the operator, is also reflected in pace.

The two phenomena requiring evaluation. Consequently, it may be seen that the evaluation or rating of performance (as is correctly done in many rating procedures) may be reduced to a judgment of not more than two items: (1) observed pace, and (2) job difficulty. In the typical time study procedure, the time study observer first judges (2), performance for the job (as required by the definition of standard time in use), and then judges (1), observed pace, against this imagined concept. This was discussed with respect to subjective rating, and while it is theoretically correct, the practical difficulties were enumerated.

Basis of objective rating. What makes a more reliable time study rating procedure possible is, first of all, the realization that the difficulty of the job and its effect on maximum possible pace does not need to be judged but may be reduced to objective terms, based on observable phenomena, and reduced to tabular form as a function of strength required, amount of body used, degree of dexterity, and the like. Thus, the sole remaining phenomenon to be judged in rating is observed pace, or rate of activity, or actual speed of movement as compared with a pace required for standard performance. Although the desired pace is implied in the definition of standard time, this is still subject, if left in this form, to subjective variation. As will be seen later, the procedure may be altered so that standard pace may be set up in objective form so that the two items that the time study observer must compare can be concrete, and any number of observers may make an appraisal of performance as related to standard, without recourse to any imagined concepts. This produces a remarkable increase in the reliability of the time standards.

Steps in objective rating. What is proposed is again a two-step rating procedure, but the steps are in the reverse of the conventional order. This is called *objective rating* and consists of the following steps:

1. The rating of observed pace against an objective pace standard, which is the same for all jobs. This standard is embodied in a film.

In this rating, no attention whatsoever is paid to job difficulty and its limiting effect on possible pace; hence, a single pace standard may be used instead of a multiplicity of mental concepts.

2. The use of a *difficulty adjustment*, consisting of a percentage increment, added after the application of the numerical appraisal from step 1 has been used to adjust the original observed data. This percentage increment is to be taken from experimentally determined tables of the effect of various observable factors that control the exertion required at a given pace. Such tables have been developed and appear later in this chapter.

Such a procedure offers a vastly improved probability that time studies rated with this procedure will more consistently represent a single concept of standard time than those adjusted by subjective procedures.

The application of such an improved procedure requires some preliminary activity before any time studies are made. However, this preliminary activity need be performed but once, and the subsequent details of taking time studies are less complex than with subjective rating procedures.

The pace standard. A word of caution—the concept of a standard pace must be fully understood. Experiments have shown (as perhaps one might expect) that rating with a wrong concept of pace may lead to gross errors in rating.

The concept of pace, as used, requires some understanding of the basic mechanics of human motion. Very early experimentation indicated that the hand or arm, when moving, accelerates for a relatively constant portion of the distance traveled, moves at constant velocity for another relatively constant portion, and decelerates for another relatively constant portion.[18] These proportions do not appear to be affected by the pace.[19] The percentage of time spent in acceleration, constant velocity, and deceleration with free movement appears to be in the neighborhood of 40, 20, and 40 per cent, respectively.[20] Now, one of the basic laws of physics is the equation: force = mass × acceleration. In respect to human motions, this may be interpreted as: the application of muscular force to a body member (or eye, and the like), with other variables constant, will produce an acceleration proportional to the force. It would, therefore, seem reasonable to define *pace or rate of activity as the rate at which muscular force is applied to the creation of body,*

---

[18]Wayne Deegan, "The Development and Use of a New Technique for Measurement in Time and Motion Study," M.S. thesis, University of Iowa, 1935.

[19]M. E. Mundel (with R. M. Barnes), "Studies of Hand Motions and Rhythm Appearing in Factory Work," *University of Iowa Studies in Engineering, Bull. 12*, 1938, p. 10.

[20]Deegan, op. cit. The discussion that follows holds true no matter what these percentages actually are.

arm, hand, or finger movements, disregarding the effect of job difficulty (mass constant), and hence as the observed rate of acceleration of the body member doing the work.

The preceding information has been given to explain a reasonable, yet at first glance peculiar, phenomenon. If the time for a given distance of body member movement, such as a movement of the arm about the elbow, is determined, a good deal less than twice the time will be required for twice the distance of movement at the identical pace or rate of activity or required rate of application of muscular force, assuming unhindered movement, all three being synonymous. If equations for distances are set up as a function of time, it will be found that for a given rate of acceleration, *a*, the time will increase as the square root of the ratio of distances with an increase of 100 per cent (double original) distance accompanied by approximately a 41 per cent increase in time. The distance–time relationship is shown graphically in Figure 20.11.

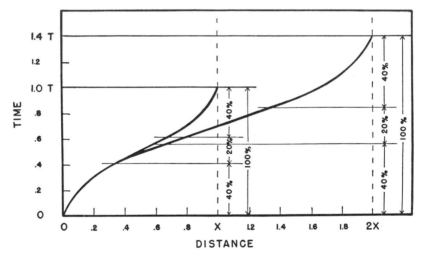

Figure 20.11. Distance versus time curves for movements of *X* and 2*X* distance at a given pace.

What is looked at to rate pace. Consequently, it may be seen that the average velocity of a body member movement is not the correct guide to use to rate the pace, but what must be rated is the rate of acceleration near the inception of the movement and the rate of deceleration near the conclusion of the movement. It is also worth noting that micromotion analyses usually reveal that most of the time, with an effective method (particularly on small jobs), is spent on the red-blue, or terminal, group of therbligs rather than on the green, or movement, group; therefore, considerable attention should be paid to the pace with which the small motions of these terminal therbligs are performed.

Creating an objective pace standard. An objective pace standard may be obtained in any one of four ways.

1. A GROUP MAY MAKE ITS OWN STANDARD

The time study group may make a series of films of a worker at different paces on a simple job, and ask management to select one of these as representing their concept of standard pace. It is true that judgment must be exercised here and some original error is possible, but this is not of prime importance. At least a standard of unchanging pace is set up. Also, if management wants to assume this prerogative, it should be assumed at a high-enough level. It is worth noting that the pace selected may also be jointly negotiated by labor and management, in which case the accuracy with which it represents the definition of standard is of less consequence, although this will be discovered later through experience with its use. At least it will be acceptable to both parties, and without such mutual acceptance, joint agreements concerning money per hour appear inadequate, to say the least. However, care should be exercised that any existing, generally accepted pace is not unduly disturbed. A negotiation for a standard pace should not be a subterfuge for a concealed negotiation for a wage increase or decrease.

Critics of time study will refer to the managerial decisions required in selecting this standard pace film as "value judgments" and charge that time study based on such a benchmark concept is not measurement. It is true that the pace selection is a value judgment, but so is the selection of any unit of measurement that meets sociological rather than physical criteria. However, whether or not time study is measurement depends on how the standard is *applied* rather than how it is *selected*. Figure 20.12 shows the simple operation filmed at the Patrick Cudahy plant for development into this type of benchmark.

Once the standard pace film is designated, it is desirable to prepare films showing step-by-step deviations from standard pace on the one job (step films), so as to establish markings on the scale of pace and to facilitate the rating. The films may be in loop form, that is, the front end spliced to the back so as to permit continuous projection for any period of time. Also, such films are commonly made with the frames divided into different areas, each area showing a different pace, so that a group of steps may be projected simultaneously. Such films are called *multi-image films*. A short section of such a film presently in use in a variety of industrial plants is shown in Figure 20.13. The originating company's pace values for the 12 paces simultaneously presented are given in Figure 20.14. The job filmed was shown in Figure 20.12.

Figure 20.12. "Measuring Hog Casings" (used for sausage casings) is the operation used as a benchmark job at Patrick Cudahy, Inc., Cudahy, Wis.

2. Existing Concepts of Standard May Be Used

One may accept proposals such as those of Ralph Presgrave, who, as was noted,[21] suggested a standard pace as that equal to dealing 52 playing cards into four piles in 0.50 minute or walking at the rate of 3 miles per hour. These are usable when the definition of standard is the same as the one last given, since Presgrave's concept of standard is similar to the one previously stated. If the definition is changed, the pace on the examples must also be changed.

3. A Professional Group May Set a Standard

As another procedure, a simple job may be shown to large groups of industrial engineers, and the values (corrected for different concepts of standard) averaged and used as a basis for the standard pace.[22]

---

[21]It is worth noting that Presgrave offers these as examples of *standard effort* and suggests that with practice one can determine a mental concept of standard effort on other jobs. The pace, at standard effort with Presgrave's rating system, will naturally be different with each job, depending on the difficulty of the job. The use of Presgrave's card-dealing pace proposed here is different from the *use* proposed by Presgrave.

[22]Films of such standards are in existence.

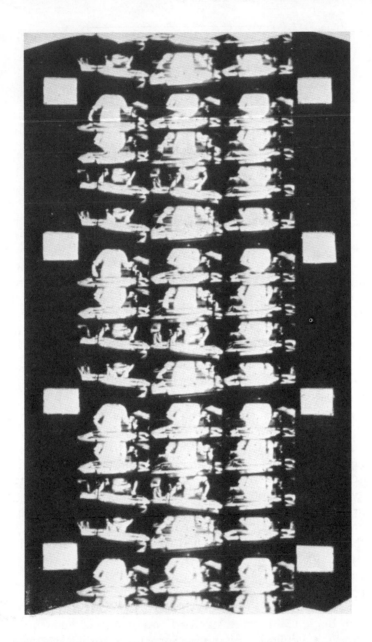

Figure 20.13. Short section of 12-image multi-image loop used in one plant. Each of the pictures shows a different pace. For a close-up of this task, see Figure 20.12.

141	132	120
100	93	82
74	67	53
44	40	36

Figure 20.14. Rating values for multi-image loop of Figure 20.13 when projected at 1,000 frames per minute.

However, there are some drawbacks to this alternative. There may be real differences in the concepts of normal within the group, and these differences will be hidden by averaging. These differences may represent valid differences in the dynamics of the situation portrayed roughly in Figure 6.1, and the reduction of these differences may create inequities.

4. ANOTHER COMPANY'S BENCHMARK MAY BE MODIFIED

As another procedure, an organization may accept a benchmark film made by another company. They may use either the values assigned by the originating organization or assign a new set of pace values. This alternative has frequently been followed by small organizations as an economical procedure. There are psychological advantages in having a benchmark pace standard made of a job in the plant, but any operation will meet the requirements of the technique of objective rating.

For example, another company in a different industry, which uses the film shown in Figure 20.13 as its benchmark, uses the values given in Figure 20.15. Note, however, that the values of Figures 20.14 and 20.15 have a constant ratio between them.

Correcting for the difficulty of a job. It is obvious that all jobs could not be performed at the standard pace, since practically all will be more difficult than the job with which standard pace is defined and, further, some jobs will be more difficult than others. Some tasks, for instance, will involve heavier parts, closer visual work, and so forth. These job differences will place different limits on the pace possible on each job with a fixed rate of exertion relative to the maximum possible on the job, but they may be objectively evaluated.

152	142	129
108	100	88
80	72	57
47	43	39

Figure 20.15. Values given multi-image film of Figures 20.13 and 20.14 for use in a different industry.

Objective evaluation of difficulty requires the prior determination of the various factors that make for difficulty on the job, evaluating their effect, and developing a *difficulty adjustment* for use in computing the standard times, so that all the standards will be consistent in regard to attainability. These difficulty adjustments have been developed and set up in tabular form.

The work of developing these difficulty adjustments is probably not fully complete; unusual conditions such as slippery floors and so forth are not covered. However, at their present stage of development they offer a much more satisfactory and reliable procedure than leaving the adjustment to the mental evaluation of the time study observer, as is inherent in subjective rating procedures. If inconsistencies appear in the results of the application of these difficulty adjustments, then the source may be traced and lasting corrections made, a course of action that is not possible with the subjective approach. The reduction of such difficulty adjustments to tabular form removes from the time study procedure additional sources of subjectively caused variation and thereby aids in meeting the basic objections to the subjective approaches. Tables of difficulty adjustments are given as Tables 20-III and 20-IV.

Using the difficulty-adjustment tables. The total difficulty adjustment for an element will be the simple sum of all the appropriate values from the scales for all the factors. (Note that the difficulty adjustments will be separate from the pace rating. For example, if the rating is 90 per cent and the total difficulty adjustment is 12 per cent, the observed time for that element will be multiplied by 0.90 and then by 1.12. The 0.12 and 0.90 cannot be added; otherwise, the actual increment would vary with the pace observed, which would not be correct.)

Table 20-III. DIFFICULTY ADJUSTMENTS FOR TIME STUDIES

Category	Description	Reference Letter	Condition	Per Cent Adjustment	Example
1	Amount of body used	A	Fingers used loosely	0	
		B	Wrist and fingers	1	
		C	Elbow, wrist, and fingers	2	
		D	Arm, etc.	5	
		E	Trunk, etc.	8	
		E2	Lift with legs from floor	10	
2	Foot pedals	F	No pedals or one pedal with fulcrum under foot	0	
		G	Pedal or pedals with fulcrum outside of foot	5	
3	Bimanualness	H	Hands help each other or alternate	0	
		H2	Hands work simultaneously doing the same work on duplicate parts	18	(Parts are "identical" with respect to work requirements)
4	Eye–hand coordination*	I	Rough work, mainly feel	0	Do not need to look other than casually
		J	Moderate vision	2	Occasional need for peripheral vision
		K	Constant but not close	4	Constant peripheral vision
		L	Watchful, fairly close	7	Foveal vision
		M	Within $\frac{1}{64}$ inch	10	Close hand sewing

Table 20-III. (CONTINUED)

Category	Description	Reference Letter	Condition	Per Cent Adjustment	Example
5	Handling requirements*	N	Can be handled roughly	0	No need to consciously control muscular forces
		O	Only gross control	1	Can squeeze or "bang" objects
		P	Must be controlled, but may be squeezed	2	Objects must not be "banged"
		Q	Handle carefully	3	Parts could be damaged by careless handling or too much pressure
		R	Fragile	5	Parts *readily* damaged by normal finger pressures
6	Weight		Identify by the actual weight or resistance	Use Table 20-IV	

*These scales could, in some cases, go higher.

369

Table 20-IV. DIFFICULTY ADJUSTMENT AS A FUNCTION OF PER CENT OF CYCLE TIME UNDER LOAD OR UNDER LOAD-CONNECTED WORK TENSION

(PER CENT OF CYCLE COMPUTED FROM RATED TIMES WITHOUT ADJUSTMENTS)

Pounds or Force Per Lift	Basic Value When Work Under Load Is 5 Per cent or Less of Cycle	Increments to Be Added to Basic Value for Load-Connected Work in Per Cent in Excess of 5 Per Cent of Cycle														Maximum Possible Total
		1	2	3	4	5	6	7	8	9	10	20	30	40	45	
1	1															1
2	2															2
3	3															3
4	3															3
5	4															4
6	5															5
7	7															7
8	8															8
9	9															9
10	11					Use basic value for all conditions up to 20 lb of lift or force per application										11
11	12															12
12	13															13
13	14															14
14	15															15
15	16															16
16	17															17
17	18															18
18	19			When adding up to get value, use the nearest whole per cent to total sum													19
19	20															20

Table 20-IV. (CONTINUED)

Pounds or Force Per Lift	Basic Value When Work Under Load Is 5 Per Cent or Less of Cycle	Increments to Be Added to Basic Value for Load-Connected Work in Per Cent in Excess of 5 Per Cent of Cycle														Maximum Possible Total
		1	2	3	4	5	6	7	8	9	10	20	30	40	45	
20	21	0.0	0.1	0.1	0.2	0.2	0.3	0.3	0.4	0.4	0.5	1	1.3	1.7	2	23
21	22	0.0	0.1	0.1	0.2	0.2	0.3	0.3	0.4	0.4	0.5	1	1.3	1.7	2	24
22	23	0.1	0.1	0.2	0.3	0.3	0.4	0.5	0.5	0.6	0.7	1.3	2.0	2.8	3	26
23	24	0.1	0.2	0.3	0.4	0.4	0.5	0.6	0.7	0.8	0.9	1.8	2.7	3.6	4	28
24	25	0.1	0.2	0.3	0.4	0.6	0.7	0.8	0.9	1.0	1.1	2.2	3.3	4.4	5	30
25	26	0.1	0.3	0.4	0.5	0.7	0.8	0.9	1.1	1.2	1.3	2.7	4.0	5.3	6	32
26	27	0.2	0.3	0.5	0.6	0.8	0.9	1.1	1.2	1.4	1.6	3.1	4.7	6.2	7	34
27	28	0.2	0.4	0.5	0.7	0.9	1.1	1.2	1.4	1.6	1.8	3.6	5.3	7.1	8	36
28	29	0.2	0.4	0.6	0.8	1.0	1.2	1.4	1.6	1.8	2.0	4.0	6.0	8.0	9	38
29	30	0.2	0.4	0.7	0.9	1.1	1.3	1.6	1.8	2.0	2.2	4.4	6.7	8.9	10	40
30	31	0.2	0.5	0.7	1.0	1.2	1.5	1.7	2.0	2.2	2.4	4.9	7.3	9.7	11	42
31	31	0.3	0.6	0.9	1.2	1.5	1.9	2.2	2.5	2.7	3.1	6.2	9.3	12.4	14	45
32	32	0.3	0.7	1.0	1.3	1.7	2.0	2.3	2.7	3.0	3.2	6.7	10.0	13.3	15	47
33	33	0.4	0.7	1.1	1.4	1.8	2.1	2.5	2.8	3.2	3.6	7.1	10.6	14.2	16	49
34	34	0.4	0.8	1.2	1.6	2.0	2.4	2.8	3.2	3.6	4.0	8.0	12.0	16.0	18	52
35	34	0.4	0.9	1.3	1.8	2.2	2.7	3.1	3.6	4.0	4.4	8.9	13.3	17.8	20	54
36	35	0.5	1.0	1.5	2.0	2.4	2.9	3.4	3.9	4.4	4.9	9.7	14.7	20.0	22	57
37	36	0.5	1.1	1.6	2.1	2.7	3.2	3.7	4.3	4.8	5.3	10.7	16.0	21.4	24	60
38	36	0.6	1.2	1.7	2.3	2.9	3.5	4.0	4.6	5.2	5.8	11.6	17.3	23.1	26	62
39	37	0.6	1.2	1.9	2.5	3.1	3.7	4.4	5.0	5.6	6.2	12.4	18.7	24.9	28	65
40	37	0.7	1.4	2.1	2.8	3.4	4.1	4.8	5.5	6.2	6.9	13.8	20.7	27.6	31	68
41	38	0.7	1.5	2.2	2.9	3.7	4.4	5.1	5.9	6.6	7.3	14.7	22.0	29.3	33	71

Table 20-IV. (CONTINUED)

Pounds or Force Per Lift	Basic Value When Work Under Load Is 5 Per Cent or Less of Cycle	Increments to Be Added to Basic Value for Load-Connected Work in Per Cent in Excess of 5 Per Cent of Cycle														Maximum Possible Total
		1	2	3	4	5	6	7	8	9	10	20	30	40	45	
42	38	0.8	1.6	2.4	3.2	4.0	4.8	5.6	6.4	7.2	8.0	16.0	24.0	32.0	36	74
43	39	0.8	1.7	2.5	3.4	4.2	5.1	5.9	6.8	7.6	8.4	16.9	25.3	33.8	38	77
44	40	0.9	1.8	2.7	3.6	4.4	5.3	6.2	7.1	8.0	8.9	17.8	26.6	35.6	40	80
45	40	1.0	1.9	2.9	3.8	4.8	5.7	6.7	7.6	8.6	9.6	19.1	28.7	38.2	43	83
46	41	1.0	2.0	3.0	4.0	5.0	6.0	7.0	8.0	9.0	10.0	20.0	30.0	40.0	45	86
47	42	1.0	2.1	3.1	4.2	5.2	6.3	7.3	8.4	9.4	10.5	20.9	31.3	41.8	47	89
48	42	1.1	2.2	3.3	4.4	5.6	6.7	7.8	8.9	10.0	11.1	22.2	33.3	44.4	50	92
49	43	1.2	2.3	3.5	4.6	5.8	6.9	8.1	9.2	10.4	11.6	23.1	34.7	46.2	52	95
50	43	1.2	2.4	3.7	4.9	6.1	7.3	8.6	9.8	11.0	12.2	24.4	36.7	48.9	55	98
51	44	1.3	2.5	3.8	5.1	6.3	7.6	8.9	10.1	11.4	12.7	25.4	38.0	50.7	57	101
52	44	1.3	2.7	4.0	5.3	6.7	8.0	9.3	10.7	12.0	13.3	26.7	40.0	53.3	60	104
53	45	1.4	2.8	4.1	5.5	6.9	8.3	9.6	11.0	12.4	13.8	27.6	41.3	55.1	62	107
54	46	1.4	2.8	4.3	5.7	7.1	8.5	10.0	11.4	12.8	14.2	28.4	42.6	56.9	64	110
55	46	1.5	3.0	4.5	6.0	7.4	8.9	10.4	11.9	13.4	14.9	29.8	44.6	59.5	67	113
56	47	1.6	3.1	4.7	6.2	7.8	9.3	10.9	12.4	14.0	15.6	31.3	46.6	62.2	70	117
57	47	1.6	3.2	4.9	6.5	8.1	9.7	11.4	13.0	14.6	16.2	32.4	48.6	64.9	73	120
58	48	1.7	3.4	5.1	6.8	8.4	10.1	11.8	13.5	15.2	16.9	33.8	50.7	67.6	76	124
59	48	1.8	3.5	5.3	7.0	8.8	10.5	12.3	14.1	15.8	17.6	35.1	52.7	70.2	79	127
60	49	1.8	3.6	5.4	7.2	9.0	10.8	12.6	14.4	16.2	18.0	36.0	54.0	72.0	81	130
61	50	1.9	3.7	5.6	7.5	9.3	11.2	13.1	14.9	16.8	18.7	37.4	57.0	74.7	84	134
62	50	1.9	3.9	5.8	7.7	9.7	11.6	13.5	15.5	17.4	19.3	38.7	58.0	77.3	87	137
63	51	2.0	4.0	5.9	7.9	9.9	11.9	13.8	15.8	17.8	19.8	39.6	59.4	79.9	89	140

Table 20-IV. (CONTINUED)

Pounds or Force per Lift	Basic Value When Work under Load is 5 Per Cent or Less of Cycle	Increments to be Added to Basic Value for Load-Connected Work in Per Cent in Excess of 5 Per Cent of Cycle														Maximum Possible Total
		1	2	3	4	5	6	7	8	9	10	20	30	40	45	
64	51	2.1	4.1	6.2	8.3	10.3	12.4	14.5	16.5	18.6	20.6	41.3	62.0	82.7	93	144
65	52	2.1	4.2	6.3	8.4	10.6	12.7	14.8	16.9	19.0	21.1	42.2	63.3	84.5	95	147
66	53	2.2	4.3	6.5	8.6	10.8	12.9	15.1	17.3	19.4	21.6	43.1	64.6	86.2	97	150
67	53	2.2	4.5	6.7	9.0	11.2	13.5	15.7	17.9	20.2	22.4	44.8	67.3	89.8	101	154
68	54	2.3	4.6	6.9	9.2	11.4	13.7	16.0	18.3	20.6	22.9	45.7	68.6	91.6	103	157
69	54	2.4	4.7	7.1	9.4	11.8	14.1	16.5	18.8	21.2	23.6	47.1	70.6	94.2	106	160
70	55	2.4	4.8	7.3	9.7	12.1	14.5	17.0	19.4	21.8	24.2	48.4	72.6	96.9	109	164
71	56	2.5	4.9	7.4	9.9	12.3	14.8	17.3	19.7	22.2	24.6	49.3	74.0	98.7	111	167
72	56	2.5	5.1	7.6	10.1	12.7	15.2	17.7	20.3	22.8	25.3	50.6	76.0	101.3	114	170
73	57	2.6	5.2	7.8	10.4	13.0	15.6	18.2	20.8	23.4	26.0	52.0	78.0	104.0	117	174
74	58	2.6	5.3	7.9	10.6	13.2	15.9	18.5	21.2	23.8	26.4	52.8	79.3	105.8	119	177
75	58	2.7	5.4	8.1	10.8	13.6	16.3	19.0	21.7	24.4	27.1	54.2	81.3	108.4	122	180
76	59	2.8	5.6	8.3	11.1	13.9	16.7	19.4	22.2	25.0	27.8	55.5	83.3	111.1	125	184
77	59	2.9	5.7	8.6	11.5	14.3	17.2	20.1	22.9	25.8	28.7	57.3	86.0	114.7	129	188
78	60	2.9	5.8	8.7	11.6	14.5	17.5	20.4	23.3	26.2	29.1	58.2	87.3	116.5	131	191
79	61	3.0	6.0	8.9	11.9	14.9	17.9	20.8	23.8	26.8	29.8	59.5	89.3	119.1	134	195
80	61	3.0	6.1	9.1	12.2	15.2	18.3	21.3	24.4	27.4	30.4	60.8	91.3	121.8	137	198

With respect to the difficulty adjustment for weight, Table 20-IV, the use is shown in the following example. Let us assume that a task *after rating* has two elements: (1) 0.15 minute, and (2) 0.10 minute. Element 2 involves motions connected with moving a 47-pound box or giving a 47-pound pull on a lever. (A spring scale should be part of the normal equipment of the time study observer so that he may determine the forces required.) Element 2 is 40 per cent of the rated cycle time:

$$\frac{0.10}{0.25} = 40 \text{ percent}$$

For 47 pounds, Table 20-IV gives a basic value of 42 per cent and an additional increment of 36.5 per cent or, in round figures, 37 per cent for the 35 per cent of the cycle *more than the minimum 5 per cent of the cycle* affected by the weight. Note that the values given for each weight permit the rapid calculation of the increment for any per cent of cycle under load (or with load-connected motions) from 5 per cent of the cycle (the minimum value), to a maximum value at 50 per cent of the cycle.

Consequently, the difficulty adjustment for weight for element 2 is 42 + 37, or 79 per cent. This would be used together with the other applicable adjustments to complete the objective rating and adjustment procedure.

The table may be used for leg lifts by using only the basic "5 per cent of cycle" value and the 10 per cent adjustment for "leg lift" of category 1.

It is further suggested that it would not be unreasonable to place an agreed-upon limit on the maximum total lift which will be permitted per day. This limit may be a function of the amount of weight per lift, the manner of lifting, and the height of the lift. Very little data exist for giving values for these variables. However, the difficulty of taking them into account in a consistent fashion with subjective rating should be obvious. Even the construction of a tentative table for use with objective rating would bring consistency into the adjustment for these factors and allow results to be observed in order to improve the tables. The use of these tentative tables would require that a dummy delay element be inserted in time standards when the use of the basic difficulty adjustment for weight would cause the standard time to call for lifting in excess of these limits.

Also, it is suggested that *carrying an object axially loaded on the body* (or held against the body) be treated as a "leg lift" rather than as an "arm" or "arm and trunk lift." A quick inspection of the muscles involved will indicate the reason for this suggestion.

As can be seen readily, some of the adjustments for weight become sizable; consequently, it is important that elements containing such activity should be kept short enough to be as homogeneous as possible.

Difficulty adjustments for unusual conditions. Table 20-III and 20-IV contain all the difficulty adjustments that are commonly used with repetitive work. However, objective rating is used with other work measurement techniques which are used with nonrepetitive work; such work situations present additional problems. Some require considerable care and study of the drawings or instructions throughout the task. This creates a slowing down beyond that caused by the physical characteristics of the work. In other cases, the worker may be working in cramped quarters (e.g., repairing tubes in a boiler sheet aboard a ship or welding inside a box girder). In such cases the cramped or limited room will slow down his motions beyond that of the repetitive worker who is usually working without such constraints. Of course, the cramped quarters may require more allowance or rest time, but the motions will also be slower, and special difficulty adjustments must be introduced to allow for this. In still other cases the worker may be working under conditions where the maintenance of his safety requires that he move slowly and carefully; he may be on a partially railed section of a ship's deck far above the ground, on a painting scaffold, and so forth. Taking such a variety of conditions into account in subjective rating seems relatively impossible.

In companies where such situations have been encountered, tentative tables of special difficulty adjustment values have been made and used. These introduce consistency into the standards in which such conditions are encountered. The initial values assigned are watched and further adjustments made until the values appear workable and acceptable.

Correcting the benchmark film. Now that difficulty adjustments have been discussed, we must return our attention to the film of pace used as an anchor for judgment. After management assigns rating values to the benchmark films, there is a modification to be made to them. Management's ratings are made subjectively when selecting the 100 per cent pace. If this pace was to be used as the 100 per cent value for objective ratings, this pace on this benchmark operation would then be rated as 100 and then multiplied by 1.0 plus the difficulty corrections resulting in calling it more than 100 per cent. This would be an error. Hence, when the 100 per cent value is selected by management, it is divided by (1 + Adj.), where "Adj." is the sum of the appropriate difficulty adjustments. This correction places the subjective rating in objective form. Such a correction is not needed if externally set, objective values for a pace loop are accepted.

Applying objective ratings and difficulty adjustments. The tapping machine time study previously shown in Figure 20.9 is shown, with the objective ratings and difficulty adjustments and with the base time computed, in Figure 20.16.

1 Engage feed	C																				
	I	20	27	23	27	27	23	22	24	25	22	23	21	24	23	23	23	25	25	24	22
2 Tap out of piece	C																				
	I	10	8	11	10	10	10	9	9	9	10	9	9	9	10	10	10	9	9	10	9
3 Release piece in truck	C																				
	I	5	5	6	6	5	7	7	6	6	6	6	5	6	6	6	6	6	7	6	6
4 Release gage on bench	C																				
	I																				
1	C																				
	I	F	23	24	23	27	27	25	24	27	20	21	24	22	21	26	23	20	18	24	20
2	C																				
	I	10	9	9	9	9	8	10	9	10	9	9	9	9	9	9	9	10	9	9	9
3	C																				
	I	5	7	6	6	5	6	–	5	5	6	6	6	6	6	7	5	6	5	6	7
4	C																				
	I										143										
1	C																				
	I	20	20																		
2	C																				
	I	9	9																		
3	C																				
	I	5	6																		
4	C																				
	I																				
	C																				
	I																				
	C																				
	I																				
	C																				
	I																				
	C																				
	I																				

RECAPITULATION

ELEMENT NO.	1	2	3	4	5	6	7	8	9	10	11	12	13	14	15	16
Amount of body	E-8	–	E-8	D-5												
Foot pedals	F-0	–	F-0	F-0												
Bimanualness	H-0	–	H-0	H-0												
Eye-hand coord.	J-2	–	J-2	J-2												
Handling Require.	0-1	–	0-1	P-2												
Weight	4-4	–	4-4	4-4												
Total time in min.	9.52	3.81	2.41	1.43												
Number of obs.	41	42	41	1												
Pro-rate divisor	1	1	1	50												
Average per cycle	.232	.091	.059	.029												
Rating	112	100	112	80												
Rated time	.260	.091	.066	.023												
Percent of cycle	–	–	–	–												
1 + Adjustments	1.15	1.00	1.15	1.13												
Base time	.299	.091	.076	.026												

SYMBOLS USED IN STUDY

112	100	91
85	81	71
65	55	49
45	41	37

Figure 20.16. Time study for tapping machine (as in Figure 20.9) with ratings and difficulty adjustments and with base time computed.

Application of Allowances

Base time, as computed by the procedures given thus far, does not include three additional groups of additives (*A*) that commonly must be included in the standard time. These are:

1. Allowance for personal time.
2. Allowance for irregular occurrences that may not have been time-studied or that cannot be prorated.
3. Allowance for machine time.

Each of these will be discussed separately. These allowances differ from the difficulty adjustments[23] in three respects: first, they are usually applied similarly to every element in the task; second, they may, in some categories, actually represent a block of time that may be accumulated from a large number of cycles to eventually provide an interval in the work spell during which the worker will not be working; and third, they relate to factors external to the job.

Allowance for Personal Time

This category is not to be confused with the catchall term *fatigue allowance*, which is frequently found in the literature and which, because of the manifold interpretations placed on the term *fatigue*, has led to a great deal of misunderstanding. It should be borne in mind that after the time study observations have been adjusted either by subjective or objective ratings (in the latter case, including difficulty adjustments), the result is supposedly a time value that would permit the operator to produce work in this time, or less, throughout the normal work spell, as far as the internal work of the operation is concerned. No allowance could be devised that would prevent the operator from becoming *fatigued* (in the common sense of the word) from the day's work, but if the concept of normal, which was used in rating, was reasonable, the fatigue (in this sense) is reasonable. The use of an additional allowance would, in subjective rating, imply that step 1 of the rating procedure was done improperly and, in objective rating, that the difficulty adjustment was inadequate.

---

[23]Whether the difficulty adjustment is made mentally, as with subjective rating, or from tables, as with objective rating, is not of concern. Allowances will be appropriate in both cases.

Providing for personal needs. In rating, no attention is given to personal needs (among other items) or to the effect of the external conditions of the operation upon personal needs. It is not usual that a person can work through a normal industrial work spell without attending to personal needs. The amount of time required for these will be affected by the conditions surrounding the work; less in comfortable, quiet surroundings, and more when they are hot, dusty, or noisy.

Commonly used allowances. Many plants follow a practice of providing for personal time during the working day in accordance with a schedule as given in Table 20-V. A means of correctly reducing the time values given in Table 20-V to a percentage figure for proper pro rata addition to the base times will be discussed later in this chapter.

Table 20-V. ALLOWANCE FOR PERSONAL TIME*

Condition	Reference Letter	Time for an 8-Hour Work Spell Divided by a Lunch Period into Two 4-Hour Work Spells (*minutes*)
Comfortable	*S*	23
Warm or slightly disagreeable	*T*	30
Hot, dusty, noisy, etc.	*U*	50
Special or unusual	*SP*	As required

*When rest breaks occur in the work spell, these may be in lieu of all or most of the personal time.

In many plants, to replace most personal time taken at will, rest pauses of various lengths are introduced into the work spells at definite intervals. An appropriate amount of time will have to be introduced into each standard so as to apportion this time to all the jobs the worker may work on during the day. In such cases, the value that would otherwise be selected from Table 20-V is replaced with the total time of these rest pauses.

Allowing for unusual conditions. A few examples of "special or unusual" conditions may be of interest. In one plant manufacturing dye transfer carbons (hectograph material), the maintenance of satisfactory industrial and public relations requires that sufficient time for washing be allowed, so that the workers do not inadvertently carry away any of the exceedingly transferable dye on their persons or clothes when leaving the plant. As a result, as much as 2 hours, or 120 minutes out of a 480-minute day, may be allowed for changing clothes when entering and for thorough washing and rechanging when leaving.

For another example, the operator of a process furnace in a pipe plant may easily be able to perform the motions of the cycle of his task in 0.25 minute, but the heat of the furnace, radiating to his body, may force him to remain idle, away from the furnace for 30 minutes out of each hour, or 240 minutes out of an 8-hour work spell.

Bargaining for allowances. Such allowances, as can readily be seen, are quite reasonably subject to collective bargaining. However, if wage incentives are in use, care should be taken that negotiations for such allowances are not confused with wage negotiations, which are something quite different. These may easily be confused if the allowances are raised beyond the point of fulfilling their purpose and instead so increase the allowed time per unit as to permit considerably more incentive increment than could be anticipated with the concept of the standard in use. This is far beyond the conceived purpose of the allowances.

Allowance for Irregular Occurrences

Although it is desirable to study and rate all irregular occurrences and prorate these so as to apportion them properly to each cycle, some occurrences are of such a nature as to make this extremely difficult.

End-of-day clean-up problem. For example, let us assume that a sewing operator must clean the sewing machine at the end of the day. Perhaps this takes 5 minutes. It is hardly proper to charge this solely to last job of the day, nor is it feasible to prorate this to a time value addable to each cycle if the machine operator works on several jobs each day (as is not uncommon) and each job has a different total cycle time. However, the cost of this 5 minutes should be distributed among the jobs. Consequently, an adjustment must be made such that a total of 5 minutes is accumulated during a day as part of the standards.

Mechanical recordings. Allowances may be made for tool and machine maintenance and the like. A mechanical or electrical recorder may be placed on the machine and the foreman or worker may note, in an accompanying log, the reason for each delay. These two records may later be analyzed to provide a basis for allowances for irregular elements.

Direct attribution to jobs preferable where feasible. It is strongly suggested, however, that where the irregular work may be directly attributed to a specific job, and a rate of occurrence during that job determined, that the time value be handled as an irregular element. Otherwise, these allowances for irregular elements may well get out of hand and eventually lead to sizable inconsistencies in the comparative difficulty of different standards. Extended time studies may be made in the course of which the work is divided only into productive work and delays. These may be made with a stopwatch or a memomotion camera. The camera offers the advantage of being able to observe more than one machine at a time, with no loss of accuracy. Using such studies as a basis, the allowances may be determined.

Making a production study. Figure 20.17 is a *production study* made, at a large printing plant, of the operation of a combiner (a machine for pasting a printed sheet to cardboard). To make this study, the time study observer was at the machine before the shift began. The supervisor and workers had been notified, on the previous day, that such a study would be made. The study was made with a stopwatch. The record was made in the following manner:

1. The basic information at the head of the sheet was entered prior to the starting of the shift.
2. The observer started the stopwatch when the shift began and recorded the following:
   a. The ending time of each period of productive work.
   b. The units produced during each period of productive work.
   c. The ending time of each delay.
   d. The reason for each delay.
3. The watch was allowed to run throughout the entire shift and all events were recorded.
4. The actual production and delay times were obtained subsequently by subtraction, and the values at the bottom of Figure 20.17 were computed.

Following the study, a delay percentage was computed for delays too small to be punched out and normally expectable in the work. Breakdowns or other delays not a regular part of the work would not, by company policy, be charged to productive time and therefore were not included in the delay allowance. Production studies are also of considerable assistance in aiding in the determination of why an operator or group of operators are not performing at standard. Inasmuch as production studies are a continuously observed, intensive sample, the exact events may be examined and the reasons for substandard performance determined. Indeed, they are more effective for this latter purpose than for determining delays, inasmuch as delay studies, in many cases, should be taken over longer periods than can be economically studied by means of production studies.

PRODUCTION CHECK STUDY (All time in minutes, readings are ending times)

Page __1__ of __1__
Dept. __16__
Product and No. __1272 PUZZLE__
Operation __COMBINE__
Machine and No. __COMBINER #7__

Study date __3/31__
Opr. and No. __PINE 817__
Work unit __SHEET__
COUNTER ZEROED AT EACH DELAY

PRODUCTIVE TIME			DELAYS			
Units made	TIME Begin 8.00		TIME			Explanation of delays (why or what done)
Count	R	T	R	T	x = Allowable	
327	805.35	5.35	806.85	1.50	X	SHEET JAMMED
740	819.16	12.31	821.16	2.00		PERSONAL
982	837.49	16.33	837.77	.28	X	OVERSIZED BOARD
544	846.83	9.06	847.39	.56	X	GLUE DRIP
163	850.10	2.71	853.82	3.72		PERSONAL
415	900.73	6.91	901.22	.49	X	SHEET FOLDED
1782	930.93	29.71	931.51	.58	X	OIL PICK UP ARMS, SQUEEKING
694	953.96	11.56	958.12	4.16		PERSONAL
603	1008.17	10.05	1008.60	.43	X	TORN SHEET
659	1019.58	10.98	1019.84	.26	X	BAD SHEETS
1019	1036.83	16.99	1037.68	.85	X	ADD GLUE
575	1047.27	9.59	1048.18	.91	X	ADJUST GLUE ROLL
256	1052.44	4.26	1053.70	1.26	x	THIN GLUE WITH WATER
694	1105.25	11.55	1107.11	1.86	x	" " " "
585	1116.86	9.75	1117.40	.54	x	SHEET JAM
283	1122.12	4.72	1124.57	2.45	x	NO PICKUP – ADJUST CUPS
169	1127.39	2.82	1128.87	1.48	x	SHEET ON GLUE ROLLER
85	1130.28	1.41	1133.45	3.17		PERSONAL
456	1141.06	7.61	1141.91	.85	x	ADJUST GUIDES TO NEW BATCH SIZE
1084	1200.00	18.09				LUNCH TILL 1 PM
337	105.62	5.62	106.45	.83	X	THIN OUT GLUE
1022	123.47	17.02	125.65	2.18		PERSONAL
675	136.88	11.23	137.60	.72	X	CARDBOARD MISFEED
332	143.14	5.54	144.62	1.48	X	ADD GLUE
1482	209.32	24.70	210.13	.81	x	ADJUST GLUE ROLL
688	221.56	11.43	222.32	.76	x	SHEET JAM ON PRESSURE ROLL
892	237.19	14.87	238.10	.91	x	REPLACE WORN PICK UP CUP
711	249.93	11.83	251.46	1.53	x	ADD GLUE AND THIN
664	302.52	11.06	303.78	1.26	x	ALIGN LOOSE GLUE GUIDE
814	317.36	13.58	321.48	4.12		PERSONAL
585	331.22	9.74	331.47	.25	x	TORN SHEET
880	346.10	14.63	347.09	.99	x	CLEAN LINT FROM PICK UP
1078	405.03	17.94	405.76	.73	x	SHEET ON PRESSURE ROLL
261	410.11	4.35	411.76	1.65	x	ADD GLUE
2618	454.30	42.54	500.00	5.70		CLEAN FOR 2nd SHIFT

1 __25,762__ Total production count
2 __427.95__ Total production time *Min.*
3 __60.2/MIN.__ Rate of production, unadjusted (1 ÷ 2)
4 __27.00__ Allowable delay time
5 __56.6/MIN.__ Adj. rate of production [1 ÷ (2 + 4)]
6 __25.05__ Unallowable delays (Personal time and clean up allowed elsewhere)
7 __53.5/MIN__ Observed rate of production

Figure 20.17. Production study in printing plant.

Using work sampling to determine allowances.   As a preferred procedure, the use of work sampling may assist in determining accurately the rate of occurrence of irregular events or delays and may more readily be extended over the necessary period of time. This approach, described in relation to methods study in Chapter 8, requires recording the events taking place during random observations of an operation and computing the percentage occurrence of any type of event as the percentage of the total observations. If the number of observations is great enough to reduce the probable sampling error to tolerable limits, then the percentage computed may be taken as representing the percentage required for this activity.

The details of making a work sampling study were described in Chapter 8. When making such a study to determine delay or other allowance percentages, there will be only two differences from the procedure previously given. These differences are:

1. The categories into which the observations will be classified must be appropriate to the concept of allowances or delays.
2. The sample size must be enough to provide an accuracy suitable for performance standards.

Table 8-I would be used to determine the size of the required study. Let us say that we wished to determine the actual per cent of time spent in tool replacement on a machine. A rough estimate places this at 5 per cent. Let us further assume that an accuracy of ± 1 per cent of total with 95/100 probability is sufficient. Table 8-I indicates that 1,900 readings are necessary. Let us assume that our random method of recording is estimated as giving us 50 recordings a day. Thus, 38 days will be required to obtain sufficient data. (It should be noted that these readings will not be other than a miscellaneous duty of the individual obtaining them unless a considerable number of work sampling projects are in progress, of which this is only a part.) To continue, let us say that at the end of 38 days we have 1,950 readings, and the percentage indicated, for the event under study, is 6. Table 8-I indicates that we must continue to accumulate readings until we have 2,260, at which time we will check again. A similar procedure would be followed with the other limits.

As an alternative procedure, the cumulative percentage may be graphed with the study continuing until a stable value is reached. The accuracy of this cumulative percentage may be checked by daily computation, as follows:

$$E_{95/100} = 2\sqrt{\frac{\bar{p}(1 - \bar{p})}{N}}$$

where

$E_{95/100}$ = value to be added and subtracted from the value of $\bar{p}$ to indicate the limits within which we have 95 chances in 100 that the true value of $\bar{p}$ lies

$\bar{p}$ = percentage of observations of any one type

N = total number of observations made in the study

Combining Personal Allowances
and Allowances for Irregular Occurrences

This is an appropriate place to examine a suitable way of making an adjustment to the base times so as to allow for the accumulation of the blocks of time required to provide for the first two allowance categories that have been discussed.

An example. Let us assume that the allowances for a sewing-machine operator for personal time and irregular occurrences not prorated to each job are as follows:

Personal—reference S	23 minutes
Irregular—daily cleanup	5 minutes
Total time to be distributed	28 minutes

In other words, if the work day is 480 minutes, it will be made up of 28 minutes of allowances and 452 minutes of work. Consequently, each expenditure of time in work must allow the worker to accumulate [(28/452) × 100] per cent more time for expending in the manner implied by the allowance categories. In this case this is 6.2 per cent.[24] In actual practice, in this case, 6 or 7 per cent would probably be used. It would appear reasonable to suggest the use of the 7 per cent (the next highest whole per cent) in such a case, because it is within the range of accuracy and does not imply a dimunition of the allowance. Undoubtedly, it would be possible to advance reasons for using the 6 per cent value. This is a matter for policy decision, inasmuch as the inherent reliability of time study could hardly support, on a logical basis, the use of a decimal percentage allowance.

Procedure to be used. As a general procedure, to combine peronal and irregular allowances into a per cent value for addition to base time, the following equation, in accord with the procedure used in the example just given, may be used.
Let

A = minutes per workday
B = minutes personal allowance per workday
C = minutes irregular allowance per workday

---

[24]See D. Anderson and A. H. Hansen, "The Right and Wrong of Time Study Computations," *Iron Age*, August 17, 1944. This article appears to be the first to propose this obviously correct procedure.

Then

$$\text{per cent allowance} = \frac{B + C}{A - (B + C)} \times 100$$

It is also suggested that the nearest higher whole per cent be used unless the the calculation gives a whole number.

Allowance for Machine Time

This third category raises certain questions that must be carefully examined. For discussion, let us say that the definition of standard time in use is: "130/100 of the amount of time required, assuming method and condition, for the typical operator expending maximum exertion on the job."

Some examples of jobs that may be encountered. The effect of the factor of machine control in such circumstances might best be understood by examining four sample jobs, as follows:

Job A: 100 per cent hand time, 0 per cent machine-controlled
Job B: 60 per cent hand time, 40 per cent machine-controlled
Job C: 30 per cent hand time, 70 per cent machine-controlled
Job D: 0 per cent hand time, 100 per cent machine-controlled

In all of the foregoing cases we may further assume that the operator is idle during the machine time, except in case D, where his work may be a feeding of the machine, work that makes but a slight demand upon him. Of course, such jobs should be considered a challenge to the method designer, who should attempt, in most cases, to find some useful activity for the otherwise idle time, but this is not always possible. Further, the desirability of maximum machine utilization or the desirability of continuous attention to the machine may outweigh the desirability of maximum labor utilization. Therefore, it does not appear unreasonable to assume that we may encounter valid instances of such jobs as have been listed.

When no machine-controlled time is in the job. In case A, if the actual standard time represented the recommended definition of standard time, the typical operator on the job would be able to exceed standard performance by 30 per cent.[25]

---

[25] With a different version of standard in use, this per cent and all others in the following discussion would change, but the implications would remain the same.

With 40 per cent of the time machine-controlled. In case B, the effect of the rest has been found to be such that it will usually permit the operator to increase his pace sufficiently, on the hand- or operator-controlled part of the cycle, above that expectable on job A. (In this case also, the expectable incentive increment would be 30 per cent without excessive exertion, provided that the pace concept used to rate the job was the same pace against which job A was rated.) To make this obvious, let us further assume that job A at the given pace (with difficulty adjustments added if objective rating was used) was 1 minute long and that job B involved the identical 1 minute of work plus 0.667 minute (40 per cent of total cycle) of rest while a machine operated.

A comparison of case A and case B yields some significant observations.

1. The maximum pace, taking into account the effect of machine-enforced rest, is higher for job B than job A even if the actual handwork is identical.

2. It is unlikely that step 1 of subjective rating will take into account the effect of the intermittent rest. To do so would require an infinite multiplicity of mental concepts. Hence, the operator will obtain the full benefits of the intermittent rest.

3. Objective rating will not differentiate between the two jobs inasmuch as all jobs are compared to a single, standard pace. The operator will obtain the full benefits of the intermittent rest.

As will be shown in a later chapter, the effect of the rest, if the operator obtains the benefit, is such that expectable production increment on job B will be the same as on job A. Hence, the standard time set by normal procedures will meet all the criteria given in Chapter 6.

With 70 per cent of the time machine-controlled. Case C raises fully the problem of machine control. In this case the operator will again, without undue exertion, find it possible, because of the larger rest, to increase his pace above even that expectable with job B. However, the fixing of 70 per cent of the cycle time will not allow him to exceed standard by 30 per cent, even if the pace required for standard was the same as on job A. This condition ensues as soon as more than 45 per cent of the cycle is machine-controlled.[26] Therefore, the indirect adjustment made, when the same pace as was used to rate job A is used to rate job C, will not fully adjust the time for job C to match the definition of standard time represented by the time standards of cases A and B.

---

[26]The data supporting this statement appear in Chapter 41.

With 100 per cent of the time machine-controlled. Case D is the extreme situation. This gets no benefit from any indirect adjustment, and the observed time is the minimum time.

Possible solutions to equalizing attainability of more than standard production. The expectable production increment over standard is greater on jobs A and B than with job C and is greater with jobs A, B, and C than with job D. What should be done, if anything, about these differences depends upon the use to be made of the standards and upon the concept of equity held; a policy decision is required.

Three basic types of solutions are possible:

1. Make no adjustment.
2. Adjust the standards so that the probable production increment over standard is related to the per cent of manual work in the cycle.
3. Adjust the standard so that the expectable production increment over standard is the same for all jobs.

Summary—Allowances

Allowances are added to all elements equally. Table 20-V is suggested with the understanding that either because of the nature of the allowance or the meagerness of data, as fits each case, some of the items of such a table may need to be recomputed or may well be negotiated before being used in an actual application.

Summary Example

A time study completed with allowances, and with the final calculations made and transferred to the front, is shown in Figures 20.18a and b. Note the two calculations at the bottom of the front of the sheet, which are designed to summarize the effect of the ratings and allowances for the time study supervisor. The "production anticipated at standard" is computed on the basis of the allowed cycle time. The "production during study" is computed on the basis of the total average time (with irregular elements prorated as necessary). These two values sum up the relationship between the observed performance and anticipated performance and are often of great aid in

ELEMENTS

No.	TERMINAL POINT	1 R	1 T	2 R	2 T	3 R	3 T	4 R	4 T	5 R	5 T	6 R	6 T	7 R	7 T	8 R	8 T	9 R	9 T	10 R	10 T	11 R	11 T	12 R	12 T	13 R	13 T	14 R	14 T	15 R	15 T
1	Release ring	7	7	33	6	62	7	86	6	15	7	43	6	71	7	98	6	24	6	50	6	79	7	16	F	44	7	71	6	98	7
2	Pick up assembly	16	9	43	10	70	8	96	10 M	52	9	52	9	80	9	7	9	32	8	59	7	89	10	25	9	53	9	80	9	8	10
3	Let go of weight	22	6	50	7	77	7	3	7	32	/	59	7	86	6	13	6	39	7	66	7	95	6	32	7	59	6	86	6	15	7
4	Let go of lever	27	5	55	5	82	5	8	5	37	5	64	5	92	6	18	5	44	5	72	6	0	5	37	5	65	6	91	5	20	5
5	Return from stockroom			180	98																										
6																															
7																															
8																															
9																															
10																															
11																															
12																															
13																															
14																															
15																															

RECAPITULATION

ELEMENTS	1	2	3	4	5	6	7	8	9	10	11	12	13	14	15
Amount of body	C-2	C-2	C-2	C-2	E-8										
Foot pedals	F-0	F-0	F-0	F-0	F-0										
Bimanualness	H-0	H-0	H-0	H-0	H-0										
Eye-hand coordination	J-2	J-2	J-2	J-2	I-0										
Handling requirements	N-0	N-0	N-0	N-0	N-0										
Weight or resistance	0-0	0-0	2·5W	5·W	W·15										
TOTAL TIME IN MIN.	.91	1.28	.92	.78	.98										
NUMBER OF OBS.	14	14	14	15	1										
PRO-RATE DIVISOR	1	1	1	1	100										
AVERAGE PER CYCLE	.065	.091	.066	.052	.010										
RATING	110	110	108	105	90										
RATED TIME	.0715	.1006	.0710	.0546	.0088										
PER CENT OF CYCLE	—														
1+ADJUSTMENTS	1.04	1.04	1.19	1.12	1.23										
BASE TIME	.0744	.1046	.0845	.0612	.0108										
1+ALLOWANCES	1.04	1.04	1.04	1.04	1.04										
ALLOWED TIME	.077	.109	.088	.064	.011										

SYMBOLS USED

C-Extra unnecessary motion
D-A.D. Dropped part
F-Unnecessary fumble
H-Unnecessary hesitation
P-Includes personal time
R-Made reject by improper work
X-Deviation from std. routine

ALLOWANCES

Personal 15 min _____ %
Clean up 3
18/(480-18)
TOTAL 4 %

LOOP VALUES		
141	132	120
100	93	82
74	67	53
44	40	36

Figure 20.18a. Completed rear of time study form for task of assemble saxophone pad (*continues*).

PRODUCTION STANDARD—OPERATION METHOD SUMMARY SHEET

Page No. 1 of 1   Department SAX

Operation and No. ASSEMBLE SAXAPHONE PAD   Product S-91 SPEC. 24382

BENCH; FIXTURE 2011   Work Unit PAD

Machine(s) and No.   Study Date 6/21

SALLY MORKIN # 5492   10⁵⁰ AM – 11³⁰ AM   Study No. S-91-2

No.	LH — Detailed Description of Operation Elements	RH	Std. Allowed Time/Unit
			Minutes
1	Get pad, line up in work area for assembling ring to pad	Get metal ring, carry to pad, assemble to pad and *release ring*	0.077
2	Hold ring and pad while right hand moistens glue and then let go	To water container, dip 4th finger, to pad, moisten glued portion with circular motion, *pick up assembly*	0.109
3	Get complete pad from fixture, set aside, get weight and place on pad, *let go of weight*	Move to lever, use heel of hand for pushing lever to open fixture, delay for left hand	0.088
4	To fixture, assemble pad into fixture, wait for right hand to close fixture	Carry pad to fixture, assemble into fixture, move to lever, move lever to close fixture; *let go of lever*	0.064
5	Exchange finished parts for stock at stockroom 50 ft. away 1/100		0.011

Production during study 211/Hr

REMARKS:

A = Pads
B = Rings
C = Fixture
D = Water
E = Work area
F = Lever
G = Operator, seated

Std. Allowed Time Minutes/Unit	0.349
Standard Hours/100 Units	.582
Standard Units/Hour	172
Observed by:	Wren
Approved by:	Brown

Figure 20.18b. (*Concluded*) Completed front of time study form for task of assemble saxophone pad.

indicating when unusual or difficult situations may arise from the installation of the standard.

Summary: Direct Time Study— Intensive Sampling

1. *Uses*:
   For setting standard times for jobs that are manual and repetitive and where the cycle time is such that a reliable sample may be observed economically by a period of continuous observation of the work.
2. *How made*:
   a. A concept of standard time must be established.
   b. The worker and supervisor should be informed that a study is going to be made.
   c. The work-unit is defined.
   d. The method is divided into elements, and the details of each element are recorded.
   e. A statistically reliable sample of time recordings is obtained.
   f. Ratings are made and applied.
   g. Adjustments and allowances are added.
   h. A standard time is computed.
   i. The standard time is validated by:
      (1) Comparison with the standard time for similar jobs.
      (2) Subjective evaluation of the credibility of the ratings.
      (3) Monitoring performance against the standard.

PROBLEMS

20.1. Prepare a time study sheet for a time study of the job of inspecting drawn copper wire as shown in:
(a) Figures 14.1 and 14.2.
(b) Figures 14.3 and 14.4.

20.2. Prepare a time study sheet for a time study of the job of assembling hand-hole cover as shown in:
(a) Figures 16.2 and 16.3.
(b) Figures 16.5a and b and 16.6.

20.3. Prepare a time study sheet for a time study of the job described in:
   (a) Problem 16.2.
   (b) Problem 16.3.
   (c) Problem 16.4.

20.4. Prepare a time study sheet for a simple job that you may have performed or any job in a shop.

20.5. Assume that the readings for an element were 0.14, 0.15, 0.14, 0.15, 0.16, 0.13, 0.15, 0.14, 0.15, 0.14, 0.16, and 0.15, all in minutes.
   (a) What can you state concerning the suitability of these values as a basis for determining an average value representing the performance observed?
   (b) What would the next step be?

20.6. Assume that the readings for an element were 0.03, 0.04, 0.05, 0.06, 0.04, 0.05, 0.04, 0.03, 0.04, 0.06, 0.05, 0.06, and 0.04, all in minutes.
   (a) What can you state concerning the suitability of these values as a basis for determining an average value representing the performance observed?
   (b) What would the next step be?

20.7. Assume that the readings for an element were 0.07, 0.04, 0.06, 0.08, 0.08, 0.04, 0.07, 0.06, 0.05, 0.06, 0.03, 0.06, 0.05, 0.06, and 0.04, all in minutes.
   (a) What can you state concerning the suitability of these values as a basis for determining an average value representing the performance observed?
   (b) What would the next step be?

20.8. Assume that the readings for an element were 0.22, 0.21, 0.23, 0.24, 0.20, 0.19, 0.22, 0.22, 0.23, 0.22, 0.24, 0.25, 0.23, 0.22, and 0.21, all in minutes.
   (a) What can you state concerning the suitability of these values as a basis for determining an average value representing the performance observed?
   (b) What would the next step be?

20.9. Set up a job that may be performed in the laboratory, record the method, and observe and record the time values for a series of cycles. Obtain enough readings to yield a reliable average and compute the average for each element. (Splicing short lengths of film makes a suitable job.)

20.10. Describe the method of rating time studies in use in the plant in which you work or have worked or in a plant with which you are familiar.

20.11. Find another time study text and report in detail on the method of rating which is proposed by the author of that work.

20.12. See the rating practice films distributed by the Society for Advancement of Management. Rate with these films and make an analysis of the results.

20.13. Films for rating practice, each 400 feet in length, and each showing 10 paces for performing an industrial job, filmed in industrial plants, may be obtained from the author.* Current titles include:

_____

*Multi-image loops are also available.

Film Number	*Title*
1	Roping Salami
2	Feeding 28-lb Tank Heads to Punch Press
3	Preassembly of Screws to Insulators
4	Set Up Wire-Bound Box
5	Lift Teletype from Mounting Bolts
6	Manual Purge of Requisition File
7	Loosen Terminal Block Support
8	Plate Radio Crystal and Check Frequency
9	Load Piece on Broach
10	Wrap $\frac{1}{4}$-cup Aluminum Measuring Cup
11	Pick Up Pressure-Sensitive Label on Carton Pack
12	Make Mold in Foundry, Element 1
13	Load 10-lb Wooden Boxes on Skid
14	Hand-Feed Wrapping Machine
15	Interleave Greeting Cards and Envelopes
16	Run Natco Multiple-Spindle Drill
17	Feed Casing-in Machine
18	Brush Small Metal Diaphragm
19	Cover Doll Boxes
20	Trim Meat from Blade Bone
21	Measure Hog Casings
22	Pull Cable Aboard Ship
23	Edge Disk in Tin Shop
24	Assemble Deck House to Coaming with Huck Bolts
25	Scribe Disk in Tin Shop
26	Push Hand Cart
27	Spot, Face, and Drill Casting
28	Assorted Pharmaceutic Operations
29	Various Machine Shop Operations

Rate a selected film. Following the analytical methods suggested by the instructor, analyze the results.

20.14. Assuming that 40 per cent of the time is spent in acceleration, 20 per cent traveling at constant velocity, and 40 per cent in deceleration, prove that with a given rate of acceleration the time for the longer movement divided by the time for the smaller movement is equal to the square root of the longer distance divided by the smaller distance.

20.15. Compute the difficulty adjustments for each element of the job of:
(a) Problem 20.1(a).
(b) Problem 20.2(a).

    (c) Problem 16.2.
    (d) Problem 16.3.
    (e) Problem 16.4.
    Note: If any additional information is required, list a reasonable assumption and proceed.

20.16. Compute the difficulty adjustments for the job studied in connection with Problem 20.9. Assume a rating of 80 for all the hand elements and compute a base time for each element.

20.17. Make a time study of a task, rate objectively, and compute the difficulty adjustments. Compute the base time.

20.18. Compute the allowance for the job described in:
    (a) Problem 20.1(a).
    (b) Problem 20.2(a).
    (c) Problem 16.2.
    (d) Problem 16.3.
    (e) Problem 16.4.
    Note: If any additional information is required, list a reasonable assumption and proceed.

20.19. What allowance would have to be added to compensate for the machine time effect on a job where the machine controlled 50 per cent of the cycle but the operator worked during the entire cycle? (Typical production expectancy is 130 per cent.) The policy is to equalize potential incentive earnings of machine and manual jobs.

20.20. Same as Problem 20.19 except that the policy is to make the per cent incentive ostensibly proportional to the per cent of cycle worked, with no consideration of the pace increase due to rest.

20.21. It is estimated that an activity takes 12 per cent of the total work day. Determine the number of observations that we would need to take, on a random basis, to measure accurately the percentage of time consumed by this activity, with the probability of 95 chances in 100 of not being in error by more than:
    (a) 1 per cent of total.
    (b) 1 per cent of 12 per cent.
    (c) 5 per cent of 12 per cent.
    (d) 10 per cent of 12 per cent.
    (e) 2 per cent of total.

21 PREDETERMINED TIME SYSTEMS

Introduction

1ST order work unit = Human Motion — Body Movements

Predetermined time systems employ first-order work-unit performance time data which have been obtained from a careful analysis of human performances. These data may be resynthesized into a standard time for a job, even when the job is entirely different from those previously studied. The presently available data may not permit this synthesis to be performed with absolute accuracy in terms of a fully consistent concept of standard. However, they produce a reasonable approximation which has considerable utility, if used with discretion.[1]

---

[1] It should be noted that most engineering techniques exhibit similar aspects. Building, bridge, road, and machine design usually employ "factors of safety," inasmuch as in the synthesis of the design, one assumes certain ideal materials or conditions and thus obtains answers somewhat less than accurate. However, if these answers are used intelligently, they

Definition. A *predetermined time system* (*PTS*) is a set of organized data for first-order work-unit standard times, representing some consistent and known concept of standard performance, together with the rules and conventions for computing and documenting a task standard time from these data. The application of a predetermined time system results in a "predetermined time standard."

Alternative definition. As an alternative definition of predetermined time systems, in the terms used in the basic equation, given in Chapter 5, for a standard time:

1. *W/C.* A count of 1 is used. The cycle of work is set up before the work; the work is modeled. It is described in terms of first-order work-units. Considerable judgment is needed to make certain that all the required motions are included in the model.
2. *W/T.* The time for each first-order work-unit is obtained from predetermined tables. Whatever the system of data employed, some judgment is necessary in selecting the appropriate time from the tables; the amount of judgment varies from system to system. Machine time is taken from similar tables for machines.
3. *M.* If the concept of normal implicit in the tables of predetermined times is different from that in use in the plant, a modifier is applied to all values. In the development of this modifier, ratings may be needed. Once this modifier has been determined, no subsequent individual judgments or ratings are needed.
4. *A.* An allowance is developed and applied in the same manner as with direct time study-intensive sampling.

Types of work with which it is used. The approach with such data primarily offers the opportunity to estimate the time for the performance of a task before it is performed; it also does away with the need for rating with each individual study. These techniques, in their basic form, are primarily applicable to repetitive work. They are an alternative approach to setting standard times for the same type of operations for which direct time study—intensive sampling is used. Of course, if part of a cycle is controlled by the use of a machine, data other than human performance time data will also be needed to complete the computation of the standard time.

---

are of great utility. We have only to look around us to see the results. However, when time standards are used as a basis for cost determination or wage payments, the tolerances to which one must work are considerably smaller than in most other design activity. Consequently, one must use even greater discretion with data such as are described in this chapter.

Uses of predetermined time systems. Using tables of predetermined times one may:

1. Compare the time for alternative, proposed methods in order to permit an examination of the economics of the proposals prior to building the equipment or prior to production runs.
2. Use the time values to construct any of the time-scale methods analysis charts as part of the previously described methods improvement procedures.[2]
3. Compute, with repetitive work, an estimate of manpower, equipment, and space requirements prior to production or prior to setting up the facilities.
4. Develop tentative layouts for assembly lines prior to their construction to minimize the amount of subsequent rearrangement and rebalancing.
5. Construct higher-order work-unit predetermined time tables.
6. Determine job time standards.

In all the above uses, the motion study techniques described in previous chapters may appropriately be employed to assist in constructing the method with the predetermined time data providing the time dimensions.

In addition to these uses, the data may provide an independent basis for checking direct time study standards. While there may at times be considerable disagreement between a direct time study standard and a standard based on predetermined times, a study of the differences may serve to increase understanding of the task and assist in determining the correct standard time.

As will be seen in Chapter 23, the data from either the application of direct time study—intensive sampling or predetermined time systems (or other work measurement techniques) may be used to build predetermined times for higher-order work-units. These higher-order data, when the values are assembled for second-order work-units, are called "standard elemental data." However, the data may also be assembled to produce a system of predetermined times for third-order work-units. This chapter, however, will cover only the use of first-order work-unit predetermined time system data to obtain second- and third-order work-unit standard times.

Limitations on the use of the technique. Eventually, predetermined time system data may serve in the development of standard times for all the uses

---

[2]See "Predetermined Time Standards," *Factory Management and Maintenance*, Vol. 111, No. 9, 1953, pp. 134–139. From very early in the use of predetermined time data, the use of such data seems to have encouraged the making of such time charts and methods improvements.

of such values, as given in Chapter 5. However, as will be indicated later, it is doubtful whether some of the presently available data permit this to be performed within the permissible tolerances for all the uses. Also, as will be seen, some types of work preclude the use of this approach.

To be applicable to a wide variety of jobs, the data must be developed in terms of very small units of work such as therbligs. This introduces new complications. The time for a therblig has been shown to be a function of:

a. Distance.
b. Complexity of action.
c. Amount of body involved.
d. Bimanualness involved.
e. Whether the use of the feet accompanies the action.
f. The eye–hand coordination required.
g. The sensory requirements.
h. The weight or resistance involved and the per cent of time involved.
i. The preceding and following therbligs as well as the context and pattern of the task.[3]
j. The direction of the movement.
k. The place of the therblig in the motion pattern.
l. The number of therbligs in the pattern and the length of time the pattern will be performed.[4]
m. The possible interactions of two variables.
n. Several other variables as yet unidentified, inasmuch as even when the effects of variables a through l are extracted, considerable variation in observed data still exists.

No predetermined time system adjusts for all the variables above.

---

[3] M. E. Mundel (with R. M. Barnes), "Studies of Hand Motions and Rhythm Appearing in Factory Work," *University of Iowa Studies in Engineering, Bull. 12*, 1938; a study of the time to carry, position, and place short cylinders in holes surrounded by varying degrees of bevel showed that not only did the positioning time vary with the bevel, as one might expect, but the transport-loaded time also varied, despite the fact that the workplace dimensions, and hence the path of transport, remained constant. This and similar experiments indicated that the time for a therblig is affected by what precedes and follows it and is not a constant.

[4] This is particularly noticeable in short-cycle jobs performed for long periods. For instance, a job may originally consist of four therbligs, *a, b, c,* and *d.* The original time may be the time for *a,* plus the time for *b,* plus the time for *c,* plus the time for *d.* As the job progresses and the operator develops familiarity with the task, part of *b* may be performed during *a* and part of *d* during *c,* thus reducing the cycle time. After a considerable number of performances, practically all of *b* may be performed during *a* and practically all of *d* during *c,* making the cycle time the time for *a* plus *c.* This phenomenon has not been well studied. Consequently, it is frequently difficult to determine beforehand where, between the two limiting conditions, the expectable method for a job will be.

Different predetermined time systems. There are many systems of predetermined time data in current use.[5] These systems differ from each other in the following respects:

1. The number of variables from the preceding list which are taken into account.
2. The manner of adjusting for these variables.
3. The assumptions concerning the independence of individual motion times.
4. The level of performance on which the time values are based, that is, normal or incentive performance (as well as the range described in Figure 6.1).
5. The manner of classifying motions, that is, therbligs (or modifications thereof), groups of therbligs, or body movements.

Each of these systems of data consists of:

1. A system of notation for describing the job being studied.
2. A set of tables of time values organized in a manner determined by the categories of the system of notation used.
3. A set of rules or conventions for using these tables in a consistent fashion. This aspect commonly amounts to a rather-good-sized book for each system.

Steps in the use of the technique. There are four steps in the use of predetermined time systems (PTS), as follows:

1. Adjusting the data to the concept of "standard" in use in the organization. This needs to be done only once for all studies.
2. Describing the job method in a manner conformant to the conventions of the particular PTS.
3. Applying the PTS values to the job, as described.
4. Applying allowances and other adjustments to the time values, as appropriate to the PTS.

Some Representative Predetermined Time Systems

It should be noted that an effective use of any system of these data depends upon a faithful following of the detailed conventions of application. Hence, the short treatment given here of the systems, described in full in

---

[5] See H. B. Maynard (ed.), *Industrial Engineering Handbook*. New York: McGraw-Hill Book Company, 1971.

other original texts devoted solely to a single system, is intended primarily as an appreciation of these systems rather than a complete treatment. A full study of a data system should be made before applying it.

Methods-time measurement. Representative of the systems that give the time for therbligs and which consider these as essentially independent values is *methods–time measurement (MTM)*.

Methods–time measurement is a procedure that analyzes any manual operation or method into the basic motions required to perform it and assigns to each motion a predetermined time standard which is determined by the nature of the motion and the conditions under which it is performed.[6]

The classification of motions used in methods–time measurement can be seen from Tables 21-I through 21-X. The system of notation used to describe any motion consists of a compound alphanumeric code consisting of a letter (or letters) to describe the motion group, followed by a number to describe the physical dimension of the motion (if applicable), followed by the "case" designator. For instance, M10A is a *Move, 10 inches long*, of a type classified as *case A; G4A* is a *Grasp, case 4A*. The time values in Tables 21-I through 21-X are TMUs, where a TMU = 0.00001 hour. These times are normal performance times without any allowance for personal

Table 21-I. REACH—R

Distance Moved Inches	Time TMU				Hand In Motion		CASE AND DESCRIPTION
	A	B	C or D	E	A	B	
¾ or less	2.0	2.0	2.0	2.0	1.6	1.6	A Reach to object in fixed location, or to object in other hand or on which other hand rests.
1	2.5	2.5	3.6	2.4	2.3	2.3	
2	4.0	4.0	5.9	3.8	3.5	2.7	
3	5.3	5.3	7.3	5.3	4.5	3.6	B Reach to single object in location which may vary slightly from cycle to cycle.
4	6.1	6.4	8.4	6.8	4.9	4.3	
5	6.5	7.8	9.4	7.4	5.3	5.0	
6	7.0	8.6	10.1	8.0	5.7	5.7	
7	7.4	9.3	10.8	8.7	·6.1	6.5	C Reach to object jumbled with other objects in a group so that search and select occur.
8	7.9	10.1	11.5	9.3	6.5	7.2	
9	8.3	10.8	12.2	9.9	6.9	7.9	
10	8.7	11.5	12.9	10.5	7.3	8.6	
12	9.6	12.9	14.2	11.8	8.1	10.1	D Reach to a very small object or where accurate grasp is required.
14	10.5	14.4	15.6	13.0	8.9	11.5	
16	11.4	15.8	17.0	14.2	9.7	12.9	
18	12.3	17.2	18.4	15.5	10.5	14.4	
20	13.1	18.6	19.8	16.7	11.3	15.8	
22	14.0	20.1	21.2	18.0	12.1	17.3	E Reach to indefinite location to get hand in position for body balance or next motion or out of way.
24	14.9	21.5	22.5	19.2	12.9	18.8	
26	15.8	22.9	23.9	20.4	13.7	20.2	
28	16.7	24.4	25.3	21.7	14.5	21.7	
30	17.5	25.8	26.7	22.9	15.3	23.2	

---

[6]By permission from H. B. Maynard, G. J. Stegemerten, and J. L. Schwab, *Methods-Time Measurement*. McGraw-Hill Book Company, New York, 1948, p. 12.

Table 21-II. MOVE—M

Distance Moved Inches	Time TMU				Wt. Allowance			CASE AND DESCRIPTION
	A	B	C	Hand in Motion B	Wt. (lb.) Up to	Factor	Constant TMU	
¾ or less	2.0	2.0	2.0	1.7	2.5	0	0	**A** Move object to other hand or against stop.
1	2.5	2.9	3.4	2.3				
2	3.6	4.6	5.2	2.9	7.5	1.06	2.2	
3	4.9	5.7	6.7	3.6				
4	6.1	6.9	8.0	4.3	12.5	1.11	3.9	
5	7.3	8.0	9.2	5.0				
6	8.1	8.9	10.3	5.7	17.5	1.17	5.6	
7	8.9	9.7	11.1	6.5				
8	9.7	10.6	11.8	7.2				**B** Move object to approximate or indefinite location.
9	10.5	11.5	12.7	7.9	22.5	1.22	7.4	
10	11.3	12.2	13.5	8.6				
12	12.9	13.4	15.2	10.0	27.5	1.28	9.1	
14	14.4	14.6	16.9	11.4				
16	16.0	15.8	18.7	12.8	32.5	1.33	10.8	
18	17.6	17.0	20.4	14.2				
20	19.2	18.2	22.1	15.6				
22	20.8	19.4	23.8	17.0	37.5	1.39	12.5	
24	22.4	20.6	25.5	18.4				**C** Move object to exact location.
26	24.0	21.8	27.3	19.8	42.5	1.44	14.3	
28	25.5	23.1	29.0	21.2				
30	27.1	24.3	30.7	22.7	47.5	1.50	16.0	

Table 21-III. TURN AND APPLY PRESSURE—T AND AP

Weight	Time TMU for Degrees Turned										
	30°	45°	60°	75°	90°	105°	120°	135°	150°	165°	180°
Small— 0 to 2 Pounds	2.8	3.5	4.1	4.8	5.4	6.1	6.8	7.4	8.1	8.7	9.4
Medium—2.1 to 10 Pounds	4.4	5.5	6.5	7.5	8.5	9.6	10.6	11.6	12.7	13.7	14.8
Large— 10.1 to 35 Pounds	8.4	10.5	12.3	14.4	16.2	18.3	20.4	22.2	24.3	26.1	28.2

APPLY PRESSURE CASE 1—16.2 TMU.   APPLY PRESSURE CASE 2—10.6 TMU

Table 21-IV. GRASP—G

Case	Time TMU	DESCRIPTION
1A	2.0	Pick Up Grasp—Small, medium or large object by itself, easily grasped.
1B	3.5	Very small object or object lying close against a flat surface.
1C1	7.3	Interference with grasp on bottom and one side of nearly cylindrical object. Diameter larger than ½".
1C2	8.7	Interference with grasp on bottom and one side of nearly cylindrical object. Diameter ¼" to ½".
1C3	10.8	Interference with grasp on bottom and one side of nearly cylindrical object. Diameter less than ¼".
2	5.6	Regrasp.
3	5.6	Transfer Grasp.
4A	7.3	Object jumbled with other objects so search and select occur. Larger than 1" x 1" x 1".
4B	9.1	Object jumbled with other objects so search and select occur. ¼" x ¼" x ⅛" to 1" x 1" x 1".
4C	12.9	Object jumbled with other objects so search and select occur. Smaller than ¼" x ¼" x ⅛".
5	0	Contact, sliding or hook grasp.

Table 21-V. POSITION*—P

CLASS OF FIT		Symmetry	Easy To Handle	Difficult To Handle
1—Loose	No pressure required	S	5.6	11.2
		SS	9.1	14.7
		NS	10.4	16.0
2—Close	Light pressure required	S	16.2	21.8
		SS	19.7	25.3
		NS	21.0	26.6
3—Exact	Heavy pressure required.	S	43.0	48.6
		SS	46.5	52.1
		NS	47.8	53.4

*Distance moved to engage—1″ or less.

Table 21-VI. RELEASE—RL

Case	Time TMU	DESCRIPTION
1	2.0	Normal release performed by opening fingers as independent motion.
2	0	Contact Release.

Table 21-VII. DISENGAGE—D

CLASS OF FIT	Easy to Handle	Difficult to Handle
1—Loose—Very slight effort, blends with subsequent move.	4.0	5.7
2—Close — Normal effort, slight recoil.	7.5	11.8
3—Tight — Considerable effort, hand recoils markedly.	22.9	34.7

Table 21-VIII. EYE TRAVEL TIME AND EYE FOCUS—ET AND EF

Eye Travel Time $= 15.2 \times \dfrac{T}{D}$ TMU, with a maximum value of 20 TMU.

where T = the distance between points from and to which the eye travels.
D = the perpendicular distance from the eye to the line of travel T.

Eye Focus Time = 7.3 TMU.

needs, fatigue, or delays.[7] They were developed from motion-picture time studies of a large number of operations.

The time values from Tables 21-I through 21-X are used in the same fashion as one uses observed values, to build numerically, rather than graphically, a simo-chart of an operation. Allowances are added to complete the standard time.

---

[7] Maynard, op. cit., pp. 5–29.

Table 21-IX. BODY, LEG, AND FOOT MOTIONS

DESCRIPTION	SYMBOL	DISTANCE	TIME TMU
Foot Motion—Hinged at Ankle.	FM	Up to 4″	8.5
With heavy pressure.	FMP		19.1
Leg or Foreleg Motion.	LM —	Up to 6″	7.1
		Each add'l. inch	1.2
Sidestep—Case 1—Complete when lead-ing leg contacts floor.	SS-C1	Less than 12″	Use REACH or MOVE Time
		12″	17.0
		Each add'l. inch	.6
Case 2—Lagging leg must contact floor before next motion can be made.	SS-C2	12″	34.1
		Each add'l. inch	1.1
Bend, Stoop, or Kneel on One Knee.	B,S,KOK		29.0
Arise.	AB,AS,AKOK		31.9
Kneel on Floor—Both Knees.	KBK		69.4
Arise.	AKBK		76.7
Sit.	SIT		34.7
Stand from Sitting Position.	STD		43.4
Turn Body 45 to 90 degrees—			
Case 1—Complete when leading leg contacts floor.	TBC1		18.6
Case 2—Lagging leg must contact floor before next motion can be made.	TBC2		37.2
Walk.	W-FT.	Per Foot	5.3
Walk.	W-P	Per Pace	15.0

Table 21-X. SIMULTANEOUS MOTIONS

Work Factor. *Work Factor* (*WF*),[8] another system, pioneered by J. H. Quick and W. J. Shea, considers motions as affected by the context of the motion sequence in which they lie. Five time values are given for each motion. The time values given in *work factor* tables are in units of 0.0001 minute and are for incentive performance but do not include allowances for delays, fatigue, or personal time. The five values given for each motion consist of a basic value for a simple minimal difficulty occurrence of a motion and four additional values for the time for the motion with one to four *work factors* of difficulty.

A "work factor of difficulty" is the presence of a manual control factor such as directional control (steer), care (precaution), change direction, or definite stop. Weights or resistances above a given level are counted as one or more work factors, depending on the motion and the sex of the operator. The steps or levels for weight appear in the work factor tables. The work factors are considered equal in their effect upon motion times, and the entry into the work factor tables is made dependent upon only the number of work factors persent.

Other systems. Numerous other systems have been proposed. Some are designed for clerical work, others for work under microscopes,[9] and so forth. The predetermined time system, whose detailed description follows, was designed for repetitive manual work.

Human Performance Times

Introduction. *Human Performance Times* (*HPT*), a system of data developed by I. Lazarus and the author, is based primarily on therbligs but considers the time value for each motion to be affected by the full nature of the motion complex in which it occurs. The time values are given for therbligs, in units of 0.00001 minute, as a function of the physical dimensions of the motions. These time values are subsequently grouped into time study elements and adjusted by means of the difficulty adjustments given in Tables 20-III and 20-IV, thus giving an almost infinite number of values for each therblig. These data and their method of use, given in this chapter, take into account, to a degree, variables a through i of the list given in the introduction to this chapter under "Limitations on the use of the technique." Variables j through n, if they enter into the problem being worked on, may

---

[8]Work Factor is the service mark (trademark) of The Work Factor Company and identifies their services and their predetermined time system, including the values and techniques of use. For a more detailed description, see Maynard, op. cit., pp. 5–65 through 5–101.

[9]For work with miniature electronic circuitry.

still be sources of error. Moreover, errors may also be introduced by improper descriptions of tasks, wrong classifications of therbligs, poor element grouping, wrong uses of the difficulty adjustments, and incorrect allowances. In short, these data do not offer an absolute solution to the problem of work measurement, but if used intelligently they may help to provide a working solution to many problems.

Developing the HPT data. In the development of these data, Lazarus collected an extensive series of time studies of industrial operations using a motion-picture camera to make the complete time study record. All the operators were experienced industrial operators. All the dimensions and features of their jobs were recorded at the time the pictures were taken. The motion pictures were taken at 1,000 frames per minute to permit later rating against a multi-image loop, as described in Chapter 20, by a group of experienced raters. The reliability of the samples of times for each therblig was checked using statistical techniques. The rated therblig times were subjected to a careful statistical analysis to determine the remaining factors which affected the required time.

The final data obtained were tested, first by reconstituting the original jobs, and second, by synthesizing a second group of industrial tasks and comparing the standards thus developed against standards set independently by stopwatch time study and objective rating. The differences between the second set of pairs of values were in the order of the usual discrepancies between pairs of independently set direct time studies—intensive sampling.[10]

Additional variables isolated. In that the data were subjected to objective rating, the time for each therblig, regardless of which job it occurred in, would be constant unless other factors or variables affected the time. The time for each therblig did, in fact, vary from job to job. Hence, an analysis was made to determine the nature and effect of the remaining variables. The statistical analysis of the therblig time data revealed the following facts[11]:

> *TE* and *TL* (*transport empty and transport loaded*)—the time for these therbligs was a continuous exponential function of distance.
> *G* (*grasp*)—the time for grasp was a discrete function of the type of grasp. Four types of grasp were isolated:
>
> 1. Contact: control gained by mere contact.
> 2. Contact pinch: control gained by contact with one or more fin-

---

[10]Standard deviation equals about 5 per cent.

[11]It should be noted that the data had been rated *objectively*. The effect of the six factors compensated for by the difficulty adjustments was therefore removed, and the statistical analysis referred to above was to find the factors causing the residual differences in time values.

gers sliding the object into a position so that the thumb could oppose them on the object.

3. Pinch: control gained by maneuvering the hand so that the thumb and one or more fingers come into opposition on the object.
4. Wrap: control gained by fingers and palm coming into opposition on the object.

P (position)—the time for position was found to be a function of the number of degrees of positioning required. A degree was defined as a restriction in a dimension or orientation which had to be provided by the operator. There are six possible sources of restriction, but no more than five were found occurring at one time. The six sources are shown in Figure 21.1.

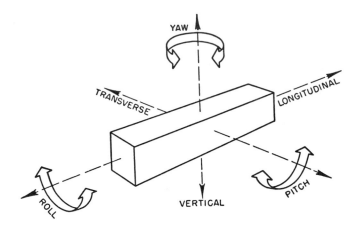

Figure 21.1. The six possible sources of restriction of position. (From Lazarus HPT.)

A (assemble)—the time for assemble was found to be a function of the distance of the restricted movement caused by the mating of the parts. *DA* and *RL (disassemble and release load)*—the time for these therbligs appeared to be constant.

General manner of use. The manner of use of the data presented later involves grouping the therbligs into elements and applying to the elements the difficulty adjustments described in Chapter 20. Thus, the method takes into account the variables that have just been listed for each therblig, those covered by the difficulty adjustments, and the effect of the context of the element the therbligs occur in, so as to produce standards fully compatible with standards set by direct time study—intensive rating.

The data developed by Lazarus were originally rated against an early multi-image loop which is not readily available. This loop was subsequently rated in comparison with the widely circulated benchmark film shown in Figures 20.12 and 20.13. The original values were then multiplied by a constant to convert them to the 100 per cent pace of the loop in Figure 20.13, so that any plant could, if necessary, subsequently convert the values of the table to their concept of normal. These adjusted data are given in Table 21-XI. The conventions for using these data are given at the bottom of the table.

How To Use HPT Data

Workplace sketch and method summary. The actual or contemplated workplace or work area should be measured and a dimensioned sketch prepared. A summary should be prepared of what is accomplished by the operation. Figure 21.2 is a plan view of the workplace showing the drill

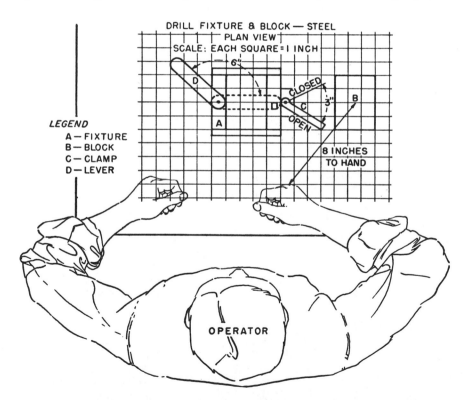

Figure 21.2. Plan view of workplace of drill press operator. (Drill and controls left out to simplify illustration.)

fixture and the piece to be drilled. (The drill and controls have been left out to simplify the illustration.) Figure 21.3 is a drawing of the fixture and part concerned. In the part of the operation treated in this illustration, the operator must *get, place, and lock the piece in the fixture.*

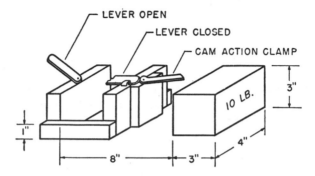

Figure 21.3. Details of drill fixture and workpiece.

HPT System Data and Conventions

Table 21-XI. HUMAN PERFORMANCE TIMES IN UNITS OF 1/100,000 MINUTE (0.00001 MINUTE)*
Basic values subject to difficulty adjustment when assembled into elements.

Transport loaded, transport empty, or for steps in walking or leg movements

Distance (inches)	Time	Distance (inches)	Time	Distance (inches)	Time	Distance (inches)	Time
1	137	13	495	25	686	37	835
2	194	14	514	26	700	38	846
3	237	15	532	27	714	39	857
4	274	16	549	28	726	40	868
5	307	17	566	29	739	41	879
6	336	18	582	30	752	42	890
7	362	19	598	31	764	43	900
8	388	20	614	32	777	44	910
9	412	21	629	33	789	45	921
10	434	22	644	34	801	46	931
11	455	23	658	35	812	47	941
12	475	24	673	36	823	48	951

Table 21-XI. (CONTINUED)

Grasp

Condition	Time
Contact	139
Contact–pinch	222
Pinch	286
Wrap	453

Position

Condition	Time
1 degree	160
2 degrees	178
3 degrees	249
4 degrees	330
5 degrees	445

Disassemble

Condition	Time
All	381

Release Load

Condition	Time
All	151

Non-therbligs

Sit	1560
Arise	1904
Eye focus	416

Assemble

Distance (inches)	Time	Distance (inches)	Time	Distance (inches)	Time
$\frac{1}{4}$	0^2	$1\frac{1}{4}$	383	$2\frac{1}{4}$	800
$\frac{3}{8}$	19	$1\frac{3}{8}$	435	$2\frac{3}{8}$	852
$\frac{1}{2}$	71	$1\frac{1}{2}$	487	$2\frac{1}{2}$	904
$\frac{5}{8}$	123	$1\frac{5}{8}$	539	$2\frac{5}{8}$	956
$\frac{3}{4}$	175	$1\frac{3}{4}$	592	$2\frac{3}{4}$	1008
$\frac{7}{8}$	227	$1\frac{7}{8}$	644	$2\frac{7}{8}$	1060
1	279	2	696	3	1113
$1\frac{1}{8}$	331	$2\frac{1}{8}$	748		

[2] $\frac{1}{4}$ inches or less of assemble is contained in position.

Conventions:
1. Measure motion paths as straight-line distances between terminal points, following location on body member that actually performs the therblig (i.e., fingertip for pinch grasp).
2. For walking, divide into 27-inch movements (steps), and use 27-inch *TE* line.
3. For twisting or cranking movements, divide peripheral distance into N motions of radius of motion length and use N times such motions for time value.
4. For a *select and grasp*, approximate by using twice the grasp time.
5. Use *RL* time even when letting go after contact grasp.
6. On foot pedals with fulcrum under foot, measure distance at ball of foot.
7. For elements that involve overcoming weight or resistance, include the motions to the lever, weight, or piece in the element.
8. Eye focus time occurs (a) when the motion cannot be completed until the eyes have assisted in performing another task which would be concurrent if the eyes were not needed; (b) when the motion cannot be initiated until the eyes have made an observation they were not free to make during the preceding movement.

*Adapted from I. P. Lazarus, "Predetermined Human Performance Time," Ph.D. thesis, Purdue University, 1952, but adjusted to Hog Casing Loop concept of normal.

Job element description. The elements of the work should be detailed in terms of constituent therbligs. A form, such as shown in Figure 21.4, may be used to list the therbligs in groups similar to typical time study elements following the criteria for elements given in Chapter 20. The conditions surrounding the performance of each therblig should be evaluated carefully and noted in the proper column in terms of the variables previously given as controlling the time for each therblig. The conventions given at the bottom of Table 20-XI should be followed closely in establishing these conditions.

From these data, a simo-chart or man and machine chart may be constructed to aid in examining and improving the method. The manner of working with such a chart is the same as when such charts are constructed from observed data.

Use HPT data table. The individual therblig times from Table 21-XI should be entered next to each therblig. This has been done in Figure 21.4 in the columns headed "Time Value." Unavoidable delays, holds, and so forth take their values from the controlling hand. The control during an element may shift from hand to hand.

Summarize. The total time for each element should be computed and transferred to a summary sheet. The recapitulation section of the rear of an objective time study form is an ideal form for this, in that it:

a. Contains room for all the necessary computations.
b. Permits assembling the data in exactly the same manner as with observed data.
c. Permits the front of the sheet to become a method record so that method and standard are preserved together as with regular, direct time studies—intensive samplings. In this manner, any subsequent processing of these data does not require a separate procedure.

In computing the element time, care should be exercised to add all controlling time values. The total time obtained should be the same as the time represented by the longest column of a simo-chart drawn from the data. Note in Figure 21.4, element 2, that the left hand controls the element time for part of the element.

Add difficulty adjustments and allowances. The elemental times should be processed subsequently, as if they were rated, observed values from direct time studies. Difficulty adjustments from Tables 20-III and IV and appropriate allowances should be used. This has been done for the two illustrative elements in Figure 21.5.

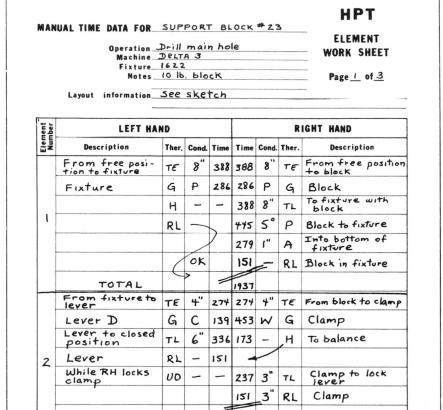

MANUAL TIME DATA FOR SUPPORT BLOCK #23

HPT

ELEMENT WORK SHEET

Operation _Drill main hole_
Machine _DELTA 3_
Fixture _1622_
Notes _10 lb. block_

Page _1_ of _3_

Layout information _See sketch_

Element Number	LEFT HAND				RIGHT HAND			
	Description	Ther.	Cond.	Time	Time	Cond.	Ther.	Description
1	From free position to fixture	TE	8"	388	388	8"	TE	From free position to block
	Fixture	G	P	286	286	P	G	Block
		H	—	—	388	8"	TL	To fixture with block
		RL	⌒		445	5°	P	Block to fixture
					279	1"	A	Into bottom of fixture
			OK		151	—	RL	Block in fixture
	TOTAL				1937			
2	From fixture to lever	TE	4"	274	274	4"	TE	From block to clamp
	Lever D	G	C	139	453	W	G	Clamp
	Lever to closed position	TL	6"	336	173	—	H	To balance
	Lever	RL	—	151				
	While RH locks clamp	UD	—	—	237	3"	TL	Clamp to lock lever
					151	3"	RL	Clamp
	TOTAL				1288			
					Note: Only two elements are shown so as to shorten the illustration.			

Figure 21.4. Use of HPT data for determining standard time for "place and lock piece in fixture." (Only two elements are shown, to simplify the illustration.)

14																
15																

RECAPITULATION

ELEMENTS	1	2	3	4	5	6	7	8	9
Amount of body	C-2	C-2							
Foot pedals	F-0	F-0							
Bimanualness	H-0	H-0							
Eye-hand coordination	J-2	I-0							
Handling requirements	P-2	O-1							
Weight or resistance	W-11	W-3							
TOTAL TIME IN MIN									
NUMBER OF OBS.									
PRO-RATE DIVISOR	FROM HPT								
AVERAGE PER CYCLE									
RATING									
RATED TIME	.01937	.01288							
PER CENT OF CYCLE									
1+ADJUSTMENTS	1.17	1.06							
BASE TIME	.0227	.0316							
1+ALLOWANCES	1.05	1.05							
ALLOWED TIME	.0238	.0143							

Figure 21.5. Use of recapitulation section of objective time study sheet to complete determination of standard time for elements shown in Figure 21.4.

It should be noted that this manner of adjusting the therblig values has the same effect as having an infinite number of values in Table 21-XI. The difficulty adjustment allows not only for all sorts of conditions of work, but inasmuch as elements rather than therbligs are adjusted, it also compensates for the context in which the therblig occurs.

Complete the method record. The method should be described in the usual manner on the front of the time study sheet and the final allowed times transferred to the front so as to complete the standard. The net result of this step will produce a sheet exactly like that of Figure 20.18b.

Conclusion. The standard time is now complete and in the same final form as when obtained with a direct time study—intensive sampling. In that the same concept of standard and the same difficulty adjustments are used that were used with objective rating (Chapter 20), HPT values may readily be mixed with standard times set in that manner.

Summary—HPT

1. *Uses:*
   a. For determining standard times for repetitive manual work:
      (1) Prior to production.
      (2) During production, without the use of a stopwatch or rating.
      (3) For contemplated alternative methods for ongoing jobs.
   b. To provide time values for all varieties of time charts used in method analysis.
   c. To assist in balancing assembly lines.
   d. For the construction of higher-order work-unit standard data.
2. *How used:*
   a. The dimensions of the workplace are determined and the work method is described in general terms.
   b. The job is described in detail by elements, with each element described in terms of its constituent therbligs. The necessary data concerning variables, in conformance with HPT conventions, are added.
   c. The HPT data tables are used to obtain therblig time values.
   d. Element time totals are obtained.
   e. Difficulty adjustments (as with objective rating, Chapter 20) are added. Allowances are also added to obtain allowed times.
   f. A method and standard time record are created, as with direct time study—intensive sampling.

PROBLEMS

21.1. Using the data of Tables 21-XI, 20-III, and 20-IV, compute the standard time for the task shown in:
(a) Figure 16.2.
(b) Figures 16.5a and b.
(c) Problem 16.2.
(d) Problem 16.3.
(e) Problem 16.4.

21.2. Make an objective-rated time study or subjective-rated time study of a simple task and then compute the time for the same task as given in Tables 21-XI, 20-III, and 20-IV. If there is a difference among the values obtained, attempt to determine the cause.

21.3. Using the data of Tables 21-XI, 20-III, and 20-IV, compute the standard time for the best method you can devise for:
(a) Problem 16.2.
(b) Problem 16.3.
(c) Problem 16.4.

21.4. In the discussion of *position*, in HPT, six degrees of restrictions were identified. However, it was stated that no more than five were found to occur at one time.
(a) Explain why this is true of most work.
(b) Describe an element in which all six would occur.

21.5. Obtain a text that describes in detail one of the predetermined time systems, and prepare a book report.

21.6. Describe any PTS in use in your plant or office.

22

STANDARD DATA SYSTEMS

Introduction

Rather than determine the standard time for each job on the basis of an individual study, standard times from a number of studies of related jobs may be organized into a data base. Using this data base, the standard times for related jobs may be constructed or synthesized. Such data bases are called *standard data*.

Synthesized standards from standard data have four general advantages over individual time studies:

1. They are usually based on more data than an individual time standard and thus have more reliability, provided that the errors in the supporting data are random rather than biased.

2. They often eliminate the need for a great many studies. However, they involve considerable work, and where only a few jobs are involved,

413

single studies are often simpler, unless the added consistency is worth the added effort.

3. They aid in estimating rates of production for price quotations on new products and in setting schedules.

4. They aid in setting up assembly or progressive work lines with a minimum of original unbalance.

Definitions. *Standard data* is the term used to refer to predetermined time data when the data are accumulated at the second-order work-unit level (or higher level), as compared with the first-order work-unit data more commonly associated with the term "predetermined time systems."

Standard elemental data system is the term used in this book to refer to the use of data bases using second-order work-unit time values for setting standard times. The source of the data may be any of the predetermined time systems, observational procedures, or any other appropriate technique.

Standard data system is the term used in this book to refer to the use of data bases, using third- and fourth-order work-unit time values, for setting standard times. The sources of the data, in addition to the sources used with standard elemental data systems, may include the use of standard elemental data.

Alternative definition. The use of standard data to set standard times, in terms of the four factors associated with the equation describing a standard time, employs:

1. W/C. A count of 1 is used; one occurrence of the work-unit for which a time is being set is *modeled* or *synthesized*.
2. W/T. Taken from tables; the data reflect the characteristics of the procedures employed in developing such data.
3. M. Implicit in the data tables; a rating is not required (except perhaps to develop the tabulated data).
4. A. Applied to standard elemental data in the same manner as with direct time study—intensive sampling. In data tables for work units larger than second-order, the incremental time for A may be included in the tabulated values.

Types of work for which standard data systems are used. Standard data systems, as one might expect, facilitate the setting of standard times prior to the inception of work. The standard time for a task, instead of requiring direct observation of the work, may be built up, or synthesized, from the existing data. In that second-order (or higher-order) work-units are much larger than first-order work-units, the work of synthesizing a standard time with such data is less laborious and less time-consuming than with first-order work-unit predetermined time systems. The synthesis, in most cases, is also easier than the making of a direct time study—intensive sampling; this cannot usually be said about the use of first-order work-unit data approaches.

As with any standard time values, the standard data must be accompanied by an adequate written standard practice and understood as representing the time expected to be taken with some specified definition of standard time. Synthesized standard times from elemental data are usually of sufficient accuracy to allow the determination of incentive rates prior to the inception of production.

Standard data systems may be developed for any type of work. The only question usually is: Is it more economical to make individual studies than to organize and use a data base?

Limitations on the use of the techniques. Synthesized time standards from standard data do not completely replace other work measurement techniques, since they are developed from a systematic accumulation of data obtained by the use of other techniques. They do, however, reduce the number of individual studies that must be made, while increasing the consistency with which the values represent a fixed concept of standard. The final accuracy of time standards set from standard data, however, is still a function of the accuracy of the compiled data, although their consistency may be (and generally is) better inasmuch as random errors tend to be eliminated.

Steps in developing standard data systems. To develop standard elemental data, it is necessary to have a series of time standards for a series of similar jobs. The time standards may be developed by direct time study—intensive sampling (Chapter 20), predetermined time systems (Chapter 21), or by means of one of the techniques discussed in chapters following this one.

To develop standard data systems the possible sources of data include standard elemental data systems. Hence, the steps listed here will be those related to the development of standard elemental data; standard data systems will be treated, in the next chapter, concomitantly with the use of an additional work measurement technique. (The work measurement technique discussed in the next chapter is one which may be used to set individual time standards, data for standard elemental data systems, or data for standard data systems.)

Therefore, it seemed appropriate in this chapter to focus solely on standard elemental data systems.

Steps in developing standard elemental data systems. The development of standard elemental data requires an analysis of the available data to determine the variables or factors that must be taken into account when applying the data subsequently to other work. The general procedure may be summarized as consisting of the following six steps:

1. Obtain elemental base times[1] for as wide a range of jobs as possible within a given group of similar jobs. The values, as has been noted, may be from direct time study–intensive sampling, predetermined time systems, and so forth.
2. Summarize the data on a spread sheet. The typical spread sheet contains columns for each element as well as columns for the values associated with the possible variables; the lines are for the data from individual studies. The elemental data and the data concerning the variables are "spread" across the lines.
3. Ascertain which elements are constant and which are variable.
4. For the constant elements, determine the average base time.
5. For the variable elements, determine the causal factors and their relationships with base time.
6. Prepare a form to facilitate the use of the standard elemental data for developing standard times on new tasks.

The application of this procedure will be explained with data from an example. Data from other more complex cases and related materials appear in later chapters.

Example—Developing Standard Elemental Data

Introduction. The steps of the procedure will be illustrated with reference to an example. The example chosen is the development of standard elemental data for the work connected with polishing and buffing brass plumbing fixtures.

Polishing and buffing are two jobs for which it is extremely difficult to set accurate and consistent time standards. On such jobs it is extremely desirable to develop standard elemental data so as to achieve uniformity of time standards. Some of the parts studied in this case and their identification numbers are shown in Figure 22.1. Drawings of some of the parts appear in Figures 22.2a and b. The standard workplace is shown in Figure 22.3.

---

[1]Standard elemental data are usually computed for base times with allowances added later so as to allow the data base to stand, even though allowances may change.

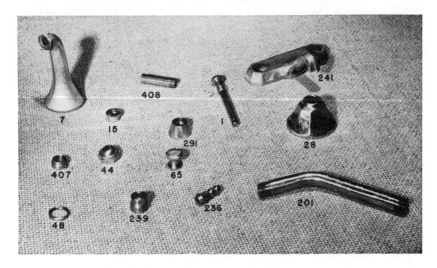

Figure 22.1. Plumbing fixture parts studied.

A motion-picture camera operating at 60 frames per minute was used to record a large number of cycles of each operation studied. Some film was also taken at 1,000 frames per minute. If in the subsequent calculations various points appeared out of line, even the ratings could therefore be reviewed or the method used on the particular job could be reexamined in detail. Only 15 rolls of film were used, since the one-per-second timing allowed extensive recording with only a small amount of film.

Obtain elemental base times for as wide a range of jobs as possible within a given group of similar jobs (step 1). The time standards used as a basis for constructing standard elemental data must have the following six characteristics:

1. Adequate written standard practice with well-defined element end points.
2. Broken into similar elements.
3. Similar methods used.
4. Similar equipment used.
5. Well-chosen elements (see the criteria given in Chapter 20).
6. Rated to a uniform rate of activity.

In that the 41 time studies made in this case were taken with standard elemental data in mind, the individual studies conformed to these criteria. Data taken from time study files, where good elemental breakdowns were not used, frequently do not.

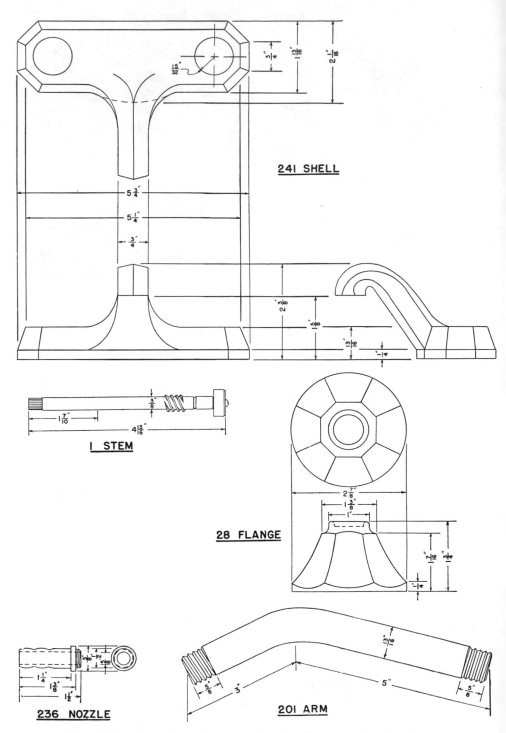

Figure 22.2a. Drawings of some of the plumbing parts (*continues*).

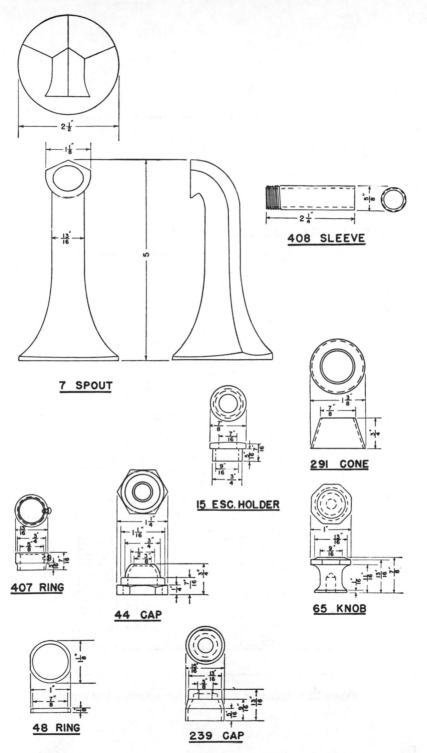

Figure 22.2b. (*Concluded*) Drawings of some of the plumbing parts.

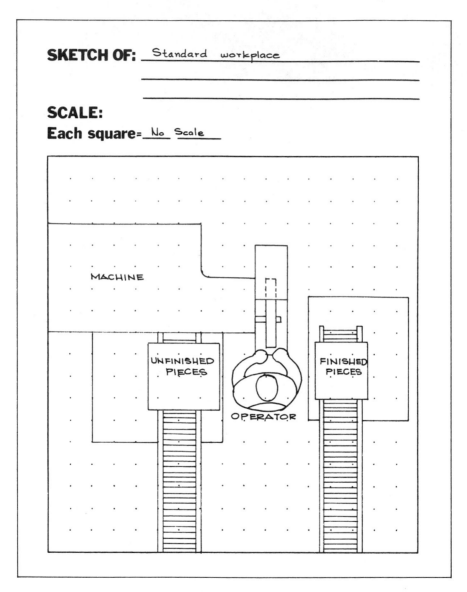

Figure 22.3. Standard workplace for polishing and buffing.

The elements used follow.

ELEMENT 1: Assemble *X* number of pieces to a tool.

Left Hand	*Right Hand*
Hold tool, *A* pieces to tool.	Pick up pieces from box, *DA* one at a time from hand *A* pieces to tool, *RL* excess pieces.

ELEMENT 2: Pick up tool and position to machine.

Left Hand	*Right Hand*
TE to tool, *G* tool, *TL* to machine area, *P* to machine.	*TE* to tool, *G* tool, *TL* to machine area, *P* to machine.

ELEMENT 3: Get handful.

Left Hand	*Right Hand*
TE to pieces in box, *G* handful, *TL* to machine area.	Hold tool or delay, *TE* to machine area.

ELEMENT 4: Get one piece and position to machine.

Left Hand	*Right Hand*
TE to pieces, *ST* and *G* one piece, *TL* to machine area, *P* to machine (can *A* to tool).*	Delay or hold tool, *TE* to machine area, *G* piece, *P* to machine.

Note: This element can have an assembly included in it if the assembly to the tool is very simple, such as assembling the number 7 spout to the finger that is used to support the bell end while buffing.

ELEMENT 5: Position one piece to machine.

Left Hand	*Right Hand*
DA one piece from handful, *TL* to machine area, *P* to machine* (can *A* to tool).	*G* one piece, *TL* to machine area, *P* to machine.

Note: This element can have an assembly included in it if the assembly to the tool is very simple, such as assembling the number 1 stem to the nail point to help support the end that is to be polished or buffed.

ELEMENT 6: Aside one piece to finish box.

Left Hand	*Right Hand*
TL from machine, *RL* piece, *TE* or delay.	*TL* from machine, *TL* to finished box and aside to box, or *DA* from tool and aside to box.

ELEMENT 7: Aside one piece to finished box and position new piece to wheel.

Left Hand	*Right Hand*
TL from machine, *RL* piece, *DA* new piece from hand, *TL* to machine area, *P* to machine.	*TL* from machine, *TL* to finished box and aside to box, or *DA* from tool and aside to box, *TE* to machine area, *G* piece, *TL* to machine, *P* to machine (can assemble to tool).*
TL from machine, *TL* to finished box. Hold tool while pieces are being *DA*.	

Note: This element can have an assembly included in it if the assembly to the tool is very simple, such as assembling the number 1 stem to the nail point to help support the end that is to be polished or buffed.

ELEMENT 8: Aside tool and *DA X* number of pieces from tool.

Left Hand	*Right Hand*
TL from machine, *TL* to finished box. Hold tool while pieces are being *DA*.*	*TL* from machine, *TL* to finished box, *RL* tool, *DA* pieces from tool, *RL* pieces.

Note: This element is used when pieces are fastened to tool by screws, nuts, and the like, and are more difficult to *DA*.

ELEMENT 9: Assemble *X* number of pieces to tool and position tool to machine.

Left Hand	*Right Hand*
Hold tool, *A* pieces to tool, *TL* tool to machine area, *P* to machine.	Pick up pieces from box, *A* pieces to tool, *RL* excess pieces, *TL* tool to machine area, *P* to machine.
TL from machine, *TL* to finished box. Hold tool while pieces are being *DA*.	*TL* from machine, *TL* to finished box. *RL* tool, *DA* pieces from tool, *RL* pieces.

ELEMENT 20: Dress wheel for regular buff.

Left Hand	*Right Hand*
Delay or hold tool, *G* buffing material, *P* to wheel, use, *RL* buffing material.	*TE* to buffing material, *G*, *TL* to machine area, *P* to wheel, use, *TL* to aside area, *RL* buffing material.

ELEMENT 21: Cut and color.

ELEMENT 22: Buff piece.

ELEMENT 30: Dress wheel for color buff (brass).

Left Hand	*Right Hand*
Delay or hold tool, *G* buffing material, *P* to wheel, use, *RL* buffing material.	*TE* to buffing material, *G*, *TL* to machine area, *P* to wheel, use, *TL* to aside area, *RL* buffing material.

ELEMENT 31: Color buff (brass).

ELEMENT 40: Dress wheel for color buff (nickel).

Left Hand	*Right Hand*
Delay or hold tool, *G* buffing material, *P* to wheel, use, *RL* buffing material.	*TE* to buffing material, *G*, *TL* to machine area, *P* to wheel, use, *TL* to aside area, *RL* buffing material.

ELEMENT 41: Color buff piece (nickel).

ELEMENT 50A: Dress wheel for polish (greaseless) (put on compound).

Left Hand	*Right Hand*
TE to machine on–off button (machine off), *TE* to machine area, *G* compound, use (repeat—machine on, then off), *TE* to machine button (machine on), *RL* button.	*TE* to compound, *G*, *TL* to machine area. Hold and use. *TL* to compound aside area, *RL* compound.

ELEMENT 50B: Dress wheel for polish.

Left Hand	*Right Hand*
Delay or hold tool, *G* polishing material, *P* to wheel, use, *RL* polishing material.	*TE* to polishing material, *G*, *TL* to machine area, *P* to wheel, use, *TL* to aside area, *RL* polishing material.

ELEMENT 51: Polish bell on No. 120 grit.

ELEMENT 52: Polish top on No. 120 grit.

ELEMENT 53: Polish edge on No. 140 grit.

ELEMENT 54: Polish all over No. 140 grit.

ELEMENT 55: Polish all over on No. 180 grit.

ELEMENT 56: Polish all over with "greaseless."

In addition to the base time data, information was obtained with respect to the weight in ounces, the area of the surface buffed or polished, and the number handled at one time, if in excess of one, for each element with each part. In that all parts were made from identical brass, no data were recorded with respect to this feature. Had the parts differed with respect to this, or any other potential variable, such data would have been recorded. As a general principle, excess data can be discarded; missing data may be hard to obtain at a later date.

Summarize the data on a spread sheet (step 2). In the example being used to illustrate the steps given, the data were transferred to a form such as the one shown in Figure 22.4. Note that the columns represent the elements and the variables; the lines represent the different parts or time studies.

Ascertain which elements are constant and which are variable (step 3). In developing elemental data for a group of jobs, the following types of elements may be encountered:

1. Constant elements, identical from job to job.
2. Variable elements, similar in motion pattern from job to job but varying in difficulty (and in time required to perform) with the size, shape, and so forth of the work. These may involve handwork with a machine if the machine work is operator-paced.
3. Machine elements, mechanically controlled by the feed, speed, length of cut, and so forth, or welding current, thickness of material, and so forth.

These three types of elements may exhibit any of the following three characteristics of occurrence:

a. Repeated the same number of times in each job.
b. Repeated a different number of times in each job.
c. Appear in some jobs and not in others.

Examination of the data on the spread sheet suggested that element 6 was a constant. (The element description is such that one would not question this theory unduly.)

Further examination of the data on the spread sheet suggested that element 22 was a variable. (The element description strongly supports this initial assumption.)

For the constant elements, determine the average base time[2] (step 4). For element 6, *aside one piece to finished box*, the average time for the values

---

[2]To keep the size of this illustration in bounds, only one element will be examined at each step of detailed analysis.

Part	Weight oz	Area sq. in	Type	Number	1	2	3	4	5	6	7	8	9	10	20	21	22	30	31	40	41	50A	50B	51	52	53	54	55	56
1 STEM	2½	2	R	9			.094		.027	.033	.056				.065		.035												
7 SPOUT	10½	7		1						.034													.110	.474					
7 SPOUT		13		1				.042															.127		.688				
7 SPOUT		2		1				.040		.036													.113			.127			
7 SPOUT		20		1				.040		.037													.129					.306	
7 SPOUT		20		1				.049		.032												1.132							.438
7 SPOUT		20	IR	1				.048		.033					.063		.993												
7 SPOUT		20	C	1							.055							.073	.180										
7 SPOUT		20		1							.052									.038	.229								
15 ESC. HOLDER	½	1½	C	9			.083		.026	.035	.053								.022										
15 ESC. HOLDER		1½	R	5½			.061		.023	.035	.048				.066		.064												
28 FLANGE	1½	2¼	IR	1										.217			.184												
28 FLANGE		10	R	1											.059		.267												
28 FLANGE		10	F	1									.075						.021										
28 FLANGE		22½	F	10						.031	.050							.068	.050										
31 O. H. SCREW	—	—		53	1.073	.033						.062			.067	.449													
44 CAP	1	50	R	20								.323	1.302		.063		1.560												
44 CAP		2½	IS	3½			.045		.023	.033								.063	.039										
44 CAP		2½		4			.055			.030	.046									.038	.067								
48 RING	—	—		30								.037	.431			.307													
65 KNOB	1½	3⅓	VIR	1								.065	.101				.439												
65 KNOB		3⅓	IS	4			.031		.023	.032	.048							.064	.057										
201 ARM	6½	10 8/10	R	1				.048							.058		.258												
201 ARM		10 2/10	R	1						.036							.207												
201 ARM		20	C	1				.049		.033									.240										
236 NOZZLE	1	2	IR	1							.054				.064		.144												
236 NOZZLE		2	IS	3			.033		.023		.055							.071	.031										
239 CAP	½	2½	C	4					.023	.036	.055							.064	.020										
241 SHELL	11½	15 9/10		1				.047		.031													.119			1.166			
241 SHELL		9 1/10		1				.042		.036													.134		.680				
241 SHELL		24 8/10		1				.041		.033												1.090							.857
241 SHELL		9 1/10	R	1				.040		.035					.054		.636												
241 SHELL		15 9/10	VR	1				.050		.037					.057		1.169												
241 SHELL		24 8/10	IS	1							.048							.062	.516										
291 CONE	1	3	R	1							.055						.107												
291 CONE		3	C	5			.040			.037	.049							.063	.031										
407 RING	½	1	IR	7			.066		.034		.048				.058		.113												
407 RING		1	IS	3			.037		.028	.031	.046							.067	.029										
408 SLEEVE	1	4½	R	5			.073		.027		.048				.065		.130												
408 SLEEVE		4½	C	6			.073		.028	.032								.065	.032										
408 SLEEVE		4½	C	10			.171		.024	.033	.055									.038	.047								

Figure 22.4. Spread sheet for polishing and buffing, giving part numbers, physical characteristics, and allowed times for each element.

shown on the spread sheet of Figure 22.4 is 0.034 minute. The reliability of this average was checked with statistical techniques and found to be within suitable limits. It should be noted that instead of 15 readings or so and one rating, as with a single study, the average standard time for *aside one piece to finished box* would, when computed in this new fashion, represent as many times this number of both readings and ratings as there are time studies. It should be much more reliable. At least the time standards for jobs for which this average value is used will be much more consistent in respect to ease or difficulty of attainment, as long as the timing and rating errors are random rather than biased.

For the variable elements, determine the causal factors and their relationships with base time (step 5)

a. Determine, on a logical basis, which job characteristic or characteristics the variable times are a function of.
b. Plot, on a graph, the time for the element against the variable or variables. (*Note:* The variables may act in simple combination or they may interact.)
c. Fit a smooth curve or curves to the points plotted.[3]

This curve (or curves) will either pass close to the points or it will not. If two variables control the values, a series of parallel curves or a curve based on a compound variable may be necessary. For example, if the time T is a function of variables A and B, the abscissa may be

$$A + B, \qquad A \times B, \qquad \frac{A}{B}, \qquad A^n B^m$$

and so forth. A considerable amount of analysis may be necessary to determine the nature of the causal relationship. Semilog and log-log graph paper are sometimes of great assistance in evaluating various hypotheses. If the curve does not fit the points reasonably well, any of the following faults may be the cause:

1. The time studies were incorrectly or inconsistently rated.
2. Other variables as well as those plotted also affected the time.
3. The method varied.
4. An incorrect variable was used.

Each possibility should be investigated and evaluated. The validity of the curve should be checked by statistical methods. Note that even if the values

---

[3]Computer routines are available to assist with this step.

fall along a smooth curve, it is still possible that two types of errors occurred that compensate for each other, although this is not likely. The variables used should be reasonable. It is recommended that in those cases where a reasonable fit with a smooth curve cannot be obtained, or the reason for lack of fit explained with certainty, the investigations and study continue until they can be. Haphazard standard elemental times are never justifiable. A good criterion is: Could this be explained logically? The careful worker may well fit his curves mathematically, but they should still be explainable on a logical basis. It is worth noting that in many cases only a straight line is justifiable and that this straight line is often fitted by the method of least squares.[4] This procedure, however, is designed to determine the straight line with the best fit and not to justify the line. The fit should appear sufficiently reasonable to permit explanation to most working groups.

Element 22, *buff piece*, appeared to be a variable. The time values from the spread sheet were plotted, on Cartesian graph paper, against the *area buffed*. The graph presented a horrifying scatter. The scatter was so bad that if one were to assume that errors in timing or rating were the source, then errors of over 100 per cent had occurred. This was hardly credible. Hence, it seemed most likely that other factors were operative. The spread sheet, the graph, and the parts drawings were examined. The following analysis resulted:

1. Part 1—stem, is perfectly circular; it presents a regular surface.
2. Part 7—spout, has a variety of surfaces; the surface is more complex than part 1.
3. Part 241—shell, has a very complex surface.

As a consequence, all the parts were classified into one of three categories: regular surface; irregular surface (subsurfaces, but each simple); very irregular surfaces (subsurfaces with compound intersections). The data points on the graph were identified and the curves shown in Figure 22.5 resulted.

d. If the variable is discrete, that is, has only definite steps between two limits, prepare a table for the range of jobs covered. If the variable is continuous, that is, has an infinite number of possibilities between two limits, prepare a graph. From the graph, prepare a table by grouping values within limits suitable for the purpose for which the standards are going to be used.

Wage incentives were in use in this plant. Hence, it was decided to develop tables that gave the time for conditions of the variable such that

---

[4]See any statistical text for the details of this method. Also, many portable calculators will do this computation as a "hard-wired" function.

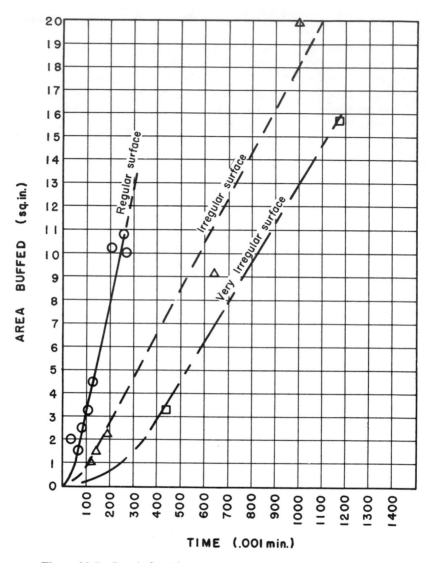

Figure 22.5. Graph for Element 22, *Buff*, in polishing and buffing plumbing fixtures. (*Note:* In actual practice a much more detailed grid was used. The more detailed grid is not shown because it does not reproduce well with normal printing procedures.)

the values would not be more than ± 2.5 per cent different from the value obtained by a direct reading of the graph. The computations for element 22, *buff*, very irregular surface, follow:

1. First, 5 per cent steps were computed for the scale of standard time and rounded to three significant digits. Starting with the lowest antici-

pated time (in minutes), this gave: 0.200, 0.210, 0.221, 0.232, 0.243, 0.255, and so forth.

2. Second, each time value was assigned a span of time from the midpoint of the range between it and the next lower value, and just below the midpoint of the range between it and the next higher value, as:

Value	Span
0.200	0.195–0.2040
0.210	0.205–0.215
0.221	0.216–0.226
0.232	0.227–0.237
0.243	0.238–0.249
0.255	0.250–and so forth

3. Third, using the graph of Figure 22.5, the values of the variable corresponding to the time span were determined and placed in tabular form as:

TIME FOR BUFF, VERY IRREGULAR SURFACE

Range of Area Buffed (square inches)	Standard Time (minutes)
0.46–0.54	0.200
0.54–0.64	0.210
0.65–0.74	0.220
0.75–0.84	0.230
0.85–0.94	0.240
0.95–1.04	0.250

Prepare a form to facilitate the use of the standard elemental data for developing standard times on new tasks (step 6). A simple typed form was prepared which had four sections. Section 1 contained space for the date, part number, and analyst identification.

Section 2 provided space for a dimensioned sketch of the part.

Section 3 contained spaces for entering data on all variables: weight, area polished, area buffed, nature of shape, number worked on together, and so forth.

Section 4 contained:

a. Twenty-five lines, one for each element. For those elements that were constant, the standard elemental time value was preprinted. For those elements that were variables, the appropriate table was referenced; a blank space was provided for entering the time value.
b. Lines to show element totals, allowances used, and final allowed time.[5]

Summary—Standard Elemental Data Systems

1. *Uses:*
   a. To reduce the time to set standard times for repetitive work.
   b. To make it possible to set standard times for jobs, prior to their inception, when they resemble other jobs previously studied.
   c. To improve the consistency of the concept of standard performance embodied in standard times.
2. *How made:*
   a. Obtain elemental base times for as wide a range of jobs as possible within a given group of similar jobs. The values may be from direct time study—intensive sampling, predetermined time systems, and so forth.
   b. Summarize the data on a spread sheet. The typical spread sheet contains columns for each element as well as columns for the values associated with the possible variables; the lines are for the data from individual studies. The elemental data and the data concerning the variables are "spread" across the line.
   c. Ascertain which elements are constant and which are variable.
   d. For the constant elements, determine the average base time.
   e. For the variable elements, determine the causal factors and their relationships with base time.
   f. Prepare a form to facilitate the use of the standard elemental data for developing standard times on new tasks.

PROBLEMS

22.1. Make a series of time studies for stripping insulation from wire, varying the size of wire (both diameter and length), its construction (solid or multiconductor), the type of insulation, and the length stripped. Develop standard elemental data for all the elements in this work.

---

[5]The concept of the form seems so simple that it appeared inappropriate to expend a page on a picture of it.

22.2. Make a series of time studies for half-round (or triangular-cross-section) sand cores ranging from $\frac{1}{2}$ inch to 4 inches in diameter and 4 inches to 14 inches in length. Develop a series of standard elemental times for the elements of these operations.

22.3. Develop the standard elemental data for elements in Figure 22.4, as assigned.

22.4. Using the material from Problem 22.3, compute the standard time for assigned operations on any of the parts shown in Figures 22.2a and b.

22.5. Design a form for use with the standard elemental data derived from Figure 22.4.

22.6. For any assigned base time of Figure 22.4 (for this problem, assume that observed and base times were alike), what range was allowable if 15 values sufficed to give 95 per cent probability that the averages were accurate to ± 5 per cent?

22.7. A series of time studies were made in a plant making steel weldments. The four elements were:

> Element 1: Get member in place in fixture.
> Element 2: Adjust horizontal position.
> Element 3: Weld.
> Element 4: Realign after welding.

The spread sheet prepared, including the selected variables, is shown in the table.

	Element 1		Element 2		Element 3		Element 4	
Study	Time (minutes)	Variable: Weight (100's of kg)	Time (minutes)	Variable: Length (cm)	Time (minutes)	Variable: Volume of Weld (cu cm)	Time (minutes)	Variable: Distance between Alignment Points (meters)
1	0.15	1.5	0.70	65	3.7	10	10.0	5.00
2	0.95	7.5	1.40	60	10.0	30	15.0	5.00
3	0.35	2.5	0.41	45	7.9	30	60.0	15.00
4	0.95	8.5	0.20	75	6.0	30	3.0	1.50
5	0.45	2.5	0.90	80	30.0	60	70.0	20.00
6	0.85	5.0	0.90	100	6.0	20	4.0	2.3
7	0.05	0.5	0.10	20	23.5	50	34.5	12.00
8	0.65	5.0	0.30	25	60.0	68	9.0	3.60
9	0.10	0.5	0.60	55	3.5	20	40.0	15.00
10	0.35	1.5	0.50	60	13.0	40	6.0	2.30
11	1.05	8.5	0.20	30	50.0	60	5.0	2.25
12	0.85	6.0	0.20	15	30.0	56	18.0	6.50

(a) Plot the variable and time on Cartesian, semilog, and log-log paper for:
1) Element 1.
2) Element 2.
3) Element 3.
4) Element 4.

(b) Select and explain the selection of the best plot for:
1) Problem a.1.
2) Problem a.2.
3) Problem a.3.
4) Problem a.4.

(c) Develop a table with a base time discrepancy of not more than ±5 per cent for the selected data for:
1) Problem b.1.
2) Problem b.2.
3) Problem b.3.
4) Problem b.4.

(d) Same as part (c) but with not more than a ±2.5 per cent base time discrepancy, for the selected data for:
1) Problem b.1.
2) Problem b.2.
3) Problem b.3.
4) Problem b.4.

22.8. In the text it was noted that polishing and buffing are jobs for which it is difficult to set accurate and consistent time standards. Why is this so?

23

DIRECT TIME STUDY— EXTENSIVE SAMPLING

Introduction

The basic difference between the technique being described and direct time study—intensive sampling is in the manner of making observations. Instead of continuously observing the task and recording the time at the end of each separated category, the task is observed at intervals, as with work sampling. However, instead of merely noting the category of work being performed, as with work sampling, the performance is rated (objectively or subjectively), and a rating is entered on the record instead of merely a check mark, except in the case of personal time or non-operator-controlled delay elements. The sampling times may be established by any of the methods described in Chapter 8. It is worth noting that a whole series of jobs may be observed in sequence, and a whole group of standards worked on simultaneously. Separate category lists would be used for each job or group of jobs.

The name *direct time study—extensive sampling* has been chosen on the following basis. The technique employs *direct* observations. The term *time study*, synonomous with work measurement, means that it is used to set time standards. Finally, an *extensive sample* of work is studied relative to the small sample of direct time study—intensive sampling.

Definition. Direct time study—extensive sampling is a procedure for setting standard times wherein the observations are made, as with work sampling, at intervals[1] (rather than continuously) over an extensive period. Each observation, as with work sampling, is classified into a category. However, the categories other than idle, rest, and so forth, are the second-, third-, or fourth-order work-units associated with the activity observed. Idle, rest, or delay categories are kept separate. The per cent of total observations represented by any one category is taken as representative of the per cent of the total time attributable to that category. Because the work is not observed continuously, a total work time and a work count to apply against the work time must be acquired by some means other than direct observation. Ratings may be applied to the observations of physically active categories. Allowances may be added as with direct time study—intensive sampling.

Other names for the technique. The technique is also referred to, in the literature, as *work sampling work measurement* and *standards by work sampling*. Further, in that the observed data are rated and time spent on activities other than direct work are kept separate, and if the method is described in detail, standards set by this technique may also be referred to as "engineered standards."

Alternative definition. As an alternative definition, in terms of the factors of the basic equation for a standard time:

1. W/C. Obtained from production records covering the period during which the observations are made.
2. W/T. Observations are made of employees at intervals. The per cent of the total observations represented by any one category is taken as representing the per cent of the total work time spent at that category. The validity of this may be tested by statistical means; it is a function of the per cent of total time represented by the category and the total number of observations.
3. M. A rating may be made by any one of the standard methods and applied to physically active categories. Because each observation of an active category is rated, there is a tendency for random rating errors to be removed; bias-type errors, if present, remain.

---

[1]Random, or at intervals other than random, as determined by some sampling plan.

4. *A*. As with direct time study—intensive sampling, a percentage, determined by policy, is added; the extensive sampling study itself may also provide some additional information relating to an appropriate value of *A*.

Types of work for which it is used. While the technique may be used to study the same types of jobs for which direct time study—intensive sampling is used,[2] its greatest field of use is for the study of nonrepetitive jobs, or jobs with too much internal variation, or too long a cycle for convenient study by means of intensive sampling. Also, standard data for second-, third-, or fourth-order work-units may be determined conveniently and directly with extensive sampling.[3] The method record, when the study is used to determine third- or fourth-order work-unit standard times, is usually not as fully described as with direct time study—intensive sampling.

The technique is particularly useful for setting standard times for work such as nonrepetitive office activities, the activities in analytical laboratories, maintenance, custom manufacturing and custom assembling, and so forth. Shipbuilding, electric switchgear assembly, heavy engine assembly, custom assembly of motors with generators and pump accessories, aircraft overhaul, flight line maintenance, and plant maintenance are also typical of the variety of work for which the technique has proved useful for setting standard times to assist in maintaining managerial control.

Steps in the use of the technique. There are eight steps in the use of the technique, as follows:

1. Define the standard of measurement so as to provide a basis for determining the modifier *M*. As with direct time study—intensive sampling, this definition needs to be established only once for all studies in a plant.
2. Create categories for observations.
3. Arrange for work count and work time data for the period of the study.
4. Design materials for recording observations.
5. Determine sample size, sampling pattern, and observation times.
6. Observe and make appropriate recordings.
7. Tabulate and evaluate the observations.
8. Apply allowances (as with direct time study—intensive sampling).

---

[2]This approach permits a standard time to be set without the use of a stopwatch; some regard this as an advantage.

[3]It is for this reason that this chapter appears after the chapter on standard data rather than subsequent to the chapter on direct time study—intensive sampling.

The details of these steps will be presented, together with an example of an application of the technique.

Example—The Metal-Clad Assembly Department of the S & C Electric Co., Chicago[4]

Introduction. The S & C Electric Company manufactures large, heavy-duty electrical switchgear. The typical unit consists of one or more bays (cabinets), each bay being about 3 by 3 by 9 feet, although these dimensions vary from unit to unit. A typical bay is shown in Figure 23.1, and a series of bays composing a complete unit are shown in Figure 23.2. The final assembly department has several unique aspects that differentiate it from the normal type of activity for which time standards are commonly engineered. These differences are:

1. The work is not repetitive, and the summing up of the "units of output" (5th-order work-units) is not a simple counting of the finished pieces. The units are individually designed to meet customer requirements.

2. The details of the work methods used by the employees, on each different task, are not economical to describe; hence, there is difficulty in defining the "standard man-hours of work input" wherein both the method and the degree of diligence of the employee is fully defined.

3. The employees work as a group on a considerable number of units simultaneously, shifting back and forth and helping each other as needed. It is difficult to get accurate time-keeping charges against each unit.

Most of the characteristics of the work are also typical of the types of work indicated as the area of application of direct time study—extensive sampling. Further, it should be noted that with such work, some type of standard data is a necessity; any standard set on individual jobs would otherwise be "after the fact." After-the-fact data would hardly serve the uses of work measurement, as they were indicated in Chapter 5.

Hence, this illustration may be taken as being representative of the special field of application of the technique. Its use as an alternative to direct time study—intensive sampling, although not specifically discussed in this chapter, should also be clearly discernible from this example.

---

[4]This case was worked on by R. Amstutz, A. J. Mandele, and M. A. Ottow of the Industrial Engineering Division of the S & C Electric Co., and the author. The data are reproduced with the kind permission of the S & C Electric Co.

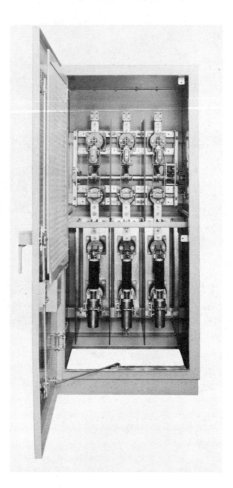

Figure 23.1. Sample bay from assembly department of S & C Electric Co., Chicago. Each bay made is custom-built and different from all other bays. (Courtesy of S & C Electric Co.)

Define the standard of measurement (step 1). The performance of this step is exactly the same as when performed with direct time study—intensive sampling. Therefore, the details will not be repeated in this chapter.

Create categories for observations (step 2). It would appear that the only repetitive aspects of the work situation are the specific fourth-order work-units, such as *door installed, panel installed, screen installed, PCS switch installed, MAG switch installed, fuse mounting installed, (x) feet of bus duct*

Figure 23.2. Complete unit from assembly department. (Courtesy of S & C Electric Co.)

installed, and so forth. These work-units meet the three criteria given in Chapter 9:

1. They fit into a clear hierarchy of countable, convertible units of quantification from objective to workload.
2. They permit a meaningful forecast of the workload to be made in terms related to the required manpower.
3. A firm relationship of the work-unit to the required manpower resources can be established.

It should be noted that the development of suitable categories for observation is an application of work-unit analysis, as described in Chapter 9. In addition, criterion 2 must be understood to include the meaning, "the work count of each work-unit must be discernible from the product drawing prior to production."

In addition, it should be noted that if a set of work-units is devised which meets the first two criteria, but none of the work measurement techniques currently in use is applicable, then either a modification of existing techniques or a new technique is needed. The work-units "speak for the situation"; the work-measurement-technique selection or design is the problem of the analyst.

In the application being examined, various amounts,[5] or quantities, of these components (4th-order work-units) appear in each assembly. However, for any component, the work of installation (and consequently, the time required at a given level of effort) varies from assembly to assembly, depending on the location of the component in the assembly. To make a sufficient number of direct time studies so as to permit the making of tables of time values, for the installation of each component, with the time required being given as a function of all the factors affecting the difficulty of installation, would be an extremely lengthy and uneconomical task, to say nothing of the labor to apply such data. Similar difficulties would beset an attempt to create performance standards for unit *acts* and apply these to set standard times.

The solution to the problem lies in considering the time *for a* (*specific*) *component installed* as being best represented by the rated average of a sizable sample of such specific component installations. If the sample is a sufficiently random mix of installations, it may be used to predict realistically and accurately the time for subsequent *groups* of installations of the subject component. In short, an extensive sampling time study is suggested, with each standard time developed representing an average standard time for *a type of component installed* (4th-order work-unit).

Specifically, with respect to the work done in the final assembly department of the S & C Electric Company:

1. A complete list of bay components (as assembled in the final assembly department) was prepared. This list contained every *thing* that might cause work on a bay in the final assembly department. This is the list of productive categories for the sampling study by tasks. A section of this list is shown in Figure 23.3.

2. A specific 4th-order work-unit was designated for each component or category so as to provide a basis for counting. These work-units were chosen so that their count indicated the relative amount of work and also so that they could be counted from the design drawings.

Note: In the case being discussed here, the selection of the method of counting was a relatively simple problem. This is not always so. In many

---

[5]In some cases, with some bays, the amount is zero.

BAY COMPONENTS - METAL CLAD ASSEMBLY DEPARTMENT						
Component	Work unit	Initial work count	Work units done prior	Work units added	Work units left	Net Work Count
Doors	Door	16	8	60	11	57
Panels	Panel	28	27	74	10	65
Indoor roofs	Roof	18	8	59	7	62
Screens	Screen	12	6	43	9	40
Front handle	Handle	16	5	55	14	52
Gasketing, rubber	Foot	336	120	1766	224	1758
GBS switches	Pole	24	21	120	17	106
Potheads	Pothead	11	9	47	16	33
Lights	Light	9	7	39	11	30
Line bus	Foot	192	64	1387	119	1396
Install and wire interlock	Interlock	8	6	49	11	40

Figure 23.3. Part of the task list, by components, for assembly of bays. (Courtesy of Industrial Engineering Dept., S & C Electric Co.)

applications of the technique of direct time study—extensive sampling the success or failure of the work measurement effort hinges on the design of a satisfactory way of performing this step. Reference should be made to the three criteria of satisfactory work-units given in Chapter 9.

Of course, in addition to the product-oriented fourth-order work-units, spaces were provided on the sampling form for idle, absent, personal, avoidable delay, as well as some blank lines for any unanticipated categories.

The categories selected for recording the observations are identified as fractions of the total work time and referred to in computations as p_1, p_2, and so forth. As a general principle, excess categories can be combined with other categories after the study is completed, but categories not separated in the study cannot later be isolated. Hence, if a category appears possibly useful, it should be separated when the study begins.

Arrange for work count data and work time data for the period of the study (step 3). In making a direct time study—extensive sampling, the work is not observed continuously. Hence, there is a need to arrange for the obtaining of the two data items, the work-unit count and the work-time data, from sources other than the sampling obervations. The work-unit count is the number of times the work-units being studied are produced during the extensive period studied. The work-time data refer to the number of hours worked by the employees who are producing the work-units. These new items will be used in the calculations. It is vital to arrange for a source for these before the observations begin.

In the example:

1. A set of drawings was obtained for all bays that were in the assembly department and a work count made for all work-units of all bays on the assembly floor (see Figure 23.3, "Initial work count").

2. Just prior to beginning the study, an inventory was made of all work-units completed on all bays in the assembly department (see Figure 23.3, "Work-units done prior"). When this inventory was deducted from the count obtained in step 1, the remainder represented the work-load (work count by 4th-order work-units) remaining on bays already on the floor.

3. Arrangements were made to obtain the work-unit count of all bays released to the assembly department during the study period, to add these to the work count from step 2 (see Figure 23.3, "Work-units added").

4. At the conclusion of the 5-week sampling period an inventory was taken of the bays on the assembly floor to determine the work count of tasks not completed at this cutoff point in the study. This was deducted from the work count obtained from step 3 to give a net work count for work performed during the study period.

Design materials for recording the observations (step 4). The simple form, shown after use in Figure 23.4, was devised. Such a form requires hand posting and tallying to summarize the results. However, because of the limited scope of this study, this was cheaper than devising and using machine-readable[6] forms.

On larger studies, such as with shipbuilding, where 1,400 observations per day may be obtained for months, direct, machine-readable forms may be much more advantageous.

Determine sample size, sampling pattern, and observation times (step 5). To maximize the number of observations for any given period of time, and with observer manpower limited, one observer was assigned to each of the two shifts; the observers were instructed to make sampling rounds as frequently as possible, with random breaks for the observers. The time each round was made was recorded, as shown on Figure 23.4. A 5-week period was calculated, as described in Chapter 8, as necessary to obtain a sufficient number of observations. A 5-week period was selected, in preference to a month, so that the "Work done prior" and the "Work-units left" work

---

[6] "Machine-readable" means "can be entered into a computer directly, without hand keying, punching, or posting."

PERFORMANCE SAMPLING DAILY DATA SHEET   ASSEMBLY DEPT.

Observer A. Mandell   Number of employees 27

Task	7:32	7:59	8.23	9.08	9:40	10:11	10.33	10:50	11:17	11:29	11:41	1.08	1.22
						Time of observation							
Doors	80 85 70	80	80 80 75	80	80 90	80							
Panels	90	90	90		90								
Indoor roofs									70		75	70	65
Screens	75	90		75	75	75							
Front handle	75						80	80					
Gasketing, rubber	80	80						80					

Task	7:32	7:59	8.23	9.08	9:40	10:11	10.33	10:50	11:17	11:29	11:41	1.08	1.22
GBS switches	70 75 70										80 70	80	
Potheads		80	75	70	70							70	
Lights												80 70	
Line bus					80		80 90	90 80 55	80	80			

Task	7:32	7:59	8.23	9.08	9:40	10:11	10.33	10:50	11:17	11:29	11:41	1.08	1.22
Read plans	✓	✓						✓	✓				✓
Personal	✓✓		✓✓			✓			✓		✓		

Figure 23.4. Part of daily sampling sheet for sampling time study by tasks, assembly department, S & C Electric Co. (Courtesy of S & C Electric Co.)

counts (see Figure 23.3) could be obtained over the preceding and final weekends.

Prior to the 5-week observation period, a "dry run," or test study, was carried on for 1 week to familiarize the observers, test the recording forms and procedures, and to accustom the workers to being observed.

Observe and make appropriate recordings (step 6). Figure 23.4, as has been noted, is one page of the multipage data recording form used. Approximately 500 to 700 observations were made per day. This is the product of the number of workers in the assembly area, multiplied by the number of rounds per day. Note, on Figure 23.4, that some observations show more than one rating in an observation space. This could refer to either two or more men on a task on one bay or several men doing the same thing on different bays. These were not differentiated. A list of the workers, by clock number, was employed on each round, so that the category "absent from work areas" could be properly entered.

In this study, as is typical of such studies, the observer when making observations proceeds as with work sampling, identifying the category, but the observation time is slightly longer so that the observer may rate the performance.[7] Only delays are recorded in the same manner as normal work sampling; the reason for not rating these should be obvious. However, each irregular element, whose time depends upon worker diligence, should be listed as a separate *p*, and observations of this activity should be rated.

Various statistical controls may be used to check on:

1. The acceptability of each day's sample.
2. The reliability of the average $\bar{p}$, or per cent of time spent on each category, as indicated by all the samples available at any particular point in the study.

Tabulate and evaluate the observations (step 7)

Apply allowances (step 8). If the categories of a direct time study—extensive sampling are 2nd-order work-units and objective rating is used, the difficulty adjustments are computed at this point for each category in the same manner as with direct time study—intensive sampling. If the categories are 3rd- or 4th-order work-units and objective rating is used, the difficulty adjustments are computed for the average or typical conditions accompanying the work, except for the adjustment for weight. For weight adjustments, only the 5 per cent or basic values are used, with the weight

---

[7]Objective rating is particularly helpful here, in that only one concept of normal speed need be used and the rater does not need to "reset his sights" for each observation; the observation time per observation may be held to a short interval.

being taken as the average of weights lifted. If there is great variation with respect to weight or force overcome during the category, the average may be obtained by recording the necessary information during the sampling period. (Information concerning unusual difficulty adjustments may be obtained in the same way.) Hence, under some circumstances, each observation may lead to a recording of the category, the rating, and the information relating to the difficulty adjustments.

At the conclusion of the study, the values for each $\sum n$ (the number of a particular category of observation) and the $\sum N$ (the total number of all observations) are tabulated. (If other data related to the difficulty adjustments have been recorded, they must also be tabulated separately for each category.) The ratings for each category are tabulated separately. The average rating, $\bar{R}$, for each category may be computed as

$$\bar{R}_i = \frac{\sum R_i}{\sum n_i} = \frac{\text{sum of all ratings of category } i}{\text{total number of observations of category } i}$$

The standard time for each category, including allowances, may be computed by means of the following formula:

$$\text{ST}_i = \frac{(T \times 60 \times \bar{p}_i) \times [\bar{R}_i \times (1 + \text{Adj.}_i)] \times (1 + \text{All.}_i)}{\text{W/C}}$$

where

    ST_i = standard time for category i, minutes

    T = work time covered by the study, usually from time cards or job cards, hours

    60 = conversion of T to minutes

    $\bar{p}_i$ = per cent of time spent at category i

    $\bar{R}_i$ = average rating for category

$(1 + \text{Adj.}_i)$ = 1 + difficulty adjustments for category i (if ratings are subjective, this term will disappear)

$(1 + \text{All.}_i)$ = 1 + allowances for category i

    W/C = work count; number of times the work-unit was performed (a source for this, to match the period of the study, must be arranged for prior to starting the study)

The basis for the formula is simple:

    $T \times 60 \times \bar{p}_i$ = time attributable to work-unit i, minutes

$\bar{R}_i \times (1 + \text{Adj.}_i)$ = correction for diligence of worker and difficulty of task using objective rating [with subjective rating, this term is $(\bar{R}_i)$]

    $(1 + \text{All.}_i)$ = allowances for the category

The product of all the terms above is the standard time for all the performances of work-unit i, in minutes; dividing by the work count puts this on a one-performance basis, giving a standard time per work-unit, in minutes, as with direct time study—intensive sampling.

In the example being followed, the standard time for each work-unit was computed by means of the formula given previously, except that they were expressed in man-hours instead of man-minutes. The T used was the total of all direct labor hours of work expended in the department during the 5-week, two-shift, study period. A complete set of standard data was developed.

Develop material to make use of the standard data. Figure 23.5 shows sections of the two-page, standard time application sheet for applying the standards. This sheet may be used as follows:

STANDARD TIME APPLICATION SHEET – METAL CLAD ASSEMBLY

CDA _____ CUSTOMER _____ SO _____ DATE _____

ITEM	UNIT OF MEASURE	QUANTITY	STD. HOURS*	STD. HRS ALLOWED
Move to assembly area	Bay	———	.4002	———
Doors	Door	———	1.0820	———
Indoor roofs	Roof	———	.3358	———
Gasketing, rubber	Foot	———	.0266	
PCS switch fused	Switch	———	.6416	
Line bus	Foot	———	.0659	
Flexible connectors	Foot	———	.0122	
SUB-TOTAL *				———
* Includes 10% allowance				
Stock up tools, sharpen,study prints			4.8%	
TOTAL ALLOWED HOURS				———

Figure 23.5. Sections from the standard time application sheet for assembly of electric switchgear units. (Courtesy of Industrial Engineering Dept., S & C Electric Co.)

a. To determine the assembly time when bidding, so as to estimate an assembly cost.
b. To "load" or schedule work into the assembly department.
c. To credit work done in assembly, in standard labor-hours, to compare to actual labor-hours, so as to measure operating productivity, and so forth.

Inasmuch as the work units are not credited as "performed" until a unit is completed and moved from the department, a 4-week, moving average is used to measure productivity and smooth out the fluctuations. It should be noted also that the use of a 4-week, moving average:

1. Increases the size of the mix used to measure performance.
2. Gives a weekly measure of performance.

By doing this, the reliability of the measure is enhanced, and yet values of performance are obtained at suitable intervals.

Evaluation of results. The data were economical to develop and economical to use, and brought reasonable control to a nonrepetitive situation. Greater productivity and better scheduling, pricing, and use of floor space resulted from the application of these standards. It is to be noted, however, that the methods are not described in detail. When a considerable number of component design changes or mounting method changes take place, either a complete new study will need to be made or a series of direct time studies— intensive sampling of the new components or mounting methods will be needed. The standards described are not fully "engineered," but they may be used for almost all uses of time standards, including wage incentives, if a proper plan is used.

Summary: Direct Time Study— Extensive Sampling

1. Uses:
   a. As an alternative technique in situations where direct time study— intensive sampling is applicable.
   b. Nonrepetitive jobs that have a finite variety of subcycles or subtasks, but in which the repetition of the subparts has no regular pattern.
   c. Long-cycle work, too lengthy to economically study with direct time study—intensive sampling.
   d. To determine standard data.

2. *How made:*
   a. The standard of measurement must be defined. This is done once for all jobs in a plant.
   b. Create categories for observation.
   c. Arrange for work-count and work-time data for the period of the study.
   d. Design materials for recording the observations.
   e. Determine sample size, sampling pattern, and observation times.
   f. Observe and make appropriate recordings.
   g. Tabulate and evaluate the observations.
   h. Apply allowances. *Note:* If the study was made to develop standard data, forms for the application of the standard data will be developed subsequent to step h.

PROBLEMS

23.1. In the S & C Electric Company example cited, an employee list was used to make certain that the correct number of "absent" observations was recorded. What would be the effect of an error in the count of this category if the error was:
(a) Too high a count?
(b) Too low a count?

23.2. Do all the preparatory work and complete planning for a direct time study—extensive sampling, by tasks, for:
(a) A garage attendant.
(b) A maintenance employee.
(c) A janitor.
(d) A housewife.
(e) A restaurant cook.
(f) A roving inspector.
(g) A gas station attendant.
(h) Any selected nonrepetitive job.
(i) Any designated office group.
(j) A member of any designated athletic team, while in competition.

23.3. For any of the situations in Problem 23.2, plan a direct time study—extensive sampling, by 4th-order work-units.

23.4. Make a direct time study—extensive sampling, by elements, of a repetitive task. (To permit a group to do this in 1 hour, have similar elements and sampling times stratified by minutes or half-minutes and randomized by seconds, and pool the group data.) A film loop may be used and the accuracy of the sampled data verified by comparison with the actual time distribution on the film.

23.5. What is the probable nature of the distribution of values of true, individual standard times for any 4th-order work-unit on the bays examined in the S & C Electric Company problem?

23.6. Under what circumstances could the pre-start and post-study inventory of work-units, used in the S & C Electric Company problem, been unnecessary?

23.7. For custom assembly work, provide, in narrative form, guidelines for adjusting the length of a direct time study—extensive sample as a function of the length of time the custom units are worked on in the department studied.

23.8. A film for practicing rating under conditions simulating direct time study—extensive sampling may be obtained from the author. The film is titled, "Rated Sampling No. 1." There are 50 scenes on this film, taking in a wide variety of jobs, ranging from stirring paint to shifting heavy containers. Each scene is only 10 seconds in length and is preceded by a 7-second title. All the scenes were taken in an industrial environment.

24

THE FRACTIONED
PROFESSIONAL ESTIMATE

Introduction

In Chapter 12, reference was made to projects and to the achievement of a *status* by performing *activities*. Implicit in the discussion was the concept that the activities are not repetitive. However, they are describable and are within the "state-of-the-art" of some field of science or technology. Similar activities are found in the assignments of industrial and government product designers, management staff, audit staff, writers, and so forth.

This chapter addresses one technique for setting standard times for the types of work enumerated. It should be noted that this type of standard time can be employed for all the uses listed in Chapter 5, although it is seldom used as a basis for incentive wages.

Many of the work-units associated with the work under consideration are services, clearly of such a nature that a wide range of work time accompanies the differentiation of outputs which are quite satisfactory from those which are superlative. It is not usually recognized that substantive outputs display similar characteristics but economic pressures keep the criterion of "quite satisfactory" more obvious to us. Hence, the wide range of possible work time is not a unique feature of the work discussed in this chapter; it is not necessarily an impediment to setting standard times.

Definition. A *fractioned professional estimate (FPE)* is a standard time set by one or more individuals, knowledgeable in the subject matter of the work-unit (or work-units), who list the components of the work-unit as a first step. The listing must be in discrete, homogeneous[1] steps (lower-order work-units), sufficiently small so that an estimate of the time required to perform each separate step may be made with reasonable accuracy, but of a size appropriate to the scope of the job. The key word is "reasonable." In this case, reasonable means sufficiently close to expectable fact that the corrective action (of the subsequent steps of managerial control) is minor. It is worth noting that gross errors of estimate will be detected and can be corrected after a small lapse of time. Of course, to make an FPE, experience in actually performing either the work-unit, or work-units like it, is desirable in addition to knowledge in the subject-matter area.

The making of an FPE includes the making of visible, documented assumptions concerning all the steps in the final work-unit, as well as all the factors affecting the work time. This listing facilitates the employment of the details of experience for the conversion of these steps to a quantitative statement of the resources required. Further, the FPE constitutes a recording of the "know-how"; it also pinpoints steps where technological change would be particularly effective.

Other names for the technique. A fractioned professional estimate, in some areas of application, is called a *technical estimate*, a *technical plan*, or a *research or design protocol*.

Alternative definition. An FPE, in terms of the four factors associated with a standard time, has the following characteristics:

1. W/C. A count of 1 is used; a model of the work-unit is constructed in terms of separable, smaller work-units.
2. W/T. An estimate is made of the time required for each of the smaller work-units by people knowledgeable in the subject-matter area. Experience with the work-unit or similar work-units is highly desirable.
3. M. Implicit in the estimates; unity. The time assigned is the actual expected performance time.

---

[1] Describable under a single action verb.

4. *A*. May either be implicit in the estimates, or an allowance may be added after the estimates are completed.

Types of work for which it is used. The FPE is the principal technique used in setting standard times for activities such as those analyzed with the network diagrams explained in Chapter 12. It also is an ideal technique for setting standard times for workers in research and development laboratories, or for foremen and other supervisors. It can also be used for salesmen. It is applicable to many government field workers, such as the field veterinarians of the Animal Health Service of the U.S. Department of Agriculture, mine inspectors of the U.S. Department of the Interior, soil conservation field representatives of the U.S. Department of Agriculture, auditors of the U.S. Department of Defense, and so forth. It has been applied to product design engineers, industrial engineers, management analysts, tool designers, and so forth.

In the areas of activity that have been enumerated, time standards are set, for the most part, by a judgmental process[2] partially similar to the FPE but differing in one vital way: the components of the work-unit assigned as a task usually are not separated. Hence, these details are not recorded.

Advantages of the FPE. The FPE introduces the following advantages:

1. The basis of the estimate, the lower-order work-units, can be examined and discussed.
2. The accuracy of the estimate can be checked, part by part.
3. The subsequent correction of the estimate may be performed with respect to either the contents or sequence of the anticipated lower-order work-units or the anticipated time required for their performance.
4. The details of the "know-how" are recorded.
5. The need for technological change can be pinpointed.
6. The effect of technological change can be pinpointed
7. Participation in making the estimate can be obtained on a meaningful basis.
8. The estimates permit more effective control because of their more adequate size.

For these reasons, the use of FPEs is an important way of providing standard times for facilitating managerial control in areas of activity long considered as not readily amenable to real managerial control.

---

[2]If they are set at all.

The FPE also replaces, to a large extent, the concept of the manning pattern. A *manning pattern* results when gross historical data are used to determine how many of one type of employee (such as personnel interviewer) are needed per other type of employee (such as employee hired), or when the number of designers is computed on the basis of the number or value of designs, and so forth.

Limitations on the use of technique. In order to make a realistic FPE, some experience with the activity is necessary, Further, some considerable motivation is required if the final time standard is to be realistic. Conversely, undue optimism must be avoided.

Steps in the use of the technique. There are only three simple steps in the making of an FPE:

1. The objective of the project (the status to be achieved) must be defined.
2. The component lower-order work-units, of a size appropriate to the work, must be listed.
3. From experience, a realistic time must be assigned to each lower-order work-unit listed in step 2.

The details of the technique will be examined with the aid of two cases. One case will relate to a design department in a manufacturing plant. Because of the large number of details that would ensue if the entire case was examined, only enough will be examined to illustrate the principles employed. The second case will relate to a professional task, which is repeated at irregular intervals and which has unique variations occurring in each repetition.

Case I—A Design Department

Introduction. The company was engaged in the design, manufacture, and construction of radio and TV stations.[3] The particular difficulties encountered, which triggered this investigation, were the failure to make installations on schedule, chaotic production schedules, high costs, and a large number of test-erection defects. ["Test-erection defect" was the designation given when mating parts (e.g., antenna sections, racks, and so forth) did not mate properly, necessitating either reworking the parts, remanufacturing, or building special mating conversions.]

---

[3] Name withheld at company request.

One easily remedial source of the difficulty with meeting delivery schedules was traced to the sales department. They were promising delivery dates without checking on required lead time for design or factory production schedules. However, the design department claimed that they could not schedule the design work; it was innovative; the time for a design was not really predictable. Further, they indicated that they had both "good" and "excellent" designers; they all produced satisfactory designs, but the excellent engineers took much less time than the "good" ones. In addition, the design department laid much of the blame for late designs at the door of the research department. The design department claimed that a major source of delay was late data from the research department.

However, everyone agreed that the situation needed remedy. The sales personnel were instructed to obtain design delivery dates and factory schedule estimates before promising delivery. Attention then shifted to the design department's claim that they could not schedule their work. The approach of the FPE was selected as the only approach suitable for examining the possibilities of setting standard times for this type of work.

Defining the status to be achieved on each design (step 1). The desired status was defined as a complete set of drawings suitable for use in the factory and during erection, representing a station meeting the performance specifications of the sales contract and *available on an agreed-upon date.* (The italicized portion represents the "sticky" aspect, as noted in the preceding two paragraphs.)

Listing the lower-order work-units (step 2). The analyst, the chief design engineer, and an "excellent" designer met. It was tentatively agreed that seven different groups of design problems could be examined: AM radio stations, FM radio stations, FM-stereo radio stations, VHF black-and-white TV stations, VHF color TV stations, UHF black-and-white TV stations, and UHF color TV stations. Other factors affecting design details and sequence were, with respect to eventual location of the station, climate, terrain, unusual but expectable weather and earth-movement events, social stability, reliability of power supply, and so forth. Finally, design details were affected by desired range, operating stability, local maintenance capability, specific channel, and so forth.

A VHF black-and-white TV station design was selected as a test case. A completed design was used as an information source. After considerable discussion, the steps in the design sequence were agreed upon, as follow.[4]

---

[4] As noted earlier, to reduce the detail, only a few will be given.

Step	Title and Contents
1	*Prepare preliminary information* Receive the assignment; obtain a copy of the sales contract, site map and data, and the country electrical code. *Obtain a hard-cover bound notebook from the supply room. Check the site data for completeness and the code data for completeness and recency. Request additional data required from Sales and/or the company library.*[5]
2	*Check frequency, power, and range data requirements* *Check company handbooks for completeness of research data on signal-generation, wave-guide, and radiator design for the given frequency and range. Search for power and range research data. If the needed data are available, transfer data to the project notebook; if not available, forward a "Research Request Form 710" through the chief engineer, one for each type of data needed.*
3a	*Design power source* If the station is to be self-contained, design the power source. *All calculations are recorded in the notebook. List the salient features and installation dimensions in the project notebook.*
3b	*Design power interface* If the station is to use available power lines, design only the main switch gear. If the station is to have its own power source, do this step after 3a. *All calculations are recorded in the notebook. If the country electrical code is not available, apply one of anticipated stringency. List, in the notebook, all decisions and locations of equipment so affected. List the salient features, dimensions, access requirements, and so forth in the notebook.*

Subsequently, "good" engineers were interviewed. Their description of the various steps, which were much harder to elicit, did not include the italicized portions of the steps given. As a result, their discussion of the time for individual steps indicated that they felt the time required was not only very variable but unpredictable. They claimed that they were frequently held up unpredictable lengths of time:

---

[5]The portion of the steps that appeared in the "excellent" engineers steps but not in those of the "good" engineers have been italicized. This could not be done, of course, when this first FPE was made. The italics, which were added later, are included in this illustration to facilitate discussion later in the chapter.

1. Waiting for sales data.
2. Waiting for climatic data.
3. Waiting for electrical code data.
4. Waiting for research data.
5. Having to recalculate various design aspects.
6. Having to search through the drawing department to obtain final drawings in order to design mating parts, and so forth.

It should be obvious that some of the differentiating characteristics of the "excellent" engineers included:

1. A more complete concept of a sequence of steps of design, although it had never been documented.
2. The use of a notebook, per project, rather than bits and scraps of paper, to keep all project data and computations readily available for reference, checking, or changing.
3. An anticipation of potential delays caused by lack of data and the forwarding of requests for such data as far as possible ahead of need.

Assigning a realistic time to the steps (step 3). When the final lists of lower-order work-units were completed, the experienced engineers, with guidance from the chief engineer, assigned, from their experience, a typical time for each step. They took both easy and unusually hard jobs into account to assign an average which they felt they could, in the long run, meet. A required understanding was that the engineers would not be expected to meet the standard time with each occurrence of each step.

Additional problem and solution. The setting of the FPEs overcame the feeling that standard times were not feasible. At this point the design department raised its final objection: "You can set standard times. We agree. However, you cannot use them to schedule our work. We are frequently interrupted by requests to assist sales, design conversions for stations previously erected by the company, help on installation or station run-in difficulties, and so forth. We cannot predict when these will occur."[6]

To overcome the difficulty cited, the work records of the design group were examined. It was found that sick and annual leave amounted to 14 per cent of their payroll time. Time on emergency work and out-of-office work (sales meetings, travel to sites, and so forth) took another 30 per cent of their time. Hence, only 56 per cent of their payrolled hours could be

---

[6] I have repeated this statement in full in that it is almost inevitably offered. I have not yet encountered a real problem area where it was not given.

considered available for planned design work. Schedule loadings were made at 56 per cent of design manpower to determine design delivery schedules.

Results. By employing the FPEs, with the loading and scheduling methods indicated, the following results were achieved:

1. Design output rose 25 per cent. This was based on a tally of a past year's designs, grouped into the seven types and weighted by the appropriate FPEs, compared to a year under the new procedure.

2. On-time delivery of designs rose from 10 per cent to 95 per cent. This is, of course, in part due to the restriction placed on the former free-promise policy of Sales. However, the new reliability of promises appeared to offset the handicap of the restriction.

3. Factory productivity rose 30 per cent as a result of the elimination of late designs, which had previously kept production chaotic.

4. Test-erection defects were reduced by 85 per cent. This is ascribable to the use of the notebook and recording data therein.

5. Overtime was essentially eliminated in the research department.

6. The relationships among sales, design, production, erection, and research noticeably improved. Of particular note was the better morale in the design department.

7. Emergency site work began to fall toward the end of the year. A full evaluation of changes in this area could not be made during the first year in that most designs, particularly during the first 10 months, had originated during the use of the previous undermanaged procedure.

8. Indoctrination time to bring new engineers, with other design experience, into full station-design productivity was reduced by 75 per cent.

An examination of the advantages of an FPE, as previously stated, will indicate why these results should not be considered surprising.

Case II—The Use of the FPE To Assist Managerial Control in the Animal Health Division, Agricultural Research Service, U.S. Department of Agriculture

Introduction. The Animal Health Division (AHD) represents a particularly complicated problem for managerial control, as control was defined in Chapters 1 and 6. The resources for the AHD are appropriated by the Congress as 14 separate funds. The use of each fund is restricted (with some small exceptions) to the work of controlling, eliminating, or preventing either a specific disease or a specific group of diseases of animals, or for the maintenance of humane methods of handling animals. The work, however, is carried out under 50 state/federal agreements, or agreements covering an area. The total territory covered consists of the 50 states, Puerto Rico, and the Virgin Islands. The work necessary to achieve the program goals is shared between federal and state employees in a different manner in each area covered by an agreement. Hence, although managerial control must begin with the usual identification of the hierarchy of work-units, (Chapter 9), the conversion of these work-units to a statement of required resources must be performed on an area-to-area basis before being aggregated into a national budget.

The work-unit structure. The work-units, at the measurable level, consist of services rendered by veterinarians, epidemiologists, diagnosticians, livestock inspectors, and various support personnel, in assisting in diagnosing animal diseases of economic importance, in making trace-backs and traces forward to determine and eliminate the sources of the disease, in maintaining a monitoring service to make certain that a disease that has been eliminated in an area is not reintroduced, and so forth. The list of work-units is far too extensive to reproduce in its entirety. This illustration will, therefore, be limited to an example selected to demonstrate the method of making and using an FPE.

A sample 5th-order work-unit and its parts. For instance, one of the 5th-order work-units produced in most of the areas served is: *A healthier animal population maintained by attention to direct control and elimination of hog cholera in swine.* Typical of the 4th-order work-units produced in the field as component parts of this 5th-order work-unit are[7]:

10901. *Diagnostic referral completely processed.*
10902. *Epidemiological study completed.*
10903. *Garbage feed premise maintained under inspection.*

Requirements for managerial control. Managerial control requires (among other things) that:

1. A forecasting system be developed to project the number of each work-unit that will be required in an area.
2. A coefficient (standard time) be developed to permit each individual work-unit (or any number of it) to be converted to a statement of the amount of resources required.

---

[7]I have retained the numbering system of the AHD. The list has been greatly foreshortened.

3. The forecasting system be checked by comparison of the actual events to the anticipated events and the actual use of resources be compared to the planned use of resources.

Normal biometrical methods make item 1 feasible; the same approaches facilitate the performance of part of item 3. Various information flow systems are required, of course, but these only require the application of well-known data-collection technology. However, to achieve the remainder of the items, some type of standard time is necessary. It is conceivable that the necessary data could be collected by a self-reporting system. However, such systems require an enormous effort on the part of all people in the system, an enormous data-reduction procedure, and still fail to give results that have the advantages of the FPE (as those were listed at the beginning of this chapter). It should also be obvious that other work measurement methods listed in preceding chapters are not applicable here.

The project team and plan.[8]   A project team for the initial effort consisted of an industrial engineer, a thoroughly knowledgeable veterinarian from the central staff, an administrative clerk, and the Federal-Veterinarian-in-Charge from the "pilot" state. It was anticipated that if a standard time for all work-units could be completed for a pilot state, the task of replicating the effort in other states would be greatly reduced; the methods, the procedures, and even the format of the FPEs would be available to them. The additional states would need merely to adapt them to the peculiarities of their territories.[9]

Developing the FPEs.   As a first step, the project group developed the details of the hierarchy of work-units referred to earlier. Subsequently, they developed an FPE for each 4th-order work-unit in terms of 3rd-order work-units. Part of such an FPE for work-unit 10901, *Diagnostic referral completely processed*, is shown in Figure 24.1. Where the word "standard" appears in the columns "Factors affecting time," the task was repeated in so many FPEs that a standard time was agreed upon for use in all FPEs. The FPE shown in Figure 24.1 shows only 10 of the 21 steps associated with the total FPE for the work-unit.

All the FPEs for all the 4th-order work-units were assembled by 5th-order work-units for convenience of use.

Meeting other control needs.   Subsequently, a work-unit forecasting system was devised for each 4th-order work-unit and the standard time used to

---

[8] It should not boggle the mind to realize that an FPE had previously been made for this entire project.

[9] This is true in general. However, a few states have required outputs not found elsewhere; in such cases, they would need to originate an FPE for these exceptions.

BASIC TIME ESTIMATE DATA FROM PROFESSIONAL EXPERIENCE

| 5TH ORDER WORK UNIT CODE 109 | 5TH ORDER WORK-UNIT DESCRIPTION Healthier swine with respect to HOG CHOLERA |
| 4TH ORDER WORK UNIT CODE 1091 | 4TH ORDER WORK-UNIT DESCRIPTION Diagnostic referral completely processed |

ACTION	MODIFIER	SUBJECT	SPECIAL FACTORS AFFECTING TIME	TIME PER OCCURRENCE IF PRORATA	CATEGORY OF PERSONNEL DIRECTLY INVOLVED			
					HOURS Fed. Diag.	HOURS St. Diag.	HOURS Fed. Vet.	HOURS St. Vet.
01. Receive phone call	from office	a suspected case of hog cholera	STANDARD		.17	x*	N/A	
02. Make arrangements by phone	with vet., diag-nostician or state employee to farm	to accompany to premises	Average 3 calls		.50	x	.17	x
03. Drive car			Average trip (outgoing)		1.50	x	.75	x
04. Discuss herd his-tory	at farm with farmer		Average		.50	x	.50	x
05. Change clothing	at farm	to work clothes	STANDARD		.17	x	.17	x
06. Inspect and observe	ante-mortem at farm	swine herd	Average herd size		.42	x	.42	x
07. Take temperatures and collect blood samples	at farm with help	swine	10–20% of average herd; 5–6 blood samples		1.00	x	1.00	x
08. Make post-mortem and collect tissue	at farm, with help	swine, complete necropsy	2 carcasses		1.00	x	1.00	x
09. Fill out form	at farm	ANH 13-7	Added time during post-mortem; STANDARD		.17	x	.17	x
10. Follow-up dispo-sition of carcasses	at farm	carcasses	By owner		.75	x	.75	x

*Indicates same time if done by alternative employee

Figure 24.1. Part of the FPE for the 4th-order work-unit, "Diagnostic referral completely processed."

convert this to a statement of the amount of manpower needed to produce the required number of work-units during the year for which a budget was being estimated. This is called the *budget year (BY)*. A tabulation of the standard times for the 4th-order work-units comprising the 5th-order work-unit, *A healthier animal population maintained by attention to direct control and elimination of hog cholera in swine* is shown in Figure 24.2. A form showing the use of the standard time and the forecast of the anticipated number of the 4th-order work-unit 10901, *Diagnostic referral completely processed*, is shown in Figure 24.3. Note the accounting for the split of work between federal and state employees for this state.

Results. Using such procedures, as well as other work measurement methods for other parts of the AHD, it is possible to greatly reduce the speculative elements in the budget and greatly increase the effectiveness of managerial control. The use of the FPE makes possible the economical application of work measurement to areas of activity usually thought of as not only uneconomical to measure but also outside the range of feasibility.

Summary—The Fractioned Professional Estimate

1. Uses:
   a. To set standard times for:
      (1) Project-type work.
      (2) Nonrepetitive tasks that are describable and within the state-of-an-art.
      (3) Long jobs which have some unique features each time they occur and which only occur at intervals.
2. How made:
   a. Define objective; the final desired work-unit is defined.
   b. Describe parts; the lower-order work-units are listed together with factors affecting the time to complete them.
   c. Assign times; using experience (and with managerial guidance), a time is assigned to each lower-order work-unit, which, on the average, could be achieved with a reasonable, acceptable level of performance diligence.

PROBLEMS

24.1. With respect to Case I, indicate which of the advantages of an FPE, as given earlier in the chapter, relate to the results achieved. Explain which features of the FPEs were related to each of the results.

STANDARD TIME DATA SHEET FOR BUDGETING AND REPORTING	STATE Maryland	DATE LAST UPDATED Current fiscal year

5TH-ORDER WORK UNIT CODE 109

5TH-ORDER WORK UNIT DESCRIPTION: A healthier animal population maintained by attention to direct control and elimination of HOG CHOLERA in swine (Fund 400)

4TH-ORDER WORK UNITS (COMPONENTS OF OUTPUTS)	WORK COUNTS FOR REPORTING AND ESTIMATING	MAN-HOURS BY CATEGORIES OF PERSONNEL						REPORTING TIME VALUES
		Fed. Diag.	St. Diag.	Fed. Vet.	St. Vet.	Fed. Insp.	St. Insp.	
01. Diagnostic referral completely processed	Referrals completely processed	11.43	x	8.35	x			19.78
02. Epidemiological study completed	Studies completed including all premises	49.22	x					49.22
03. Garbage feed premise maintained under inspection	(1)Per premise visit (2)Per district mon't./mo (3)Wash.gbge.ctr./week					1.78 8.00 28.00	x x	1.78 8.00 28.00
04. Quarantined herd administered because of garbage feed lot problem	Quarantine completed and removed					9.11	x	9.11
05. Quarantined herd administered other than garbage feed premise problem	N/A in Maryland							
06. Herd depopulated to remove focus of infection	Herd depop'ed. appraised Premise cleaned and disin. Quarantine released			18.25 1.50	18.17 x	10.75 7.00		29.00 7.00 1.50
07. Survey activity performed	N/A in Maryland							
08. Concentration point and interstate movement monitored	21-day quarantine released					2.17		2.17
09. Minute-man activities completed	Per presumptive or infected premise; Per man of 2-man team					15.60		15.60

x = Alternative person who may perform work; use same time as counterpart

Figure 24.2. Standard times for all field-produced 4th-order workunits comprising the 5th-order work-unit, "A healthier animal population maintained by attention to direct control and elimination of HOG CHOLERA in swine."

REQUIRED RESOURCES COMPUTATION SHEET

STATE	DATE
Maryland	Budget year

5TH ORDER WORK-UNIT CODE 109

5TH ORDER WORK-UNIT DESCRIPTION
Healthier swine with respect to HOG CHOLERA

4TH ORDER WORK-UNIT CODE 1091

4TH ORDER WORK-UNIT DESCRIPTION
Diagnostic referral completely processed

WORKLOAD GENERATING SOURCE
Private practice vets., owners, state inspectors, etc., requesting diagnostic assistance

PERSONS DIRECTLY INVOLVED
Fed. Diagnostician, State Diagnostician, Fed. & State Vet.

PERSONS PROVIDING SUPPORT

HOW DIRECT WORK IS DIVIDED	WORKLOAD SPLIT	Fed. Diagnostician – 67%; State Diagnostician – 33%
	INTERNAL SPLIT	Fed. Vet and State Vet assist the diagnosticians 80% of time; 50% split each

OTHER DATA AFFECTING PRORATA DISTRIBUTION OF TIME VALUES

	DESCRIPTION	Diagnostic referr.			Positive case			
NUMBER OCCURRENCES IN YEAR	FISCAL YEAR	BY	+1	+2	+3	BY	+1	+2	+3
	WORK COUNT	24				1			

NUMBER OF SIGNIFICANCE TO FEDERAL EMPLOYEES BY CATEGORIES

NUMBER OF SIGNIFICANCE TO STATE EMPLOYEES BY CATEGORIES

FEDERAL REQUIRED RESOURCES COMPUTATIONS:

10901 Time = No. of D.R. x hrs/D.R. x Fed share
Diagnostician
$$24 \times 11.43 \times .67 = 184 \text{ hrs}$$

10901 Time = No. of D.R. x hrs/D.R. x Fed share split x Fed share assist
$$24 \times 8.35 \times .50 \times .80 = 80 \text{ hrs}$$

STATE REQUIRED RESOURCES COMPUTATIONS:

10901 Time = No. of D.R. x hrs/D.R. x State share
Diagnostician
$$24 \times 11.43 \times .33 = 91 \text{ hrs}$$

10901 Time = No. of D.R. x hrs/D.R. x State share split x State share assist
State Vet
$$24 \times 8.35 \times .50 \times .80 = 80$$

VETERINARIAN	80
LIVESTOCK INSP.	
EPID. OR DIAG.	184
PORT INSPECTOR	
INVESTIGATOR	

VETERINARIAN	80
LIVESTOCK INSP.	
EPID. OR DIAG.	91
INVESTIGATOR	

Figure 24.3. Forecast of work-units and required resources for the 4th-order work-unit, "Diagnostic referral completely processed."

24.2. A staff analyst is unable to give a list of lower-order work-units associated with a contemplated project.

(a) Give the possible reasons for this situation.

(b) Suggest remedies for each potential reason.

(c) If the analyst is really unable, comment on probable success. Explain fully.

24.3. (a) Define "pure research."

(b) Is pure research, as defined, subjectable to an FPE?

24.4. If an FPE is somewhere between the time needed for an acceptable output and a superb one, describe the behavior pattern needed to conform to the requirements of managerial control.

24.5. Develop an FPE for the following:

(a) Preparing a term paper.

(b) Completing a shopping trip to the supermarket.

(c) Completing registration.

(d) Having a parking ticket processed.

(e) Preparing a bundle of laundry for the laundry service.

(f) Changing from regular tires to snow tires.

(g) Washing and polishing a car.

(h) Determining the number of different kinds of people that pass a given place at different hours of the day.

(i) Determining the details of the flow of traffic at a simple street intersection.

(j) Determining the details of the flow of traffic at a complex street intersection.

(k) Determining the average number of pairs of shoes examined by the typical woman purchaser in a shoe store.

(l) Determining the amount of time spent, per person, at the return desk at the department store.

24.6. Determine an FPE for any of the intermittent or project-type assignments that you are likely to receive in the near future.

24.7. Design a simple reporting form and system for use by a group for reporting their use of time and the status of projects with respect to the FPEs orginally set for the projects.

25

TIME STANDARDS BY FIAT

Introduction

The concept of a standard by fiat is a necessity if we are to think of the cycle of managerial control as being applicable to all the manpower resources of an organization. It is also useful in directing our attention to the areas where change is feasible rather than expending fruitless efforts on work measurement.

Definition. A time standard by fiat is one that is implicit in the design criteria used to design the work-unit or output. (The word *fiat* is used to indicate that the standard is by decree rather than by measurement.) This is the simplest type of work measurement. Of course, it is not always applicable.

There are many areas where a failure to recognize the existence of this type of standard has caused the expending of much fruitless effort on the wrong aspects of managerial problems. For a simple example, let us take the case of an in-plant seminar with 10 days of all-day meetings. The standard time for the seminar leader, *for leading this seminar*, is 10 man-days. (This example is so simple that it may sound inane. However, it will be shown that the same simple concept has great applicability to problems that appear to be hopelessly complex if viewed from a wrong aspect.) Of course, the 10 man-days does not include the time to plan the seminar; the planning time might well be the subject of an FPE, as those were described in the preceding chapter. However, in the case being discussed, if one wishes to change the manpower resource required to lead the seminar, the problem to tackle is the design of the larger work-unit, *seminar completed*. The problem is one of changing the design of the output rather than one of complex work measurement.

Types of work for which it is used.   An organization may wish to divide a country into territories or regions to facilitate administrative activities. It may wish to have a regional director (of sales, service, and so forth). It may also want an assistant regional director to make certain that there is always someone readily available in the region to perform the assigned functions. It is important to recognize that the first and most important problem is the selection of the basis for dividing the country into regions. What basis is to be used? What are the design criteria? Subsequent to this, an FPE approach may be used to determine whether a regional director can do all the things he is expected to do. Perhaps the use of the FPE will show that the work of two regions could be combined under one regional director. However, this combining would result in a change in the design criteria for a region, and in a change in the design criteria for a regional director. These are not work measurement problems.

In a chemical plant it may be desirable to have an individual to watch certain gages and controls *in case something happens*; then he is there to react. The placing of a man at such a post is the result of design criteria; it is not a simple work measurement problem. The design criteria may not, of course, be correct. However, the recognition of the area of examination is a first step toward the development of a solution. Perhaps the real risk of something happening should be evaluated. Perhaps the "watcher's" span of attention (the number of gages watched) can be increased without increasing the risk; perhaps the possibility of simultaneous emergencies is too high to do this. These are not work measurement problems as we have defined them. They concern the design of the output, that which has been called design criteria. Management technologies other than those described in this book must be used. Methods studies involving mathematical models may be undertaken to improve the design of the output. The subsequent standard time will be essentially by fiat, as the term has been used.

Case—The U.S. Secret Service

Introduction. Let us assume that we are attempting to determine the total number of men required for the Secret Service of the United States. If we were to make a complete work-unit analysis of this organization (see Chapter 9),[1] one of the 5th-order work-units would be: *A guarded White House.* Within this 5th-order work-unit one of the 4th-order work-units would be: *A guard post maintained 24 hours per day all year.*

The work-unit design problem. The determination of the number of such guard posts needed is a design problem; it is a function of the design criteria of a guard post. It is not a work measurement problem. The work measurement problem concerns only how many man-years are needed per year to maintain each such post.

For instance, the design criteria for such posts may be: "All avenues of approach to the White House are to be under the simultaneous observation of at least two posts at all times; each post must be under the simultaneous observation of at least two other posts; the posts must be unobtrusive."[2]

Some may question the validity of using design criteria of the type given. However, this use would appear to be no more arbitrary than putting one spare tire in a automobile, two ice-cube trays in a refrigerator, and so on. One may object and say that these substantive design criteria are based on market surveys, consumer-use reports, and so on. However, surely one should anticipate that the Chief of the Secret Service uses a parallel experience-study procedure for developing his design criteria. The similarity should be obvious. Note also that the selection of two ice-cube trays per refrigerator affects the amount of manpower resources needed to make refrigerators, although additional work measurement studies are needed to determine the effect. In the case of the White House guards, the effect is merely more immediately discernible and less subject to modification once the design criteria have been fixed.

Work measurement by fiat. With respect to the White House guard posts, the work measurement problem concerns only the determination of

---

[1]Such an analysis has been made and used for budgeting. For obvious reasons it cannot be given here.

[2]For obvious reasons the statement is hypothetical, only in general resembling the actual criteria. For a similar reason the data given later are fictitious substitutes for the real data.

the amount of staff resources required to produce a unit of output, *One guard post maintained 24 hours per day all year.* Predetermined time systems cannot be used. Extensive or intensive sampling studies would shed no light on the problem. The appropriate approach is much simpler.

Let us examine the actual work measurement problem. Each 4th-order work-unit requires 365 (days) $\times$ 24 (hours/day) = 8,760 hours of direct "on-post" time each year. A guard works 8 hours per day, 5 days a week, $52\frac{1}{4}$ weeks per year, or 2,090 hours per year. However, as in most such cases, one cannot expect all of these hours to be available for the direct production of "on-post time." The 4th-order work-unit has associated with it other tasks besides being "on-post." For instance, experience may indicate that 14 per cent, or 293 hours[3] per year, will be lost because of sick and annual leave. This must be taken into account when computing the staff resources needed per 4th-order work-unit produced. There are additional losses created by additional design criteria associated with the 4th-order work-unit. For instance, it may have been decided that it is desirable for the guards to maintain proficiency in the use of small arms and that 2 hours per month (another design criterion with a fiat-type, but verifiable, effect) should be devoted to this. There will be similar factors. A table may be constructed as follows:

Total payroll time		2,080 hours/year
Less:		
Sick and annual leave at 14 per cent	(293)	
Small-arms proficiency training		
2 hours/month $\times$ 12 months	(24)	
Physical fitness		
1 hour/day $\times$ 251 working days	(251)	
First-aid refresher courses		
8 hours/course $\times$ 1/year	(8)	
Lectures on handling the public		
8 hours/lecture $\times$ 2/year	(16)	
Total deductions from available time	(592)	
Time available for actual guard-post work		1,488 hours/year

Hence, the work measurement coefficient equals

$$\frac{8{,}760 \text{ hours/post/year}}{1{,}488 \text{ hours on-post time/person/year}} = 5.89 \text{ staff-years/post}$$

---

[3]The use of fictitious numbers does not affect the validity of the example.

The personnel required in man-years (MY) for all the guard posts would be the number of such posts multiplied by 5.89. Essentially, the standard is by fiat, being computable from the design criteria. It is to be noted that this situation, properly analyzed, is one of relative simplicity.

It should further be noted that the 5th-order work-unit, *A guarded White House*, will contain other 4th-order work-units. If these are to be aggregated, each 4th-order work-unit must be subjected to an adequate time study or work measurement procedure, as the terms are used in this book. Each 4th-order work-unit may need a different procedure, but each must be subjected to work measurement before being aggregated to determine the total manpower resources needed for the 5th-order work-unit. Unfortunately, all are not as simple as the one given. However, the one given would not be simple if the concept of design criteria and a standard by fiat were not separated and identified.

Summary—Time Standards by Fiat

1. Uses:
    Applied to jobs where the manpower is a function of only the job design; where additional effort would not alter the time required.
2. How made:[4]
    a. The time required for the daily duration of the job is determined.
    b. The time available per day, from an appropriate type of employee, is determined.
    c. Using a and b, the number of needed employees is computed.

PROBLEMS

25.1. Assume that the employees of the Secret Service are assigned a 30-hour work week. How many guards are required per post?

25.2. For the Secret Service, draw a graph with the ordinate showing guards per post and the abcissa showing the hours of work per week from 45 to 24 in appropriate steps.

25.3. A hospital has put ward clerks into each 80-bed ward to relieve the professional personnel of all nonprofessional activity. The ward clerks post records, keep the bed roster, and so forth, as well as provide the capability to type patient insurance forms, and so forth, as needed and when needed by the patients. A

---

[4]Time, in some cases, is more conveniently computed per job, per month, or per year.

rough work sampling survey has found that the ward clerk is only working 45 per cent of the 8-hour tour of duty. Comment on:

(a) A proposal has been made to apply a PTS to the work of the ward clerk to more accurately determine the actual workload.

(b) The earned hours (3.56) and the actual work hours (8) are being added to the earned hours and worked hours of the total hospital staff to compute hospital efficiency.

(c) It is proposed to assign two wards to each ward clerk.

(d) It is proposed to add medical records typing to the ward clerk's duties so as to give them 8 hours of work.

25.4. A supermarket, to expedite checkout of a reasonable number of customers, has installed an "express line" limited to customers with not more than 10 items of purchase.

(a) What type of time standard is required to determine the necessary staffing?

(b) Describe the decision-making process which might lead to two such lines. Carefully separate the output design aspects from the work measurement aspects.

25.5. (a) List five instances of common occurrences where the staffing is in terms of a fiat time standard.

(b) In each of the situations in part (a), indicate the variables that may cause the situation, at some time in the future, to be other than a fiat manning.

26 TIME STANDARDS BY MATHEMATICAL ANALYSIS

Introduction

There are many areas of activity where the classical procedures of time study (direct time study—intensive or —extensive sampling or PTS) are not applicable. For this reason previous chapters have dealt with the use of the fractioned professional estimate and the standard time by fiat. Standards by mathematical analysis represent another alternative to the use of the classical techniques.

Definition. Statistical data, obtained over a period of time concerning the production of outputs and the use of manpower resources, is subjected to mathematical analysis to determine the relationship between required manpower resources and outputs. The result is *a standard time derived by mathematical analysis*. The necessary data may be available, in some cases, from existing records; in other cases, special data-collection systems need to be designed and used.

470

A variety of techniques. The available mathematical techniques are numerous. In some procedures additional judgment may be introduced to alter or modify the data so that values approaching engineered standard times may be obtained. The mathematical techniques usually enable one to determine time standards faster than with any other type of approach. In some cases they are the only readily feasible methods. The use of such techniques does not, of course, preclude possible subsequent refinement with other more detailed techniques, when feasible. As with direct time study—extensive sampling, the most difficult problems encountered in the use of the mathematical techniques frequently concern the determination of the work-units and how to count them.

Other names for the techniques. In much of the literature the types of standards discussed in this chapter are referred to as statistical standards. In that the work *statistical* is commonly associated with the theory of probability, and in that probability is not necessarily computed with statistical standards, the title *mathematical time standards* or *time standards derived mathematically* would seem to be a more appropriate way of referring to these types of standard times.

Alternative definition In terms of the factors constituting the equation of a standard time, time standards by mathematical analysis employ the following:

1. *W/C*. The number of work-units produced may be taken from historical records; as an alternative, a data-gathering system may be introduced to obtain these data over a selected period of time. The period of time needed for gathering data depends upon the mix contained in the work-unit work counts.
2. *W/T*. The work time may be taken from past payroll or attendance records or gathered in connection with the data-gathering system related to the work-units or outputs.

Note: With some mathematical techniques, the simple division of the work time by the work count (as with other work measurement techniques) may appear to be replaced with a much more complex mathematical analysis, depending on the relationship between the units used to count the work and the work time. This comes about because, in some cases, the work time associated with a variety of work counts may be aggregated. The mathematical techniques are employed to determine how much of the total time to divide by each separable work count; the basic equation still holds.

3. *M*. Depending on the mathematical technique chosen, the value of *M* may be taken as 1 (past performance is assumed to be satisfac-

tory), or judgment may be used in a systematic fashion so as to introduce an *M* greater or less than unity.
4. *A*. If historical data are used, the value of *A* is usually implicit in the data; if a special data-gathering technique is employed, the data may be adjusted to correct the actual *A* to some desired value.

Types of work for which the techniques are used.   The basic field of use is for gross standard times and for situations similar to those where the FPE is used but where the nature or variety of work-units and the quantity (work count) make the FPE relatively inconvenient or infeasible.

Use in computing gross standard time.   A foundry may use past data to compute a mathematical standard for the number of man-hours per ton of castings produced. A change in the number of man-hours per ton may be caused by a variety of factors. The change may be due to a different work pace; to more- or less-difficult castings; to a change in the size of castings, requiring more or fewer castings per ton; to a change in the amount of delays caused by the equipment; to more or less supervision; to changes in the efficiency of scheduling; or to unbalanced changes in different directions in several of these factors. In addition, the factors cited are by no means a complete list.

Consequently, if performance deviates from the simple mathematical standard, remedial measures are not specifically indicated, and additional data must be obtained in order to determine the necessary action. This is in comparison with performance measured against standards where the method and the work-unit are more specific and most of the effect of factors outside the control of the group being measured are removed from the measure of performance of the group. However, the simple mathematical standard does measure the overall effect of all the factors affecting the work force, and thus has considerable management utility if used with a great deal of discretion. In particular, all downward deviations should not immediately be ascribed to poor performance from the work force; all improvement should not be ascribed to better management.

Use when a great variety of work-units is encountered.   For examples where the variety of work is too great to make the use of the FPE practical, one need not look beyond the work of patrol police officers (including the FBI and the field forces of the Secret Service), billing and purchase order clerks, insurance account data processors,[1] and so forth. In such cases the mathematical procedures are more complex than those used to set gross standard times.

Mathematical standards derived by more complex methods have been used with considerable success for some but not all of the common uses of

---

[1] Preparing data for computers, and so forth.

work measurement.[2] They are frequently basic measures, which may be supplemented eventually with standards for smaller work-units. However, it should be pointed out that there are situations, particularly where the final work-unit has almost infinite variety, in which the mathematical standards are the only standards feasible on an economic basis, and their wise use is often of tremendous assistance to management. Particularly with certain types of service work-units, the use of mathematical techniques appears to be the only feasible method for work measurement. However, these techniques are not the simple ones. Also, the use of newer work measurement techniques, such as those described in Chapters 22 through 25, permits many applications of more detailed work measurement to be made in situations previously deemed uneconomical to measure except with mathematically derived standard times.

The uses of mathematical standards. In summary, mathematical standards, correctly derived, may be used for almost all of the normal uses of standard times. In some cases they are fully comparable to standard times derived by other methods. Of course, in some cases, the details of the manner of use need to be altered. For instance, if the mathematical standard for a foundry is man-hours per ton of castings,[3] it would probably not be proper to compute an individual's pay on this basis. However, a whole plant's productivity wage increment may be based on a measure of this general type, if the product mix is stable. However, in most cases, a simple measure of this type is seldom feasible; a better system of work-units is usually needed.

Further, it should be noted that the application of a mathematical standard frequently may require the assumption that past practice was not unusual in any respect and that the performance indicated may be expected in the future. When the work performed is a service, and when the work situation is relatively stable, this may be a tenable assumption. Hence, much government budgeting employs mathematical time standards. As an alternative to accepting past performance, gross assumptions as to the quality of the total of past performance may be made and used to modify the results. Whichever assumption is made, deviations from some mathematical standards are often hard to interpret. There is often no way of separating the amount of deviation ascribable to factors subject to the control of the working group, whose performance is measured by the standard, and the amount ascribable to factors outside their control. In such cases, even when the performance meets the mathematical standard, the significance is not always clear. Inasmuch as mathematical standards lack most of the basic controls

---

[2]Such standards came into wide use in government operations between the years 1917 and 1949 because of the congressional prohibition, during this period, of direct time study.

[3]If the mix of the size of castings is not a stable statistic, this will not be a good work-unit.

of the other work measurement approaches, we have no way of knowing whether changes both in the factors controlled by the group and in factors outside of their control counterbalance each other. This defect is primarily associated with the simpler techniques; when an adequate system of work-units is employed, these difficulties may be overcome.

The steps in the use of the technique (compared with other techniques). The comparison is as follows:

Steps in Other Techniques	Steps in Computation of Mathematical Standard
1. Define standard of measurement.	Past practice (or some gross modification of it) accepted as typical performance.
2. Describe product, equipment, and record method for detailed operation. (Not fully implemented with some techniques.)	No method record made. Only rough descriptions of the work-units are made, and a variety of similar work-units may be lumped together. However, in many modern applications this "lumping" is avoided.
3. Observe time taken by an actual operator, or postulate such time values.	Production and work-time records are used.
4. Rate or relate observed performance to definition of standard, or select a suitable estimate.	An arbitrary mathematical basis is used to determine the relationship between production and man-hours.
5. Apply allowances of a known amount.	Usually assumed to be included in gross production and work-time records.

Limitations on the use of the techniques. The primary advantage of standards derived by mathematical procedures is the rapidity with which they can be determined for most situations. Therefore, the development of such standards often precedes the employment of other techniques. Speed of development is the trade-off gain, accompanied often by loss of detail. The mathematical standards serve as stopgap standards until accurate measurements may be developed. (However, there are situations wherein the mathematical techniques appear to offer the only feasible approach.)

The usual form of a mathematically derived standard time is a standard expressed in *man-hours per work-unit*, as with standard times developed with other procedures, but the work-unit, instead of being a second- or 3rd-order work-unit, is frequently a work-unit encompassing all the activity required to produce a 4th-order work-unit, or the entire processing of a product, or a large phase of the processing, or the processing of a variety of products with some similarity. However, this is not an inherent limitation in the techniques available. Standards for smaller work-units may be derived by mathematical procedures. In some cases the mathematical approach may be the most suitable of all of the work measurement methods.

An examination of the basic steps in mathematically deriving a standard, as compared with the steps of other work measurement techniques, will reveal the reasons for the rapid computation of the statistical standard as well as its inherent deficiencies. However, if these deficiencies are understood, the standards may have considerable utility.

A type of gross unweighted mathematical standard is often hidden under the names *manning ratio* or *staffing pattern*. Staffing patterns were mentioned in Chapter 24; it was further indicated that in many cases an FPE could be developed more usefully to set a standard for many situations currently covered by staffing patterns. A staffing pattern was defined as follows: "Ratios of one type of employee to another type are used to set a 'standard' for manning." For example, a plant may have seven tool designers and 300 direct machine operators. The staffing pattern for tool designers is 1 per 42.9 workers. These gross measures are often used as rules of thumb to compare various periods of activity or different plants. There are great dangers in their use. The danger increases as the relationship between the two groups used in the ratio becomes more tenuous. For instance, it would seem more reasonable to accept a ratio of chair assistants per dentist, or tool-crib attendants per machine operator, than the example given previously of tool designers per machine operator. In either case, however, the use of an FPE approach seems more realistic unless a quick temporary rule of thumb is desired. Using such ratios to compare from plant to plant is even more dangerous. The assumptions concerning the similarity of the factors affecting the ratio are too large for ready acceptance. At the very most, such ratios may have some value in assisting in evaluating a change from period to period within a plant, or the difference between similar plants. If, in such cases, an investigation is made concerning the reason for the change in values within the plant, or the reasons for the differences between plants, some additional useful information may result.

Advantages and limitations, summarized. Most work with mathematically derived standards has suffered from either the use of an inadequate system of work-units for describing the outputs, or from the use of unreasonable assumptions in manipulating the data. In many cases, both types of errors have been made. With an adequate system of work-units and with adequate mathematical procedures, excellent results can be obtained; situations that defy the setting of standards by other methods may be handled, in some cases, quickly and with ease. Some procedures may involve much detail, but they enormously extend the range of work to which work measurement may be applied to usefully serve management control systems.

As indicated, the procedure for determining a mathematical standard is relatively simple. A mathematical relationship is determined between man-hours expended and the work-units produced. Often, the records necessary for the computation of such a standard are already available as part of the organization's normal production and payroll records. However, this is not always true of indirect work or service activities. These are seldom counted in a manner appropriate for even mathematical techniques. If service activities are represented by a suitable hierarchy of work-units, the mathematical techniques may be highly effective procedures. As was indicated in Chapter 9, work-unit analysis may be a natural starting technique.

Case I—The Simple Computation of a Gross Mathematical Standard

Introduction. A gross mathematical standard is one wherein the work count of a single work-unit is used to quantify the output. It gives an overall measure reflecting the results of all the factors affecting the work situation. This procedure is the least applicable of all the procedures given in this book (with due allowance for the range of application of each) for the derivation of work measurement values for all the uses of managerial control. The error is usually in the assumption that the single work count is an effective indicator of the required human resources.

For an example of the procedures, prior to a discussion, suppose that the production and payroll records of a foundry gave the following data (the numbers have been rounded for convenience):

Month	Direct Labor (hours)	Castings Produced (tons)
January	4,800	550
Febuary	3,300	480
March	3,900	350
April	3,000	240
May	1,000	150
June	1,200	350
July	3,200	390
August	4,500	490
September	4,000	590
October	6,200	840
November	4,600	700
December	6,300	640

Different methods of computing the gross mathematical standard. Three methods of computing gross mathematical standards are in common use. They are:

1. *The standard based on the weighted average.* From the data given, the total direct labor-hours for the 12 months were 46,000; the total tons of castings produced was 5,770. For the period, 7.96 direct labor-hours were expended per ton of castings produced. This standard reflects not only the monthly figures but gives more weight to the months with more tons or more direct labor-hours.

2. *The standard based on the unweighted average.* If we compute the direct labor-hours per ton for each month and average these figures, we obtain a standard of 8.11 direct labor-hours per ton. This standard gives each month an equal weight, regardless of tons produced or hours consumed.

3. *The standard based on a fixed measure other than the average.* Taking the hypothesis that work previously performed without a standard time to work against is performed in an inferior fashion, as far as diligence of work and supervision are concerned, some applications of gross mathematical standards employ measures such as the third quartile as the basis of the standard. To determine such a measure for the data previously given, it is necessary to compute the direct labor-hours per ton for each month and then place these in an array as follows:

Direct Labor-Hours per Ton	*Month*
3.42	June
6.57	November
6.66	May
6.78	September
6.88	February
7.38	October
8.20	July
8.73	January
9.18	August
9.84	December
11.13	March
12.50	April

The third quartile is defined as a value below which are three-quarters of the values. In this case, the third quartile would be between 6.78 and

6.66, or 6.72 hours per ton. In an isolated application such a procedure may be justifiable, but when several such standards are in use in different parts of one plant, such a procedure introduces new, questionable features. It has already been noted that the mathematical standard confuses factors under the control of the group (whose performance is measured against the standard) with factors outside their control. If the data for one standard have more dispersion than the data for a second standard, we have no way of knowing whether the difference in variation is attributable to the first work group or not, or what factor caused the additional variation. The use of the third quartile, or any similar measure, to set the standard will result in a standard farthest from the average for the group of data with most dispersion. Consequently, it may be more difficult to attain this standard than another standard for another task whose data showed less dispersion. It will also be more difficult to administer such a set of standards in a uniform fashion, and their already limited utility will be further reduced.

Utility and limitations of such gross standard times. The value of a standard computed by one of the gross computational methods lies in the simplicity of the calculation; this may be a small value if the results are not valid. Reasons for suspecting the validity are often numerous. First, if the data given, such as were used in the preceding example, are the only data, nothing is known about the quality of performance implicit in the data and in the resultant standards. This certainly detracts from their usefulness. Second, if nothing is known concerning the manner in which manpower was allocated during the data-base period, we may have even less use for the resultant standard. If monthly manpower was allocated on the basis of gross outputs scheduled, or if the gross outputs were scheduled on the basis of the monthly direct labor-hours avaliable, the computed standard may be merely a recomputation of the prior assumption; this would hardly appear useful. It is not uncommon to find either such a circular computation or computations approaching such nonsense.

Third, and most important, the gross computational methods assume that gross outputs, in the manner used to count them, are directly related to direct labor-hours required. As noted in Chapter 9 when discussing hierarchies of work-units, and as noted earlier in this chapter, this is not usually a valid assumption. Further, using the gross computational techniques, particularly methods 1 and 2, it is not known how well the computed standard really fits the data.

As will be seen later, almost any other method of computing a mathematical standard, other than the three described here, appear more useful. Certainly, more information than mere gross data seems necessary if the standard is to have any credibility. However, the methods described are employed, and their defects must be understood if one is going to suggest alternative procedures.

Case II—The Computation of a Gross Mathematical Standard by Curve Fitting or Simple Regression

Definitions. *Curve fitting* is the development of a curve for the plot of a number of data sets when each data set represents a work-count and a work-time value.

Simple regression is the development of a statement describing a straight-line relationship among data sets, such as described above, by the method of least squares.

Computing the standard time. For an example, let us use the same data as were used with the discussion of the simple computation of a gross mathematical standard. Figure 26.1 is the graph of the data previously given, with a trend line drawn. The trend line would be taken as a standard. In this case, the standard would be variable and would be a function of the tons of castings scheduled.

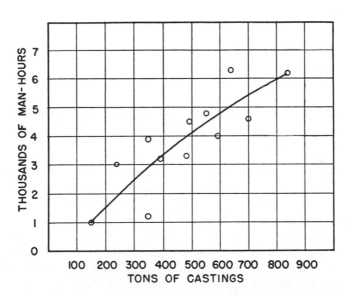

Figure 26.1. Man-hours of direct labor expended per month in a foundry versus tons of castings produced per month. A trend line has been drawn.

Assumptions.   To fit a trend line, some assumptions must be made concerning the probable nature of the relationship (i.e., straight line, hyperbola, and so forth). Just what relationship is shown in Figure 26.1 is not easily determined. In some cases, plotting the data on log or semilog graph paper assists in formulating some hypothesis concerning the nature of the relationship. Whatever relationship is finally assumed, the reliability of the relationship should be evaluated by statistical means to avoid making the standard reflect a chance or spurious relationship. (The subject of statistical reliability of relationships is beyond the scope of this book, and reference should be made to adequate material on this subject before endeavoring to apply this procedure. Particular reference should be made to "curve fitting," "goodness of fit," and "correlation.")

Fitting curves by visual methods; "eyeballing."   As an alternative to mathematical analysis and curve fitting, a visual method may be employed to fit a credible curve to the data points. The criteria would be credibility and a sufficiently good fit such that all data points are either within ± 5 per cent of their equivalent value on the curve or a valid reason for the lack of fit, which does not conflict with the theory of the curve, has been ascertained. Such an approach may require the introduction of additional work-units; if this increases validity, it seems appropriate.

Fitting a straight line.   As an alternative to fitting a curve, as in Figure 26.1, a straight line may be fitted to the data by the method of least squares. The use of this approach involves two assumptions, which are not always valid:

1. A straight line is the best explanation of the relationship between the two variables.
2. Large deviations from the line (the distance from a data point to the fitted line) should be given greater weight in locating the line than small deviations.

Defects in the standard times so far computed.   In either case, curve or straight line, the plot of Figure 26.1 reveals two facts not revealed by any of the gross computational procedures. It reveals that the fitted curve and the actual data have many discrepancies; some of these discrepancies are sizable. For instance, at 640 tons of castings per month, the actual direct labor-hours used exceeded the amount explained by the standard by approximately 1,300 hours, or 26 per cent. Also, at 350 tons per month, one data set exceeded the amount explained by the standard by approximately 900 hours, or 30 per cent; with another data set for the same tonnage, the hours used were below the amount suggested by the trend line by 1,800 hours, or 60 per cent.

It would appear most reasonable to assume that there are additional sources of variation or additional variables that have not been accounted for, and that the trend line shown has severely limited utility. If additional data are not available, it does not appear that such a trend line could serve well the purposes of managerial control.

Improving the utility of computed standard times. Let us assume that more than a trend line of a gross mathematical standard of low utility is desired and that additional data are available or can be collected. Further, let us assume that we wish to apply curve fitting in a useful fashion. Our first change must be to discard the concept of a single work-unit and find a series of work-units that are more appropriate, as was described in Chapter 9.

Let us assume that an analysis of the outputs or shipments of castings reveals that when the castings are separated by factors affecting the amount of time required per casting weight, there appear to be at least three different kinds of castings (three different work-units):

Work-unit 1. Castings ranging from 1 to 10 pounds, without cores.
Work-unit 2. Castings ranging from 1 to 10 pounds, with cores.
Work-unit 3. Castings ranging from 11 to 50 pounds, without cores.

Sorting the data by work-unit categories. A new table may be constructed as follows:

		Castings (*tons*)		
Month	*Direct Labor (hours)*	*Work-Unit Type 1*	*Work-Unit Type 2*	*Work-Unit Type 3*
January	4,800	400	30	120
February	3,300	80	95	305
March	3,900	200	89	61
April	3,000	100	95	45
May	1,000	50	16	84
June	1,200	0	28	322
July	3,200	250	23	117
August	4,500	250	85	155
September	4,000	300	23	267
October	6,200	450	51	339
November	4,600	300	45	355
December	6,300	450	79	111

Even a cursory analysis of the data indicates that the mix of kinds of castings is fluctuating; that the single work-unit did not provide a reliable statistic. Let us array the table by direct labor-hours, to obtain:

		Castings (tons)		
Month	*Direct Labor (hours)*	*Work-Unit Type 1*	*Work-Unit Type 2*	*Work-Unit Type 3*
December	6,300	450	79	111
October	6,200	450	51	339
January	4,800	400	30	120
November	4,600	300	45	355
August	4,500	250	85	155
September	4,000	300	23	267
March	3,900	200	89	61
February	3,300	80	95	305
July	3,200	250	23	117
April	3,000	100	95	45
June	1,200	0	28	322
May	1,000	50	16	84

As a side note, the data in this table reveal the incorrectness of the third quartile method for the gross standard; the mix at the third quartile level is not a representative mix; a standard set on this mix will not predict manpower requirements.

Fitting a curve to the segregated data using simple procedures. A straight line may be fitted to the detailed data in the table by simple means. Equations may be written for the first 3 months as follows:

$$6,300 \text{ hr} = 450X_1 + 79X_2 + 111X_3$$
$$6,200 \text{ hr} = 450X_1 + 51X_2 + 339X_3$$
$$4,800 \text{ hr} = 400X_1 + 30X_2 + 120X_3$$

where

$X_1 =$ standard time per ton for castings of type 1 (unknown)

$X_2 =$ standard time per ton for castings of type 2 (unknown)

$X_3 =$ standard time per ton for castings of type 3 (unknown)

Values for X_1, X_2, and X_3 may be computed from these simultaneous equations by the simple algebraic procedure of determinants.

A weighted value of the outputs may now be computed for each month by multiplying the outputs by their respective computed times. A new graph may be drawn of the computed direct labor-hours versus the actual hours. If a good fit is not obtained, a new computation should be made, or in that three adjoining points were used, a curve may be detected and fitted by visual or mathematical means. If the curve is poorly defined, three close data points (in this case, January, November, and August; or February, July, and April) may be used to get time values to assist in plotting the points in a manner that better defines the curve.

Making additional decisions. If the function appears to be a straight line, but the points do not fit the line well, three new data sets should be selected. If the relationship is found to be a straight line, but the computations from the first three points do not give a good fit, the months selected for the new computation require that some additional assumptions be made. For instance, if it does not seem reasonable to assume that the pace of work ever exceeded a reasonable pace, the data from the three lowest points (with respect to the first line) would be used to set up new equations.

As an alternative, if it can be assumed that the pace, from month to month, varied from above normal to below normal, the data from points which tended to place the line so that an equal number of points were above and below it would be used for the three new equations. In some cases facts, such as the knowledge of much idle time during a particular month, may be ascertained, and such points allowed to fall well off the curve. (Such points would, of course, not be used to compute the values of standard times.)

If a good fit on a believable basis cannot be obtained, the investigation must continue until explanations of the data points seriously off the line (more than say ± 5 per cent) can be made, or the system of work-units expanded to enable the close fitting of a line or curve.

Summary. The curve-fitting methods that have been described have much utility. However, in many cases, when the work-unit system is expanded to its true size so that the categories are relatively homogeneous, the number of work-units may exceed the number of data sets that can be obtained. In such cases, the mathematical technique described is not applicable; more complex mathematical approaches must be used; such techniques are described in later chapters.

The Use of Specialized Data-Collection Methods and Mathematical Techniques for Deriving Standards More Detailed Than Gross Standards

Introduction. In the cases of mathematically derived standards which were discussed to this point, when there were deficiencies in the standards, the usual cause was the inadequacy of the work-units used in the work-counting system. Other defects were in some cases indicated, but the foregoing was basic. As noted in Chapter 9, the problems of "what to count" and "how to count" must be solved before one can select and apply a work measurement technique.

In addition to the types of work situations described to this point, situations will also be encountered in which the work-unit list is large, the work is performed in scattered locations, and observations may not be feasible. Such characteristics often accompany service-type work-units. However, work measurement data are needed to effect managerial control; some special techniques must be developed and applied.

In such cases it is not likely that historical data containing work counts of a suitable list of work-units are available, although such data are occasionally found. In such instances, standard times may be computed with great rapidity. However, in most cases, some sort of data-collection system must be designed and put to use.

Self-reporting. In many cases a self-reporting system may be used in which the people doing the work report their use of time and the work completed. These data are then used to determine a standard time for each work-unit identified. Obviously, a consistent set of work-units must be used. Obviously, there are risks concerning the accuracy of reporting and of errors in the tabulation of such data. Experience suggests that the reporting of the use of time and the completion of work-units usually should be restricted to a base period used to establish the time standards. The base period should be as short as possible, yet extensive enough to provide a sample of the mix of work. Subsequent reporting may be confined to reporting the work-units produced. Attempts to maintain a continuous reporting of both time use and work performed frequently deteriorate in accuracy unless the reporting group fully believes in the need for such data on a continuous basis. In all cases, the need to demonstrate the necessity for the data collection to those reporting is a prime requisite. The more professional the group, the more the demonstration of need is feasible, and the more true participation for the necessary duration can be obtained. As an alternative to complete self-reporting, sampling may be used by attaching data-collection forms to a sample of the work, such as every tenth purchase requisition, every fifth payroll record, and so forth. It should be obvious that such methods are only feasible where some document or object is a constant concomitant to the work.

Types of work for which self-reporting is probably the only realistic method. One of the typical places where self-reporting is one of the few feasible approaches is with "undercover" work. To attempt to observe such work would obviously be impossible. Also, maintenance work, sales work, purchasing offices, and so forth may at times be advantageously studied with this technique. The technique can be substituted for the use of the FPE approach when some real amount of repetition of the work-units can be expected, or when the number of work-units is extremely large. Various data-gathering and data-reduction techniques may be employed to reduce the cost of, and the time required for, setting such standards.

In some situations, difficulties have been encountered in the use of self-reporting work measurement techniques. These difficulties are usually attributable to three types of causes. First, the system of work-units may have been inadequately developed; the work-units did not meet the criteria for well-chosen work-units. The usual defects are that they are usually neither mutually exclusive nor all-inclusive; they do not relate to the objectives of the organization. Second, the purpose of the self-reporting may not have been made clear. Indeed, in many cases, a system for use of the data may not have been contemplated or designed. It is hard to get effective cooperation in reporting under such circumstances. Third, inadequate time may have been spent in training the people to properly and accurately self-report. Such systems are frequently referred to as "garbage-in" systems. If these deficiencies and their causes are understood and avoided, much effective use may be made of procedures such as those described in this section.

Case III—Specialized Data Collection to Set Standards for Lawyers

Introduction. The Office of the Solicitor, U.S. Department of the Interior, presented a complex work measurement problem for which the self-reporting approach was appropriate. The discussion that follows will present the analysis of the problem situation which led to the selection of the technique. This will be followed by a short discussion of the actual methods used, the standard times obtained, and the use made of these data.

The Office of the Solicitor provides legal services to the Secretary, the Deputy Secretary, the Under Secretaries, and the program bureaus and offices of the Department. It is divided into 11 subject-matter offices located in Washington, nine area offices, and 20 subarea offices, a total of 29 locations in the United States. It employs approximately 210 professional employees. Services are provided to all 36 program offices and bureaus of the Department of the Interior in their dealings with the private sector, the Indians, the states, the Congress, the Executive Office, and in certain foreign problems dealing with fish, water, and so forth. Typical of the program bureaus served are the Bureau of Mines, the Office of Oil and Gas, the Bureau of Indian Affairs, the National Park Service, and the Bureau of Commercial Fisheries.

The work-unit structure. The 8th-order work-unit of the Office of the Solicitor is the same as the 8th-order work-unit of the Department of the Interior; the Solicitor only serves to facilitate it.

The 7th-order work-unit is the totality of advice, opinions, briefs prepared, and so forth, as characteristic of the output of lawyers.

It seems convenient to think of the 6th-order work-units as being 36 in number, with each described as: "A [specific office or bureau of the Department] served for 1 year."

Note, it is the bureaus or offices that produce the real outputs in a variety of ways. Hence, it seems appropriate that the workload forecasting and work measurement system of the Office of the Solicitor should be related to the individual offices and bureaus that generate the workload. Also, service functions are frequently undermanned. If the required resources of the Office of the Solicitor were computed from the total of the individual workloads generated by the offices and bureaus, any overall shortage of legal manpower could be reacted to by changing the allocations of legal manpower to the program offices or bureaus, with some ability to anticipate the consequences.

The special problems of selecting the 5th-order work-units. The selection of the 5th-order work-units presented a special problem. Legal assistance takes many forms, such as undocumented advice, aid in negotiating, reviewing for legality, or documented assistance such as interpretations, opinions, drafts, and so forth. However, if the outputs were divided by kind of service, the residual mix of subject matter in any of these 5th-order work-units would constitute a heterogeneous mix that would be difficult, if not impossible, to relate to the variety of outputs of each office or bureau served. Hence, the 5th-order work-units were defined as "A matter [of a specific subject variety] processed." (A *matter*, briefly, is a document or other communication requiring a formal professional legal response considered completed at the time no further work can be done.)[4]

Fifty-nine varieties of matters were separately identified by *subject* so as to allow relating matters to the variety (caused by the external society) in the mix of outputs for the offices and bureaus served. These 59 categories were devised as a result of many discussions between the analysts and the attorneys. Numerous test periods were used to make certain that the selected categories were such that matters could be so classified, and that the list of

---

[4]"A case completed" might appear to be the obvious output, but if legal aid is effectively applied, cases may not even develop, only matters will. Further, cases constitute too heterogeneous a mix; some last for 100 years. Large cases have many matters; small cases few. Hence, matters seemed a more appropriate and convenient concept for counting. The brief description of a matter given here has been condensed. The work-unit, in actual practice, was defined much more fully to assure uniform counting of outputs produced.

categories was all-inclusive and mutually exclusive, a basic requirement for work-unit structures. *All-inclusive* means that all matters could be classified; *mutually exclusive* means that a matter would not be counted in more than one category. When the work-unit structure was completed [with the work-unit structure being defined (see Chapter 9) as the hierarchy of work-units from objective to tasks] a special work measurement procedure was designed and applied and the standard times used for budgeting, productivity measurement, and ongoing managerial control.

Preliminary analysis of the problem.   From the preceding brief material, let us define the basic workload forecasting and work measurement problems before examining further details. First, no data existed that would permit a forecast of the workload; it had never been counted or quantified as matters in the fashion proposed. Second, the associated activities were performed in 30 locations (29 subarea offices and Washington). Third, with 36 6th-order work-units (offices and bureaus served) and 59 5.5th-order work-units, (types of matters), there are potentially 2,124 separate 5th-order work-units, if a 5th-order work-unit is "All of the tasks associated with the production of one matter of one kind of a 5.5th-order output for each 6th-order work-unit." Hence, if the time associated with a performance of a particular 5th-order work-unit is affected by the location in which it is performed, then there are potentially $2,124 \times 30 = 63,720$ work-units to forecast and for which work measurement coefficients must be developed.

Self-reporting was selected.   For the reasons given in the preceding paragraph, a self-reporting type of approach was selected. Such an approach could be carried on over an extensive period so as to cover a representative sample of the mix of outputs; it would not require the addition of a large number of analysts; it would not require observers who would need to be able to tell what an attorney was working on by observation (if this is at all possible); such an approach would allow both a work count (as a basis for future forecasting) and an attributable time for each 5th-order work-unit in each separate office to be accumulated simultaneously. Of course, such an approach would not permit purification of the time data by extracting any unreported idle time from the reported data, or by judging pace; reported data would have to be accepted at face value. Values could not be generated independently for M and A in the basic equation of a standard time. An assumption would have to be made that adequate values were included in the self-reported data and that these values were not excessive. Obviously, the quality of the data would reflect the attitude and ability of those being measured. Hence, adequate steps were taken to assure effective participation of all professional staff. The need for the study was demonstrated; the backing of the study by the Chief Solicitor was made clear.

Devising a self-reporting system; codings. To facilitate the reporting of the work count for each work-unit, a coding system was devised. Although a six-digit code was used, the last two digits were for the collection of information not directly related to the work measurement study. Hence, they will not be discussed here.

The first two digits were used to identify the *program served;* the second two digits indicated the *subject-matter nature* of the "matter."

Category I: coding by program area (first two digits).[5] When the activity or matter concerns several bureaus or offices, an appropriate summary number is to be used; for example, if the problem concerns 51 and 53, use 50; if the problem concerns 19 and 43, use 10; if 25 and 61, use 10.

10–. Office of the Secretary
  18–. Office of Water Resources Research
  19–. Job Corps

20–. Office of the Assistant Secretary for Fish, Wildlife, and Parks
  22–. Bureau of Commercial Fishing
  25–. Bureau of Sport Fishing and Wildlife
  28–. National Park Service

30–. Office of the Assistant Secretary for Mineral Resources
  31–. Geological Survey
  32–. Bureau of Mines
  34–. Office of Geography
  35–. Office of Coal Research
  36–. Office of Minerals and Solid Fuels
  37–. Oil Import Administration
  38–. Office of Oil and Gas

40–. Office of the Assistant Secretary for Public Land Management
  41–. Bureau of Indian Affairs
  42–. Office of Territories
  43–. Bureau of Land Management
  48–. Bureau of Outdoor Recreation

50–. Office of the Assistant Secretary for Water and Power Development
  51–. Bonneville Power Administration
  53–. Bureau of Reclamation
  56–. Southeastern Power Administration
  57–. Southwestern Power Administration
  58–. Alaska Power Administration

60–. Office of the Assistant Secretary for Water Pollution Control
  61–. Federal Water Pollution Control Administration
  65–. Office of Saline Water

80–. Office of the Solicitor

90–. Office of the Assistant Secretary for Administration
  91–. Office of Management Operations
  92–. Office of Survey Review
  93–. Office of Budget
  94–. Office of Management Research
  95–. Office of Personnel Management

---

[5] From "Workload Analysis Reporting System," Office of the Solicitor, U.S. Department of the Interior.

Category II: coding by type of work (second two digits). These classifications are grouped by kinds of matters having some general relationship to assist in locating them and are not necessarily strict subdivisions of the general group heading.

INTERNAL, *Office of the Solicitor*

–00. Personnel actions, including supervision, work assignment, work organization, improvement study, etc., internal to the operation of the Office of the Solicitor

PERSONNEL, *other than Office of the Solicitor*

–10. Organization and individual duties

–11. Safety problems or reports

–12. Speeches or publications by employees

–13. Employee claims for loss, damage, or injury

–14. Coordination between bureaus or offices

–15. Conflict-of-interest issues or opinions

–16. Other employee problems or matters, other than classified above, including internal regulations

–17. Internal reports, other than safety

–18. Reports of the department or branches or offices

–19. Improvement studies

CONTRACTS–RATES, *including documents, letters, discussion, advice, formats, opinions, negotiation, litigation, etc., excluding patents and copyrights*

–20. Power development

–21. Power other than above

–22. Water use, irrigation

–23. Water user, other than irrigation

–24. Construction, other than power

–25. For research and development, other than management

–26. For goods or services other than above

–27. Rate and fee development

LEASES, *including claims, permits, and licenses, but excluding patents and copyrights*

–30. Lease, oil and gas

–31. Lease, other minerals

–32. Lease, other than above

–33. Permits, all

–34. Licenses, all

PATENTS AND COPYRIGHTS

–40. Applications

–41. Acquisitions

–42. Licenses

–43. Search

–44. Not otherwise classified

LAND–Ownership, *mining claims, grants, entries, etc.*

–50. Acquisition by exchanges

–51. Acquisition by gift or donation

–52. Acquisition by eminent domain

–53. Acquisition by purchase

–54. Grants or patents

–55. Entries, all aspects

–56. Transfer

–57. Mining claims

–59. All other

LEGISLATION–POLICY, *including congressional, state, proclamations, policy statements, and departmental regulations for external application*

–60. Congressional legislation

–61. State legislation and regulations

–62. Proclamation

–63. Departmental policy, not concerning personnel

–64. Departmental regulations for external application

TORTS; CLAIMS (*other than land and not related to contracts*)

-70. Tort, trespass
-71. Tort, all other
-72. Discretionary irrigation
-73. For loss or damage, other than employee

INDIANS

-80. Probate
-81. Tribal laws and customs
-82. Jurisdiction over, other than tribal
-83. Rights, privileges, and responsibilities other than above

-84. All other

MISCELLANEOUS

-90. Rights and privileges, other than Indian, and not classifiable into any previous category
-91. General correspondence, not classifiable into previous categories
-92. Grants and loans
-93. Mixed batch of various letters or documents
-94. Water pollution abatement
-95. Water quality standards
-99. Not otherwise classifiable

Self-reporting; forms. In addition to the coding details already given, each attorney and all clerks who were assigned the task of collecting the data were given a uniform set of instructions and a uniform set of forms. The form for the collection of the work-unit work-count information (by location) is shown in Figure 26.2. The form for reporting the use of time

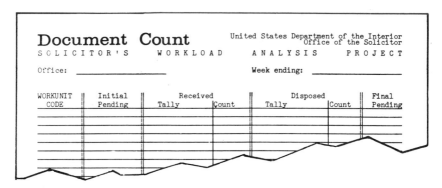

Figure 26.2. Form for use in the Office of the Solicitor for collection of work count of work-units completed in a location.

(by individual attorney by work-unit category) is shown in Figure 26.3. This form, used daily by the attorneys, produced a considerable amount of data. Hence, the format was designed to be "machine-readable"; the data could be converted directly to punch-card or magnetic-tape record.

The amount of data collected and processing the data. The data were collected for a 6-month period so as to provide data from a representative mix of work. The data were processed in the following manner. The data

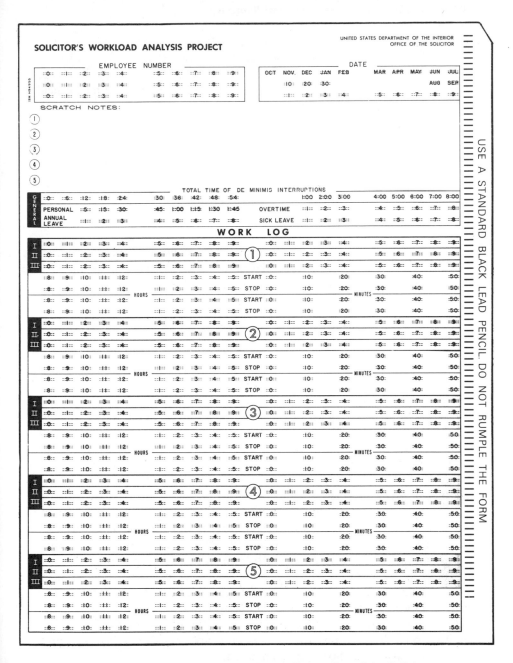

Figure 26.3. Form for use by attorneys in the Office of the Solicitor for reporting use of time during data-collection period.

contained in the Work Log and Document Count forms were converted to punched cards. The combined inputs of matters processed and related work times were matched and merged into a tape. The tape was processed to produce printed summary data, and data sorted by divisions and by field offices. These data provided, for each location, the average time standards[6] for the various matter codes processed. In addition, time standards were established for matters regardless of which bureau or office created the workload. Each location was given a complete set of time standard times, consisting of a mix of the three following types:

1. A time standard for each *kind of matter* (category II of the classification), developed without regard to which bureau or office it concerned.

2. A time standard for each *kind of matter by program area* (categories I and II of the classification), developed where there were 10 or more occurrences of the matter within the Office of the Solicitor.

3. A time standard for each *kind of matter by program area by division or office of the Office of the Solicitor*, developed where that division or office had 25 or more occurrences of that kind of matter.

Each division and office was given the time standards of type 3 that it generated during the study. Where there were less than 25 occurrences, the time value used was type 2. If no value existed in the type 2 accumulation, the time value of type 1 was used. In this way every division and office had the best available time standards for any type of work that might occur. It is to be noted that the time standards derived from those matters occurring frequently in a division or office constitutes the great bulk of its work; that the infrequent use of the balance of the time standards had only a minor effect in its periodic reports.

Three exceptions to this method in selecting time values were made. Time values of type 1 were used for code 91, General Correspondence; code 93, Mixed Batch of Documents; and code 99, Not Otherwise Classifiable.

A nonmatter factor was developed which reflected time expended that was not picked up in the standard times (e.g., time expended but no matter recorded, and similar recording errors). Since no office differed from another in this respect by more than 1 or 2 per cent, and since these differences did not appear to be statistically reliable, a common nonmatter factor was applied to all divisions and offices as part of an allowance. (This is related to A in the basic equation for a standard time.)

---

[6]These were called "weights" in the actual study, a term the lawyers preferred.

Standard times. In this manner, a work measurement coefficient (standard time) was developed, for each 5th-order work-unit for each office, for use with a forecasting system related to the need for assistance generated by each type of work in each office or bureau served. A sample of the work measurement data is shown in Figure 26.4.

WEIGHTS FOR MATTERS
OFFICE OF THE SOLICITOR
U.S. DEPARTMENT OF THE INTERIOR

PROGRAM AREA CODE - 43 - Bur. of Land Mgmt.
FOR USE IN OFFICE 06 - PUBLIC LANDS
EDITION 1

Matter code	Weight	Matter code	Weight	Matter code	Weight	Matter code	Weight
4300.	10.30	4326.	5.17	4350.	5.10	4380.	1.51
4310.	4.39	4327.	2.43	4351.	3.68	4381.	3.72
4311.	1.88	4330.	5.97	4352.	2.88	4382.	2.91
4312.	16.50	4331.	8.10	5353.	4.32	4383.	8.14
4313.	2.57	4332.	0.74	4354.	6.31	4384.	3.00
4314.	3.72	4333.	3.55	4355.	17.54	4390.	2.79
4315.	2.57	4334.	12.76	4356.	1.37	4391.	1.64
4316.	10.23	4340.	5.56	4357.	14.93	4392.	3.24
4317.	0.40	4341.	1.00	4358.	4.71	4393.	4.63
4318.	1.25	4342.	1.48	4360.	2.16	4394.	4.78
4319.	54.58	4343.	0.34	4361.	2.75	4395.	7.25
4320.	5.32	4344.	8.91	4362.	5.14	4399	3.03
4321.	0.50			4363.	2.32		
4322.	2.36			4364.	13.66		
4323.	4.09			4370.	2.13		
4324.	1.51			4371.	10.74		
4325.	2.33			4372.	2.51		
				4373.	5.02		

Figure 26.4. Time standards for one program branch served by one location of the Office of the Solicitor, U.S. Department of the Interior.

Using the standard times. Subsequently, a forecasting system and a work-performed and backlog reporting system (by work-units) were developed and put in use. These systems, together with the work measurement data, make it possible for the Solicitor to:

1. Objectively estimate the professional manpower required for future periods of time (zero-base budgeting).
2. Allocate professional manpower to locations.
3. Temporarily redeploy manpower to locations where undesirable backlogs occurred.
4. Evaluate the degree to which productivity is maintained or improved.
5. Evaluate the timeliness of service rendered (backlog analysis).

These, as have been frequently noted, are the managerial uses of work measurement data. For situations such as the Office of the Solicitor, the self-reporting approach appears to be almost the only feasible approach. The data were used over a period of years as a basis for management decisions.

Over a period of years productivity increased by 25 per cent. The value of the increased productivity, in salaries to lawyers alone, exceeded $1,500,000 per year.

Summary—Time Standards by Mathematical Analysis

1. *Uses:*
   a. For setting standard times when observational techniques are inconvenient, uneconomical, or infeasible; further, where the variety of work-units is too great to use the FPE and where fiat standards are not applicable.
   b. For setting standard times for 4th-, 5th-, 6th-, or 7th-order work-units.
2. *How made:*
   a. A suitable work-unit structure should be delineated. A decision must be made concerning the order of work-unit to be subjected to work measurement.
   b. Past data concerning output and work time are used if available in a suitable form. If not available, special data-collection systems must be devised and used. Data may be collected on a sampling basis, or a work period may be fully reported. The data-collection period is usually extensive.
   c. The work-count and work-time data are manipulated by mathematical procedures to produce a standard time. The most common method is

$$ST_i = \frac{W/T_i}{W/C_i}$$

   ST_i = standard time for work-unit i
   W/T_i = total work time for work-unit i, for the period for which data were collected
   W/C_i = total work count for work-unit i for the period covered by W/T_i

PROBLEMS

26.1. The table lists the data obtained from the shipping department of a plant, for 15 working days, selected at random from the year's record (numbers rounded).

Man-hours	Items Picked, Packed, and Shipped
340	1,600
450	2,900
380	2,400
560	3,800
390	2,100
220	1,900
200	1,200
650	3,400
340	2,100
400	2,400
500	2,700
440	2,400
510	3,100
540	3,100
650	3,600

(a) Compute a gross mathematical standard for this activity.

(b) Within what limits would you expect this standard to predict performance? How reliably?

26.2. (a) If the policy of the plant is to use the third quartile of the arrayed performance effectiveness as the standard, using the data from Problem 26.1, what is the standard time?

(b) How realistic do you feel this value is?

(c) Why?

26.3. Develop standard times for the tabular data given, for the foundry example, cited is this chapter, sorted into three types of work-units. Compare these to the standards derived earlier in the chapter by gross techniques.

26.4. Plot the data from the computations of Problem 26.3. Design and compute a table for determining the required monthly manpower, between units of 1,000 to 10,000 man-hours per month, as a function of the work mix of orders to be scheduled that month.

26.5. Design a self-reporting data-collection system for:

(a) Determining whether the semester-hour values assigned the courses in a university are correct.

(b) Determining the realism of an intracity bus schedule.

(c) Determining the effectiveness with which a staff (of any assigned kind) is being employed.

(d) Determining the workload in the office of the Dean of Student Affairs.

(c) Determining delay time in a factory shop.

26.6. The time for a work-unit consists of a setup time and a performance time. The setup time is approximately four times the performance time, but neither are known. Develop a table for converting the work counts of the work-unit, within the range 1 through 25, to a linear function of time required.

26.7. The time to assemble a large ship section appears to increase with the square root of the tonnage of the section. Prepare a table for converting a work count of tons assembled, in the range 1 to 150 tons, by tons, to a linear function of the time required.

26.8. (a) Develop a presentation to be made to the group who are going to self-report the necessary data, using the system designed in response to Problem 26.5(a), which will maximize sincere participation and accuracy of data.
   (b) Same as Problem 26.8(a) but with reference to Problem 26.5(b).
   (c) Same as Problem 26.8(a) but with reference to Problem 26.5(c).
   (d) Same as Problem 26.8(a) but with reference to Problem 26.5(d).
   (e) Same as Problem 26.8(a) but with reference to Problem 26.5(e).

26.9. (a) In the system designed in response to Problem 26.5(a), would you use the entire student body? Describe the logic supporting your stand.
   (b) In the system designed in response to Problem 26.5(a), would you keep any data separate from other data? Describe the logic supporting your stand.
   (c) Would the decisions made with respect to Problems 26.9(a) and (b) affect the presentation prepared in response to Problem 26.8(a)? Describe any anticipated changes in detail. Describe your supporting arguments.

26.10. Which of the work-units, if any, in the work-unit structure of Figures 9.2, 9.3, 9.4 and 9.5 could be measured conveniently using mathematical procedures? Support your conclusions with a full argument.

part four

APPLICATION OF MOTION
AND TIME STUDY TECHNIQUES

27

POLICIES FOR THE USE OF MOTION AND TIME STUDY

Introduction

In Chapter 6 it was stated:

In the widespread use of motion and time study in an organization there are certain relatively routine activities that must be performed. The means of carrying on these routine activities should be formally stated. A group of policies is necessary to achieve consistency of actions. Formal procedures are necessary to routinize the carrying out of the policies.

Policies are statements of the aims to be sought in handling recurring types of situations. They are the rules by which the organization functions and are vital in all phases of plant activity. Procedures are the details of the methods to be employed to achieve these aims.

For the routine motion and time study work to be effective, the policies and procedures should cover at least the following seven items:

1. What does a standard time represent?
2. Who shall determine the standard method?
3. How will the standard method be made regular practice?
4. Who shall determine the standard time and how will it be determined?
5. Under what conditions may a standard time be changed?
6. How will production be reported?
7. If a wage incentive plan is in use, what are the rules covering its use?

The policies and procedures for items 2 through 7 can most readily be examined after the detailed techniques of motion and time study have been presented. However, the first question, "What does a standard time represent?" is basic to all techniques. A standard time is the final value resulting from both motion study and time study.

Hence, the significance of standard time was examined in detail in Chapter 6. This chapter will consider items 2 through 7. An example of a time study manual, giving detailed procedures, will be given in Chapter 46.

Formalizing the Standard Method

Responsibility for determining the standard method. In some plants the work of developing standards is divided between two groups: one, a methods group for developing standard methods, and two, a time study group for developing standard times. In simple assembly industries, these two functions are likely to be grouped together. As the nature of the product and the production sequence increase in technical complexity, these functions will be spread among more and more groups. In industries involving a great deal of machining, a group called "tool engineers" or "manufacturing engineers" may take over many of the methods functions. In organizations with service outputs, the motion and time study activities may be carried on by line people aided by staff. (Motion and time study procedure are for the use of such groups, no matter what they choose to call themselves.) Regardless of whether there is one group, two, or more, or what they are called or choose to call themselves, the final responsibility for determining methods must be placed somewhere in the organization in such a way that standard times will accompany only the methods they were designed for, so that there will be a constant striving for better methods, and also so that the two will complement one another.

Sample statement of responsibility. The standard process sequence as well as the standard practice for each job in the sequence will be specified by the chief manufacturing engineer, who shall consult with foremen, department superintendents, quality department, design engineers, tool engineers, plant engineers, and time and motion study engineers, but retain final responsibility for the selection of the methods. (Naturally, in a large plant, more and more subordinates will do the actual work.)

In many plants, the jurisdiction of the chief manufacturing engineer may be different from that suggested here, depending on the individual, the size of the plant, product, process, and plant custom. In some cases the chief manufacturing engineer will be called the plant superintendent, and so forth. Titles vary widely. Suggestions for changes in methods may develop in many places in the management of an organization. Motion and time study workers may suggest tool or design changes, design engineers may suggest process changes, or quality control workers may suggest new motion patterns. It is important that an individual or group making such suggestions report to a central source to[1]:

1. Request permission to proceed with an innovation.
2. Provide a record of what is changed.
3. Provide material to guide other groups in achieving similar results. (The larger the plant, the more important this item becomes.)

In the interest of saving the time of the executive who reads it, as well as adequately presenting the material, the report may well be constructed around a suggested outline.

Sample procedure relating to suggestions for changing methods. The following outline is to be used as a guide in reporting suggested changes to the chief manufacturing engineer. It is suggested that as far as possible a proposal for change be built around a letter having four sections, as follows:

A. *Identification* (use interoffice form):
   1. Date.
   2. From.
   3. To.
   4. Subject.

B. *Advantages and disadvantages*: Under this head the reporting person should summarize what would be accomplished by the change, the estimated saving in money, scrap, or increase in output or quality, as appropriate; the cost of the change, and any training problems involved, and so forth.

C. *Exhibits*: Supporting data that are attached to substantiate the claims under item B should be listed here and indexed, but appended after

---

[1]This is separate and distinct from employee suggestion plans.

item D. Insofar as possible, standard charts (as described in Chapters 7 through 19) should be used to increase intelligibility.

D. *Specific proposal*: All that has to be done to put the proposal into effect should be listed here, so that if permission is granted, not only does the executive know exactly what he is agreeing to, but also, so that if the proposal is accepted, it may be achieved rapidly. In the case of a change already made, all that has been done should be summarized here.

Making the standard method regular practice.   As was established in the chapters on time study, the determination of standard time is preceded, usually, by the determination and recording of a standard practice. For adequate performance to be obtained in a production shop, regardless of the purpose of the time study, the standard method, in the form of a *written standard practice (WSP)*,[2] as well as the standard time, must be supplied in some manner to the operator. Methods that have been carefully developed must be used if their value is to be obtained. The WSP giving the method to be followed on a job should be designed, as was noted earlier, for use in connection with time study. The instructional material for use by the operator or for use by a group leader, foreman, or special instructor may well be an adaptation of this. The more continuous the production, the more likely that these two versions of the WSP will exist. The economic reasons for this should be obvious. In a jobbing shop, with extremely short runs, the economical WSP for operator use may be merely a copy of the written standard practice from the time study. It is often merely a list of steps, a blueprint, and a list of tools, but this may be improved upon. In better-managed jobbing shops it is desirable to design general routines for each type of job and merely add specific details to them to adapt them to particular jobs. WSP and supervision are a good means of consistently obtaining effective, low-cost methods. Where large numbers of people perform the same operation, a very detailed instruction sheet may be prepared.

Photographic versions of instruction sheets at some plants use small 35-mm pictures or Polaroid pictures[3] to show the steps described on the instruction sheet. Actual photographic prints are pasted on the instructions to save printing costs. If much detail is required, pictures are cheaper and more effective than lengthy descriptions.

This critical aspect of methods and standards work is worthy of considerable discussion. In all too many plants, carefully developed time stan-

---

[2]Also referred to as Standard Operating Procedure (SOP), or Instruction Sheet.

[3]A regular 35-mm still camera takes about 20 to 36 pictures on a roll, so these illustrations are not excessively expensive. The quality of Polaroid and Kodak Instamatic pictures has so improved since their introduction and they are so simple to take that these appear to be effective methods.

```
                        WRITTEN STANDARD PRACTICE
                           The Perfect Circle Co.
     SECOND INSPECTION                                    No. SI 14
                                                          Issue 2
                                                          Sheet 1 of 3
     OPERATION:  Gap Gage Earthworm Tractor 5 3/4" Rings  3-6-
                              EQUIPMENT

     1 Sizer Ring Gage                      1 V Trough
     1 Set Feeler Gages                     1 V Trough (non-sectioned)

                             REQUIREMENTS

        Operator will wear gloves when handling rings.

        All pans of rings will be lifted by service men.

        Pan ticket will remain in pocket on side of pan,
        except when being read by operator.

        Operation will be performed by checker.

        Ring gages will be checked at the beginning of each shift by
        a checker.

        A ticket showing ring gage size, variation number, feeler
        gage number, operator's clock number and date will be filled
        out when ring gage is checked and will be kept with gage
        while it is in use.

        Feeler gages will be kept at the Foreman's desk and will be
        returned to desk at end of working period.

                            SPECIFICATIONS

        Gage Size:  5.747              Gap Clearance:  .011 - .021

        Rings will be gaged on a percentage basis.  Operator will gage
        every 8th ring, beginning with bottom ring and ending with top.

        If any ring is found out of limits, operator will check 8 rings
        above and 8 rings below place from which reject was removed
        until all rings not within limits have been rejected.

                            SECURE RINGS

     1. HAVE SERVICE MAN PLACE ENTIRE TEST of rings at back of bench in a
        horizontal position.

                            CHECK SET-UP

     2. CHECK PAN TICKET  to see that previous operation has been completed.  Check
        number of rings in pan against count shown on ticket.  (Count remnant row
        and add to number in full rows).  If counts do not agree, have floor clerk
        make necessary corrections.
```

Figure 27.1a. Written standard practice for operation (*continues*). (Reproduced by courtesy of D. C. Parsons, Methods and Standards Superintendent, Perfect Circle Co. Richmond, Ind.)

dards are poorly applied because of the inadequateness of the method information supplied to the shop. When the standard practice is not transmitted to the workers, deviations in methods commonly ensue. Also, when methods and standards are changed, the failure to detail the methods often

```
SECOND INSPECTION          WRITTEN STANDARD PRACTICE       No. SI 14
                               The Perfect Circle Co.       Issue 2
                                                            Sheet 2 of 3
    OPERATION:  Gap Gage Earthworm Tractor 5 3/4" Rings     3-6-

                                GAGE RINGS

     3.  REMOVE EVERY 8TH RING  from first row of pan, beginning with bottom
         ring and ending with top.

     4.  PLACE RINGS ON BENCH  in front of first row with gaps toward operator in
         the order removed from pan.

     5.  REMOVE SPECIFIED NUMBER OF RINGS  from remaining two rows in the manner
         described in Steps #3 and #4.

     6.  MOVE GAGE  to a position at left end of first pan.

     7.  PICK UP FEELERS AND HOLD  them between thumb and index finger while
         gaging rings.  (When placing ring in gage or removing ring from gage,
         hold feelers in palm of hand.)

     8   PICK UP ONE RING  from first stack on right approximately 1" on right
         of gap with thumb and index finger of right hand.

     9.  BRING RING TO A POSITION OVER GAGE.

    10.  GRASP RING  on left of gap with left index finger and thumb, placing
         middle fingers on top of ring and thumbs on face and corners about ½"
         on either side of gap.

    11.  TILT RING SLIGHTLY AND SLIP BACK OF RING INTO GAGE.

    12.  INSERT REST OF RING INTO GAGE  by squeezing thumbs together toward gap.

    13.  ROLL THUMBS OFF RING  to release points slowly.  Do not flip points
         into gage.  Be sure back of rings is pressed firmly into gage.

    14.  TEST GAP  by placing Go feeler between points, starting at inside
         diameter and bringing feeler toward operator through gap.  Go feeler
         should pass between points freely.  If feeler passes through gap with
         a slight drag, ring should be rejected.

    15.  PLACE NO GO FEELER BETWEEN POINTS,  starting at inside diameter and
         bringing feeler toward operator through gap.  The feeler should either
         refuse to pass or should fit very tightly.  Do not force feeler between
         the points.

    16.  REMOVE RING FROM GAGE  by closing gap with thumb and index fingers of
         each hand and lifting ring.  Do not flip points out of gage.

    17.  PLACE IN NON-SECTIONED TROUGH ON LEFT OF GAGE AND AT SAME TIME REPEAT
         STEP #8,  if ring is within limits.

         a. If ring is not within limits, place in sectioned trough on left of
            gage in section marked for that type reject.
```

Figure 27.1b. (*Concluded*) Written standard practice for operation.

allows the change in standards to be misinterpreted as a "speed-up," and poor labor relations or grievances may needlessly result. Part of a fully detailed job instruction sheet appears in Figures 27.1a and b.

The instruction sheet used to accompany each standard issued at Patrick Cudahy, Inc., of Cudahy, Wisconsin, has been instrumental in reducing labor disputes. This instruction sheet is reviewed by the time study man and the departmental union steward. The steward signs the original copy, indicating that he has read and understood the sheet. (He does not have the choice of approval or disapproval but is encouraged to indicate whether any job steps have been neglected.) A sample of such an instruction sheet is shown in Figure 27.2.

C.B.CO.1124 B   PRODUCTION STANDARD--JOB SUMMARY SHEET

PAGE NO. 1 OF 3   DEPT. Process Ham   STUDY DATE 2-16-   STUDY NO. 3-6-111

OPERATION Bone 4½-5½ S.S. Picnics -Skin On   JOB   Bone Picnics

MACHINE - - - - - - --   MACHINE NO. - - - - - -AUX.EQUIP. - - -   Job Symbol 150-4

NO.	DETAILED DESCRIPTION OF JOB ELEMENTS		STD.ALLOWED TIME/UNIT
	LEFT HAND	**RIGHT HAND**	
1.	Grasp picnic from conveyor or from pile at side of operator. Slide picnic on cutting board to position in front of operator - face side up.	Hold knife.	.0371
2.	Pick up steel from cutting board - Hold steel - aside steel to cutting board.	Slide knife blade across steel. (3 strokes)	.0408
3.	Smooth out skin around shank end - pull up on skin during cut - use skin as handle to roll picnic over so skin side is up. Roll picnic back so face side is up - aside skin to conveyor.	Slit skin on shank end - cut skin away from skin side of picnic complete skinning by cutting away skin from shank end on face side.	.2786
4.	Hold picnic down on cutting board.	Make horizontal cut at side of picnic length of the arm bone to the area in the center of the picnic near the bone.	.1390
5.	Hold picnic - grasp and hold bone at center area. Hold bone at blade end, until bone is free of picnic.	Guide knife along contour of arm bone, around center area of bone until entire center area is free of picnic. Cut away meat at blade end of arm bone. Cut away meat from shank end of arm bone until bone is completely free of picnic.	.5877
6.	Aside bone to top conveyor - hit counter on return motion to record number of picnics boned.	Aside picnic to bottom conveyor.	.0370

REMARKS

		Cwt.	STD.ALLOWED TIME MIN/UNIT	1.1762
CONVERTED TO			STD.HRS./ 100 UNITS	1.9603
BASIS UNIT Wt. Per Piece		4.83 Lbs		
STD.ALLOWED TIME MIN/UNIT		.2435	UNIT HRS.	51.0
STD.HRS./100 UNITS		.4058	OBSERVED BY A.D.	
UNITS /HRS.		246.4	APPROVED BY	

Figure 27.2. Job summary sheet from Patrick Cudahy, Inc., Cudahy, Wis. (Courtesy of C. J. Allen, Chief Industrial Engineer.)

A similar instruction sheet in use at a printing and children's game plant is given in Figure 27.3. At this plant, many of the games are assembled by sequential crews (line work), and the individual standards are summarized in a crew or process standard as shown in Figure 27.4. Note that delays caused by unbalance in the line are indicated at each appropriate

---

PRODUCTION STANDARD—OPERATION METHOD SUMMARY SHEET

Page No. 1 of 7   Department Box

Operation and No. Wrap box, bottom   Product Sewing card 4600

size 8½ x 8-1/8 x 1-1/16   Work Unit Game

Machine(s) and No. A-1   Study Date 12/8/

Study No. 551208-01

No.	Detailed Description of Operation Elements	Std. Allowed Time/Unit
1	Reach 18" to bottom wrap, grasp sheet, move 18" to and place on feeding table, help center box on bottom wrap, carry to machine plunger, fit on plunger and release. Reach 18" to supply of stayed bottoms, grasp one and carry to and center on positioned bottom wrap, pick up and carry to, fit on and release on machine plunger.	.0446
2	Reach to right and move stack of stayed boxes with both hands into position to feed single boxes to positioning area. NOTE: Supply girl puts boxes into original position from truck.	.0008
3	Delay element. Load glue pot, load wraps. Reduced to a per cent as it occurs 1 time per hour and added as an allowance - 2.5%	

REMARKS:		
	Std. Allowed Time Minutes/Unit	.0454
	Standard Hours/ 100 Units	.0756
	Units/Hour	1320
	Observed by:	*G. Kaplan*
	Approved by:	

Figure 27.3. Operation standard sheet from printing plant. (Courtesy of G. Kaplan, Western Printing and Lithographing Co., Racine, Wis.)

PRODUCTION STANDARD
JOB SUMMARY

Page No. 8 of 8   Study Date 12/8/

Department Box   Study No. 551208, 1-7

Product and No. Sewing Cards, 4600   Work Unit Game (#4600)

Operation(s) and No.(s) Complete assembly

Total Operations

Operation	Std. Hrs./ 100 Units	Units/Hr./ Operator	Crew	Hrs./ 100 Units Delay Time/ Operator
1. Wrap box bottom	.0756	1320	1	.0049
2. Assemble platform and tray	.0761	1315	1	.0044
3. Accordian fold six card insert and				
insert in game	.0712	1385	1	.0093
4. Select, fold and insert six strands	(.0932			
of yarn	(.0932	1070	2	.0678
5. Wrap box cover	.0805	1245	1	------*
6. Close box and stack on table	.0438	2260	1	.0367
7. Supply two wraps, trays, cards; load				
skid with finished boxes	.0592	1790	1	.0213
*Pacing Operation	.0805	Totals .5928	8	.1444

		REMARKS:
Crew Efficiency	92 %	
Std. Hrs./ 100 Units/Crew	.6440	
Crew	8	
Std. Hrs./ 100 Units/Oper.	.0805	
Units/Hr./Crew Crew	1245	

Daily std. 7.5hrs x 1245 = 9320

T32

Figure 27.4. Crew summary of standard operation sheets. (Courtesy of G. Kaplan, Western Printing and Lithographing Co., Racine, Wis.)

position. Transmitting information concerning these delays to the shop is excellent preparation for future changes when work may be added to these stations *without increasing the time allowed*. The workers' occupying positions containing delays are given to understand that at some time in the future, when better methods are devised, this delay time may be assigned to productive work. Discussing this, when the standard is issued, greatly facilitates future understanding.

The policy for WSP should place responsibility for their creation, state the requirements for them, their authority, and when they are to be available with respect to the inception of production.

Sample policy for WSP. The chief manufacturing engineer will develop the WSP, when possible, prior to the inception of production.[4] If this is not possible, a tentative method will be arranged in consultation with the departmental foremen, time study man, tool engineer (these two men may be one and the same), and others, as necessary. The WSP shall be followed except where obvious errors exist, in which case the chief manufacturing engineer shall be notified immediately. The chief manufacturing engineer will develop and apply improved methods whenever economically feasible.

The WSP shall be of two forms: one form for the process sequence, the other form for giving, for each step in the sequence, the list of tools and equipment, the specifications for the job, and complete details for performance. (They should follow the sample WSPs given as examples in Figures 27.1 through 27.4. This sample of illustrations has been given in order of increasing formality of presentation. Also, as an additional illustration, Figure 24.1 represents the equivalent of the industrial WSP for the very different output, *Healthier swine by diagnostic referral*, completed.) It is to be noted that the process standard is usually done well in most plants but that the operation standard is frequently skimped.

Formalizing the Standard Time

The meaning of standard time. Since standard time is to be the unit of measurement, it must be defined and the definition generally known throughout the plant. Samples, as well as the reasoning behind the development of one definition of standard time, were discussed in detail in Chapter 6. Consideration should be given to union participation in its formulation, to help develop a socially acceptable definition. The definition may well be embodied in a multi-image, pace-film loop, as described in Chapter 20.

Responsibility for determining the standard time and method of determination. The time study rating and allowance procedures should be specified, with the deviations that will be allowed in unusual cases. This is the

---

[4]This is frequently delegated to the time and motion study group, who must usually obtain the participation of other affected parties in the development of the WSP.

only way to obtain a uniform time study practice. This uniform practice may be changed form time to time, but a good uniform practice ensures the use of the best knowledge obtainable at any time. The practices outlined in Chapters 20 through 26 may be followed. The procedure should indicate the assignment of responsibility for the supervision of the following and the detailed criteria for acceptable technical performance with respect to[5]:

A. The nature of the method record.
B. The manner of obtaining the work time and work count.
C. Number or basis of number of readings to be taken with direct time study.
D. Permissible use of standard data.
E. Manner of handling irregular elements.
F. Basis of rating and standard difficulty adjustments.
G. Standard allowances.
H. Necessary level of accuracy.
I. The manner of reporting allowed times and standard methods as necessary to the payroll department, schedule department, shop supervision, and the like.
J. A method of handling grievances related to standards.

A manual of time study procedure. Because the bulk of an example of this type would throw this chapter out of balance, the details of such a manual appear as Chapter 46.

Changing a standard time. Properly set standards may be guaranteed against revision except in specified cases, rather than whenever actual production gets out of line with expected production. Poorly set standards require constant revision and lead to industrial chaos. An organization must realize that unusual workers will exceed standards by considerable amounts. Recognition of this fact is particularly important if incentive wages are used, such as piece rates or time standards with pay in terms of standard hours of work done. (The latter is the same as a piece rate, except that it is more convenient, since it allows changing basic pay with automatic adjustment of the incentives, without complete retyping of the rate book.)

Maximum individual production tends to give lowest cost of manufacture. If rates are continually revised downward merely because unusual workers exceed them, a ceiling is in effect placed on earnings, and workers soon become aware of this. Of course, it is undesirable to have whole groups of workers making excessive earnings because of improperly set rates. They

---

[5]The U.S. Air Force considered these items of such great importance, with respect to contract costs, that it issued MIL-STD (Military Standard)-1567 (USAF) setting policies for these for certain classes of contractors.

create friction with other groups. Such situations are best avoided by adequate time study procedures. When workers are aware of ceilings on earnings, they restrict production, which raises costs. If rates are to be effective, or if standards are to be used as real goals, they must be sufficiently reliable to be guaranteed, except for changes in the basic conditions.

Sample policy concerning changes in the standard time. The company will not change any regular production standard, regardless of how high earnings over day rate may be, unless there has been a significant change in method affecting the standard by 5 per cent or more or where an error has been made in arithmetic computation or in typing or posting of a production standard. A method change means a change in speeds, feeds, design, material, quality requirements, jigs, fixtures, dies, machinery, number on crew, or a different method or process of doing the work, including different hand and body motions or crew balance. This would naturally change the way of performing the work, requiring either more or less time, and would naturally require a different production standard. However, only that part of the production standard affected by the method change will be modified. The records in connection with any change, because of arithmetic, typing, or posting errors or method change, are always available for review by an affected employee or his authorized representative.

Many minor improvements are made from time to time that individually would not increase or decrease the time required to any appreciable extent. However, over a period of time, several of these minor changes may accumulate and materially affect the time required. From time to time, standards will be reviewed, and, on jobs where several of these changes combine to produce a 5 per cent or greater change in the time required, the time standard will be changed.[6]

Reporting Production

Purpose of reporting production. To control activities properly, it is necessary to know how much of each individual's time is expended per unit on each operation (if the individual's work is separately identifiable), or how much group time is expended (if the production of the group is kept in a common pool). The production record is an aid in properly attributing cost and is a necessity if incentive wages are used.

Procedures for reporting production. Methods of performing this function vary widely, depending on the industry. This variation is so wide that

---

[6]The extreme importance of an adequate WSP should be apparent here.

no single sample illustration can be offered as typical. The complete design of an integrated system is a considerable task, and process chart—combined analysis may well be used in its formulation.

In general, production-and-time-expended reporting procedures may be divided into five general categories:

1. Reports on job tickets.
2. Reports on daily time cards.
3. Reports on "travelers" (sheets accompanying the work).
4. Supervisory reporting:
   a. By the job.
   b. By the work order.
5. On-line reporting.

All procedures are not applicable in all situations; their feasibility is affected by the nature and duration of specific assignments. All systems require cooperation between workers and managers in order to obtain accurate and timely information. If the production and time report is designed to serve as a basis for incentive wages, great care must be taken to keep the system free of erroneous data.

Reports on job tickets. When several jobs or batches of work are accomplished each day, it is not uncommon to find each operation reported on by an individual job ticket showing worker identification, number of pieces finished, start-time, and stop-time. Auxiliary procedures are used to account for "machine downtime" or other delays not built into the standard time. In some cases such systems have been used when a single job, without subordinate work-unit counts, lasts for a number of days. However, in such cases, without adequate supervision, it is not unusual to find workers keeping "banks" of completed work tickets "in their pockets." This enables them to have a means of reporting a steady flow of work accomplished despite an occasional bad day. They may also use the "bank" to provide an occasional "lazy day." Such abuses in reporting have a serious, deleterious effect on efficient production control. Certainly, data from such situations could not be used to set standards by mathematical procedures, as described in Chapter 26.

Reports on daily time cards. As an alternative to job tickets, the worker may be required to submit a production and time use report each day. While such a system does not eliminate the possibility of the same abuses as with job tickets, it does not equally motivate the abuses. Abuse requires a deliberate false report and some self-generated records of production. The employee who would withhold job tickets will not as willingly create false reports.

Reports on "travelers." A *traveler* is a report that flows through the shop with the work; it moves with the work. Its use tends to reduce the abuses discussed earlier. The difficulty that arises is caused by the different lengths of time the work may be in the shop; this causes difficulty in creating periodic reports on individual productivity.

Supervisory reporting: by the job. Essentially, this procedure is the substitution of a supervisor-generated report for the worker-generated job ticket or time card.

Supervisory reporting: by the work order. In some industries a work order may refer to a job on which many workers in a department may spend many hours over a period of time ranging up to a month (longer periods are not unknown). In such cases, the supervisor will list, against each worker's name entered on the work order, the hours spent by each worker each day on that work order. Such systems are designed to relieve the worker of clerical work, and to employ the greater responsibility of the supervisor to achieve accuracy.[7]

On-line reporting. On-line reporting refers to situations in which communication "terminals" are used to report finished work and receive assignments from a work control center. The terminals may vary in complexity from a voice communication system to a computer terminal. Such systems facilitate rescheduling of work, taking hourly performance into account.

Wage Incentive Plans

Wage incentive plans were introduced in Chapter 5 as being supported by standard times. Wage incentive plans may be divided into two major categories. One, measured day work, is that in which the worker, to remain on the job rolls, is required to show an average performance (over various periods of time) equal to some per cent of standard performance.[8] The second category consists of plans that reward workers in proportion to their performance. In the most primitive of these, the pay is strictly a function of work done. In the most common U.S. plans, the worker is guaranteed a base rate per hour, with the incentive above the base rate related directly to

---

[7]The author has encountered cases where the supervisor reported hours of work on such sheets in conformance with planned hours rather than work time actually used. In one case, to balance output with hours used, nonworked jobs were also reported as worked; these were later reported as "lost jobs."

[8]Standard performance is usually interpreted to mean performance in the standard time.

production. However, it is not uncommon to find the guaranteed base rate different from the incentive base rate used with the per cent of standard performance to compute earnings.

Mathematical Description of Basic Incentive Plans

Let
E = earnings for the period (day, week, etc.)
H_A = actual hours worked
H_E = hours earned; note that
$$H_E = N \times \text{ST} \qquad \text{where} \quad N = \text{number of pieces completed}$$
$$\text{ST} = \text{standard time per piece}$$
R_H = guaranteed hourly rate
R_I = incentive base rate

Measured day work

$$E = H_A \times R_H$$

and discipline is exerted unless

$$H_E = H_A \times (\text{some per cent of standard performance})$$

Primitive plans

$$E = H_E \times R_I$$

Guaranteed plans; base rate and incentive rate are the same.   If $H_E > H_A$

$$E = H_E \times R_H$$

If $H_A > H_E$,

$$E = H_A \times R_H$$

Plans with different guarantee and incentive base rates.   If $H_E > H_A$,

$$E = H_E \times R_I \quad \text{or} \quad H_A \times R_H \qquad \text{whichever is greater}$$

If $H_E < H_A$,

$$E = H_A \times R_H$$

Other wage incentive plans.   Some plans involve reduced incentives once the standard performance is reached. The reduction may be by a fixed percentage or a gradual reduction such that a ceiling of earnings is eventually reached; additional production does not produce additional earnings.

In contrast, where fixed costs are high, relative to labor costs, an incentive plan may be used which increases the incentive earnings at a faster rate than the per cent performance attained. For a detailed description of the multiplicity of plans available, reference should be made to a specialized text or to an industrial engineering handbook.

Factors affecting the acceptance of incentive plans. The most important factor seems to be the consistency with which a given effort will maintain a relatively constant level of earnings. This concept of consistency should be thought of as extending to all members of the group rather than merely to an individual. Hence, all the details of the methods and time study procedure manual in use should be directed at standard times embodying a consistent requirement of exertion for attainment.

Selecting and installing a wage incentive plan. Labor unions have, from their inception, held that wages and working conditions were subject to collective bargaining. The introduction of a wage incentive system certainly is included in this concept. However, the details of the bargaining procedures that are involved are beyond the scope of this book. Suffice it to say that advantages must accrue to both management and labor or not much will be accepted by either party.

"Runaway" incentive plans. In some cases, because of either improperly set or maintained standard times, incentive systems "run away"; they produce wages incompatible with profitable, competitive operation, or they create capricious differences among workers' wages.

Remedies, as with installation, involve agreements between labor and management. Remedies range from a reestablishment of all standard times without special compensation, to *buy-back*. Buy-back may refer to paying the workers for the incentive loss, for some agreed-upon period, as a price for the new standards. Buy-back may also refer to the disestablishment of individual incentive payments with all earnings against the existing standard times distributed in a more acceptable fashion (than may exist) to all workers. It should be obvious that each situation must be treated as a separate, special problem.

Summary

A general discussion of policies and procedures was given. However, as noted in Chapter 2, people tend to disagree with what they do not understand. Hence, to have effective use of policies, it is necessary to have understanding of them. To achieve understanding, training programs may be necessary.

Training in motion and time study. Inasmuch as the procedures of motion and time study are not commonly understood, particularly in the newer areas of application, in-plant training courses or conference sessions for all groups affected are highly desirable. With motion and time study regarded today as a tool to be used by most of the divisions of an organization, a means of distributing the knowledge is often required for its effective use. A staff specialist frequently assumes the responsibility for the educational training of the organization, although this is one of the most common instances of the utilization of outside help. Certainly, at the very least, the help of the most properly oriented and knowledgeable group should be sought.

Designing the training program. Any training program for an organization should be designed specifically to meet its needs. It is not possible, therefore, to lay down a recommended educational program. In each case there will be differences in:

1. Typical problems created by outputs and processes.
2. Educational level of the trainees.
3. Objectives of the training.

Basic principles of effective training programs. However, the basic principles embodied in successful programs will be applicable. They are as follows:

1. *The course should be given on company time during regular working hours.* If the program is to be considered as dealing with material of direct interest to the organization and the individual, this is an absolute necessity.

2. *Sessions of 3 hours' duration, with breaks on the hours, are about a maximum, if the purpose is merely to update the person's knowledge.* Two hours is about a minimum in such cases. However, the higher the educational level of the group, the longer the daily training session may be. One meeting a week is usually sufficient. This is not the case, though, when the individual is fully relieved of his duties during a training period and expected to fully use the materials discussed. In such cases the whole working day may be used.

3. *The activity should be varied.* Lecture, discussion, demonstration, and practical work should be interspersed.

4. *Extensive use should be made of visual aids and demonstrations.* During the early part of the course there is usually considerable advantage in obtaining visual aids from outside sources to increase the "psychological distance" and objectivity of the trainees. As the course progresses, the emphasis may swing to in-plant illustrations.

5. *Participational teaching should be used.* In general, this may be obtained by having the group read the topic for a session prior to the meeting and attempt to apply the material discussed, after the meeting. Thus, after the first meeting, the conference leader may answer questions on the new topic, give a short lecture on additional aspects, show the application of the material, and have the group work a problem with it. Subsequent to this, the conference leader should have the group discuss their applications of the topic of the preceding session.

6. *The reading material should be geared to the educational level of the trainees.* Particular note should be taken that the educational level of the typical industrial employee is rapidly rising in the United States and the material should not be insultingly simple.

7. *The conference leader should be skilled at such activity and should have firsthand knowledge of the material.* He should be in a position that will aid him in commanding the respect of the group.

8. Last, but not least important, *the training program must be given the effective and visible support of an executive sufficiently high in the organization so as to encourage fully effective participation in the training.*

PROBLEMS

27.1. (a) If a plant that you are familiar with does not have a definite policy for its routine motion and time study activities, prepare one.

    (b) If the plant has a definite policy for its routine motion and time study activities, constructively criticize it.

27.2. (a) Outline a course of study to acquaint a group of foremen from one plant with motion study procedures.

    (b) Outline a course of study to acquaint a group of foremen from one plant with time study procedures.

27.3. Prepare a methods proposal summary for the two methods shown in Figures 17.3 and 17.4.

27.4. Prepare a methods proposal summary for the two methods shown in Figures 17.9 and 17.10.

27.5. Prepare a procedure proposal summary and supplementary sheet for the two procedures shown in Figures 15.3 and 15.4.

27.6. If the labor rate is $4.95 an hour and the machine hour rate is $48.50 per hour and if the method has been changed from that shown in Figure 17.3 to that shown in Figure 17.4, prepare a work measurement proposal summary such as would accompany the necessary change in the standard.

27.7. Take a job in a plant, do a complete motion and time study analysis, install your proposal, and prepare an adequate report on the results.

27.8. Examine an employee suggestion system and describe its interaction with motion and time study policies and activities.

27.9. A plant, using wage incentives, evaluated the profitability of all the jobs worked on each pay period by each individual worker. It compared the cost (wages plus overhead) to sales price; conclusions were that the profit was *good*, *acceptable*, or *undesirable*. In order to motivate the workers, one of these words, as applicable, appeared on each weekly pay check. Discuss in detail:

(a) Possible effects of this practice.

(b) Desirability of this practice.

27.10. A wage incentive plan is

$$E = H_A \times R_H + 0.50(H_E - H_A)$$

Plot the per cent earnings from

$$H_A = 2H_E \text{ to } H_A = 0.5H_E$$

27.11. Same as Problem 27.10, except

$$E = H_A \times R_H + \frac{H_A}{H_E}(H_E - H_A)R_H$$

27.12. Same Problem 27.10, except

$$E = H_A + R_H + \frac{H_E}{H_A}(H_E - H_A)R_H$$

part five

ADDITIONAL DETAILS
FOR SERIOUS STUDENTS
AND PRACTITIONERS

28   NOTES ON POSSIBILITY GUIDES

Introduction

The two cases that follow were deliberately chosen from areas other than manufacturing. They were selected to suggest how the possibility-guide approach may be used no matter what the output or field of activity.

Case I concerns a section in the main office of a large shipbuilding concern. Case II concerns a sales counter in a department store.

Case I—Main Office of a Large Shipbuilding Concern

Introduction.   The main office employed 850 people. The company was building a new plant in another location. They felt that it was an ideal time to move any unnecessary employees from the home office to the new plant. The problem was: Which, if any, employees were unnecessary?

To resolve the problem a work-unit structure was developed for each section of the main office. Simple mathematical techniques of work measurement were used to determine the staff resources associated with each work-unit. Each work-unit was then scrutinized with the aid of possibility guides. Various innovations were selected. The work measurement data were used to convert the innovations into staff available for other work. A total of 165 employees were transferred to the new plant. The illustrative material that follows concerns only one section of 22 persons.

The work situation. The group studied was concerned with organizing the ship-launching ceremonies and coordinating related activities. The 5th-order work-unit 0101, *Ship launching ceremony completed*, included the following list of 4th-order work-units:

010101 Names and addresses of 1,500-person guest list on addressing-machine plates, updated

010102 Invitations prepared and mailed (addressing machine was very old and was frequently complained about)

010103 Acceptances received and party size estimated

010104 Presents for guests selected and bought (approximately 1,000)

010105 Shipyard carpenters notified to build grandstand of suitable size

010106 Arrangements to have path to shipway cleared and made safe, completed

010107 Catering for party arranged
and so forth.

The analysis. Figure 28.1 is a preliminary possibility guide made with respect to the 5th-order work-unit. Figure 28.2 is the detailed possibility guide. Note that class 1 changes were not listed on the preliminary possibility guide; larger changes were the main objective.

It should be noted that the ship-launching ceremony had been initiated when the yard was new and was launching one ship a year. At the time the study was made, they were launching one per month.

Innovation. Possibility 1 was discarded in that the owner of each new ship wanted a ceremony. Possibility 2 met this constraint while providing great savings. Further study showed other benefits; the guests (vendors, government officials, etc.) invited each month were glad to be relieved of any obligation to attend. This change reduced the workload of the group by 21 full-time employees.

BASIC CHART FORM

Prelim. Pos. Guide **Type of chart**   _12-B_ **Department**

_____ **Original or proposed** _J. M._ **Chart by**

Ship launch. ceremony **Subject charted** _6/2_ **Date charted**

Suggestion Number	Class of change	Description	Other classes affected
1	5	Eliminate party	4, 3, 2, 1
2	5	Reduce guest list	4, 3, 2, 1
3	4	Use stock presents	3, 2, 1
4	4	Use closed circuit TV with guests in town hall	3, 2, 1
5	3	Computerize invitation list	2, 1
6	2	New addressing machine	1
7	2	Prefabricate grandstand	1

Figure 28.1. Preliminary possibility guide for _Ship launching ceremony completed._

POSSIBILITY GUIDE

File Number 12-13-3
Analysis by J. M.
Date 6/2

Name of Operation _Ship launching ceremony_

Operation number _____

Class of Change	Hand and Body Motions	Tools, Workplace and Equipment	Process	Product Design	Raw Material
1	Not considered				
2	Reduce work by on man-work	New addressing machine			
	Reduce carpenter work	Prefabricate grandstand			
3	No addressing work	Eliminate addressing machine	Computerize invitation list		
4	Eliminate personal shopping	Add to computer inventory routine	Stock presents as other inventory	Use stock presents	
5	Most jobs disappear	Eliminate list of addresses, etc.	Use stock gifts except for adml owner; grandstand on flatbed truck	Party only for adml owner and the guests	Reduce guest list
	No work	No equipment	No process	No ceremony	Eliminate party

Figure 28.2. Detailed possibility guide for _Ship launching ceremony_ completed.

521

Case II—Sales Counter in Department Store; Women's Nylon Stockings

Figures 28.3 and 28.4 are the preliminary and detailed possibility guides for the sale of a pair of women's nylon hose in a large midwestern department store. The class 5 suggestion was selected as most feasible. Subsequent detailed product and man analysis proved that it was most desirable, and it was installed with considerable labor and cost savings.

Summary

In each case cited it will be noted that the situation examined merely provided information with respect to the status. The analyst, using the checklist of Chapter 7, had to provide alternatives to the status. However, this was not an uncontrolled generation of ideas; the checklist provided a general guide to alternatives.

PROBLEMS

28.1. The police of a town near a recreational area operate a road block at night on weekends. The purpose is to reduce drunken driving. All cars are stopped during the hours of 9 P.M. to 5 A.M. and the drivers scrutinized while still seated behind the wheel. If drunkenness is suspected, the drivers are requested to walk on a painted line. Those who fail to show a steady gate are made to take a balloon-type drunkenness test.
(a) Prepare a preliminary and detailed possibility guide for this situation.
(b) Select and defend an innovative possibility.

28.2. A U.S. professional technical society is planning a program of certification for its members. Initial certification will be by test. Recertification will also be by testing at 5-year intervals. Tests will be given at 60 locations in the United States.
(a) Prepare a preliminary and detailed possibility guide with the object of reducing the cost of the certification program.
(b) Select and defend an innovative possibility.

28.3. In school, courses in the humanities are designed to "broaden" technical students.
(a) Prepare a preliminary and detailed possibility guide for this situation. The criterion is "more effective broadening."
(b) Select and defend an innovative possibility.

28.4. A series of convenience stores have been subjected to frequent armed robberies. The object is to reduce the frequency of robberies, the loss of money, and the hazard to the employees.
(a) Prepare a preliminary and detailed possibility guide for this situation.
(b) Select and defend an innovative possibility.

BASIC CHART FORM

Prelim. Pos. Guide Type of chart   _Hosiery_ Department

_____ Original or proposed _Ann W._ Chart by

Sale of nylons Subject charted   _12/4_ Date charted

Suggestion number	Class of change	Description	Other classes affected
1	5	Purchase stockings prepackaged; 1 pair per pack	4, 3, 2, 1
2	4	Prepack in receiving, 1 pair per pack	3, 2, 1
3	3	Sell from samples on counter, sell hose without showing actual pair to customer	1
4	2	New packing desk for silk and nylon hose counter instead of department desk for packing	1
5	2	New cash register, one for each counter	1
6	1	Teach salespersons easy way to wrap hose	

Figure 28.3. Preliminary possibility guide for sale of pair of nylons in a department store.

POSSIBILITY GUIDE

Name of Operation Sale of pair of women's nylon hose   **File Number** WW-H-H

Operation number   **Analysis by** Ann W.

   Date 12/4

Class of Change	Hand and Body Motions	Tools, Workplace and Equipment	Process	Product Design	Raw Material
1	Teach salespersons easy way to fold and wrap hose				
2	Less walking	Cash register for each counter			
3	Much less walking	Packing bench for hose counter; Sample display unit; See above; suggestion also	Sell from samples on counter without showing actual pair to customer		
4	See above; also less handling	See above; packing at receiving	See above; also prepack in warehouse	Prepack in receiving; one pair per package	
5	See above	See above on Line 3	See above on Line 3	Give customer prepack without opening	Buy prepacked nylon stockings

Figure 28.4. Detailed possibility guide for sale of pair of nylons in a department store.

29 NOTES ON WORK ACTIVITY ANALYSIS AND WORK SAMPLING

Introduction

The two cases that follow were selected for two reasons. One, as with the illustrations in the previous chapter, was to demonstrate how the approaches may be used in areas of activity other than manufacturing. The second was to illustrate data-gathering techniques additional to those described in Chapter 8. Case I concerns maintenance work in a large hotel. Case II concerns part of a study made which eventually resulted in a large increase in the productivity of dentists.

Case I—Electrical Maintenance Department in Hotel (Work Activity Analysis)

Introduction. In this instance the data for a work activity analysis were available in the form of job tickets used to assign jobs to the workers and to record the use of their time. A job ticket existed for each task assigned.

Analysis. The job tickets were classified and summarized both by individuals and for the electrical maintenance group as a whole.

One category that appeared in the summary was of particular interest. Replacing burned-out hall light bulbs took an average of 9 hours and 43 minutes of work per day. Further investigation revealed that there were 856 hall lights; the lights burned 24 hours per day; the average life of a light bulb was 56 days. (The last figure was obtained from the light bulb manufacturer.) The fixtures were one-bulb fixtures.

Lights were replaced when room maids reported outages in the part of the halls associated with their assigned rooms. The task, each time it was performed, required the electrical maintenance worker to get a job ticket, obtain a light bulb from the storeroom, take out a ladder, use the freight elevator to reach the floor, locate and replace the burned-out bulb, and so forth.

Critiquing the data. Questions such as question 3 of the checklist for work activity analysis led to a complete replanning of this phase of the electrical work. First, the hotel management realized that since 15.35 lights, on the average, were being replaced at irregular intervals during each day, it was most probable that there were always some lights out. Second, they decided that an outage of 1 per cent was tolerable.

Innovation. Reference to a mortality curve for light bulbs, of the type being used, indicated that if a fresh set was installed at one time, 1 per cent would burn out in 28 days. Calculations showed that if 9 hours and 43 minutes were not used per day in relamping, the total time accumulated in 28 days was approximately 263 man-hours, or approximately 39 hours more than the total work time of one man. The largest portion of this time could be eliminated if once every 28 days an electrician was assigned to the task of changing all hall lamp bulbs whether they were burning or not. The economy of discarding the half-consumed light bulbs was easily determined. The bulbs cost 20[1] cents each, new, and at the end of 28 days the remaining total lamp-bulb value was only $85.60. With a large fraction of 263 labor-hours to save at $7.00 per hour, the advantage of cyclic lamping was obvious.

Results obtained. Using a cyclic lamping procedure once every 28 days in place of replacement of outages, and taking advantage of a similar procedure for ballrooms and public rooms, one electrician was released. Additional advantages were obtained in terms of more freight elevator time available for other functions, and better hall lighting.

---

[1] Wholesale price; unfrosted bulb.

Case II—A Study of Dental Activity
(Work Sampling)

Introduction. The Milwaukee Dental Research Group,[2] an informally organized group of dental practitioners, decided to attempt to ease the strain of dentistry by applying motion and time study to their work. They felt they were overtired at the end of a workday; that their fees were high but income was not satisfactory. In order to determine what types of activity would offer the greatest potential for improving their productivity, as well as improving the physical aspects of work, a work sampling study was made by several members of the group.

Data gathering by self-sampling. The observation times were randomized, by means of method 4 described in Chapter 8, and the observations were made by either the dentist or the chair assistant. Observations were made by entering a work code on a 3- by 5-inch card, one card for each observation. The cards were tossed into a box for sorting later. An interval alarm clock was used to indicate the exact reading time, and this clock was reset after each reading as indicated by the next card in the day's deck.

Data obtained. A typical summary for a dentist with one chair assistant is as follows:

Location and Activity		Per Cent of Total Time
In operatory		
Total		77.7
Preparing, placing, finishing, trimming, mixing, and polishing fillings	25.6	
Making examinations	8.6	
Scaling and cleaning teeth	7.6	
Adjusting, cementing, and trying dentures	7.1	
Seating and excusing patient	5.5	
Taking and carving wax impressions	3.7	
Talking to patient	2.9	
Taking, checking, and explaining roentgenograms	2.9	
Giving prophylaxis	2.1	
Giving fluoride treatment	2.1	

---

[2]Led by David A. Hoffman, D.D.S.

Location and Activity	*Per Cent of Total Time*	
Placing base	2.1	
Applying silver nitrate	1.8	
Giving local anesthetic	1.8	
Extracting teeth	0.8	
Holding celluloid strip	0.3	
Adjusting head rest	0.3	
Treating perodental pocket	0.3	
Placing cavity liner	0.3	
Relieving toothache	0.3	
Waiting for anesthetic to take effect	0.3	
Preparing penicillin	0.3	
Mixing alginate	0.3	
In office		
Total		17.7
Personal and idle	8.4	
Writing and going over records	3.7	
Phone	1.8	
Financial work and mail	1.0	
Talking to assistant and patient	0.8	
Talking to assistant	0.8	
Dictating instructions	0.3	
Checking roentgenograms	0.3	
Signing cards	0.3	
In laboratory		
Total		3.7
Working on prosthetics	3.4	
Waxing impression trays	0.3	
Out of office		
Total		2.7
Attending meetings	2.6	
Total		100.7

Initial conclusion. All the samples gave this same general form, indicating that:

1. The operatory was the major time consumer.
2. The flow of work and the work external to the mouth were the critical elements.

Subsequent detailed analysis. Therefore, a technique that preserved a record of the flow of work and gave the movements of the dentist as a whole, together with the activities of the assistant, if there was one, seemed most appropriate for making a more detailed study. Memomotion study was therefore selected and further studies using this technique were made and are described in Chapter 38.

Other analyses of the work sampling data. However, a different summary of the work sampling data permitted additional conclusions to be drawn. A comparison of four types of practices sampled gave the following tabulation:

RESULTS OF RANDOM SAMPLING DURING OFFICE HOURS GIVING TIME
DISTRIBUTION IN PER CENT OF DAY

	Dr. A, No Assistant	*Dr. B, 1 Assistant*	*Dr. C, 2 Assistants*	*Dr. D, 3 Assistants*
Productive time				
1. Restorative	20	29	$35\frac{1}{2}$	27
2. Prosthetics	4	7	7	$19\frac{1}{2}$
3. Oral examination and diagnosis	6	11	10	15
4. Oral surgery	$7\frac{1}{2}$	1	$\frac{1}{2}$	$5\frac{1}{2}$
5. Miscellaneous*	$3\frac{1}{2}$	2	4	4
Total	41	50	57	71
Nonproductive time				
6. Necessary general obligations	1	$5\frac{1}{2}$	$8\frac{1}{2}$	$5\frac{1}{2}$
7. Dental service delegatable	14	26	24	$16\frac{1}{2}$
8. General business	13	$5\frac{1}{2}$	4	—
Total	28	37	$36\frac{1}{2}$	22
9. Personal business	5	12	1	—
10. Idle	26	1	$5\frac{1}{2}$	7
Total	31	13	$6\frac{1}{2}$	7

*Local anesthetic averages $2-2\frac{1}{2}$ per cent.

Use of the additional analyses. These data, together with data concerning work count or volume of work performed in each office, was of great value to the group in examining the most desirable type of operation for each practice.

An Additional Method of Gathering Data

P. Timoshenko of the U.S. Veteran's Administration was confronted with the problem of gathering data on a large, central, claims administration office. The work-units used as categories for observations were of a nature such that they could not be identified by an observer. As a consequence, Timoshenko arranged to have the phone bells activated to give a distinctively, different from normal, ring at random intervals. At each such ring each person in the study made an entry on a data sheet for the day, one sheet being kept by each person. Various tests of the data indicated a high level of accuracy.

Summary

As was noted earlier, in Chapter 8, both work activity analysis and work sampling are diagnostic techniques. They assist in making a preliminary analysis in situations where a great number of 3rd-, 4th-, or 5th-order work-units are produced in a random sequence. The illustrations given in this chapter indicate that the effective use of these techniques is not restricted to office, manufacturing, or warehouse operations.

PROBLEMS

29.1. (a) If, in Case I, the new work time for a complete cycle of hall relamping is 16 hours, what is the per cent saving of time?

(b) What is the per cent saving of money?

29.2. Prepare and present a report, designed to be given to the hotel manager, to obtain approval of the cyclic hall lamping procedure.

29.3. In Case I, reference was made to the use of a *similar* (to the cyclic hall lamping) procedure for ballrooms and public rooms. In what ways would such a procedure be similar; in what ways different?

29.4. If a cyclic lamping procedure is used for florescent lights, under what circumstance would spot replacement still be required?

29.5. If, in Case I, the hall fixtures, which had one lamp bulb, were replaced with two-bulb (reduced watts per bulb) fixtures, what period of time could elapse between cyclic lampings?

29.6. Compute the direct time savings or loss under the conditions of Problem 29.5 as compared with the new method installed in Case I.

29.7. Plan a self-sampling study for your use during your work time.

29.8. With respect to the procedure used in the Veteran's Administration claim office, what activities must Timoshenko have performed prior to the study?

29.9. Prepare a presentation, for use in a hypothetical situation similar to the situation of Case II, designed to assist in obtaining cooperation.

29.10. Plan a set of categories and a procedure for use by a homemaker to determine time usage.

30 NOTES ON WORK-UNIT ANALYSIS

Introduction

This chapter presents information with respect to three important aspects of work-unit analysis. These are:

1. *Properly identifying outputs.* With service activities, the outputs to be counted and subjected to work measurement must be selected with great care. They are most frequently quite different from what at first appear to be the outputs.

2. *Statistical interfaces.* Peculiar quantitative relationships are frequently encountered between orders of work-units. These relationships, called *statistical interfaces*, must be thoroughly examined.

3. *Types of work-units.* Various work-units have different characteristics with respect to the feasibility of various work measurement techniques, the method of developing a work count, and the relationships among

work count, standard time, staff resources required, and productivity. Each work-unit must be identified with respect to *type* to make effective use of a work-unit structure.

Properly Identifying Outputs

Introduction. In Chapter 26, in the examination of the work of the Office of the Solicitor, U.S. Department of the Interior, it was noted that at first glance the work of the lawyers seemed to involve *cases*. It was shown, with respect to workload, that this would have been a very defective measure. This difference between that which appears as an obvious measure and that which is a good measure is a rather common characteristic of service activities. It is not infrequent to find overly simplistic but inappropriate measures in use.

Counting outputs for the U.S. Meat and Poultry Inspection Service (MPI).[1] Previous to the application of work-unit analysis, budget planning procedures were based on either an empirical relationship between total agency staffing and the number of meat-packing plants needing service, or an empirical relationship between total staffing for the agency and the total number of pounds of all types of meat inspected. Needless to say, these two methods did not produce compatible answers. Neither method led to a management control system; neither method represented a procedure for accounting for all the factors affecting the required manning and funding. Further, and this is most important, although "meat-inspected" is one dimension of the output, and perhaps the most obvious one, it has been found that the workload of inspection, as a total, is relatively *independent* of the amount of meat produced and inspected. "Meat-inspected" is an obvious but not a useful way to describe the outputs when one is concerned with determining staff requirements.

Applying work-unit analysis. To apply the work-unit-analysis approach of Chapter 19:
The 7th-order work-units were described as:

1. Official meat-packing-establishment inspection services provided for 1 year.[2] (Paid for by public funds.)

---

[1] To reduce the size of the illustration, only a part of the meat-inspection activities will be examined, as well as only a part of the work-unit structure.

[2] Note, as was suggested in Chapter 9, the use of the past-tense verb to assist in reducing the confusion of outputs with activities.

2. Nonofficial meat-packing or meat-products-establishment inspection services provided for 1 year. (A reimbursed activity.)

Outputs within each of these two work-units were more homogeneous than when not separated because they are funded differently. This is the beginning of a replacement for the natural division supplied by the engineering drawings of the hardware-producing organization. To continue with the analysis:

The programs (6th-order outputs) or groupings of like outputs were listed as follows[3]:

0101 Inspection services provided for 1 year where the plant workload requires less than one inspector.

0102 Inspection services provided for 1 year where the plant workload requires exactly one inspector.

0103 Inspection services provided for 1 year where the plant workload requires more than one but less than six inspectors.

0104 Inspection services provided for one year where the plant workload requires six or more inspectors per establishment.

0201 Government product specification examinations performed.

Within each of the 6th-order work-units 0101 through 0104, the natural 5th-order work-unit would appear to be *A plant [of that type] serviced*, but a count of these plants cannot be added within a program or from program to program to give a number with a fixed relationship to staff required.

Additional division is desirable in this case between the 6th-order and the 5th-order. This additional, needed subdivision is encountered with many government work-unit hierarchy problems. The list of Table 9-I was given only as a guide, not as a limiting list. In this case, the lower-order work-units of program 0101, "Inspection services provided for 1 year where the plant workload requires less than one inspector," can be usefully divided into two subgroups, which are not units of final output. These work-units, designated as 5.5th-order work-units, are:

010101 Inspection services provided for 1 year where the workload requires less than one inspector but the plant is in an isolated location where no other work exists for the inspector, or where, because of the nature of the work, the inspector cannot leave during the working day.

---

[3] This division is not arbitrary. Each group of plants requires a different amount of inspection staff per plant. Further, the number of plants in each group is changing independently and responds differently to economic factors; each 6th-order work-unit needs a different forecasting system.

010102 Inspection services provided for one year where the workload requires less than one inspector but the plant is of such a nature or so located that the plant may be combined with other plants into a multiplant assignment, using one inplant inspector.

Within each of these 5.5th-order work-units we have 5th-order work-units: *A plant [of a type] serviced.*

The 4th-order work-units, the intermediate products required to produce *A plant [of a type] serviced* will be a mix from the following list (which is the equivalent of the "parts list" in manufacturing; the situation resembles "custom" manufacturing):

-----01 Cattle slaughter inspected in-plant.
-----02 Hog slaughter inspected in-plant.
-----03 Sheep slaughter inspected in-plant.
-----04 Sausage manufacturing maintained in compliance in-plant.
-----05 Cured product manufacturing maintained in compliance in-plant
-----06 Inspection supervision provided.
-----07 Inspection results appraised.
-----08 Inspection methods designed.
-----09 Laboratory support provided, and so forth.

(This list has been greatly shortened for simplicity's sake. A complex service may have as many "parts" as a complex manufactured product.)

Summary. It should be obvious that the list of outputs above is quite different from the older, incorrect, albeit simpler methods of counting *pounds of meat inspected* or *number of plants maintained under inspection.* However, the new procedure will provide a meaningful basis for predicting and controlling staff resources.

Statistical Interfaces

The outputs of a grant-awarding organization. The National Institute of Allergies and Infectious Diseases (NIAID) of the National Institutes of Health, among other activities, makes grants to external organizations for research.[4]

---

[4]The information presented in this section was extracted from a report prepared under Contract HEW-OS-72-179 with the U.S. Department of Health, Education, and Welfare.

The four areas in which research grants are made are:

1. Allergic and immunological diseases.
2. Bacteriological and mycological diseases.
3. Virological and rickettsial diseases.
4. Parasitic and anthropod-borne diseases.

Hence, one 6th-order work-unit would be:

0101. Research grants awarded and managed for allergic and immunological diseases.[5]

Further, a 5th-order work-unit under 0101 would be:

010101. A project grant made and supported for research on an allergic or an immunological disease.

Problems in counting in a meaningful fashion. It should be noted that this discussion does not concern the actual grant money; this can be quite meaningfully added from grant to grant to obtain a true total. This discussion concerns counting the grants to obtain a number or numbers meaningful with respect to the staff needed by NIAID.[6] The counting problem is complicated by the nature of the constituent 4th-order work-units of any one grant. The 4th-order work-units are:

01010101 Areas for emphasis within program selected.
01010102 Program opportunities and objectives delineated and announced.
*01010103 Application received and reviewed; project to support selected.
*01010104 Grant award prepared and issued.
*01010105 Grant managed.
01010106 Program accomplishments analyzed and evaluated.
01010107 Coordination, with other federal and nonfederal offices, agencies, and institutions, maintained.

An examination of these work-units will reveal that only the three marked with asterisks are related to a count of the 5th-order work-units in 0101.

Statistical interface, defined. The 4th-order work-unit 01010101, *Areas for emphasis within program selected*, will be performed whether 10 grants

---

[5]The first 01 of 0101 indicates the 7th-order work-unit category, *Research grants made and supported*.

[6]Note the similarity of this aspect of counting with the MPI illustration that preceded this discussion.

or 100 grants are to be made. The amount of work time required is not a function of the count of 5th-order work-units supported. On the other hand, the work time needed for 4th-order work-unit 01010105, *Grant managed*, will be a function of the 5th-order work-unit count.

The 4th-order work-units and the 5th-order work-units of this work-unit structure do not have a constant, simple, one-for-one relationship: hence, the term "statistical interface" to refer to the relationship between the 4th- and 5th-order work-units.

Summary. In situations such as was described, it should be obvious that any useful work measurement effort must be performed at the 4th-order work-unit level; that the work time associated with a 5th-order work-unit will be an empirical value, correct only for a specific period of time. Further, it should be obvious that the recognition and delineation of the statistical interface is crucial to successful managerial control, as control was defined in Chapter 1.

Types of Work-Units

With respect to the use of work counts, standard times, and feasible work measurement methods, work-units may be usefully classified into types or categories. Each of these categories will be examined in detail. (It should be noted that these five types may also occur in manufacturing. However, the framework provided by the drawings and the bill of materials has such a strong influence that the problem of handling these different types of work-units has never, in manufacturing activities, been a problem of any magnitude.)

A linear work-unit. A *linear work-unit* (*LIN*) is one for which the required total work time (using a constant method of production) is a direct function of the work count and the work count is a discrete variable. For example, if a linear work-unit has a standard time of one hour and 10 units are to be produced, 10 hours will be required.

A curvilinear work-unit. A *curvilinear work-unit* (*CLIN*) is one for which the required total work time increases with the work count but not in direct proportion. For example, if a work count of 1 takes one hour, a work count of 10 may take 5 hours; there may be an "economy of scale." The standard time may be expressed by a mathematical formula relating to the work count or by a table giving the standard time for each possible work count, by integers or by ranges.

A level-of-effort work-unit. A *level-of-effort work-unit* (*LE*) is characterized by the following:

1. There is no objective-oriented, lower-order, work-unit work count.
2. Additional performance, beyond an assigned amount, would not contribute to the achievement of the organization's objectives.
3. There usually is not a clear indication of the completion of the work-unit.

The 4th-order work-unit of NIAID, 01010101, *Areas for emphasis within program selected*, is a typical example. Let us assume that in a base year[7] it was satisfactorily performed and required 700 hours of effort. More time or more production of the work-unit would not have produced additional value. We can also presume that it should take the same time the next year.

The work count of such a work-unit may be assigned as 1, 3, or 4, depending on whether an annual, a one-third-year, or a quarterly computation is desired. If only a 1-year computation is desired, the work count is 1; the standard time per work count is 700 hours. If a one-third-year computation is desired and the annual work count is designated as 3 (to avoid fractions), the standard time per work count of 1 is 233 hours. If a work count of 4 is used, the standard time per work count of 1 is 175 hours.

If the organization learns to perform the work-unit more rapidly, and uses the time to produce other outputs, they will appear more productive; this seems correct. If they use more time, and therefore fail to produce other outputs, they will appear less productive; this also appears to be correct.

It should be noted that a level-of-effort work-unit is *not an open account* to which time may be charged willy-nilly. There is a fixed value assigned the work. Additional performance requiring additional time does not add to the earned or produced hours of the organization.

A work-unit with a fiat standard time. These fiat work-units were described in detail in Chapter 25. It is important to recognize work-units with such standard times as being different from other types; much needless use of other work measurement techniques may be avoided. However, fiat work-units may also be LIN, as with the White House guards of Chapter 25; they may also be LE, as with a sales manager of a territory.

A work-unit that is a hybrid; it has both linear and level-of-effort components. An example of a *hybrid* (*H-LIN-LE*) type of work-unit is the 4th-order work-unit of NIAID, 01010102, *Program opportunities delineated and announced*. The *delineation* is linear with respect to the number of different

---

[7]A year selected as a starting point for future budgeting and productivity measurement.

program opportunities to describe; each requires a separate effort.[8] *Announcement*, on the other hand, which involves mailings and discussions at various meetings, is level-of-effort; the same mailings are made and the same meetings attended regardless of the number of opportunities.

A work-unit that is linear but limited.   These (*LIN-LIM*) work-units are frequently found in the subordinate outputs of an organization. For instance, a major manufacturer may feel that having executives, engineers, or scientists present papers before learned societies or "sit" on industry or international panels is important with respect to the company's public image. Hence, one 5th-order work-unit of a staff group may be *Learned appearance made*. However, no more than 10 per year may be desired. For convenience of control (and based on past experience), a standard time of 40 hours per *Learned appearance made* may be established. However, in any "earned hours" or productivity computation, no more than 10 per year would be counted.

Summary.   Each work-unit in a work-unit structure must be identified with respect to type (as defined in this section), as well as by other characteristics, in order to effectively select work measurement procedures, work counting methods, and procedures for using the data in managerial control and productivity measurement systems.

Summary

Various aspects of work-units have been examined. The purpose has been to provide an understanding of the relationships among work-unit types and work measurement procedures with respect to feasibility or special requirements.

PROBLEMS

30.1. (a) Would you anticipate that the money spent on the control, elimination, or prevention of a disease by the U.S. Public Health Service would be a function of the incidence of the disease in the United States?

   (b) Support your position by listing some of the work-units that would be involved.

---

[8]In actual fact, discontinued opportunities must also be counted in that a notice of discontinuation and the reasons therefore must be prepared.

30.2. Would you anticipate that the work time required by a checker in a super-market would be a function of one or more of the following:
(a) Dollar volume checked?
(b) Number of customers checked?
(c) Number of items checked?
Support your position; add work-units if you think it necessary.

30.3. What is an appropriate procedure for quantifying the work of a:
(a) Grade school teacher?
(b) High school teacher?
(c) Technical school teacher?
(d) College teacher?
(e) Graduate college teacher?

30.4. What is an appropriate set of work-units for your work?

30.5. Describe, in detail, an appropriate work measurement procedure for the work-unit or work-units of:
(a) Problem 30.1.
(b) Problem 30.2.
(c) Problem 30.3.
(d) Problem 30.4.

30.6. A worker is using an overhead-track cutoff saw for cutting pallet face boards 3 feet long from random-length boards ranging from 10 to 16 feet in length. The elements are:
(a) Get one length of random lumber and bring to saw.
(b) Cut first end square.
(c) Cut off one 3-foot length; repeat as possible.
(d) If remainder is over 2 feet but less than 3 feet long, place in special "save" pile.
    1) If remainder is under 2 feet, place in scrap bin.
(e) Change saw blade after approximately 5,000 pieces.
(f) Make out lot ticket for each 1,000 pieces.
(g) Blow off sawdust after approximately 3,000 pieces.
(h) Replenish random-length boards in 200 board lots. Let
    T_1, T_2, etc, $=$ time for each element listed above
        $n_1 =$ average number of pieces per random-length board
        $n_2 =$ number of pieces meeting the criteria for element (d)
        $n_3 =$ number of pieces meeting the criteria for element (d1)
    1) If the 3rd-order work-unit is described as *Cut all 16 pieces required for one pallet*, describe the statistical interface between the 2nd-order work-units given and this 3rd-order work-unit.
    2) Describe the statistical interface between the 3rd-order work-unit and the 4th-order work-unit, *One pallet completed.*

30.7. The political reporter for the newspaper, the *Washington XXX*, has the assignment of tracing down all rumors concerning potential political appointments.
(a) If the 4th-order work-unit is *An appointment recognized prior to its*

official announcement, what is the statistical interface between this 4th-order work-unit and its constituent 3rd-order work-units?

(b) What is the nature of the statistical interface between the 4th-order work-unit produced by this reporter and the 5th-order work-unit, *The Washington political scene fully reported*?

30.8. Classify the following work-units by types as described in Chapter 30 and defend your position:

(a) A term paper written.

(b) A wedding attended.

(c) One's own high school graduation attended.

(d) Annual plans made.

(e) A house bought to live in.

(f) A vacation used.

30.9. Suggest an appropriate work measurement procedure, and defend your position, for the work-unit or work-units of:

(a) Problem 30.8(a).

(b) Problem 30.8(b).

(c) Problem 30.8(c).

(d) Problem 30.8(d).

(e) Problem 30.8(e).

(f) Problem 30.8(f).

30.10. Prepare a work-unit structure for the work of your branch.

30.11. Prepare a work-unit structure for the work of your division.

30.12. Prepare a work-unit structure for the work of your service.

30.13. Propose a work measurement procedure or procedures for use with:

(a) Problem 30.10.

(b) Problem 30.11.

(c) Problem 30.12.

31

NOTES ON PROCESS CHART— PRODUCT ANALYSIS

Introduction

This chapter, like the two that preceded it, illustrates how the basic technique described earlier (in this case in Chapter 10) also has great utility in the service environment. In addition, this chapter describes the method of charting the manufacturing process with a multiple-component product. Finally, the potential relationship between process chart—product analysis and mathematical models will be described.

Case I—A Use of a Process Chart to Rearrange and Relocate the Steps in a Service Process

Introduction. This case is drawn from a service industry to illustrate further the use of process chart—product analysis. Figures 31.1a and b are a process chart—product analysis from a large midwestern department store.

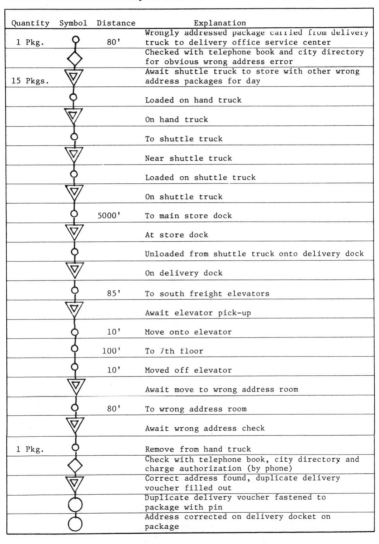

BASIC CHART FORM

Proc.chart-Product anal. **Type of chart**   Delivery   **Department**

Original _____ **Original or proposed**   K. Baker   **Chart by**

Redelivery W.A. package **Subject charted**   4/1 _____ **Date charted**

Quantity	Symbol	Distance	Explanation
1 Pkg.	○	80'	Wrongly addressed package carried from delivery truck to delivery office service center
	◇		Checked with telephone book and city directory for obvious wrong address error
15 Pkgs.	▽		Await shuttle truck to store with other wrong address packages for day
	○		Loaded on hand truck
	▽		On hand truck
	○		To shuttle truck
	▽		Near shuttle truck
	○		Loaded on shuttle truck
	▽		On shuttle truck
	○	5000'	To main store dock
	▽		At store dock
	○		Unloaded from shuttle truck onto delivery dock
	▽		On delivery dock
	○	85'	To south freight elevators
	▽		Await elevator pick-up
	○	10'	Move onto elevator
	○	100'	To 7th floor
	○	10'	Moved off elevator
	▽		Await move to wrong address room
	○	80'	To wrong address room
	▽		Await wrong address check
1 Pkg.	○		Remove from hand truck
	◇		Check with telephone book, city directory and charge authorization (by phone)
	▽		Correct address found, duplicate delivery voucher filled out
	◯		Duplicate delivery voucher fastened to package with pin
	◯		Address corrected on delivery docket on package

Figure 31.1a. Process chart—product analysis for original method of handling wrongly addressed packages in a department store (*continues*).

BASIC CHART FORM

PCP Page 2 of 2 **Type of chart** ______ **Department**

______ **Original or proposed** ______ **Chart by**

Redelivery W.A. Pkg. **Subject charted** ______ **Date charted**

Quantity	Symbol	Distance	Explanation
	o		Package placed on hand truck for correctly addressed packages
20 Pkgs.	▽		Await pick-up to elevators
	o	80'	To south freight elevators
	▽		Await elevator
	o	10'	Moved onto elevators
	o	100'	To street floor
	o	10'	Moved off elevator
	o	70'	To delivery department; west building
	▽		Await delivery route check
	○		Marked for route identification; duplicate delivery voucher pulled
	o	20'	To shuttle truck
	o		Loaded on shuttle truck
	▽		On shuttle truck
	o	5000'	To service center
	▽		At shuttle dock
	o		Unloaded on shuttle dock
1 Pkg.	o	90'	Moved to proper delivery bin and stacked in bin
	▽		Awaits next regular delivery when it will be listed on log
	o	12'	Moved to truck
	o		Loaded on truck
	▽		On truck
			SUMMARY
○ - 3	o - 25	▽ - 17	◇ -2 Distance 10,797'

Figure 31.1b. (*Concluded*) Process chart—product analysis for original method of handling wrongly addressed packages in a department store.

They show the sequence of work on a wrongly addressed package, of which delivery could not be made. The possibility guide in this case suggested a class 4 change; making wrongly addressed packages correctly addressed packages at first addressing.

Preliminary innovation. A campaign was undertaken with the sales force to cut down the incidence of wrong addresses on sales slips. It is worth noting that during the year checked there were 4,796 wrongly addressed customer packages in this store, a number that was not considered excessive for a store of its size.

Further critique. In addition to the campaign to cut down wrongly addressed packages, the process chart was subjected to the same treatment detailed with respect to a manufactured item, as in Case II, Chapter 10, but using a different field of knowledge. This resulted in the proposed chart shown in Figure 31.2, which was the method eventually adopted.

Summary of improvements. In summing up the advantages, in addition to those indicated at the bottom of Figure 31.2, the analysts, who were employees of the store, stated:

1. There will be considerable improvement in customer service resulting from the saving of time and handling between the service center and the main store.

2. There will be a great saving in distance traveled, movement, and delays of packages, which will result in less possibility of breakage and loss. To put it briefly, we will save roughly 9,600 *package-miles* of transportation per year with an accompanying reduction of work and loss hazard.

3. There will be a reduction in traffic for the shuttle truck and freight elevators. There will also be some reduction in handling time by personnel at the service center and the main store.

Case II—Charting a Multiple-Component Product

Introduction. This is not a complete case study. The sole purpose of this case is to demonstrate an additional charting technique.

The charting technique. Figure 31.3 is part of an extensive chart of the entire manufacturing process of a molded cork product and its corkboard center. Note that separate columns are used, not only for each part of the product, but for the molds used to form each part. If problems larger than this are encountered, it is obvious that the use of a form should be abandoned and a large sheet of drawing paper be used.

BASIC CHART FORM

<u>Proc.chart-Product anal.</u> **Type of chart** <u>Delivery</u> **Department**

<u>Proposed</u> **Original or proposed** <u>K. Baker</u> **Chart by**

<u>Redelivery W.A. package</u> **Subject charted** <u>4/10</u> **Date charted**

Quantity	Symbol	Distance	Explanation
1 Pkg.		80'	Wrongly addressed package carried from delivery truck to delivery office service center
			Check with telephone book, city directory and charge authorization (by phone)
			Correct address found
			Duplicate delivery voucher filled out
			Duplicate delivery voucher pinned to package
			Address corrected on delivery docket
			Placed in canvas truck
20 Pkgs.			Await distribution of all wrong addressed packages to delivery bins
1 Pkg.		90'	Moved to proper delivery bin and stacked in bin
			Await next delivery when it will be listed on log
		12'	Moved to truck
			Loaded on truck
			On truck

SUMMARY AND RECAPITULATION

Symbol	Original	Proposed	Saved
◯	3	2	1
○	25	5	20
▽	17	5	12
◇	2	1	1
Distance	10,797'	182'	10,615'

Figure 31.2. Process chart—product analysis for proposed method of handling wrongly addressed packages in a department store.

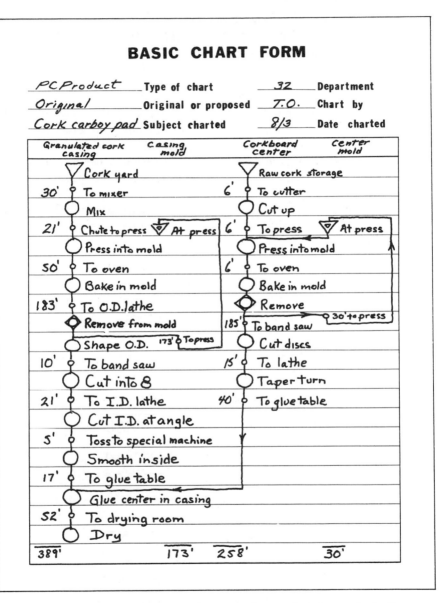

BASIC CHART FORM

PC Product ____ Type of chart ____ _32_ ____ Department
Original ____ Original or proposed ___ _T.O._ ___ Chart by
Cork carboy pad Subject charted ____ _8/3_ ____ Date charted

Granulated cork casing	Casing mold	Corkboard center	Center mold
▽ Cork yard		▽ Raw cork storage	
30' To mixer		6' To cutter	
◯ Mix		◯ Cut up	
21' Chute to press ▽ At press		6' To press ▽ At press	
◯ Press into mold		◯ Press into mold	
50' To oven		6' To oven	
◯ Bake in mold		◯ Bake in mold	
183' To O.D. lathe		◈ Remove	
◈ Remove from mold		185' To band saw 30' to press	
◯ Shape O.D. 173' To press		◯ Cut discs	
10' To band saw		15' To lathe	
◯ Cut into 8		◯ Taper turn	
21' To I.D. lathe		40' To glue table	
◯ Cut I.D. at angle			
5' Toss to special machine			
◯ Smooth inside			
17' To glue table			
◯ Glue center in casing			
52' To drying room			
◯ Dry			
__389'__	__173'__	__258'__	__30'__

Figure 31.3. Part of process chart—product analysis for method of manufacture of molded cork product.

Case III: A Use of a Process Chart—
Product Analysis That Led to the Creation
of a Mathematical Model[1]

The graphic model. Figure 31.4 is the process chart—product analysis for part of the manufacture of a pressed-aluminum article. This particular product was made during only a small part of the year on presses and polishers that were parts of machine lines set up to manufacture other products, which were made in great volume during the rest of the year. Note the various quantities produced by each machine and the various trucking times given at the bottom of the chart. (These trucking times were determined by means of time studies.)

This analysis was made because of confusion in trucking from the presses to the polishers. The analyst wished to determine the most efficient hourly trucking pattern.

The mathematical model. A mathematical model of the situation may be made to assist in the solution of a problem of this type, as follows:

Let the quantity of skid loads of parts going from press A to polisher 1 be represented by the letter D, from A to polisher 2 by E, and so forth, as follows:

From Press	To Polisher 1	2	3	4
A	D	E	F	G
B	H	I	J	K
C	L	M	N	O

The press-to-polisher trucking times in minutes per load are as follows:

From Press	To Polisher 1	2	3	4
A	2.2	3.5	4.1	5.2
B	2.7	1.2	4.4	3.2
C	1.2	1.6	2.1	4.3

[1] The numbers involved have been altered to facilitate presentation.

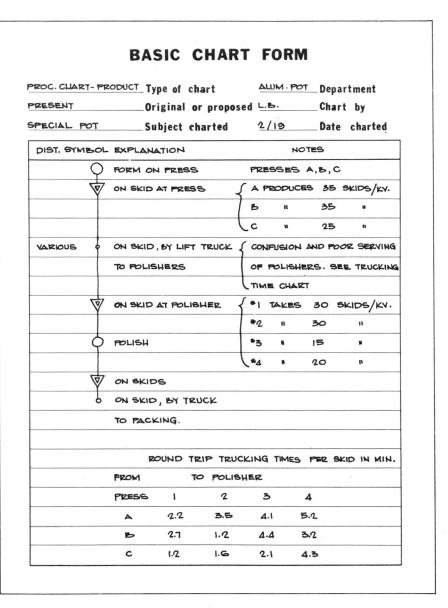

Figure 31.4. Process chart—product analysis of part of a manufacturing sequence.

Let the total trucking time (which we wish to minimize) be represented by T.

Taking the information from the process chart, we may now write the following equations:

$$D + E + F + G = 35$$
$$H + I + J + K = 35$$
$$L + M + N + O = 25$$
$$D + H + L = 30$$
$$E + I + M = 30$$
$$F + J + N = 15$$
$$G + K + O = 20$$

$$2.2D + 3.5E + 4.1F + 5.2G + 2.7H + 1.2I + 4.4J + 3.2K + 1.2L$$
$$+ 1.6M + 2.1N + 4.30 = T$$

Solving the mathematical model. This system of equations, a mathematical model of the situation, may be solved by linear programming methods[2] to produce $T = 211.5$ minutes, its lowest possible value, when:

$$D = 30, \quad E = 0, \quad F = 0, \quad G = 5, \quad H = 0, \quad I = 20, \quad J = 0, \quad K = 15,$$
$$L = 0, \quad M = 10, \quad N = 15, \quad O = 0$$

Summary

A process chart–product analysis may be used effectively to study the flow of any object, even when the object is being subjected to a service rather than being manufactured.

The form of a process chart–product analysis alters when a multiplicity of objects are recorded on a single chart. As was noted in Chapter 10, the information is the important aspect; the chart serves only as a convenient recording device; it may need to be adapted for problems with unique requirements.

Some process chart—product analyses may lead to the development of mathematical models to aid in finding a solution. Common models involve linear programming, queuing (waiting line), or simulation.

---

[2]For methods of solution, see any text on linear programming.

PROBLEMS

31.1. The following article appeared in the Newsletter of the Transportation and Distribution Division of the American Institute of Industrial Engineers:

SERVICE TRACKS INCREASE LOCOMOTIVE RELIABILITY AND AVAILABILITY
Raymond P. McFarland
Senior Mechanical Engineer
International Engineering Company

The basic function of the Mechanical Department in any railroad is to maintain the rolling-stock equipment to achieve a high level of reliability and availability within reasonable economic constraints. Therefore, the facilities, methods, and procedures of the Mechanical Department must be geared to meet this requirement. This is particularly true for the high-investment, high-revenue-generating equipment such as diesel-electric locomotives.

In the last 10 years, some American railroads have applied industrial engineering analysis to the locomotive service and trip inspection process and found they cause excessive out-of-service time. Since locomotive servicing and the trip inspection are daily processes, the total impact on the railroad is highly significant if this time can be greatly reduced. In the past, fueling, sanding, and the trip inspection have been accomplished at three separate locations requiring individual locomotive moves to each work station. It was common for the total locomotive servicing and inspection cycle to take up to 4 hours when done sequentially.

The concept of a Locomotive Service Track, where all three functions (fueling, sanding, and the trip inspection) are done simultaneously, has reduced the servicing and inspection cycle to 1 hour or less. In terms of one locomotive over a 1-year period (assuming a road locomotive has to be serviced and inspected every day), an additional $1\frac{1}{2}$ months of locomotive availability time for transportation service can be achieved. While the author realizes that the actual recovery of $1\frac{1}{2}$ months of locomotive time is not totally achievable, the addition of more locomotive availability time can make the difference between meeting train schedules or delaying trains.

To some shippers the adherence to reliable schedules is one of their primary criteria for selecting a certain type of transportation mode. In high-volume terminals the turnaround time for locomotive "consists" is essential to an effective dispatching operation. A Locomotive Service Track that keeps locomotive out-of-service time to a bare minimum is one of the key ways progressive railroads are meeting tight train schedules.

The Locomotive Service Track must be individually designed to meet the specific transportation requirements in a high-volume terminal. A high-volume terminal is any location that must fuel more than 50 road locomotives per day. The major factors that affect the Locomotive Service Track design are:

1. The maximum number of locomotives to be fueled on a peak day.
2. The maximum number of locomotives to be fueled in a peak hour.
3. The maximum number of units in the locomotive "consist."
4. The maximum quantity of fuel and sand to be dispensed in the peak hour.
5. The configuration and location of the fuel and sand inlets for the locomotives being serviced.
6. The physical dimension (length, width, and height) of the locomotives being serviced.
7. Nominal capacities of the fuel tank and boxes of the locomotives being serviced.

Once these basic data are known, the plan layout of the Locomotive Service Track can be prepared based on certain parameters such as:

1. The cost of waiting time and how much lost time can be tolerated.
2. The normal and peak departure and arrival times for trains at the terminal.
3. The staffing level and hours of operating of the facility.
4. The pumping rate for fuel.
5. The sanding replenishment rate for the sanding system.

Besides simultaneously fueling, sanding, and trip inspecting of the locomotive "consist," the Locomotive Service Track is designed to provide the capability to add engine lubricating oil, treated engine cooling water, and traction motor support bearing oil. All services are manifolded with conveniently located outlets so the service can be rendered no matter where the locomotive is located on the Locomotive Service Track. The Locomotive Service Track is provided with a between-rail pit, with convenient service outlets and lights, so the truck, traction motor, and running gear can be properly inspected and serviced.

The Locomotive Service Track is also designed so that minor repairs, such as changing filters, brakeshoes, or traction motor brushes can be done while the locomotive is being serviced and inspected. Spare parts are conveniently stored on the Locomotive Service Track for easy application to the locomotive. The philosophy is to do as much work on the locomotive as possible at the Locomotive Service Track, to keep from sending the unit to the shop for minor repairs. This is especially true if the locomotive is not immediately scheduled for another trip. The minor repair work is done at the Locomotive Service Track as long as it does not interfere with other servicing or cause delays to other locomotive "consists."

The Locomotive Service Track is designed to allow the workforce to concentrate on the servicing and inspection process by providing the resources they need to do the job. Time is not lost traveling to and from the warehouse or supply house, because all necessary supplies are either piped to or stored at the Locomotive Service Track.

Not only does the Locomotive Service Track increase availability of units, but it also increases the reliability of units. This is true for two reasons:

1. The Locomotive Service Track is designed to enhance the trip inspection process, so more defects are seen and corrected before the locomotive is released to the Transportation Department.
2. Since all the necessary supplies are readily available, they are more likely to be applied when needed.

Where the daily volume of road locomotives is high, 100 road units or more, the use of the locomotive wash facilities can greatly aid the inspection process. Also, a large Locomotive Service Track should have a spectrometer to analyze a lube oil sample each time the locomotive arrives at the Locomotive Service Track. The spectrometer analysis must be completed before the unit leaves the Locomotive Service Track, so if something is wrong, the unit can be diverted to the shop for repair.

Increased locomotive reliability and availability are the two benefits to be gained by constructing a Locomotive Service Track. A properly designed Locomotive Service Track can achieve these goals only if good industrial engineering analysis has first been applied to the problem. The potential benefits more than repay the capital expenditure needed to construct a Locomotive Service Track. Most profitable class I railroads have proved this by constructing integrated service and inspection facilities at their major terminals.

(a) Prepare an appropriate set of process chart—product analysis for the original method described in the article.
(b) Prepare an appropriate process chart—product analysis for the improved method described in the article.

31.2. Same as Problem 31.1 but with respect to an article assigned by the instructor.

31.3. (a) Draw a process chart—product analysis for a hog slaughter plant. (You do not need to visit one.) Account for:
1) Head.
2) Ribs.
3) Tenderloin.

4) Front butts.

5) Feet.

6) Ears.

7) Liver.

8) Hams.

9) Bellies.

10) Viscera.

(b) In what important way does the chart drawn differ from the chart of Figure 31.3?

31.4. Same as Problem 31.3(a), but with respect to an automobile wrecking and stripping company.

31.5. Prepare a paper describing, in simple terms:

(a) Linear programming.

(b) Queuing.

(c) Simulation.

31.6. Solve the equations of Case III, Chapter 30, to determine:

(a) The most efficient hourly trucking pattern.

(b) The amount of trucking time needed per hour.

(c) The maximum number of trucks that will be in the department at any time. (Defend your solution.)

31.7. Same as Problem 31.6 but with data furnished by the instructor.

32

NOTES ON HORIZONTAL BAR TIME CHARTS

Introduction

With service outputs one frequently encounters a large variety of problems involving timing and strategy that are different from the example given in Chapter 11. The horizontal time bar chart is a useful aid, in such instances, for examining alternative procedures. For the most part, problems of this type may have widely varying criteria of success.

One will find a modified form of horizontal bar time charting, with special symbols, described in the literature under titles such as *management by objectives*, *management by results*, and so forth. However, rather than getting into this extensive subject, this chapter will examine a basic service problem with alternative strategies.

Settling Automobile Insurance Claims
Involving Personal Injuries

Figure 32.1 is a general presentation of two alternative procedures (or strategies) for adjusting an automobile accident insurance claim relating to a personal injury. Method A was the method originally in use. For accidents of different severity, the average amount of negotiator and legal time and the cost of such time could be determined as well as the average amount of benefits paid. Thus, total direct costs could be determined. However, with a service output, direct costs do not tell the whole story. The quality of the service has many indirect effects.

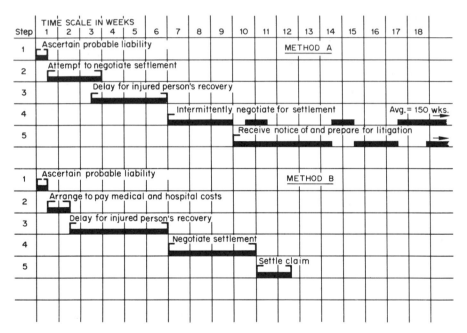

Figure 32.1. General presentation of two alternative procedures for adjusting an automobile accident insurance claim relating to a personal injury.

An alternative method. Method B of Figure 32.1 represents an alternative procedure reflecting the philosophy of checklist questions 4 and 5 of Chapter 11. The change in strategy is based on the assumption that if the initial determination of liability indicates that it is probable that the company will be obliged to remunerate the injured person, a prompt and unquestioning initial payment of the hospital bills will produce the following benefits:

1. No time will be spent in negotiating during the period of the person's recovery, during which negotiations may be slow and difficult.

2. A more amenable attitude toward final settlement on the part of the complainant much earlier in the total sequence. This will produce savings in negotiating time and in final payments in that final payments might otherwise include the litigant's cost of litigating.

3. An improved insurance company "image" leading, perhaps, to more clients.

Evaluating the alternative. In the case above so many intangible factors were involved that the problem of which alternative is the best could only be determined by testing method B. However, a test of method B yielded data that permitted the conversion of the theoretical diagram (or model) of method B to a statement of the real world. The value of method B could be determined.

The role of the chart. In the preceding illustration it was not meant to imply that the motion study technique of the horizontal time bar chart solved the problem. The technique was merely a vehicle to assist in examining alternatives; a method of using symbols to facilitate thinking and discussion. This is the basic function of most of the charting methods described in this book.

Summary

As with other motion study techniques, a criterion of success is necessary if one is going to generate alternative procedures. The horizontal bar time chart resembles a process chart, in general nature, but does not employ fixed classifications of steps, is usually applied to more involved and longer processes, and has a large field of utility with complex services.

PROBLEMS

32.1. It was traditional for one cabinet department of the United States government, when acquiring land by right of eminent domain, to first offer the landowner approximately two-thirds of the market value. As a consequence, in almost all cases, this started a series of exchanges, by mail, with the landowner's lawyer.

After approximately 6 months of correspondence, the department prepared a brief for the U.S. Department of Justice. Discussions between the two departments, with respect to the brief, took 3 more months. The Department of Justice then undertook condemnation proceedings via the courts. During the 2-year delay, the cabinet department and the landowner's lawyer continued bargaining, finally reaching approximately the original market value of the land. About 20 per cent of cases were settled in this manner, out of court. In many of the remainder, the court awarded more than the original market price; the land had increased in value during the delay.

(a) Under what economic conditions would the strategy employed have been a good one?

(b) Prepare a horizontal time bar chart of the strategy.

(c) Prepare an alternative strategy that you think is superior. State the criteria of preference. Present your alternative as a horizontal bar time chart.

32.2. Make a horizontal time bar chart of the development and issuance of a new regulation, for application to industry, by a U.S. regulatory agency or commission.

32.3. Make a horizontal bar time chart of the process to convert a country using English measures to metric.

32.4. University engineering extension services, and some professional societies, offer continuing education credits (CEUs) for attendance at seminars and workshops. Let us assume that society X has "certified" a person, Y; that Y needs 200 CEUs in the subsequent 5 years to maintain his certification. Prepare a detailed horizontal bar time chart for Y to maintain certification.

32.5. (a) Prepare a horizontal bar time chart for the completion of your education.

(b) Prepare a horizontal bar time chart of an alternative strategy that may have advantages and disadvantages.

(c) Discuss the preferability of one of the alternatives.

32.6. Make a horizontal bar time chart of a strategy for introducing frozen foods into a developing nation where supermarkets have been in common use for only 2 years.

33   NOTES ON NETWORK DIAGRAMS

Introduction

As was noted in Chapter 12, network diagrams are frequently employed on large-scale, long, and complicated projects. A discussion of the details of such charts and the computer routines used to develop reports of progress to management would be out of place in this book. However, to give more "feel" for the technique, illustrations of suitable size have been selected for this chapter.

Case I—The Use of a Network Diagram to Plan
Ship Erection Activity in a Shipyard

Terminology for the problem area.   Modern ship erection work employs a block system. Large sections of a ship, referred to as blocks, are assembled apart from the shipway (or graving dock) proper. Subsequently, the large blocks are lifted to the shipway by large cranes, aligned with the blocks already in place, tack-welded and guyed, and then welded to the other blocks to become an integral part of the ship. A simplified view of a ship showing how the hull is divided into blocks is shown in Figure 33.1. The broken lines

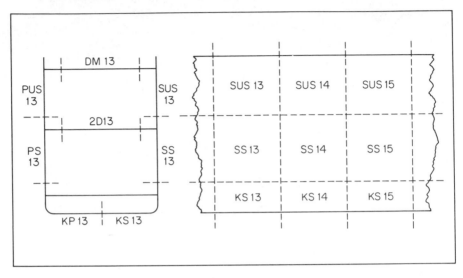

Figure 33.1. Simplified view of a ship, showing how the hull is divided into blocks.

show the division into blocks. In identifying the blocks in the figure, the following system of notation was employed:

$$KS13 = \text{keel block, starboard, ship section 13}$$
$$KP13 = \text{keel block, port, ship section 13}$$
$$SS13 = \text{side block, starboard, ship section 13}$$
$$PS13 = \text{side block, port, ship section 13}$$
$$SUS13 = \text{upper side block, starboard, ship section 13}$$
$$PUS13 = \text{upper side block, port, ship section 13}$$
$$2D13 = \text{second deck block, ship section 13}$$
$$DM13 = \text{main deck block, ship section 13}$$

Other ship section blocks are similarly identified but have different numbers; to simplify the illustration, the engine room blocks, deckhouse blocks, stern blocks, bow blocks, and so forth, have been left out of this example.

The erection process basic network diagram. As one might anticipate, there are a large number of dependencies in the ship erection process. A network diagram such as appears in Figure 33.2 is commonly employed to

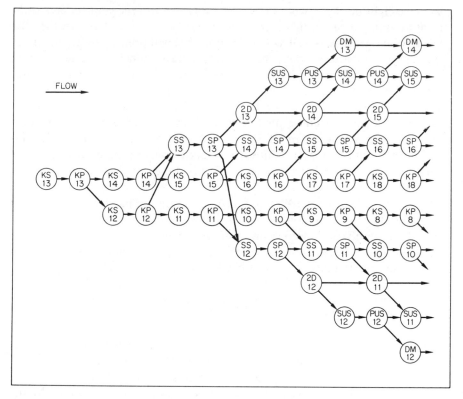

Figure 33.2. Network diagram commonly employed to set forth the basically feasible ship erection process.

set forth the basically feasible erection schedule. The network diagram of Figure 33.2 differs from the network diagrams of Figures 12.1 and 12.2 with respect to the conventions employed. The lines are used to indicate only the dependencies; the circles indicate both the activities and the status that results from the activities. As will be seen later, this modification increases the utility of this network diagram for the manner in which it is used. Further, it greatly simplifies the actual charting and reading.

Adjusting the network for feasibility. In most cases, particularly with ships larger than the yard was originally planned for (and this is a common event), the network diagram of Figure 33.2 must be modified or converted to a production schedule so as to meet the criterion: *control the impact of the project on the organization.* The reader will note that the right side or later phase of the network diagram indicates periods when the freedom exists, within the dependencies, to work on eight blocks simultaneously.

To have eight blocks to work on within 1 day requires the yard to have sufficient assembly areas so that eight blocks may be assembled and readied for lifting when needed, as well as a plate shop that will prepare sufficient plates for such assemblies. Further, the yard must have sufficient crane capacity for lifting and holding the blocks in place for a sufficient time for the fitting, guying, and tack-welding to be performed until the time when the crane may safely disengage. With the increasing size of ships, a shipyard seldom has sufficient facilities to utilize the freedoms shown on the basic network diagram. The network diagram of Figure 33.2 must be converted to a schedule within the limitations of the shipyard organization. Hence, each circle on a network diagram as shown in Figure 33.2 is dimensioned in terms of such items as the following:

1. Plate-shop support-hours for preparation of parts.
2. Assembly area needed.
3. Assembly time needed and number of men.
4. Crane lift time until safe disengagement is possible.
5. Shipfitters time after lift of block and number of men.
6. Tackwelders time after lift of block and number of men.
7. Welders time and number of welders after crane disengage, before next dependent block may be lifted and fitted, and so forth.

Planning the final network diagram. Using the dimensions above and a knowledge of the yard capability, the network diagram is redrawn with all work to be performed on successive days aligned into successive vertical columns, with the totals of each dimensional aspect shown under each column. The chart is further modified to equalize the workload per day as much as possible. The resultant network diagram becomes a tentative production schedule. Taking into account the anticipated delays from inclement weather, the schedule is checked against the delivery requirements. Further modifications may be required. The final network diagram becomes the basis for the actual scheduling and control activity.

Case II—The Use of a Network Diagram to Plan a National Project

The problem area. The U.S. national system of interstate and defense highways is a federal–state cooperative program designed to provide a coast-to-coast, border-to-border, 41,000-mile system of ultramodern controlled-access highways. The interstate highway system was designed to effectively handle the 1975 volume of traffic. Naturally, all the construction work could not be undertaken simultaneously. The impact on employment, the amout of equipment, reinforcing rod, concrete, and so forth, required would have dislocated other activities in the society. Further, the impact of such large simultaneous expenditures would have dislocated the economy.

Using a network diagram. Therefore, a network diagram, such as employed in the shipyard in Case I, had to be constructed (with each section of highway treated as a block), and a master plan evolved to fit the total interstate highway project into the nation's society and economy. The procedure was similar to that used in the shipyard. The variables, of course, were more numerous and somewhat more difficult, in some cases, to dimension accurately. Subsequently, each section of the highway had to be planned in detail. In this subsequent planning, a network diagram such as shown in Figures 12.1 and 12.2 was more appropriate. The technique needed to be adapted to fit the problem, rather than vice versa. This basic fact is too often ignored. Many varieties of network diagrams may be employed, depending on the circumstances.

Summary

A variety of formats may be used for network diagrams; the user should adapt the technique to needs. Network diagrams, properly employed, provide a valuable planning aid on complex projects where many special constraints may apply.

PROBLEMS

33.1. (a) Prepare a network diagram for the situation described in Problem 32.1.
  (b) Which technique do you feel is preferable for this problem, horizontal bar time charts or network diagrams? Why?

33.2. Same as Problem 33.1, but with respect to Problem 32.2.

33.3. Same as Problem 33.1, but with respect to Problem 32.3.

33.4. Same as Problem 33.1, but with respect to Problem 32.4.

33.5. Same as Problem 33.1, but with respect to Problem 32.5.

33.6. Same as Problem 33.1, but with respect to Problem 32.6.

34

NOTES ON PROCESS CHART— MAN ANALYSIS

Introduction

In Chapter 14, where process chart—man analysis was introduced, it was noted that four types of jobs might be encountered:

1. The work has a single, repeated cycle.
2. The work is cyclic, but there are several cycles performed with different frequency.
3. The work varies from cycle to cycle.
4. The work is such that there is no cycle or pattern.

Chapter 14 discussed work of type 1. It was suggested that type 4 jobs were best examined with the aid of work sampling or work activity analysis (Chapter 8).

This chapter is concerned with jobs of types 2 and 3. One illustration of each type will be presented. The examples have been deliberately drawn from activities other than manufacturing so as to demonstrate the wide range of applicability of the technique.

Case I—Sacking Fish-Counter Ice (Work with Subcycles)

This application of the process chart—man analysis was made in a large ice plant. Fish-counter ice is crushed ice used in supermarkets with fresh fish counters. It is prepared in a room cold enough to prevent clumping of the small pieces. It is sacked from a hopper into 50-pound sacks which are wire-tied at the top. The ice is then held in a storage room below melting temperature.

Original method. The original method is shown in Figure 34.1 and has four subcycles, *A*, *B*, *C*, and *D*. The filling of the sack *A* is performed on each of 10 bags before the worker performs *B*, which occurs only once for 10 bags. The worker then performs *C* 10 times, once for each bag, and the *D* once for all 10 bags.

Changes in charting to show the subcycles. An additional column has been added to the chart to provide a place for identifying the subcycles and to give their frequency of occurrence. Two distance and time columns are used, one to give the time for each step and one to give the prorated distance or time per bag. The times are in seconds and could be estimated from the sweep-second hand of a wristwatch.

Improved method. The proposed method, shown in Figure 34.2, took 26 per cent less time, or, in other words, increased production 35 per cent. This proposed method required the following changes. The workplace had to be rebuilt with a supply bin for bags on the right of the hopper mouth, a support in the center for the bag being filled, and a place on the left for tying bags, with a holder for tie wires and a shelf for the tier. Also, the hopper opening had to be adjusted to fill the bag more slowly and cut off when full; the worker's mittens had to be replaced with gloves, to permit him to work with them on.

Summary. It is worth noting that although the improved method in this case reduced the job to a simple cyclic task, this is by no means necessary to the improvement of all work with subcycles.

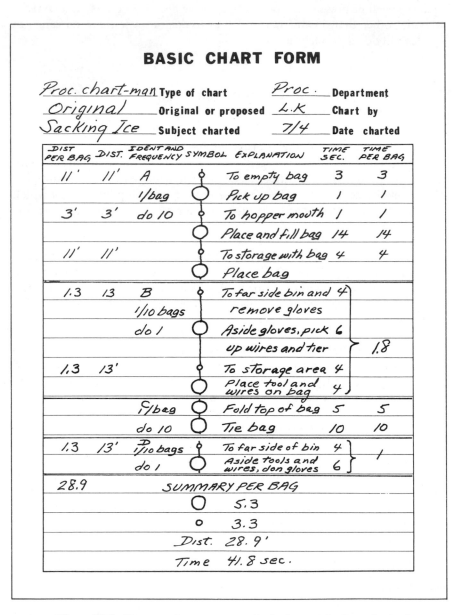

Figure 34.1. Process chart—man analysis for original method of sacking fish-counter ice.

BASIC CHART FORM

Proc. chart-man Type of chart   _Proc._ Department

Proposed Original or proposed   _L.K._ Chart by

Sacking Ice Subject charted   _7/4_ Date charted

DIST.	SYMBOL	EXPLANATION	TIME IN SEC.
11'		To hopper mouth	3
		Aside full bag	2
		Place empty bag	4
		Fold top of bag	5
		Tie bag	12
11'		To storage area with bag	4
		Place bag	1
			31

SUMMARY AND RECAPITULATION

	Proposed	Original	Saved
◯	5	5.3	.3
○	2	3.3	1.3
Dist.	22'	28.9'	6.9'
Time (sec)	31	41.8	10.8

Figure 34.2. Process chart—man analysis for proposed method of sacking fish-counter ice.

Case II—Salesclerk
(Work Varies from Cycle to Cycle)

The chart is only of one variation. This study was made with a process chart—man analysis, despite extreme variation from cycle to cycle. Because the exact order and amount of action of the women's sweaters salesclerk were subject to customer wish or whim, the process chart—man analysis, shown in Figures 34.3a and b, of a salesclerk selling a cardigan was merely representative of one type of performance.

Improved method. Application of the checklist suggested several changes. More care was exercised to keep a complete assortment of sizes, forward at the counter, in order to reduce trips to the stockroom and to have clean stock on hand. A supply of boxes and bags was kept at the sweater bar to eliminate the trips to a wrapping desk. The splintery ladder was replaced with a safe ladder. Increase in sales per unit time, more rapid customer service, and less absence of the salesclerk at the sweater bar were all advantages connected with the improved process chart—man analysis of the sweater clerk shown in Figures 34.4a and b.

Summary

Two additional types of work have been examined. Some changes in charting procedure were shown. Further, it was indicated that a job does not need to be repetitive to be charted, examined, and improved.

PROBLEMS

34.1. The following was the routine used by a worker loading refrigerators onto a trailer truck which moved them from the plant to the company's warehouse. The layout is shown in Figure P34.1.

    The loader walked 5 feet to the trailer at the edge of the dock, pushed the opening bar across the side door of the truck to the right, and then with the assistance of a hand truck located 5 feet away picked up a steel loading plate right at the edge of the dock and pushed it partly into the truck trailer so that the plate bridged the gap between the dock and the side of the trailer. The loader then pushed the empty hand truck to the conveyor line, 10 feet away, picked up a crated refrigerator with the hand truck, and pushed the refrigerator into the trailer. He then returned the 16 feet to the conveyor line and got

BASIC CHART FORM

Process chart — man **Type of chart**   383-#138   **Department**

Original   **Original or proposed** S.W   **Chart by**

Sweater sale   **Subject charted**   9/14   **Date charted**

DIST.	SYMBOL	EXPLANATION
	▽	Waits to greet customer (from counter)
	◯	Greets customer, inquires and shows black cardigan
	◯	Gets black pullover from display case
	◯	Shows to customer
	◯	Puts cardigan on to display; too small for custom.
	◯	Takes cardigan off
	◯	Replaces cardigan in display case
90'	○	Walks to stockroom
5'	○	Climbs ladder (splintery and unsafe)
	◯	Gets box
	◯	Takes out 1 cardigan of larger size
5'	○	Climbs down ladder
90'	○	Returns to customer
	◯	Shows cardigan to customer
12'	○	Walks around counter
	○	Gets whisk broom
12'	◯	Returns to customer
	◯	Brushes lint off sweater
	◯	Hands brush to customer at customers request
	▽	Waits while customer brushes sweater
	◯	Accepts brush; lays on counter
	▽	Customer agrees to buy
	◯	Asks if "charge of cash" [charge]
12'	○	Walks to end of sales counter
	◯	Gets sales book
12'	○	Returns to customer
10'	○	Walks to shelf

Figure 34.3a. Process chart—man analysis for typical original method of selling a cardigan (*continues*).

	◯	Looks for another cardigan (not there)
10'	ᴑ	Returns to customer
	◯	Writes sales check and enters on book index
	◯	Hands ballpen to customer
	▽	Waits while customer signs
	◯	Takes ballpen; turns book around
	◯	Tears out sales check set; picks up charge card
15'	ᴑ	Walks to charge verifier
	◯	Enters charge card number
	▽	Waits for response
	◯	Enters response on sales check
	◯	Separates sales check set; original in box
15'	ᴑ	Returns to customer
	◯	Picks up cardigan
15'	ᴑ	Walks with customer (for box for cardigan)
	◯	Lays sweater with sales check on wrap desk
	◯	Thanks the customer
15'	ᴑ	Returns to counter
	▽	Waits for next customer

		SUMMARY
		Dist.walked. 318'
	◯	26
	ᴑ	14
	▽	6

Figure 34.3b. (*Concluded*) Process chart—man analysis for typical original method of selling a cardigan.

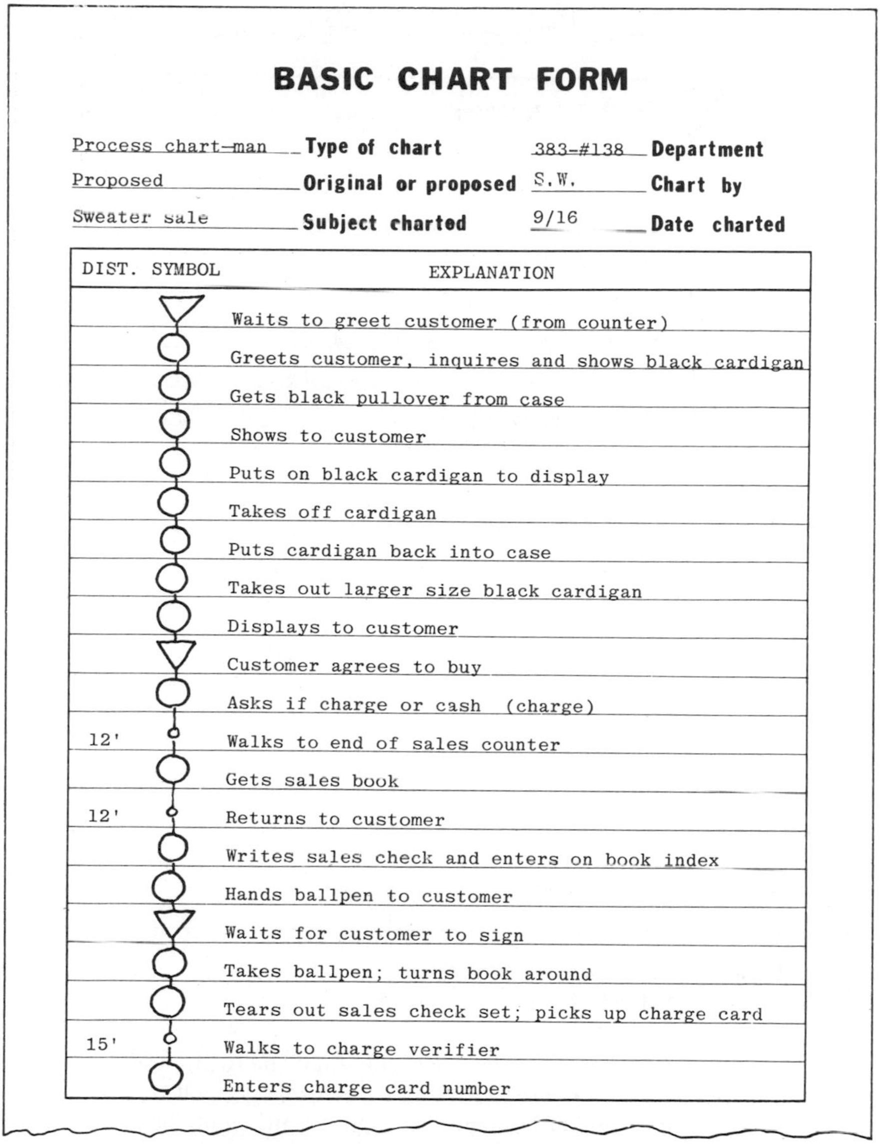

BASIC CHART FORM

Process chart—man __**Type of chart**   383-#138 __ **Department**

Proposed _____**Original or proposed** S.W. _____ **Chart by**

Sweater sale _____**Subject charted**   9/16 _____**Date charted**

DIST.	SYMBOL	EXPLANATION
	▽	Waits to greet customer (from counter)
	◯	Greets customer, inquires and shows black cardigan
	◯	Gets black pullover from case
	◯	Shows to customer
	◯	Puts on black cardigan to display
	◯	Takes off cardigan
	◯	Puts cardigan back into case
	◯	Takes out larger size black cardigan
	◯	Displays to customer
	▽	Customer agrees to buy
	◯	Asks if charge or cash (charge)
12'	○	Walks to end of sales counter
	◯	Gets sales book
12'	○	Returns to customer
	◯	Writes sales check and enters on book index
	◯	Hands ballpen to customer
	▽	Waits for customer to sign
	◯	Takes ballpen; turns book around
	◯	Tears out sales check set; picks up charge card
15'	○	Walks to charge verifier
	◯	Enters charge card number

Figure 34.4a. Process chart—man analysis for typical improved method of selling a cardigan (*continues*).

another load. This procedure was repeated 21 times before the trailer was almost fully loaded. At the end of the next trip, the loader went to the loading plate and with the help of his hand truck picked up the plate and moved it about 2 feet back on the dock so as to allow more space to stand refrigerators

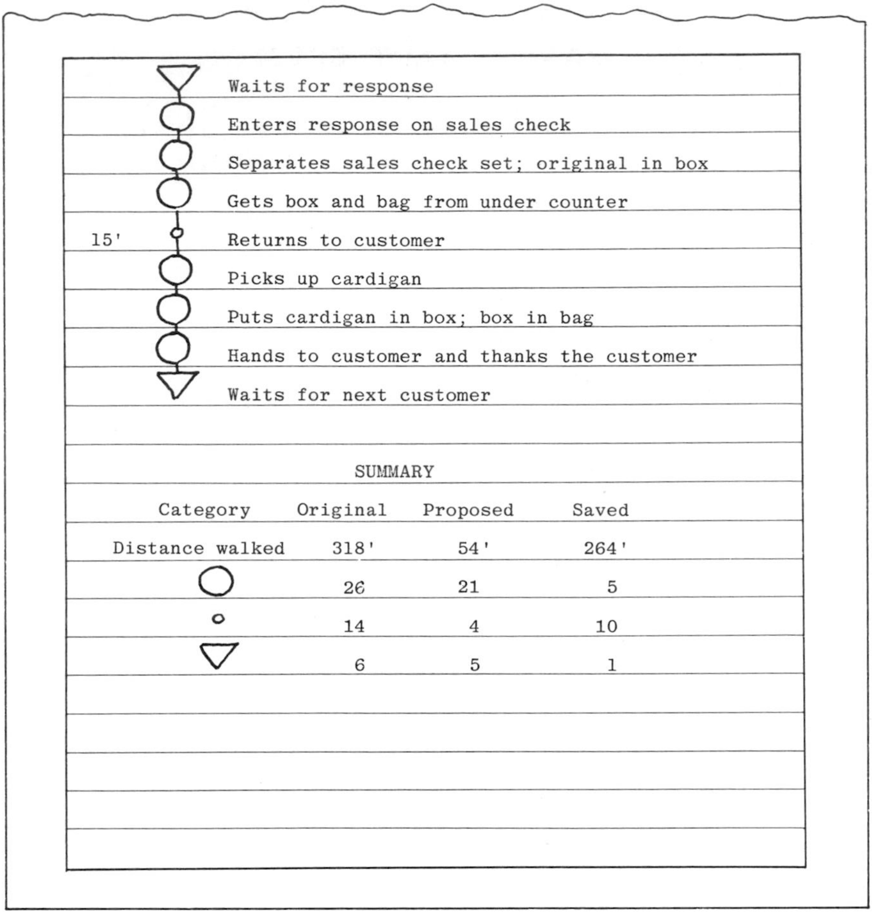

Figure 34.4b. (*Concluded*) Process chart—man analysis for typical improved method of selling a cardigan.

within the truck. He then loaded the twenty-third and twenty-fourth refrigerators. After the twenty-fourth refrigerator was loaded, he pulled back the loading plate with the aid of his hand truck, pushed the empty hand truck to its designated spot 5 feet away, returned to the trailer, closed the side bar on the trailer, and signaled the driver that his load was ready to move.

(a) Prepare a process chart—man analysis and flow diagram for the man completely loading a trailer.
(b) With the aid of the checklist, prepare a process chart and flow diagram for a better method, not exceeding a class 2 change. The driver of the truck will not handle refrigerators. The warehouse must be used, as the plant does not contain enough storage space to hold seasonal accumulation.

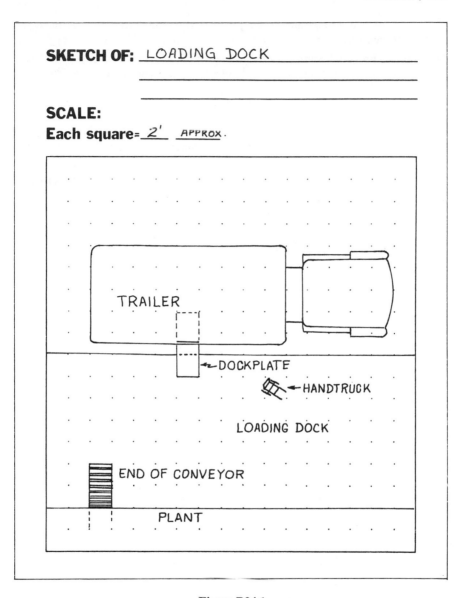

SKETCH OF: LOADING DOCK

SCALE:
Each square= 2' APPROX.

TRAILER

←DOCKPLATE

←HANDTRUCK

LOADING DOCK

END OF CONVEYOR

PLANT

Figure P34.1

(c) List the information you would add to your checklist and for each item indicate the checklist question number and letter under which you would file it.

(d) List all the advantages and disadvantages of the proposal, including all estimated savings under advantages and all estimated additional expenditures under disadvantages.

(e) What would logically be the job that you would also study at the same time as the one given here?

34.2. (a) Visit any industrial plant and observe the activities of any person whose job is such that he moves from place to place during his work. Prepare a process chart—man analysis and a flow diagram of his activities.

(b) With the aid of the checklist, prepare a process chart—man analysis and a flow diagram for a better method.

(c) List the information that you would add to your checklist and for each item indicate the checklist question number and letter under which you would file it.

34.3. Same as Problem 34.2, but for the parcel-post counter at your local post office.

34.4. Same as Problem 34.2, but for the clerk in a large store.

35

NOTES ON INFORMATION FLOW ANALYSIS

Introduction

This chapter introduces a type of chart designed to assist in the analysis of more complex problems than were treated in Chapter 15. Three different methods of charting are given.

The Problem Situation

The subject of the charts is the following procedure taken from a large manufacturing plant.

Activity at the home office. When a purchase order is received in the home office in New York City, it is checked by a sales agent. After initialing, it is passed to the office manager at the next desk, who places it in the work-basket of one of four typists. After a sales order and duplicate are typed

575

from it, the typist places the purchase order and the two copies of the sales order in the out-basket, whose contents are periodically taken by the office manager. The office manager checks the typing and puts the original purchase order and original sales order into one out-basket for periodical filing by purchaser, and the duplicate sales order into another basket for eventual enveloping in a window envelope, which is sent to the proper manufacturing plant. One of these plants is in Gary, Indiana.

Activity at the Gary plant. At the Gary plant, the mail clerk routes such sales orders to the production order department. The production order clerk automatically types the information from the sales order onto a production order, making an original and three carbons, and assigns a production sequence number from the work log. The typing is self-proofed. A batch of the first copies and the original related sales orders are sent to the plant manager's office. At the end of each day, the plant manager's secretary checks these against a telegraphic summary of orders sent from New York. The first copy and the sales orders are filed in the plant manager's office against the chronological sequence on the telegraphic summary.

The original production orders are sent in batches via plant mail to the accounting office, where they are held for the completion of the order, being filed by date of receipt in the accounting office.

Copy 2 of the production order is sent, in batches, to the materials control section, where a typist automatically makes three copies of a stores issue order and files the production order by its production sequence number.

Copy 3 of the production order is sent, in batches, to the forge shop foreman, who holds it in an open (unorganized) file until copy 2 of the stores issue order arrives. The production order sequence numbers are matched and the mated production orders and stores orders are placed on the desk of the shop production control clerk.

After performing other duties, the foreman, once each day, checks the production control clerk's batch of production orders and stores issue orders against the copies of the production orders on the telegraphic summary in the plant manager's office.

Too much for previous simple forms. The forms previously used for charting would obviously be inadequate to chart the numerous people and forms involved in this procedure.

New Formats for Charting

A ruled sheet, 17 by 23 inches with lines, six to the inch, as shown in Figure 35.1, is commonly used to chart situations of this complexity, although a blank sheet may be used. However, the rulings are of great as-

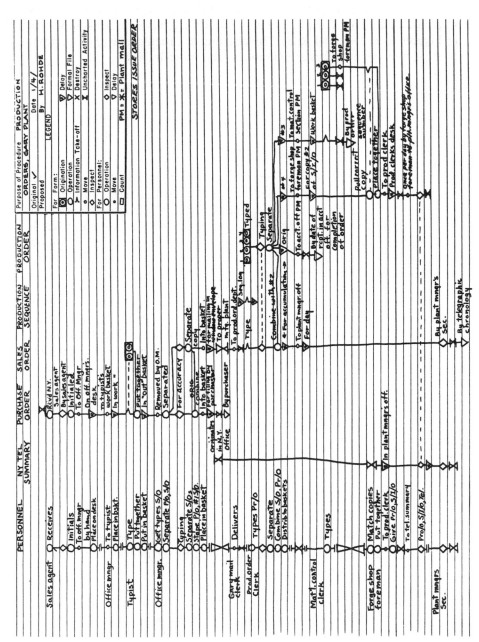

Figure 35.1. Process chart—combined analysis of production order procedure.

sistance in keeping actions that are occurring at the same time horizontally aligned on the chart. The procedure previously given is charted on this form in Figures 35.1, 35.2, and 35.3. (The charting has been confined to a corner of the form to permit a larger and more legible reproduction.)

Charting by forms with all human activity in one column. In Figure 35.1 the same general scheme as was used in Figure 15.3 has been followed, with all the activity for each form kept in a single column. A column for personnel activity has been added to show who performs each action and what tasks each person concerned has in respect to this procedure.

Charting flow by organizational division. In Figure 35.2, the personnel column has been deleted. The chart has been divided vertically to represent various divisions of the organization. The flow line for each form shows how and when the form moves from division to division. It also shows what is done with the forms in each division. Although it is a little harder to follow all the steps for an individual form, the function of the form is usually shown more clearly in a chart of this type.

Charting for organization and people involved. In Figure 35.3, personnel columns have been added to the general form of the chart used for Figure 35.2. Thus, this chart shows not only "who does what," as does Figure 35.1, but keeps clear the part played by each form in each division of the organization, as in Figure 35.3.

Advantages. The chart formats of Figures 35.2 and 35.3 generally become more useful as the procedure being analyzed becomes more complex. It is instructive to compare the information derivable from the purely verbal description of the process to that gained from study of Figures 35.1, 35.2, and 35.3.

Summary

It is with complex procedures that one may fully appreciate the part played by the graphic analysis technique in placing the information in a form that permits intelligent study.

The value of a chart, of the types shown, for designing a procedure prior to its inception should be obvious. With the aid of the chart, effective participation of all those concerned may be more easily gained because of the greater ease of indicating how their work relates to the total goal. The items in the checklist of Chapter 15 may be more fully employed to assist in developing an economical procedure. Indeed, it is hard to imagine how an effective procedure with many ramifications can be designed without some aid of this type.

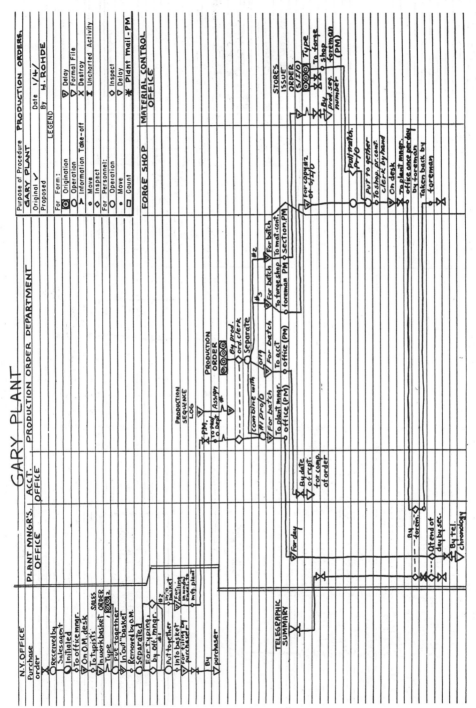

Figure 35.2. Process chart—combined analysis of production order procedure, columnated by divisions of company.

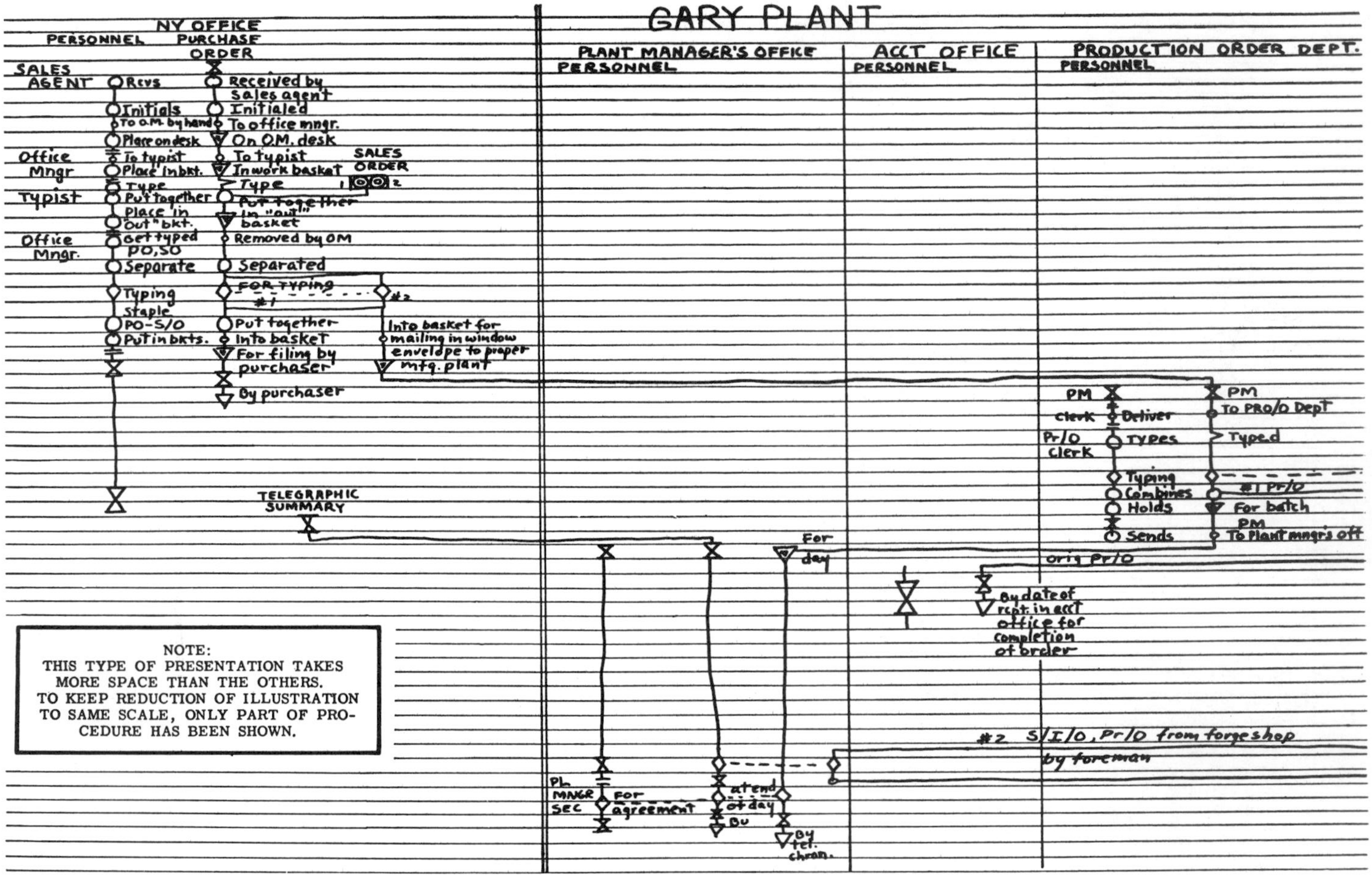

Figure 35.3. Process chart—combined analysis for production order procedure, columnated by divisions of company and with personnel columns.

PROBLEMS

35.1. Using the material of Problem 15.6:
- (a) Prepare a process chart—combined analysis for procurement of class B merchandise. (See Figures 35.1, 35.2, and 35.3 for formats.)
- (b) Using the same general charting pattern so as to make the two charts comparable, prepare a process chart—combined analysis for the procurement of class C merchandise.
- (c) Which is the better procedure? Justify your selection.
- (d) Under what conditions is it desirable to separate the merchandise into classes for separate ordering as this chain does? Explain your answer.
- (e) Prepare a proposed improved procedure, for each class, and present it as a process chart—combined analysis.
- (f) List the information you would add to your checklist and for each item indicate the checklist question number and letter under which you would file it.

35.2. (a) Follow a form used by an organization you are familiar with and prepare a process chart—combined analysis for the work involved.*
- (b) Prepare a process chart—combined analysis for a better method of performing the work.
- (c) Redesign the form or forms involved in this procedure if they require redesign.

  *Note: The instructor may, alternatively, assign students to examine such procedures as the following:
  1) A purchase requisition procedure.
  2) The procedure used to pay student help.
  3) A library procedure.
  4) Charge-account procedures in local stores.

35.3. Prepare an improved procedure to replace the one charted in Figures 35.1, 35.2, and 35.3. Indicate why you consider your proposal an improvement. (If you need any additional information, make a reasonable assumption, list your assumption, and proceed.)

36 NOTES ON OPERATION CHARTS

Introduction

This chapter deals with three subjects. The first is the method of charting an operation with subcycles. The second is a modified form of an operation chart used as an operator instruction sheet. Finally, additional material is provided with respect to the problem of adjusting jobs to the physical dimensions and the capability of people.

Work with Subcycles

The original method. Figure 36.1 is the original operation chart for drilling three holes in each of two ends of a hinge channel. The channel and two sets of holes drilled are shown in Figure 36.2. A sketch of the workplace is shown in Figure 36.3.

582

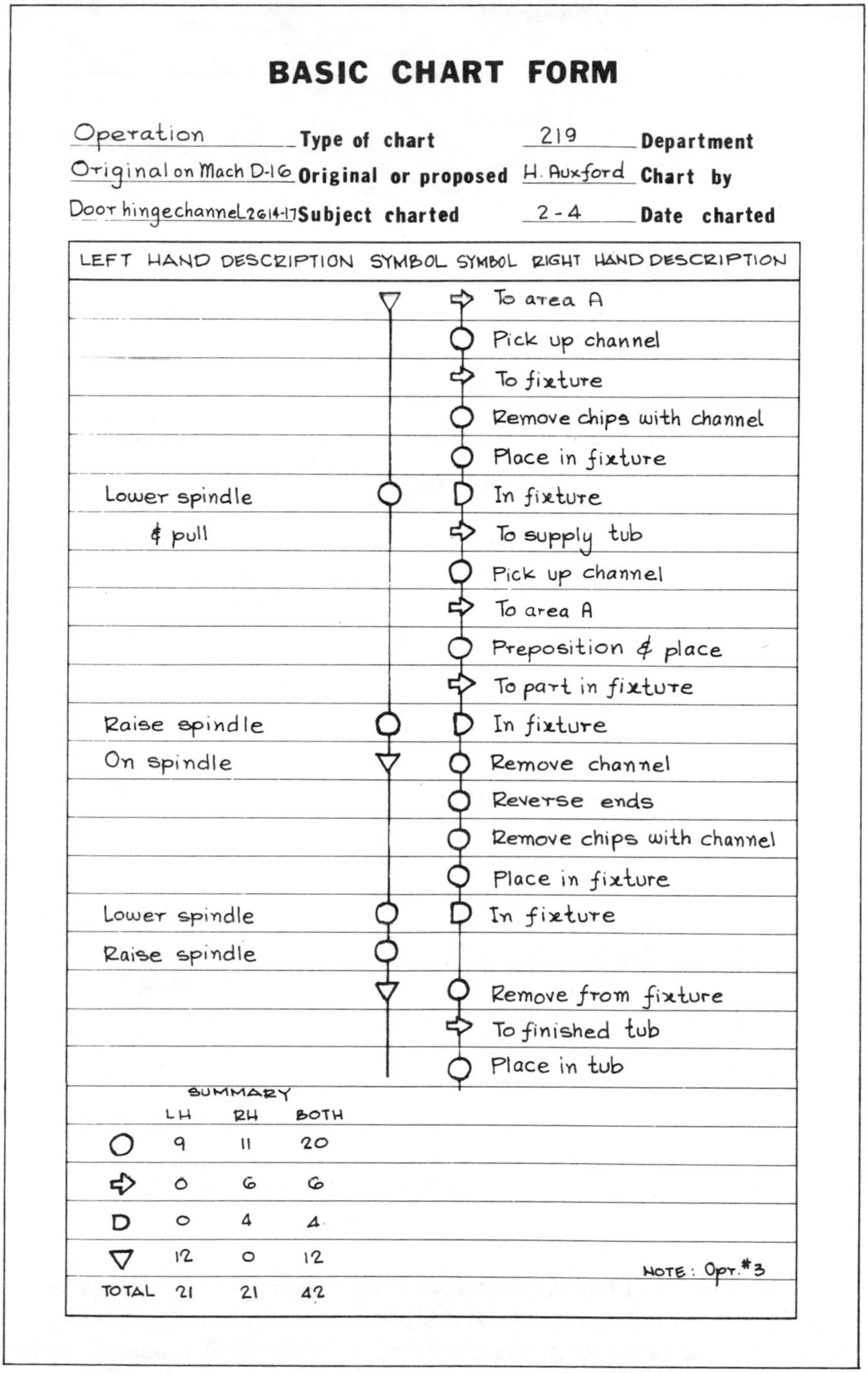

BASIC CHART FORM

Operation _____ Type of chart ___219___ Department

Original on Mach D-16 Original or proposed H. Auxford Chart by

Door hinge channel 2614-17 Subject charted ___2-4___ Date charted

LEFT HAND DESCRIPTION	SYMBOL	SYMBOL	RIGHT HAND DESCRIPTION
	▽	⇨	To area A
		○	Pick up channel
		⇨	To fixture
		○	Remove chips with channel
		○	Place in fixture
Lower spindle	○	D	In fixture
& pull		⇨	To supply tub
		○	Pick up channel
		⇨	To area A
		○	Preposition & place
		⇨	To part in fixture
Raise spindle	○	D	In fixture
On spindle	▽	○	Remove channel
		○	Reverse ends
		○	Remove chips with channel
		○	Place in fixture
Lower spindle	○	D	In fixture
Raise spindle	○		
	▽	○	Remove from fixture
		⇨	To finished tub
		○	Place in tub

SUMMARY

	LH	RH	BOTH
○	9	11	20
⇨	0	6	6
D	0	4	4
▽	12	0	12
TOTAL	21	21	42

NOTE: Opr.#3

Figure 36.1. Original operation chart for drill holes in hinge channel.

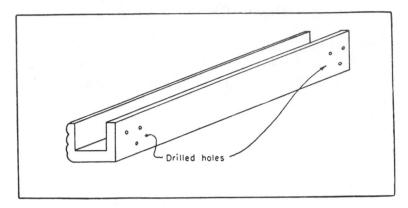

Figure 36.2. Hinge channel.

The improved method. The checking of the original operation chart with the checklist of Chapter 16, with particular reference to those questions aiding in the application of principles *B* and *C* (at the top of the checklist), resulted in the improved method shown in the operation chart in Figure 36.4. An air cylinder was attached to the drill press, an air jet was used to blow away chips, and the cyclic nature of the task was changed. A sketch of the workplace is shown in Figure 36.5.

The new method has subcycles. Particular attention is called to the method of charting used in Figure 36.4, which shows a task with several subcycles performed with different frequency.

The effect of subcycles. It should be noted that in this case the introduction of subcycles improved the task and was partly responsible for the increase in production of 138 per cent. In other cases we have seen, as in Chapter 34, Case I, Sacking Fish-Counter Ice, the elimination of subcycles was desirable. Hence, no general rule can be given with respect to the desirability of introducing or eliminating subcycles.

Operation Charts Modified for Use as Instruction Sheets

Many plants, with highly repetitive work, design their operations before they begin to design workplaces and tools. The workplaces and tooling are subsequently designed to facilitate the desired methods. A modified version of an operation chart describing the method is eventually fastened to each

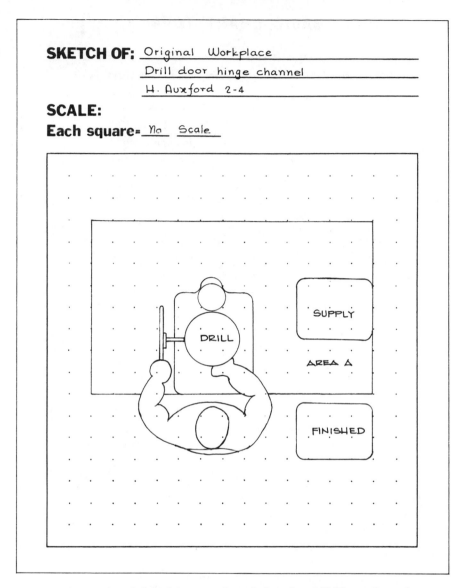

SKETCH OF: Original Workplace

Drill door hinge channel

H. Auxford 2-4

SCALE:

Each square= No Scale

SUPPLY

DRILL

AREA A

FINISHED

Figure 36.3. Original layout of workplace for drill hinge channel.

workplace to serve as an operator instruction sheet. A gross breakdown (see the footnote to Table 16-I) is used. Symbols are not employed inasmuch as "flagging" the types of steps for improvement is not required for this use. The type of chart used is illustrated in Figure 36.6.

BASIC CHART FORM

Operation _____ **Type of chart**   219 _____ **Department**

Proposed for Mach D-16 **Original or proposed**   H. Auxford **Chart by**

Door hinge channel 2614-17 **Subject charted**   2-4- _____ **Date charted**

QUAN.	LEFT HAND DESCRIPTION	SYMBOL	SYMBOL	RIGHT HAND DESCRIPTION
1/PC. 20 Times	For right hand	▽	⇨	To supply tub
	To channel in fixture	⇨	○	Pick up channel
	Remove from fixture	○	⇨	To fixture
	To table	⇨	○	Place in fixture
	Turn end & place	○	D	Piece in fixture (Foot advances drill)
1/20 pcs. 1 Time	To area A	⇨	⇨	To area A
	Help R.H.	○	○	Pick up 20 channels
	To area B	⇨	⇨	To area B
	Help R.H.	○	○	Place 20 channels
1/PC. 20 Times	For R.H.	▽	⇨	To area B
	To channel in fixture	⇨	○	Pick up channel
	Remove from fixture	○	⇨	To fixture
	To "finished" tub	⇨	○	Place in fixture
	Place in tub	○	D	Piece in fixture (Foot advances drill)

NOTE: Opr. #3

SUMMARY & RECAP PER PIECE

		Improved			Original			Saved	
	LH	RH	BOTH	LH	RH	BOTH	LH	RH	BOTH
○	4.1	4.1	8.2			20			11.8
⇨	4.1	4.1	8.2			6			-2.2'
D	0	2	2.0			4			2.0
▽	2	0	2.0			12			10.0
TOTAL	10.2	10.2	20.4			42			21.6

Figure 36.4. Improved operation chart for drill hinge channel.

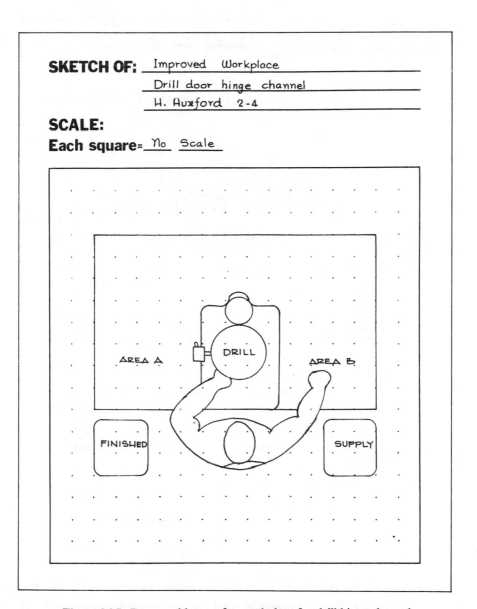

SKETCH OF: Improved Workplace
Drill door hinge channel
H. Auxford 2-4

SCALE:
Each square= No Scale

AREA A   DRILL   AREA B

FINISHED   SUPPLY

Figure 36.5. Improved layout for workplace for drill hinge channel.

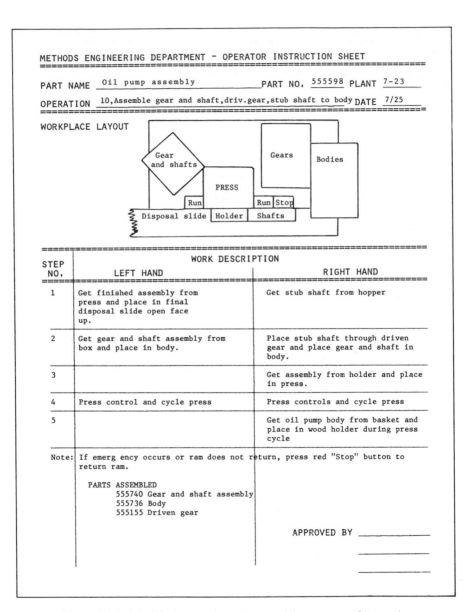

METHODS ENGINEERING DEPARTMENT - OPERATOR INSTRUCTION SHEET
==

PART NAME ___Oil pump assembly_____ PART NO. _555598_ PLANT _7-23_

OPERATION ___10,Assemble gear and shaft,driv.gear,stub shaft to body_ DATE _7/25_
==

WORKPLACE LAYOUT

Gear and shafts

Gears   Bodies

PRESS

Run

Run Stop

Disposal slide | Holder | Shafts

==

STEP NO.	WORK DESCRIPTION	
	LEFT HAND	RIGHT HAND
1	Get finished assembly from press and place in final disposal slide open face up.	Get stub shaft from hopper
2	Get gear and shaft assembly from box and place in body.	Place stub shaft through driven gear and place gear and shaft in body.
3		Get assembly from holder and place in press.
4	Press control and cycle press	Press controls and cycle press
5		Get oil pump body from basket and place in wood holder during press cycle

Note: If emergency occurs or ram does not return, press red "Stop" button to return ram.

        PARTS ASSEMBLED
            555740 Gear and shaft assembly
            555736 Body
            555155 Driven gear

                                APPROVED BY _____

                                            _____

                                            _____

Figure 36.6. Modified operation chart used for operator instruction sheet at workplace.

588

Adjusting Jobs to People

Areas of reach. In Figure 16.4, the dimensions of the normal work area were given on a horizontal plane. It should not be too difficult to translate the diagram shown into a three-dimensional concept in which the arcs of reach shown become hemispheres. If we think of the individual being free to turn, the arcs of reach become spheres of reach.

Use of restricted space. Where the space in which people work is severely limited, as in airplane cockpits or galleys, or in working spaces on naval vessels, plan-view templates of people may be used to see if they can fit into the space and do their work. Figure 36.7 shows such templates in place on a tentative layout of a working space on a naval vessel.

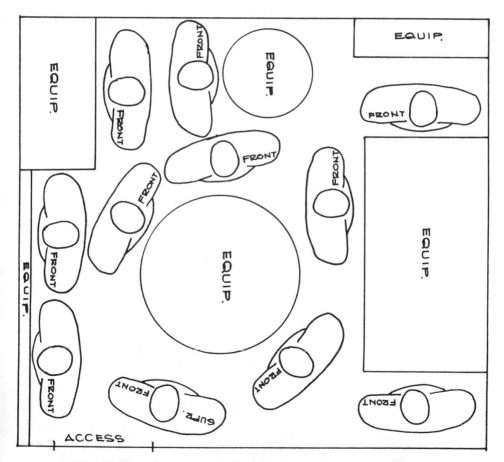

Figure 36.7. Layout of working space with human plan-view templates.

Sequence of reach and move body. Consideration of the areas of reach should include consideration of the sequence of reaches. Merely because a control object is within the normal work area does not mean that it is at the most convenient location. Where the worker reaches from, as well as where the worker will need to reach next, must also be considered.

Body moves must be considered in the same fashion as body member moves. Further, when areas such as the one shown in Figure 36.7 are under consideration, any movement of any one individual must be considered with respect to the activity of all the others in the compartment.

Lifting weight or overcoming force

An unprecedented entry of women into jobs formerly held by men has resulted from governmental regulations designed to eliminate discrimination in job placement based on a person's sex. Employers now find themselves faced with a major reevaluation of their workplace designs, job duties, physical-effort demands of the job, and time standards as a result of women entering jobs formerly held by men.[1]

The amount of force required of the worker, male or female, should be carefully reviewed against the limits given in Table 36-I.

Table 36-I. MAXIMUM WEIGHTS (1b) ACCEPTABLE TO VARIOUS PERCENTAGES OF THE MALE AND FEMALE POPULATIONS WHILE LIFTING A 19- BY $13\frac{1}{2}$- BY $5\frac{1}{2}$-INCH TOTE BOX*

		90%	75%	50%	25%	10%
Shoulder height to	Industrial men	29	39	49	59	68
arm reach	Industrial women	24	26	29	32	35
	Housewives	13	15	18	21	23
Knuckle height to	Industrial men	34	43	53	62	71
shoulder height	Industrial women	25	29	34	38	42
	Housewives	16	18	21	23	26
Floor level to	Industrial men	37	45	54	63	70
knuckle height	Industrial women	28	33	37	42	47
	Housewives	14	17	21	24	27

*S. Snook and V. Giriello, Maximum Weights and Work Load Acceptable to Female Workers, *Journal of Occupational Medicine*, Vol. 16, No. 8, 1974, pp. 527–534, as quoted in Purswell, op cit. For more detailed information, reference should be made to E. McCormick, *Human Factors in Engineering and Design*. New York: McGraw-Hill Book Company, 1976; A. Chapanis, *Ethnic Variables in Human Factors Engineering*. Baltimore, Md.: The Johns Hopkins University Press, 1975.

[1]J. L. Purswell, "Designing Jobs for Women: Productivity Implications," *Proceedings, American Institute of Industrial Engineers 1976 Systems Engineering Conference*, p. 108.

Other physical considerations. *Organoleptic* is a word with which the workplace designer should become familiar. It means "relating or making an impression upon one or more of the organs of special sense." While it is usually used with respect to the tasting of food, it is a handy term to refer to all the aspects of a job for which the worker must use his senses.

For instance, in one automobile parts supply plant, special problems were created by the testing of horns on the assembly line. In that the noise was excessive, an enclosure was built over part of the line to house the test employees, who wore ear protectors. The enclosure became a location of illicit smoking and loafing. Both these problems and the organoleptic problems were solved by removing the enclosure and furnishing each test station with a sound-absorbent box. During the test the horn was placed in the box, the lid closed, and the horn activated. A faint noise, detectable from the outside, indicated a properly noisy horn.

Problems of visual discrimination and tactile considerations are all part of workplace design. Posture must also be considered as well as articulation with respect to the use of tools.[2]

Behavioral considerations. In various cultures various jobs have had special stigma attached to them. Typical of such jobs were slaughtering animals for meat, tanning leather, and handling human manure. In modern cultures jobs that call for independant thinking tend to have a higher status than jobs which are purely repetitive.

In the view of the behavioral sciences, short-cycle repetitive jobs are "boring," although there is much debate about this, a debate that has been going on since the early 1900s. One fact which surfaced early was that "boring" was a subjective phenomenon; a job was boring only if so perceived by the worker.

Behavioral research has more recently been focused on "job enlargement" and "job enrichment," two terms for changing jobs into something other than short, repetitive cycle tasks. Social reaction to such changes seems to be based on personal reactions similar to the perception or lack of perception of boredom. Further, it is well recognized that long cycles take longer for the acquisition of skill than short cycles. With short-lived jobs, job enlargement may be counter to raising productivity.

In conclusion, it would appear that jobs should be designed with the specific workforce in mind, although their participation, which may elicit the "Hawthorne effect" described in Chapter 2, may well be more powerful than all other factors.

---

[2]See McCormick, op. cit.

Summary

Three additional aspects of jobs performed at one place have been considered. First, methods of charting work with subcycles were examined. Second, additional, necessary physical considerations were discussed. Third, behavioral aspects were briefly examined.

PROBLEMS

36.1. Prepare an instruction-sheet-type operation chart for the job charted in:
   (a) Figure 16.2.
   (b) Figures 16.5a and b.
   (c) Figure 36.1.
   (d) Figure 36.4.

36.2. Prepare a report on a reading (between assigned dates) from the periodical literature on:
   (a) Boredom.
   (b) Fatigue.
   (c) Job enlargement.
   (d) Job enrichment.
   (e) Job status.
   (f) Behavioral science.
   (g) Motivation.

36.3. Describe jobs wherein the lack of acuity with respect to hearing would not constitute a handicap.

36.4. Same as Problem 36.3, but with respect to vision.

36.5. Same as Problem 36.4, but with respect to lower body mobility.

36.6. Same as Problem 36.4, but with respect to finger dexterity.

37 NOTES ON MULTIPLE-ACTIVITY CHARTS

Introduction

This chapter deals with the conversion of a man and machine time chart into either a multiman and machine time chart or a man and multimachine time chart. The conditions under which such charts are useful will be examined.

Multiman and Machine Charts

A one-man-and-machine time chart for making, polishing, wiping, and packing pots was shown in Figure 17.6. In order to increase production, a crew size change may be contemplated.

593

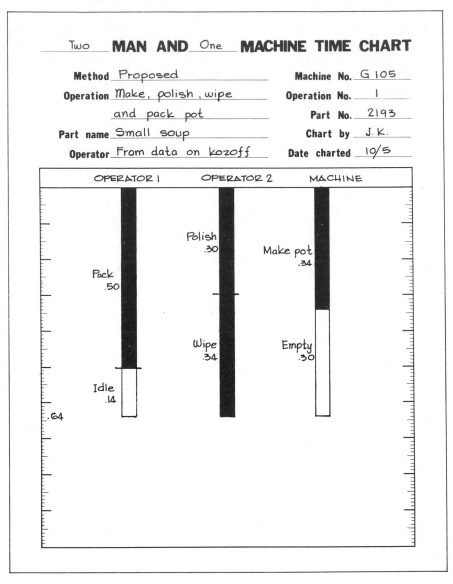

Figure 37.1. Multiman and machine process time chart for two-person crew on polish, wipe, and pack pots.

Charting the alternatives. Figure 37.1 is a two-person multiman and machine process time chart, and Figure 37.2 is a three-person multiman and machine process time chart for the same job: making, polishing, wiping,

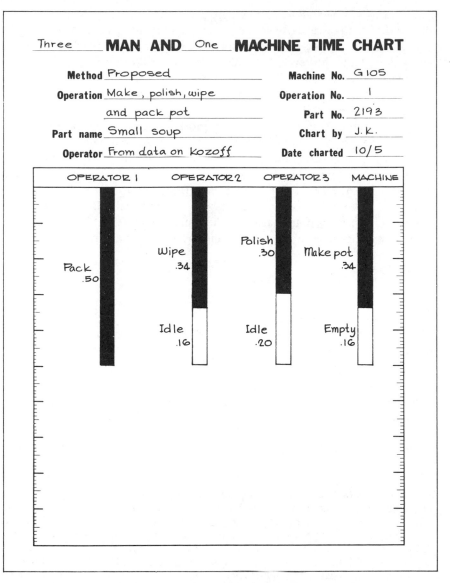

Three   **MAN AND** One **MACHINE TIME CHART**

Method Proposed

Operation Make, polish, wipe

and pack pot

Part name Small soup

Operator From data on Kozoff

Machine No. G 105

Operation No. 1

Part No. 2193

Chart by J. K.

Date charted 10/5

OPERATOR 1   OPERATOR 2   OPERATOR 3   MACHINE

Pack .50

Wipe .34

Polish .30

Make pot .34

Idle .16

Idle .20

Empty .16

Figure 37.2. Multiman and machine process time chart for three-person crew on polish, wipe, and pack pots.

and packing pots. The method of charting is that used on the one-man chart. Attention is called to the manner in which the machine and operator time chargeable to each pot vary from method to method.

Picking the best method. The most desirable solution is, of course, a function of the pot capacity of the plant, the amount of labor available, the demand for the product, and the cost relationships between machine time and labor time. Note, however, that this technique makes possible a good estimate of the effects of different-size crews, and that it also provides the material for instructing the crew in the proper distribution of work, once the crew size is determined.

The solution that would be found most desirable in the case would be only for one type of pot and one set of values of the economic variables. All these may change from time to time. Also, since the making, polishing, wiping, and packing time are all functions of the size and shape of the pot, and so forth, the best solution will vary from pot to pot as well as with changes in the basic economic factors previously listed. However, the man and machine process time charts and the multiman and machine process time charts provide an easy means of comparing and evaluating the possible solutions.

The Use of the Man and Machine Process Time Chart for the Analysis of a Man and Multimachine Assignment

A more complex situation arises when one man must operate several machines and the job involves moving from machine to machine as well as working at each machine. Jobs of this type are usually most usefully analyzed with a time chart.[1] Here it is usually desirable to analyze the worker as a single unit to avoid confusing the data by overdetailing. Hence, it is custom-

---

[1] In some cases a mathematical model may be evolved. See W. D. Jones, "Mathematical and Experimental Calculation of Machine Interference Time," *The Research Engineer* (Georgia Institute of Technology), 1949, pp. 9–10; R. N. Fetter, "The Assignment of Operators to Service Automatic Machines," *Journal of Industrial Engineering*, Vol. 6, No. 5, 1955, pp. 22–30; R. F. Lomicka and J. M. Allderige, "Mathematical Man–Machine Analysis," *Journal of Industrial Engineering*, Vol. 8, No. 3, 1957, pp. 157–164. (These last two articles have additional bibliographies.) See also K. R. Baker, "Priority Dispatching in the Single Channel Queue with Sequence-Dependent Setups," *Journal of Industrial Engineering*, Vol. 19, No. 4, 1968, pp. 203–206; C. R. Glassey, "An Algorithm for a Machine Loading Problem," *Journal of Industrial Engineering*, Vol. 18, No. 10, 1967, pp. 584–588. A single facility with a multiplicity of user problem is described in J. Goldman and H. A. Knappenberger, "Simulation of Operating Room Scheduling Policies," *Proceedings, 19th Annual Institute, American Institute of Industrial Engineers*, 1968, pp. 148–152.

ary to use one column for the operator, with a process chart type of break-down, and one for each machine.

Different objectives are possible. There can be four different objectives with problems involving the proper number of machines that an operator should run:

1. To arrange the method and number of machines so as to reduce the operator's delays to the minimum required for rest and personal time. There may be considerable machine delay.

2. To arrange the method and number of machines so that the machine delay is reduced to the minimum required to provide the operator with rest and personal time, during which the machine may be unattended. There may be considerable other operator delay time.

3. To arrange the method and number of machines so that the combined operator idle time and machine idle time give the lowest cost of operation.

4. To arrange the method and number of machines so that both operator and machine idle time are reduced to the minimum required to provide the operator with rest and personal time.

Selecting an alternative. Type 4, although usually desirable, is seldom feasible. The problem in most cases therefore becomes a matter of deciding which of the other types of solutions is the most desirable in view of existing conditions. If the plant has excess capacity, it is often best to keep actual costs to a minimum by reducing operator delays to a minimum—solution 1—although this may mean considerable machine delays. If the machine overhead is high and capacity is a limiting factor, an arrangement that keeps the machines busy but makes, perhaps, poor use of the operator's time may be desirable—solution 2. If the direct labor cost and machine overhead rates are somewhat similar, the third type of sloution may be the same as one of the first two, although it may shift from one to the other as the cost picture changes. With complex cycles, it may be distinctly different from any other solution, and quite desirable, if minimum cost without maximum possible production from the equipment is a tolerable situation.

Using mathematical models. If a mathematical model of the man–machine situation is created, it may be used to find any of these solutions. A mathematical model permits direct determination of the optimum arrange-ment after the nature of "optimum" is stated. The first three objectives which have been stated may be described mathematically. The fourth is a fortuitous situation. It should be noted that the logical procedure stated in Chapter 3 will still be followed, although the step of analysis and the subsequent steps will be performed with a mathematical model as opposed to a schematic model. However, the schematic model or man and machine time chart may well be an important step in creating the mathematical model; further, in many cases, the schematic model is sufficient to permit the development of a satisfactory solution.

Common characteristics of good solutions. In all solutions, the most efficient integration of the man work cycle into the machine work cycle is usually desirable. The process chart—man analysis checklist may aid in obtaining this. Preparation of loads and performance of inspection during machine running time, as in the case of the grinding of engine bearings (Figure 17.4), are also usually attributes of efficient man and multimachine setups. Easier methods of performing the hand part of the job done at each workplace, through class 1 or 2 changes, are usually helpful, particularly when they affect what is done during machine downtime.

Charts and mathematical models compliment each other. However, changes of the preceding, more detailed types of improvements are usually achieved through further analysis by other techniques of various steps in the sequence. This last statement should be carefully considered. Changes of this type are not suggested from a mathematical model of the situation, yet they may have a vital bearing on the final effectiveness of the man–machine arrangement. The analyst who proceeds to the more sophisticated techniques of mathematical models should ascertain that the model represents the embodiment of all reasonable class 1 and 2 changes and that the work of creating the model is compatible with the economics of the problem. Waiting-line or queueing problems involving machine servicing depend upon accurate values for "time to service." Certainly, all improvements possible through class 1 or moderate class 2 changes should be taken into account. In many cases, the economics of class 2 changes that affect the "time to service" may be more fully evaluated with the aid of a queueing model. The two types of techniques, schematic and mathematical, complement each other and should not be thought of as alternatives.

An example. Figure 37.3 is a man and two-machine process time chart for the job of edge-turn piston rings on lathe. Two machines are used for maximum machine output. A typical work cycle is charted. Note that considerable operator delay is present, but that the machines are kept busy all the time except when they are being unloaded or loaded. The only means of increasing production from these machines, other than higher feeds or speeds,

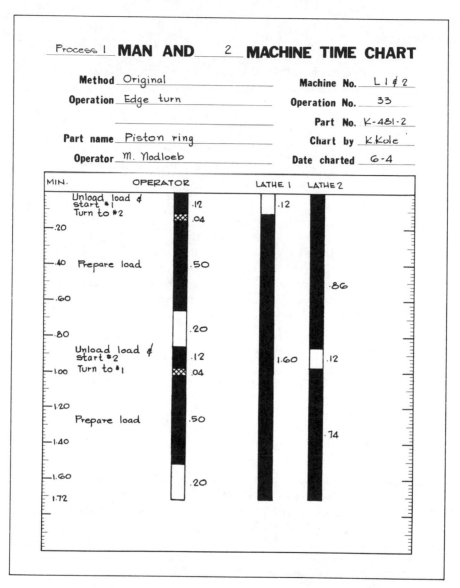

Figure 37.3. Man and two-machine process time chart for edge-turn piston rings.

is through a facilitation of the machine unloading and loading. A better detailed operation chart of this step in the work sequence would aid in the search for a better method.

Summary

The man and machine process time chart and the multiman and machine process time chart are used to determine:

1. Where methods changes achieved with the aid of other analysis techniques would be of value.
2. What, in view of existing production requirements and cost relationships, is the most economical man and machine work pattern to follow?

PROBLEMS

37.1. Use the data of Figures 37.1 and 37.2. Construct a graph using the ordinate for total cost per pot. Use the abcissa for machine hour rate, covering the range from $1 to $20 per hour by 50-cent increments. Plot the cost curves for each of the following costs of direct labor:
(a) $1.00 per hour.
(b) $2.50 per hour.
(c) $3.50 per hour.
(d) $4.50 per hour.
(e) $5.50 per hour.
(f) $6.50 per hour.
(g) $7.50 per hour.
(h) $8.50 per hour.
(i) $9.50 per hour.
(j) $10.00 per hour.

37.2. Create a table for selecting the lowest cost arrangement for each combination of machine hour and labor hour costs.

37.3. Draw a one-man and machine process time chart for the job shown in Figure 37.3. Use the data from this new chart and Figure 37.3. Use the ordinate for the total cost per load of piston rings. Use the abscissa for the machine hour rate, covering the range from $1 to $50 in $1 increments. Plot the cost curves for each of the following costs of direct labor:
(a) $1.00 per hour.
(b) $2.50 per hour.
(c) $3.50 per hour.
(d) $4.50 per hour.
(e) $5.50 per hour.
(f) $6.50 per hour.
(g) $7.50 per hour.
(h) $8.50 per hour.

(i) $9.50 per hour.

(j) $10.00 per hour.

37.4. Create a table for selecting the lowest-cost arrangement for each combination of machine hour and labor costs.

37.5. If a robot machine server costs $500, at what points of machine hour and labor hour costs does the robot become economical in:

(a) Problem 37.1?

(b) Problem 37.3?

38

NOTES ON MICROMOTION AND MEMOMOTION ANALYSIS

Introduction

To keep Chapter 19 to a reasonable size, the details of the mechanics of filming or taping were not given there; they appear in this chapter. Also, in keeping with the concept of these chapters, referred to as notes, additional examples of memomotion study are given so as to provide an expanded view of its use. Three examples are provided. The first concerns a three-person and machine activity in a large steel foundry. The second concerns maid service in a hotel. The final example concerns an extensive study made of drugstore managers.

The Mechanics of Using Films and Video Tapes

A description of the phases of recording and transcribing, using film or video-tape recording, follows. The description is general; no particular level of detail and no particular graphic presentation is predicated.

Equipment for filming. The following equipment is needed for making motion-picture films:

1. The camera should be for 16-mm or Super-8 film. A fast, wide-angle lens should be available, e.g., $f1.2$, 12 mm for the 16-mm camera or better. The camera should be electric-motor-driven. Reliable speeds of 60, 100, 1,000, and 1,440 frames per minute should be available.

2. A tripod is highly desirable. A tripod helps get better pictures for little extra money. It does away with jerkiness often caused by unsteady hands. If a trpiod is used, it should have a pan-and-tilt head.

3. An exposure meter is a necessity. Any of the electronic exposure meters presently available are suitable for assisting in determining the proper exposure. A regular exposure meter, rather than one specially set up for cine work, is to be preferred because of the odd exposure intervals used with memomotion speeds.

4. A timing device is optional unless the camera speeds are not truly reliable. Then the timing device becomes a necessity. The device may be a clock like the one shown in Figure 38.1, with 100 divisions on its

Figure 38.1. Microchronometer.

face and a large hand making 20 rpm or 10 rpm, and a small hand making 2 or 1 rpm, depending on the speed of the large hand. Since slow-motion pictures at 32 frames per second are sometimes necessary instead of the normal 16 frames per second, the faster clock is desirable. The clock is usually referred to as a microchronometer. If an electric-drive camera with a synchronous (constant-speed) motor is used, the microchronometer may be dispensed with, although it may be left in the picture in order to:

a. Be sure the workers know that time is recorded. Values surreptitiously obtained have no advantage and create mistrust.
b. Be proof of the continuousness of the film.
c. Identify, at any future time, the speed of picture taking.
d. Measure the length of any missing section if the film is inadvertently damaged in use.

Current practice attempts the use of motion pictures with as little disturbance of normal work conditions as possible. Since the placement of a timing device on the workplace can be a distraction, the use of these devices is decreasing.

Special lights are usually not required since the introduction of modern high-speed lenses and films. Indeed, modern lenses and films permit the making of good-quality motion pictures in almost any location where there is enough light for the worker to see what is being worked on. The use of high-intensity photographic lights in the factory is a source of great disturbance. The glare is not only annoying to the operator, but it makes the operator the focus of attention. Certainly, with special lighting, the conditions filmed are different from normal. The employment of available-light photography also reduces the setup time for picture taking. Equipment without lights greatly reduces the bulk that must be brought to the job, making it possible for one man to carry, set up, and employ the equipment.

Equipment for video-tape recording. Video-tape recording differs little from film recording. A special camera is used, and this camera is connected to a special tape recorder.[1] The equipment is handled in a manner similar to that used with film. A tripod is necessary. A monitor replaces the exposure meter; all the human problems remain the same.

Making the film record. Filming for data recording involves only six simple rules:

1. Obtain the cooperation of the operator and his foreman. In a union shop it may be worthwhile obtaining the cooperation of the steward. No one should have anything to fear from pictures, since they are merely a more reliable means of studying what would otherwise be visually studied with less accuracy. When filming large crews, subsequent analysis of the film may be much easier if the workers are issued inexpensive T-shirts with large numbers on the front and back. If so, adequate explanations for the shirts should be made well beforehand and all questions answered.

---

[1] I have not given make or model numbers of suggested equipment because of the extreme rate of change and improvement taking place in commercially available equipment.

2. Place the camera as close to the action of the job as possible, but be sure the view of the camera includes all the activities wanted for analysis. Set the lens focus and check the view through the finder. It should be borne in mind that the films are to gather data and not to represent an artistic effort; they should be taken with as little fuss as possible.

3. Select the speed of picture taking and set the camera speed control or motor drive accordingly.

4. Get the best exposure possible. Quality films are not a requisite, but better exposures make the analysis easier. With modern high-speed films, surprisingly little light is required. Exposure instructions accompany cameras, film, and exposure meters. Basically, the length of exposure is controlled by the number of frames per second, in a manner explained in the instruction book accompanying the camera. Usually, 8 frames per second give an exposure of $\frac{1}{15}$ second, 16 frames, an exposure of $\frac{1}{30}$ second, etc. Only a few cameras are at all adjustable in this respect, and the adjustment usually permits cutting down the exposure but not increasing it. The exposure meter has a calculator attached to it that takes into account:

a. The exposure interval.

b. The sensitivity of the film, which is given either on the film carton or on a folder in the film carton.

c. The amount of light on the subject. (Pictures for data should usually be exposed for the darkest part of the picture *in which detail is wanted*.) The use of available light is usually accompanied by more uniform lighting than with photographic lighting; this simplifies the exposure problem.

   The calculator on the exposure meter is used to determine the correct lens aperture for any conditon of these three variables.

5. If you use a microchronometer, make sure it is in the picture as much as possible. (It is preferable that it be in all the time; otherwise, the counter on the projector will have to be used to interpolate.) However, the use of a synchronous motor drive without a microchronometer means one less feature of the workplace to fuss with. This contributes to faster filming.

6. Make a record of exposure data and pertinent job information. This record permits discovery of the reasons for poor exposures, if they occur, and the application of corrective measures on future films. It also provides a place, identifiable against the film, to record important job information that does not appear on the film. An example of such a record appeared as Figure 19.1.

Making the video-tape recording. The procedure for video-tape recording is the same as filming except that different taping speeds are not used.

The equivalent of the effect of different camera speeds is imposed via the analysis equipment.

Film analysis. The films must be processed after they are taken. Inasmuch as motion-picture film processing is available on a one-day or same-day basis in most major cities, the film is rapidly available for analysis. Plants located in cities without processing service can usually arrange for mail service from the nearest processor; a three-day turnaround time is usually the maximum. (When time was of the essence, the author has obtained 45-minute service.)

A projector used for film analysis must have specific features somewhat different from the conventional projector. The optical system must have heat filtering sufficient to permit prolonged examination of single frames. A frame-by-frame advance of the film, forward or backward, must be convenient; the projector should also have a built-in frame counter. All of the commercially available analysis projectors have been specially designed for motion and time study work, and each offers specific advantages.

Data used with any of the analysis methods described in the preceding chapters may be taken from the film. Obtaining the exact relationship between the two hands or the members of a crew is much easier with film than with actual observation, since the film can be stopped and the action held still from step to step during the analysis. Each frame may be individually examined and notes made of the method and the time for each step. The analysis is usually made with a portable shadowbox so that normal office lighting may be maintained.

Tape analysis. Video-recorder tape is analyzed in a manner similar to motion-picture film but, instead of a projector, a closed-circuit television set is used. The tape controls must give a "stop-motion" capability and an ability to select positions for "stop-motion" at desired time intervals, as well as the ability to hold such pictures on the picture tube. All such sets provide the feasibility to change the brightness and the contrast of the image at will over a wide range; this is a feature which is superior to the motion-picture camera–film system. All the required equipment is commercially available.

Three-Man Crew Running Heavy Castings Through a Rotoblast

Introduction. A possibility guide had indicated that a class 1 or 2 change was desirable, and a study of the process chart—product analysis suggested that the operation, *shot-blast casting in Rotoblast*, was necessary. This completed step 1, aim, of the logical approach.

A camera was used to record the information. For convenience and accuracy, step 2, analysis, was performed with a memomotion camera.[2] Sufficient light from the building monitor penetrated even the murky foundry interior to do away with the need for supplementary lighting. The crew, aware that they were being photographed, were filmed at work.

Analysis. An analysis was made of the film in order to obtain data with which to construct a multiman and machine process time chart. The analysis was made with a projector equipped with a frame counter. A typical cycle was selected from those available on the film. Since the selected cycle was 439 seconds long (7.3 minutes) and there are 40 frames per foot on 16-mm film, the selected section was only 11 feet long. At normal speeds the cycle would have occupied almost 200 feet of film. The reduction in analysis time by using the memomotion film should be obvious. The selected cycle was identified for future examination by placing an ink spot on the beginning frame. The film was analyzed by noting the beginning counter reading, together with a process chart—man analysis symbol and a description for one operator, on the first line of a ruled sheet, and indexing the projector until the first frame showing a new activity appeared; thereupon noting again the counter reading, process chart—man analysis symbol, description, and so forth. This procedure continued until all the activities of the operator were detailed for the whole cycle. The film was then returned to the starting point and the second operator's activities analyzed. In this case, the film was studied four times, once for each worker and once for the machine.

Charting. Sections of the chart of the original method are shown in Figure 38.2. The sections on the chart marked "overblast" could not be determined from the film, but were found with the aid of experimental data on the minimum required shot-blasting time. In drawing the chart, these experimental data were used to determine unnecessary blasting, which was labeled "overblast." The chart could have been made without film, but the use of film assured obtaining data on all men from the same cycle, was quicker, gave greater accuracy, and provided a record to use in checking the analysis. It also helped in discussions with foremen and workers, because they needed but a short time away from their job to review a rather lengthy period of work.

Improvement. The proper multiman checklist was used on this foundry chart, and the new method, parts of which are shown in Figure 38.3, resulted. The improved method saved 35 per cent of the direct labor time per unit while increasing hourly production by 2 per cent. The new method did not

---

[2]A synchronous motor-driven camera was used.

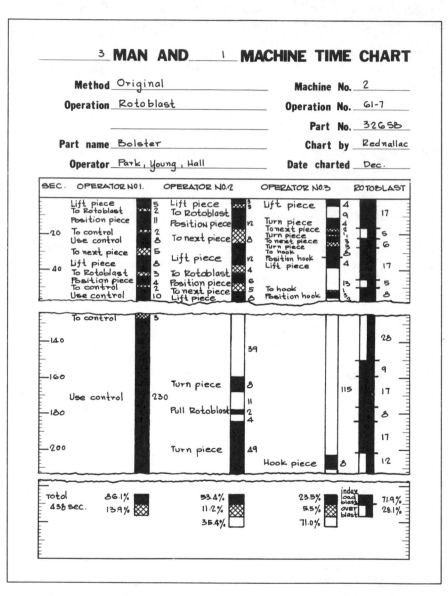

Figure 38.2. Sections of chart showing original method of Roto-blast.

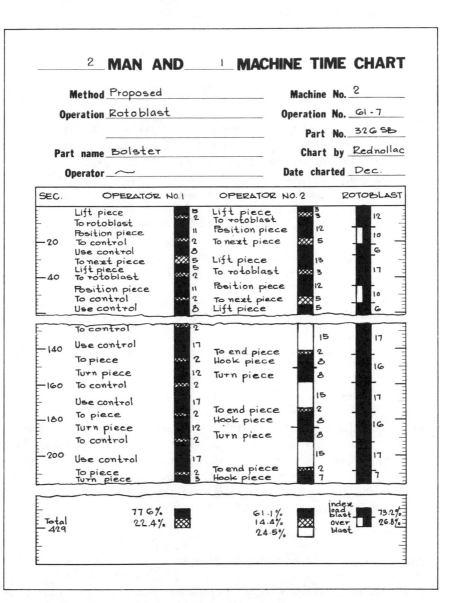

Figure 38.3. Sections of chart showing proposed method of Roto-blast.

require the crew to work faster, but made for a better distribution of the work and provided a more effective work pattern.

The memomotion film was later used in the discussions prior to the successful installation of the new method.

Maid Service in a Hotel

Introduction. A large Chicago hotel experienced difficulty in getting its maids to clean and make up a sufficient number of rooms; hence, costs were excessive. Visual recording and timing of such a task was extremely difficult, since the routine varied from maid to maid and from room to room.

The first step in reducing costs on this operation involved setting up reasonable quality requirements (class 4 changes) and adequate scheduling (class 3 changes). To obtain information from which to develop class 1 and 2 changes, a memomotion study was made of the maids' activities. The area covered in the work was large, but no one spot permitted photographing from a fixed location. Consequently, a portable setup was used with the camera mounted on a holder similar to a portable TV camera rig, which enabled the observer to carry the camera and motor drive and follow a maid.

Analysis. The analysis of this film permitted the flow path of the work to be studied, as well as the time for and nature of each step. Part of the graphic presentation of the method most frequently observed is shown in Figure 38.4, together with part of the flow path, which was also taken from the film.

Improvement. Discussion of this analysis by the hotel staff and study of the method with the checklist for process chart—man analysis led to the development of a proposed method that promised to save 34 per cent of the time. Part of this proposed method is shown in graphic form in Figure 38.5.

A trial application proved successful. Subsequently, a training manual and a training film were prepared, and the maids, in small groups, were trained in the new method and properly equipped. The memomotion film was not used in training.

Pharmacist, Work Activity Analysis

Introduction. Work activity analysis was discussed in Chapter 8. A simple example was given. In more complex cases, the task of self-recording can readily become excessive. In this case, information was desired with respect to how the pharmacist managers of drugstores used their work time.

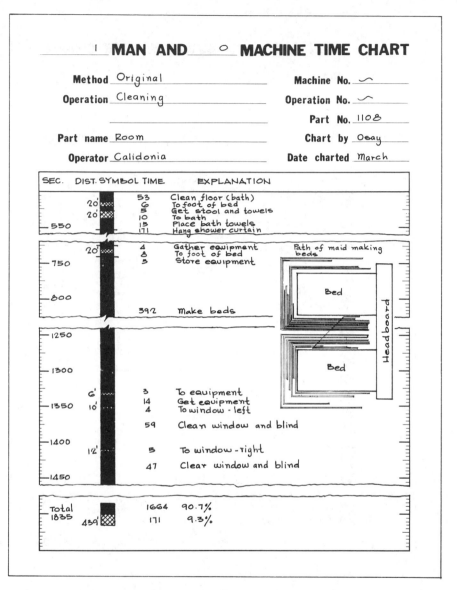

Figure 38.4. Sections of chart showing original method of maid cleaning hotel room.

Memomotion study was an ideal technique for studying this activity in that the sequence of activity was not predictable in advance. Visual recording of the events while they took place would have been extremely difficult and of doubtful accuracy. Second, using this technique the activities of the phar-

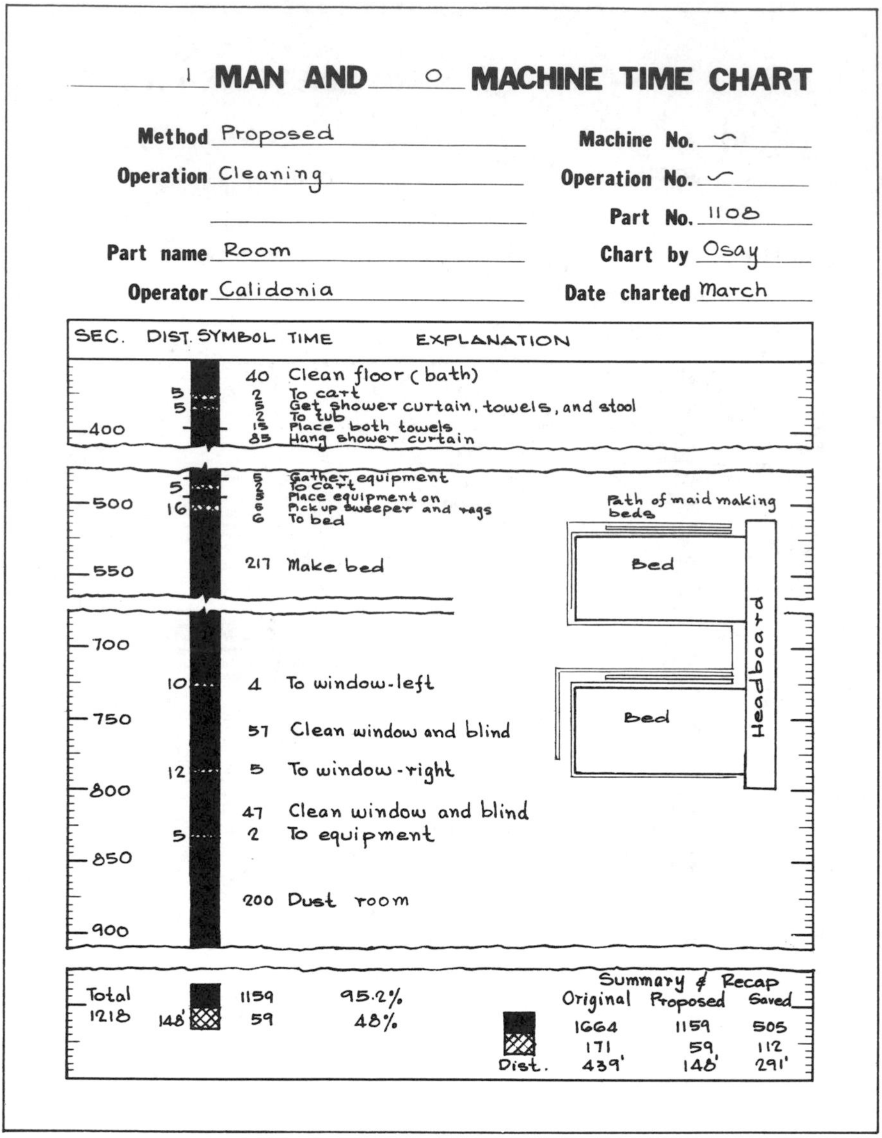

Figure 38.5. Sections of chart showing proposed method of maid cleaning hotel room.

macists in three drugstores were studied over a period running from 7:00 A.M. to 2:00 A.M. the next morning. Approximately 1,900 feet of 16-mm film were used in each store instead of the 27,010 feet that would have been required at normal film speeds, yet every second of the pharmacists' time could be grouped into activity categories that meaningfully explained their use of time. Third, a "bird's-eye" view of the activities could be gained from a rapid review of the film. Fourth, the method of analysis could be determined from a study of the film, permitting study at leisure.

The manner of using the memomotion film was predicated upon the definition of *analysis* given earlier in this book: *to break the work down into subdivisions or steps, pertinent to the job, appropriate to its scope, possessing known characteristics, or concerning whose performance information is already available.* The purpose in this study was to improve the performance of the job in terms of time without lessening the quality of the work. Therefore, in this case, it should be apparent that the steps into which the work was to be analyzed had to separate each part of the work related to a different type of accomplishment (output-oriented), as well as separating each type of man activity.

Preliminary analysis of the film. A preliminary study of the part of the film dealing with the preparation of prescriptions indicated that the pharmacist worked with the following:

1. Ingredients.
2. Tools (scales, liquid measures, etc.).
3. Finished prescriptions.
4. Labels and prescription orders.
5. Wrappings.
6. Cash.
7. Customers.

The basic information on human activity indicated that in working with the items listed a person would:

1. Work at one place.
2. Scrutinize or inspect.
3. Move from place to place.
4. Talk to customers.

Creating categories for analysis. Compounding these two groups of categories into a single one, together with a cursory review of the film, yielded the following categories into which the work shown on the film was then analyzed (the column on the left shows the mnemonic symbol used to facilitate recording):

WRL Work putting items back on shelves.
WQ Work with liquid measures.
WCA Work at cash register.

WD   Work on drugs (pulverizing, taking out of jar, etc.).

WLA Work applying labels.

WB   Work with balance and accessories.

WL   Work on labels or prescription blanks.

WR   Work wrapping prescriptions.

WA   Work putting material into prescription containers or work with prescription containers.

WC   Work counting items.

WG   Work getting down items (from shelves).

WCP Work compounding.

IC   Inspection of drug containers or contents.

IP   Inspection of prescription blanks.

IS   Inspection of shelves.

TR   Travel to and from cash register.

TL   Travel to shelves or cupboards to put material away.

TE   Travel to shelves or cupboards for material.

VC   Talk to customers.

Although this is not the only set of categories into which this work could be divided, any basis of analysis, considering the particular objectives of this study and the nature of the work, would closely resemble the list given here, although minor differences are possible.[3]

Making the analysis. After the development of this list of categories, large accounting-type work sheets were prepared. The first column was designated for a short description of the work observed, the second column for the film-analysis projector counter reading corresponding to the first frame of film of that activity, a column for the subtracted time, and a column for the symbol assigning the activity into one of the previously determined categories. To facilitate tabulation and checking, the rest of the columns of the work sheet were used to extend the subtracted times into columns for individual categories. A section of such a sheet is shown in Figure 38.6.

One of the first results of this analysis was the table shown in Chapter 18, resembling a work activity analysis summary. This table was extremely valuable in suggesting improvements in the task.

---

[3]It should be noted that a list somewhat different from the one given here was used when the analysis began. As the film analysis progressed, the list was modified in view of the details actually encountered on the film, and the list given here was the one finally used. Such a change is almost impossible if one is recording directly from observation of the actual activity. Even where the analysts have considerable experience, they usually do not anticipate all activities that will be encountered; film is of great assistance in allowing the backtracking necessary to adjust to the new groupings.

DESCRIPTION	COUNTER	TIME	SYMBOL	WRL	WQ	WCA	WD	WLA	WB	WL
Sheet 3 of 9 – Pharmacist – Film C–1–17–W										
Take phenobarb back to shelf	79307	3	TL							
Put phenobarb on shelf	79310	2	WRL	2						
To elixir of m.	79312	5	TE							
Take elixir off shelf	79317	3	WG							
Take elixir to bench	79320	3	TE							
Measure reqd. amount	79323	17	WQ		17					
Put elixir away	79340	4	TL							
Put elixir on shelf	79344	3	WRL	3						
Back to bench	79347	3	TL							
Pour elixir into Rx	79350	10	WA							
Type label, number label of Rx	79360	122	WL							122
Paste label on bottle	79482	15	WLA					15		
Wrap bottle	79497	137	WR							
To cash reg. for customer	79634	11	TR							

Figure 38.6. Section of memomotion analysis sheet of film of pharmacist's activity.

It is worth noting that prescription activity occupied only 28 per cent of the drugstore manager's time. The part of the study discussed to this point was only a subproject within the main study concerning 100 per cent of the time.

Purpose of the study. The main study required a determination of what a drugstore manager did during the day to permit:

1. The development of suggestions concerning alterations in the activity pattern to lead to more profitable operation.
2. The determination of typical duties, as practiced, in order to assist in:
a. Selecting potential managers with respect to job requirements.
b. Advising schools supplying supporting curricula of actual job characteristics.
3. An analysis of the effect of store layout and merchandise arrangement.

In order to achieve these objectives, a second group of categories was employed in the analysis of the memomotion film. It appeared to the analysts that the desirable major groupings would result from a compounding of the accounting divisions of the business and the basic human activities. Consequently, the separable parts of the work were divided into the following human activities:

—O   Do work at one place.
—T   Move from place to place.
—I   Inspect.

—U   Unavoidable delay.
—A   Avoidable delay.
—V   Talk to people.

These activities were further classified as attributable to one of the following: selling, administration, customer relations, phone, stock control, fountain and lunch counter, prescriptions, and personal. The data from an analysis of three stores and their managers are given in Figure 38.7.

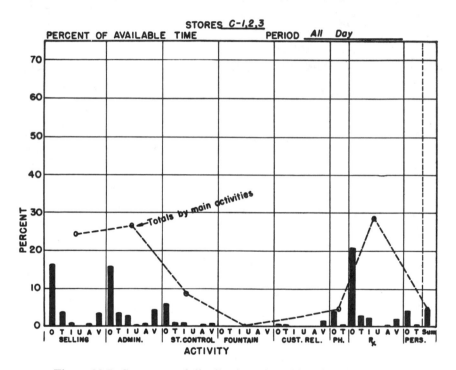

Figure 38.7. Summary of distribution of working time for pharmacist–managers of three drug stores.

Summary

Equipment lists for filming, taping, and making analyses were given. They were not extensive. A complete set can be assembled for less than the cost of one direct labor-year. Hence, the equipment should readily be cost-effective.

Memomotion was shown to have a wide range of applicability, not only to a variety of method problems but as a basic data-gathering approach.

PROBLEMS

38.1. In the summary, the filming or taping equipment was referred to as "cost-effective." Elaborate on this concept.

38.2. Assemble, from appropriate catalogs, a suitable list of equipment for filming and analysis. Identify makers and model numbers. Indicate the current price of each piece of equipment and the total price.

38.3. Assemble, from appropriate catalogs, a suitable list of equipment for video taping and analysis. Identify makers and model numbers. Indicate the current price of each piece of equipment and the total price.

38.4. With reference to Figure 38.7, suggest:
   (a) Desirable characteristics for managers of drugstores.
   (b) Necessary fields of knowledge.

38.5. Devise a set of categories for making a memomotion-based, work-time distribution study of:
   (a) A bank teller.
   (b) A bank loan officer.
   (c) A student counselor.
   (d) A meat cutter in a supermarket.
   (e) A group of typists.
   (f) A hospital nurse.
   (g) A clerk in the womens' shoe section of a department store.

38.6. In each of the circumstances above, comment on the relative desirability of the use of memomotion, work sampling, or self-reporting (as in Chapter 8).

38.7. It is desired to take memomotion pictures in the following locations:
   (a) The panchromatic film-packaging room of a film factory.
   (b) The darkened interior of a radar room.
   Describe and support your conclusion with respect to feasibility or infeasibility.

39

NOTES ON DIRECT TIME STUDY— INTENSIVE SAMPLING

Introduction

In Chapter 20, in which direct time study—intensive sampling was introduced, the procedure was described as having five steps: (1) the defining of standard time, (2) the recording of the standard method, (3) the timing of a performance, (4) rating, and (5) adding allowances. The notes in this chapter concern the last four steps.

Combining the Recording of the Details of Method with Timing

Using film or taping for recording. Film or tapes may be used as part of a direct time study—intensive sampling in two ways. Both procedures offer the advantage of providing an objective record, which may be reviewed later by anyone, including the worker, and which is not "interpreted data" as with a stopwatch record. Different element breakdowns may be made at any time, as the need arises. In addition, more smaller elements may be timed, a decided advantage, as will be seen later. Also, a whole crew may be timed simultaneously. The necessary data may be recorded on the film or tape more rapidly and more completely than with stopwatches. Indeed, most of the advantages given for photographic or electronic aids over visual method study techniques apply here also.

It is particularly suggested that disputed rates subject to a grievance or arbitration procedure may be most ideally studied with the full-camera technique as a more accurate and reliable record, since the objectiveness of the process safeguards the interests of both management and labor, if they are both concerned with accuracy, as they should be. Furthermore, rates set with these data may be discussed more factually and many grievances may be avoided. The two procedures are:

1. Partial use of camera (motion picture or video recorder).
2. Full use of camera.

With either method, as with stopwatch studies, the worker should be aware that a time study is being made. Also, all the usual data taken prior to a stopwatch time study should be recorded except the motion pattern, which may later be taken from the film or tape. No microchronometer need be used, but the worker should be fully aware that time is being recorded.

Partial use of camera. In this method, the camera is used only at the one-per-second speed or at the 100-frames-per-minute speed; hence, the camera is used only to record method and time.[1] However, once set up, it does this automatically, leaving the observer free to devote his attention fully to the task, and provides a much better opportunity to formulate a rating than when using a stopwatch. This method, however, only begins to make use of the full potentialities of the camera.

The film, when developed and ready for analysis, is put into an analysis projector with a frame counter, and, using a time study sheet, the data (which would have been recorded while the job was observed if a stopwatch was used) is entered in the blanks in a similar fashion—counter readings instead of stopwatch values being used, and an initial reading instead of the implied 0 of the stopwatch being made at the beginning of the first element. This is placed above the first *R* box. Otherwise, the procedures are identical. However, the record may be made at the analyst's own pace, small elements, down to 1 or 2 seconds identified and recorded, and the film later reanalyzed and the elements changed if this is found desirable. The data of the film are permanent and accurate. All calculations are made in a fashion identical with

---

[1] As was noted in Chapter 14, a video-tape record may be used in an equivalent fashion.

stopwatch calculations. A short section of memomotion time study film and its relationship to the time study sheet are shown in Figure 39.1 for the job of "tube cutoff." This illustration also includes the data related to the rating and the allowances.

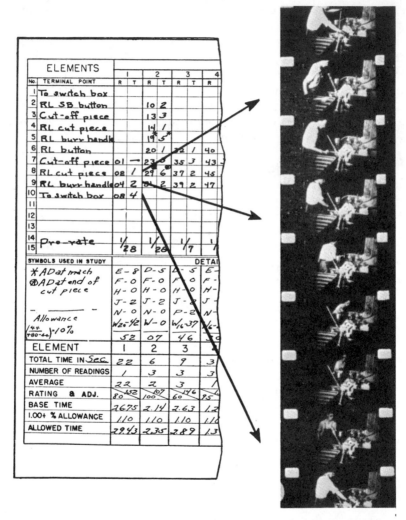

Figure 39.1. Short section of memomotion film and time study recordings made from it for operation of tube cutoff.

Studying jobs with large crews. A 25-man molding crew at the American Steel Foundries plant at Granite City, Illinois, was time-studied with the aid of memomotion camera timing. Because of the extreme area involved, the men were given numbered jerseys to wear so as to increase the ease of identifying them on the films. The gang was filmed in six sections as indicated in Figure 39.2. The films were taken at 100 frames per minute in order to obtain values similar to those obtained by stopwatch methods. The advantages of using films rather than direct observations have been summarized by C. H. Walcher, Works Manager, American Steel Foundries, Granite City, Ill., as follows[2]:

> The pictures were taken as a result of a grievance. The pictures gave us a very adequate coverage and enabled us to show all concerned parties how the rate was determined. Because the time study men were not involved in recording elemental break-downs, much more attention was given to pace determination, which resulted in better leveling factors. The pictures proved that the present rate on this job was not tight. All men on the unit were covered in one day, which represents a tremendous reduction of in-shop time to obtin the necessary information.

L. Randolph, the Chief Industrial Engineer at the plant, reports that the pictures taken have been used continually as reference points for methods descriptions. They have also, he reports, been used for method instruction with the old as well as with new employees.

Full use of camera.   This is the same as the partial procedure except that either during the study or immediately after it, the gear shift on the camera drive is used to obtain some footage at 1,000 frames per minute so as to record a reviewable version of the pace exhibited by the operator during the study.[3] This will be used for rating. Provided that care is taken to see that what is filmed at 1,000 frames per minute is similar in pace to what was filmed at 1 per second, this procedure should aid time study work as follows:

1. Allow others to check the rating.
2. Provide a record of the pace.
3. Facilitate actual rating.
4. Provide material for discussion rather than grievances.

An example of the full use of the camera.   The Meat and Poultry Inspection Program of the U.S. Department of Agriculture uses the full-film technique to set standard times for all the in-plant slaughter inspection outputs produced by the service. When new standards are needed, a small team from the central staff office in Washington, D.C., visits selected, representative slaughter establishments in all sections of the United States. They make a

---

[2]Letter to author.

[3]When an extensive series of time studies are being taken on film, some plants have found it economical to employ a second camera to do the 1,000-frame-per-minute sampling.

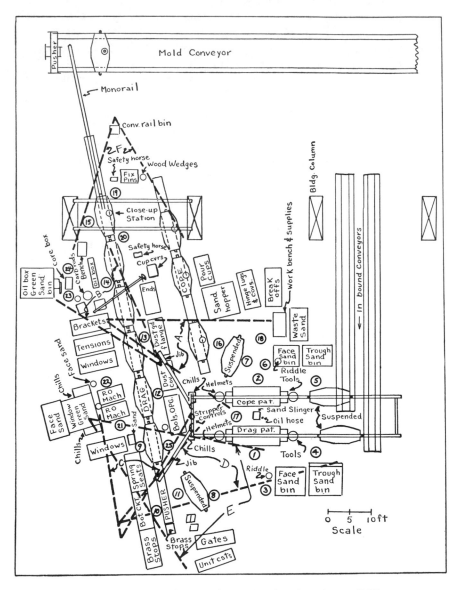

Figure 39.2. Layout of mold floor showing locations of 25-man crew (small numbered circles) and six memomotion camera locations and views (broken angles A–F), for time study timing. (Courtesy of L. Randolph, Chief Industrial Engineer, American Steel Foundries, Granite City, Ill.)

film record (at 1,000 frames per minute, because of the rapidity of the operations) of selected inspectors. The inspectors are selected from among those who perform the work correctly and maintain adequate standards of inspection.[4] The use of the full-film technique gives the following advantages:

1. Field time and the expenses associated with field time are reduced to a minimum. This more than offsets the cost of the film.

2. The methods can be reviewed in detail by all staff members concerned, to make certain that standards are set on effective and adequate methods of inspection. As a by-product, method changes may be introduced.

3. The determination of the rating is a joint effort of the entire staff. The film constitutes a record of the pace of the worker and is available for review at any time.

The standard times set for slaughter inspection are used in budget formulation for manpower resource determination and for the detailed allocation of manpower to slaughter plants serviced by the program.

A More Rigorous Method for Computing the Required Number of Readings

Basic theory. This application of sampling theory is based on the assumption that chance or random causes control the variation from reading to reading for a given element in a time study. In most cases this is a tenable assumption.

Formula (A)[5] gives a measure of the variability of data about its average. The variability is represented by σ, the standard deviation, which is expressed as follows:

$$\sigma = \sqrt{\frac{\sum d^2}{N}} \qquad \text{(A)}$$

where

$d = X - \bar{X}$ computed for each reading of the element separately before squaring and then summing

X = individual readings of an element

$\bar{X}$ = mean or average of all readings of an element

Σ = sum of like items

N = number of readings of an element

---

[4]The original standards were set in a similar fashion.

[5]A very early suggestion of a somewhat similar mathematical procedure was suggested for use without a performance rating in E. B. Royer, "How Many Observations Are Necessary in Setting Wage-Incentive Standards?" *Personnel*, May 1937, pp. 137–139.

This equation may be expressed for use on calculators as

$$\sigma = \sqrt{\frac{\sum X^2}{N} - \left(\frac{\sum X}{N}\right)^2} = \frac{1}{N}\sqrt{N \sum X^2 - (\sum X)^2}$$

From this, assuming it to represent the variability of a huge group of similar readings or the parent population (a commonly tenable assumption), another measure, $\sigma_{\bar{x}}$, the standard error of the mean (or average), may be computed, by (B), which indicates the probable variability of the averages of groups of N values of X about the obtained $\bar{X}$.

$$\sigma_{\bar{x}} = \frac{\sigma}{\sqrt{N}} \tag{B}$$

The property of this last measure is such that 95 per cent of the probable values of $\bar{X}$ (average for the element) will lie within $\pm 2\sigma_{\bar{x}}$ of the true average.

Hence, if $2\sigma_{\bar{x}}$ is equal to or less than 5 per cent of $\bar{X}$, we may say that the chances are at least 95 out of 100 that our average for the element, to which the rating will be applied, is within ± 5 per cent of the true average representing the performance we observed. If the 10 per cent criterion is used, the above may be restated by the reader, using "10 per cent" in place of "5 per cent." As was explained earlier, neither of these would seem to be an unreasonable minimum criterion to apply to our time studies.

If these conditions are not met, we may work formula (B) backward, using the σ we first obtained, setting $2\sigma_{\bar{x}}$ equal to 5 per cent of $\bar{X}$, and solving for N', which will indicate the number of readings that will probably be needed.

Indeed, it is this last property that makes this test feasible, easy, convenient, and economical to use after some additional mathematical manipulation of the formulas.

Combining formulas (A) and (B), we may state that

$$\sigma_{\bar{x}} = \frac{(1/N)\sqrt{N \sum X^2 - (\sum X)^2}}{\sqrt{N'}}$$

and setting 5 per cent of $\bar{X}$ equal to $2\sigma_{\bar{x}}$, we get

$$0.05\bar{X} = \frac{\sum X}{20N} = 2\frac{(1/N)\sqrt{N \sum X^2 - (\sum X)^2}}{\sqrt{N'}}$$

$$\frac{\sum X}{20} = \frac{2\sqrt{N \sum X^2 - (\sum X)^2}}{\sqrt{N'}}$$

and

$$N' = \left[\frac{40\sqrt{N \sum X^2 - (\sum X)^2}}{\sum X} \right]^2 \qquad \text{(C)}$$

where N' is the required number of readings.

This equation may be handled easily, even by one who is not familiar with the mathematics of its derivation.

If the analyst prefers to set his limits as 95 chances out of 100 of being within ± 10 per cent, then

$$N' = \left[\frac{20\sqrt{N \sum X^2 - (\sum X)^2}}{\sum X} \right]^2 \qquad \text{(C')}$$

Use of the equations. Equations (C) and (C') are the important ones. A small electronic-chip calculator with one memory may be used to obtain a solution to these equations without obtaining intermediate answers. Small programmable calculators are also available, which merely require the entry of the raw data. If the time values are obtained with a device like the Datamyte, shown in Figure 20.5, the accompanying software can produce the necessary answers.

It should be noted in that the units in which the element times are recorded occur in both the numerator and denominator of the expressions (C) or (C'), the decimal place of whatever unit is used in timing may be disregarded, and the values handled as whole numbers, with the resulting greater convenience.

Examples. Equation (C) or (C') in practice may be solved for an element quickly and interpreted as follows, provided that there is not a consistent pattern to the element variation. (However, in all cases, at least 10 readings should be taken as a minimum sample.)

1. *5 per cent criterion* [*Equation (C)*]. If N' is equal to or less than the number of readings recorded, then the average for that element is probably (95 chances out of 100) within ± 5 per cent of the correct representative average. If N' is greater than the number of readings taken, the study does not meet our criterion of reliability, and a new study with N' or more readings of this element should be taken and rechecked.

2. *10 per cent criterion* [*Equation (C')*]. The same as above, but replace "5 per cent" with "10 per cent."

The two illustrations that follow show how the measure and criteria developed would be applied. For example, let us say our element readings were 6, 7, 6, 8, 7, 5, 6, 8, 7, 6, 7, 6, 6, and 7 (all in 0.01 minute). $\sum X^2 = 614$ and $\sum X = 92$; consequently, $N' = 24.9$. Thus, with the real N of 14 and 25 values required to give an average with a probability of 95 chances out of 100 of being within ± 5 per cent of the correct average, we must conclude that the readings for the element are inadequate. Hence, a new study with at least 25 readings should be taken and rechecked.

For another example, let us say our element readings on another element were 10, 11, 10, 12, 11, 10, 12, 13, 10, 12, 10, 11, 11, 12, 11 (all in 0.01 minute). $\sum X^2 = 1,850$ and $\sum X = 166$; consequently, $N' = 11$. Thus, the 15 readings that were obtained give an average that has a probability of 95 out of 100, or better, of being within ± 5 per cent of the correct average; hence, the average for the element is acceptable as representing the performance observed, and a rating may be applied.

A Procedure for Developing Skill at Time Study Rating[6]

Characteristics of rating. Research involving an analysis of the performance of trained time study observers using conventional rating procedures shows that they may perform these judgments with surprising results. A study by A. J. Keim[7] showed that a trained group of time study engineers (approximately 50) rating films[8] of 57 performances at different paces using the conventional, subjective approach rated 46 per cent of the performances with less that ± 10 per cent error and 54 per cent with more than ± 10 per cent error.[9] The correct values were obtained by using the group average, cor-

---

[6]The footnotes of this section refer to rather old publications. The reason is quite simple. These data were established at that time and have not since been challenged with newer studies.

[7]M. E. Mundel and A. J. Keim, Proceedings of Second Annual Purdue Motion and Time Study Work Session, *Purdue University, Div. Tech. Ext., Misc. Bull.*, April 1945.

[8]M. E. Mundel and L. Margolin, Report of the 4th Annual Purdue Motion and Time Study Work Session, *Purdue University, Div. Tech. Ext., Misc. Bull.*, 1948. This study showed that with trained time study observers, "Ratings are more consistent . . . and . . . more accurate when made from motion pictures than when made from the actual performance of the operator . . ." (p. 5). Consequently, the data presented here are from a situation where at least a reasonably fair situation was used. However, it should be noted that in both Keim's experiment and Lehrer's experiment (quoted later), the raters were not too familiar with the jobs being rated.

[9]A 10 per cent error was defined as a rating in error by 10 per cent of the value that should have been assigned. A pace of 75 (100 being standard) could thus be rated within ± 7.5 scale points and a pace of 140 within ± 14 scale points without exceeding the ± 10 per cent error as defined.

rected for concept of standard,[10] and adjusting for known relationships between certain of the 57 performances shown. (The 57 performances were really pace variants of only 10 jobs.) Further experiments by Lehrer, with more carefully constructed films in which practically no method variation appeared from pace to pace and with extremely careful adjustment of the data to a comparable base, gave similar results inasmuch as they indicated that with 31 time study engineers observing various paces of four jobs, 19 per cent of their ratings, made in this conventional fashion, were within ± 5 per cent error, 42 per cent within ± 10 per cent error, and 58 per cent exceeded ± 10 per cent error.[11]

On the other hand, R. G. Carson reports (with respect to ratings made by six time study men on 63 films over a 6-year period): "The standard deviation of independent ratings made by men trained together on operations with which they are familiar will be approximately 6.67 per cent. Differences between the standard deviations of different men are so small they can be attributed to sampling errors."[12]

Carson attributes this greater accuracy to the uniform, planned, periodic training given his group. Also, the men were all employed in one plant and were rating films of jobs in that plant. These conditions were not characteristic of the other studies. Carson's group placed 56 per cent of their ratings within ± 5 per cent error, 74 per cent within ± 7.5 per cent error, and 87 per cent within ± 10 per cent error.[13] However, information was not given concerning the range of performances presented to the group. A reduced range would reduce the apparent error of variability, particularly if the time study men being tested had some knowledge of the approximate range covered.

The effect of excessive variation in rating. While this accuracy is surprising, considering the difficulty of the judgment involved, such results can

---

[10]Standard, as defined, may vary from plant to plant. Depending on both the definition and the scale, the typical operator with typical performance was variously defined as 75 (60 base) or 120, 125, 130 or 150, and so forth (100 base). The values actually assigned by the time study observers were corrected to the numerical value they would have assigned had they all defined typical maximum performance of the typical operator as 130, so as to make all data comparable. In this manner, erroneous conclusions arising from the use of different units were avoided. However, real differences in the concept of what could physically be expected from an incentive operator may have made the performance of these raters appear somewhat less accurate than it may, in reality, have been.

[11]R. N. Lehrer, "Development and Evaluation of a Pace Scale for Time Study Rating," Ph.D. dissertation, Purdue University, June 1949.

[12]R. G. Carson, Jr., "Consistency in Speed Rating," *Journal of Industrial Engineering*, Vol. 5, No. 1, 1954, p. 17.

[13]Carson, op. cit.

easily give rise in practice to extremely difficult situations and have led to such remarks as the following, which can hardly be disregarded.

> Obviously, if after months of negotiations and possibly strikes at great financial sacrifice to both sides, a settlement has been reached involving a 10 per cent change in the basic rates, neither management nor labor is prepared to sacrifice its respective rights to the blind operations of a technique of questionable accuracy.
>
> The use of a time study technique to set production standards whose demonstrated inaccuracy may exceed this percentage can become the source of much controversy. Naturally the demand by either side to monopolize the function arouses the suspicion of the other. Thus the solution to the basic problem of the validity of existing time study practice lies at the very heart of satisfactory industrial relations.[14]

Suspicion arises because of the ease with which the entire scale may be shifted in the course of step 1. In management's hands, this would possibly offset on all new jobs any wage increase gained, and, in labor's hands, produce opposite results. Even if this suspicion does not arise, the variation in the relative ease or difficulty of the standards possible with such procedures creates other problems. True, if the variation is random, the labor force as a group stands to gain as much as it loses, but the individual "inequities" create new pressures in the shop. It must be realized that the successful labor leader is committed to the removal of these in-group inequities almost as firmly as to the removal of what the group as a whole may feel are group "inequities."

In unorganized shops the situation is not greatly different, although grievances may not lead as rapidly to formal expressions of dissatisfaction.

Rating requires skill. It should be obvious that no matter what method of rating is used, skill at rating must be developed. To ascertain whether skill is being developed or maintained, periodic testing is necessary.

Starting a training program. A 16-mm film was made, in a plant, showing 10 paces of performing an actual job. The method used for all paces was identical. These paces were arranged in random order for presentation to the time study group. To make certain that the films were projected at the same speed at which they were taken, and thus keep random variations from being introduced by changes in projector speed, a constant-speed projector was used. The ratings were entered on forms by the 10 men. One such form with the data for one man is shown in Figure 39.3.

The ratings for the 10 men were collected and tabulated on the form shown in Figure 39.4. The sum and then the average of the ratings were

---

[14]W. Gomberg, *A Trade Union Analysis of Time Study*, 2nd ed. Englewood Cliffs, N.J.: Prentice-Hall, Inc., 1955, p. 36.

RATING OF ACTIVITY COMPARISON SHEET

No. 1
Rater JL
Date July 8
Operation Punch Press

Sequence Number	Rating of Activity in Percent	
	Rater's	Actual
1	110	
2	55	
3	130	
4	95	
5	100	
6	65	
7	112	
8	105	
9	65	
10	85	
11		
12		
13		
14		
15		
16		
17		
18		
19		
20		

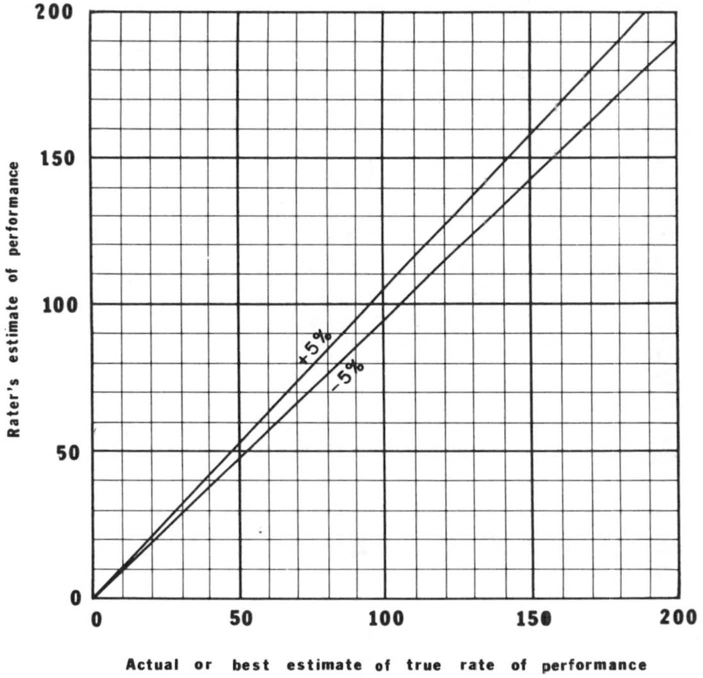

200
150
100
50
0
Rater's estimate of performance
+5%
-5%
0
50
100
150
200
Actual or best estimate of true rate of performance

Figure 39.3. Individual rating sheet with ratings of 10 films.

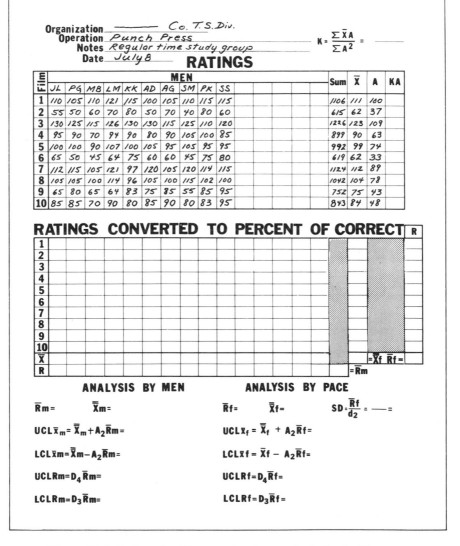

RATING ANALYSIS CALCULATION

Organization ——————— *Co. T.S.Div.*
Operation *Punch Press*
Notes *Regular time study group* $K = \dfrac{\Sigma \bar{X}A}{\Sigma A^2} =$ ———————
Date *July 8*   **RATINGS**

File	MEN														Sum	$\bar{X}$	A	KA
	JL	PG	MB	LM	KK	AD	AG	SM	PK	SS								
1	110	105	110	121	115	100	105	110	115	115					1106	111	100	
2	55	50	60	70	80	50	70	40	80	60					615	62	37	
3	130	125	115	126	130	130	115	125	110	120					1226	123	109	
4	95	90	70	94	90	80	90	105	100	85					899	90	63	
5	100	100	90	107	100	105	95	105	95	95					992	99	74	
6	65	50	45	64	75	60	60	45	75	80					619	62	33	
7	112	115	105	121	97	120	105	120	114	115					1124	112	89	
8	105	105	100	114	96	105	100	115	102	100					1042	104	78	
9	65	80	65	64	83	75	85	55	85	95					752	75	43	
10	85	85	70	90	80	85	90	80	83	95					843	84	48	

RATINGS CONVERTED TO PERCENT OF CORRECT

															R
1															
2															
3															
4															
5															
6															
7															
8															
9															
10															
$\bar{X}$															$=\bar{X}f$ $\bar{R}f=$
R														$=\bar{\bar{R}}m$	

ANALYSIS BY MEN   **ANALYSIS BY PACE**

$\bar{\bar{R}}m =$ $\bar{\bar{X}}m =$ $\bar{R}f =$ $\bar{\bar{X}}f =$ $SD = \dfrac{\bar{R}f}{d_2} =$ ——— =

$UCL\,\bar{x}_m = \bar{\bar{X}}_m + A_2\bar{R}m =$ $UCL\,x_f = \bar{\bar{X}}_f + A_2\bar{R}f =$

$LCL\,\bar{x}m = \bar{\bar{X}}m - A_2\bar{R}m =$ $LCL\,x f = \bar{\bar{X}}f - A_2\bar{R}f =$

$UCLRm = D_4\bar{R}m =$ $UCLRf = D_4\bar{R}f =$

$LCLRm = D_3\bar{R}m =$ $LCLRf = D_3\bar{R}f =$

Figure 39.4. Ratings for 10 men entered on analysis sheet with sums and averages computed and "arbitrary" ratings added.

computed for each film. Then an arbitrary set of ratings (column A) was entered on the sheet.[15] This arbitrary set was computed from the film frame count per cycle for the 10 films, each of which showed a different pace.

How the relative ratings for the ten scenes were established. The first pace (or film) was arbitrarily assigned a rating of 100. The product of the time per cycle at this pace, and a rating of 100, was computed. The ratings for the other paces were then assigned such that the cycle time for each pace multiplied by the assigned rating equaled the product obtained with the first film. These calculations produced a set of ratings correctly proportioned to the differences in paces but not representing any particular concept of standard.

Relating the relative ratings to the groups concept of standard. However, this series of ratings had subsequently to be multiplied by some constant, K, so that it would reflect the concept of normal held by the group.

This constant, K, was defined as the constant such that if the $\bar{X}$ values of ratings (group averages) were plotted against the KA values, the sum of the squares of the deviations about a 45° line would be at a minimum. Thus, the KA values would provide an ideal abscissa for graphic analysis of the ratings on forms such as Figure 39.3. Also, the KA values would represent a series of ratings, in correct proportion to the relative speeds of performance on the various films but representing the concept of normal held by the group. We might call the KA values "the best approximation of the correct ratings."

The derivation of this K is as follows. With 10 points, such as shown in Figure 39.5, the sum of the squares of the deviations, $\sum d^2$, is

$$\sum d^2 = d_1^2 + d_2^2 + \cdots + d_{10}^2$$

or

$$\sum d^2 = (\bar{X}_1 - KA_1)^2 + (\bar{X}_2 - KA_2)^2 + \cdots + (\bar{X}_{10} - KA_{10})^2$$

and

$$\sum d^2 = \bar{X}_1^2 - 2\bar{X}_1 KA_1 + K^2 A_1^2 + \bar{X}_2^2 - 2\bar{X}_2 KA_2 + K^2 A_2^2 + \cdots + \bar{X}_{10}^2 - 2\bar{X}_{10} KA_{10} + K^2 A_{10}^2$$

and setting the first derivative with respect to K equal to 0,

$$0 = -2\bar{X}_1 A_1 + 2KA_1^2 - 2\bar{X}_2 A_2 + 2KA_2^2 \cdots - 2\bar{X}_{10} A_{10} + 2KA_{10}^2$$

---

[15]If a standardized film had been used, this set of A values would be the standard, assigned ratings for the different scenes. Many such standardized films are available.

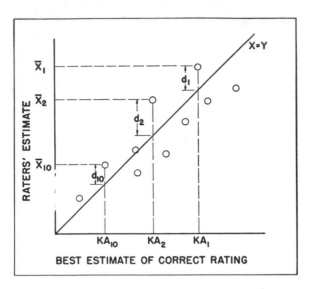

Figure 39.5. Plot of hypothetical ratings showing deviations used to derive K.

and solving for K,

$$K = \frac{\bar{X}_1 A_1 + \bar{X}_2 A_2 + \cdots + \bar{X}_{10} A_{10}}{A_1^2 + A_2^2 + \cdots + A_{10}^2}$$

or we may write

$$K = \frac{\sum \bar{X} A}{\sum A^2}$$

Equation (A) appears at the top of the sheet in Figure 39.4. The value of K is shown in the upper right-hand corner of Figure 39.6, and this K was used to compute the KA column shown in Figure 39.6. This column represents a set of ratings, properly proportioned to the true differences between the paces and such as to reflect the concept of normal held by the group.

The first critique of the skill at rating. It is worth noting that if the A values had been the KA values from another group of time study men, K would indicate the relationship between the two concepts of normal. Further, if the A column had been the KA values from an earlier set of ratings by the same group whose ratings are on the sheet, the K value would critique the stability of the concept of normal (provided, of course, that enough time had elapsed so that the correct ratings were forgotten). Films with known values which can be used as a real basis of comparison are available from a variety of sources.

RATING ANALYSIS CALCULATION

Organization ——— *Co. T.S. Div.*
Operation *Punch Press*
Notes *Regular time study group*
Date *July 8*

$$K = \frac{\Sigma \bar{X}A}{\Sigma A^2} = \frac{67180}{51942} = 1.29$$

RATINGS

Film	JL	PG	MB	LM	KK	AD	AG	SM	PK	SS				Sum	X̄	A	KA
1	110	105	110	121	115	100	105	110	115	115				1106	111	100	129
2	55	50	60	70	80	50	70	40	80	60				615	62	37	48
3	130	125	115	126	130	130	115	125	110	120				1226	123	109	141
4	95	90	70	94	90	80	90	105	100	85				899	90	63	82
5	100	100	90	107	100	105	95	105	95	95				992	99	74	96
6	65	50	45	64	75	60	60	45	75	80				619	62	33	43
7	112	115	105	121	97	120	105	120	114	115				1124	112	89	115
8	105	105	100	114	96	105	100	115	102	100				1042	104	78	101
9	65	80	65	64	83	75	85	55	85	95				752	75	43	55
10	85	85	70	90	80	85	90	80	83	95				843	84	48	62

RATINGS CONVERTED TO PERCENT OF CORRECT — R

Film															R
1	85	81	85	94	89	77	81	85	89	89				(86)	17
2	115	104	125	146	167	104	146	83	167	125				(127)	(84)
3	92	89	82	89	92	92	82	89	78	85				(87)	14
4	116	110	85	115	110	98	110	128	122	104				110	43
5	104	104	94	111	104	109	99	109	99	99				103	17
6	151	117	105	149	175	140	140	105	175	186				(144)	81
7	97	100	91	105	84	104	91	104	99	100				97	21
8	104	104	99	113	95	104	99	114	101	99				103	19
9	118	145	118	116	151	136	154	100	154	173				(136)	(73)
10	137	137	113	145	129	137	145	129	134	153				(135)	40
X̄	112	109	100	118	(120)	110	115	105	(122)	(121)				113 = X̄f	R̄f = 41
R	52	64	43	60	86	63	73	45	(97)	(103)				= R̄m	

ANALYSIS BY MEN

$R_m = 69$ $\bar{X}_m = 113$

$UCL\bar{X}_m = \bar{X}_m + A_2\bar{R}_m = 113 + .308 \times 69 = 134$

$LCL\bar{X}_m = \bar{X}_m - A_2\bar{R}_m = 113 - .308 \times 69 = 92$

$UCLR_m = D_4\bar{R}_m = 1.777 \times 69 = 122$

$LCLR_m = D_3\bar{R}_m = .223 \times 69 = 15$

ANALYSIS BY PACE

$\bar{R}_f = 41$ $\bar{X}_f = 113$ $SD = \frac{\bar{R}_f}{d_2} = \frac{41}{3.078} = 13.3$

$UCL\bar{X}_f = \bar{X}_f + A_2\bar{R}_f = 113 + .308 \times 41 = 126$

$LCL\bar{X}_f = \bar{X}_f - A_2\bar{R}_f = 113 - .308 \times 41 = 100$

$UCLR_f = D_4\bar{R}_f = 1.777 \times 41 = 72$

$LCLR_f = D_3\bar{R}_f = .223 \times 41 = 9$

Figure 39.6. Rating analysis sheet of Figure 39.4, completed.

More detailed statistical critique of the skill at rating. As the next step in the analysis of the quality of the ratings, the ratings made by each man, at each pace, and the group average for each film were converted to "per cent of correct (*KA*) rating" so as to obtain a common statistic, and these are shown in the bottom table of Figure 39.6. These values were treated by standard statistical quality control techniques as shown at the bottom of Figure 39.6.[16] In this analysis, UCL represents "upper control limit" and LCL represents "lower control limit." These are the maximum and minimum values reasonably expectable by chance alone,[17] considering the variability of the data. Values outside these limits suggest the existence of some assignable cause other than chance. Hence, a slow pace, rated higher on the average than the UCL for averages, suggests that the slow pace affected the rating error, and so forth.

The subscripts in the formulas are *m* for men and *f* for film, $\bar{x}$ for control limits of averages, and *R* for control limits for ranges; hence:

$UCL_{\bar{x}m}$ = upper control limit for men's averages
$LCL_{\bar{x}m}$ = lower control limit for men's averages
UCL_{Rm} = upper control limit for men's ranges
LCL_{Rm} = lower control limit for men's ranges
$UCL_{\bar{x}f}$ = upper control limit for group's average on a film
$LCL_{\bar{x}f}$ = lower control limit for group's average on a film
UCL_{Rf} = upper control limit for group's range on a film
LCL_{Rf} = lower control limit for group's range on a film
SD = standard deviation or limits such that:

$\bar{\bar{X}} \pm$ SD probably includes 68 per cent of all expectable ratings
$\bar{\bar{X}} \pm$ 2SD probably includes 95 per cent of all expectable ratings
$\bar{\bar{X}} \pm$ 3SD probably includes 99.7 per cent of all expectable ratings

"Out of control" and "probably out of control" values are marked with solid and broken circles, respectively. Examination of these out-of-control points in Figure 39.6 will reveal the following:

1. Different paces received a rating varying from the correct value by more than was ascribable to chance alone.
2. Slow paces resulted in loose standards.
3. Fast paces resulted in tight standards.

---

[16]The constants A_2, d_2, D_4, and D_3 are a function of sample size and may be found in any book on statistical quality control. In the case given here, the sample sizes by men and by films were the same size. This is not a necessity in making this analysis.

[17]The control limits are such that they should include approximately 99.7 per cent of all chance values.

4. The time study men tended to see all performances as too much alike.
5. Paces below 80 would frequently have been accepted but misrated.
6. Paces above 120 would frequently have been accepted but misrated.
7. The time study men agree with each other better than they agree with the facts.
8. The time study men vary in their concept of normal.
9. There is reason to suspect that some time study men are more variable than others.

A graphic analysis by individuals. The individual rating form shown partially filled out in Figure 39.3 is shown completed in Figure 39.7.

Graphing the group results. An $\bar{X}$ and R chart for the groups' ratings, by films, is shown in Figure 39.8. The group displays the same general faults as the individual plotted in Figure 39.7.

Periodic training and testing. The rating faults revealed by these analyses may be used as a guide to assist the time study group to be trained to rate more effectively. A periodic testing session of this type is almost a requisite if realistic ratings are to be obtained.

Graphic analyses of periodic training sessions. The individual $\bar{X}$ and R values from each session can be plotted on a chart similar to Figure 39.8 to show the progress of each individual. Likewise, an individual SD can be computed by taking a five-session, moving $\bar{R}$ and an appropriate d_2 so as to obtain a single figure representing individual proficiency. Such a tabulation appears in Figure 39.9. Many plants maintain data of these types as a means of assisting those charged with the responsibility for rating.

The Background of the "Difficulty Adjustments" Used in Objective Rating

In Chapter 20 it was indicated that the factors for which difficulty adjustments are added are:

1. Amount of body used.
2. Foot pedals.
3. Bimanualness.
4. Eye–hand coordination.
5. Handling or sensory requirements.
6. Weight handled or resistance encountered.
7. Other special constraints on work speed.

636

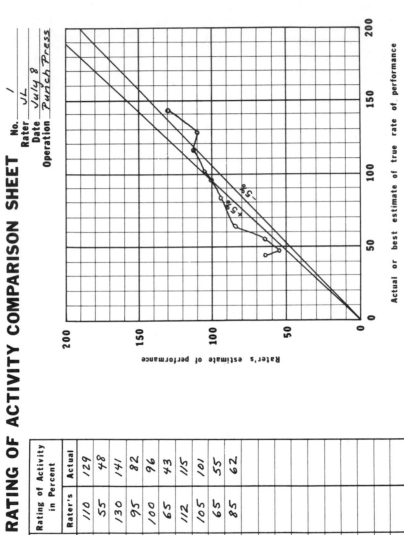

RATING OF ACTIVITY COMPARISON SHEET

No. _1_
Rater _JL_
Date _July 8_
Operation _Punch Press_

Sequence Number	Rating of Activity in Percent	
	Rater's	Actual
1	110	129
2	55	48
3	130	141
4	95	82
5	100	96
6	65	43
7	112	115
8	105	101
9	65	55
10	85	62
11		
12		
13		
14		
15		
16		
17		
18		
19		
20		

Figure 39.7. Individual rating sheet with *KA* values and plot of ratings.

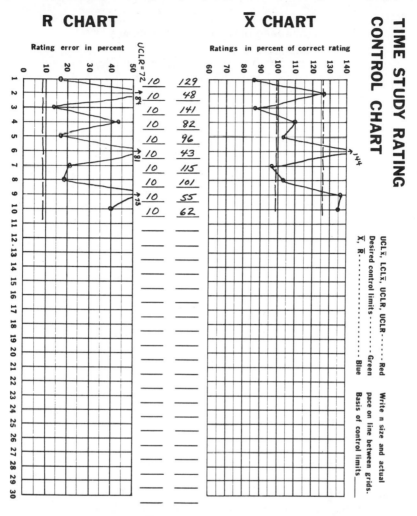

Figure 39.8. Group $\bar{X}$ and R charts of ratings, from Figure 39.6.

This seventh factor, which did not appear in the original list, has been added as the result of various situations encountered during the past 35 years' experience with extensive and varied applications of objective rating. (Many of the data sources are archival-type material. However, the data have not been challenged or replaced. For the sake of the serious practitioner or researcher, I have indicated the author and source rather than the detailed typical bibliographic footnote.)

A detailed discussion of each of these factors follows.

STANDARD DEVIATION FOR INDIVIDUALS						
NAME J. Jones						
Film No.	Date	Range	Sum of R's	Number of R's in sum	$\overline{R}$	S.D.
8	16 Dec 1958	36				
9	18 Dec 1958	45				
10	22 Dec 1958	59				
11	29 Dec 1958	40				
12	31 Dec 1958	33	213	5	42.6	13.8
13	5 Jan 1959	35	212	5	42.4	13.6
14	7 Jan 1959	Absent	167	4	41.8	13.6
15	9 Jan 1959	34	142	4	35.5	11.7
16	12 Jan 1959	41	143	4	35.8	11.7
17	14 Jan 1959	28	138	4	34.5	11.4

Figure 39.9. Record of standard deviations of ratings for an individual from successive sessions. (Adapted from U.S. Navy, Bureau of Ships, NAV-SHIPS 250–746, supplement.)

Amount of body used. Early research indicated that a 10-inch guided movement took more time when the movement was far enough away from the body to require a movement of the full arm about the shoulder than when sufficiently close to the body to permit movement primarily about the elbow.[18] The actual percentage difference in time obtained in early experiments could not be used due to certain features of the experiment.[19]

---

[18]Mundel and Barnes, University of Iowa.

[19]The operators were requested to work at a pace they thought they could keep up all day, and the full-arm movement and forearm movements were two of the four or more movements in the cycle. Besides having no check on the validity of the pace subjectively selected, there is no way of checking on the evenness or unevenness of beat possibly imposed by the subjects and thus reflected in the data in contrast to the imposed physical differences that were being evaluated.

Subsequent research was set up to evaluate separately the two types of movements.[20] Finger-actuated switches were arranged so that a series of 10-inch movements of the forearm, without wrist action, could be contrasted with a series of 10-inch full-arm movements, without wrist action. The finger switches were placed so as to prevent the operators from performing part of the movement with the wrist, inasmuch as this would reduce the amount of body used and eliminate part of the imposed differences. Actually, using a series of 10 college students, the full-arm motion (disregarding the time to actuate the switches) was found (with an accurate electrical kymograph for timing equipment) to take approximately 8 per cent longer than the forearm movement. However, it was suggested that the terminal therbligs, in practice, are usually a larger portion of the element than they were in this test situation; if they were taken as 50 per cent and assumed unchanged from forearm to forearm–upper arm movements, the obtained difference would have been 4 per cent. Actually, in the experiment, the terminal therbligs were performed more rapidly after the slower full-arm motion than after the forearm motion, possibly due to the greater overlap of finger reaction time with the arm movement. Increasing the actual terminal therblig values from the experimental data to a weight of approximately 50 per cent, the cycle would leave only a 1 per cent differential between forearm and full-arm movements.

Although the results were statistically significant, there is no indication that they were accurate to within 1 per cent of the data from a larger sample; consequently, the 3 per cent, which was the differential originally contained in the semiempirical table, Table 20-III, has not been changed.

Sekerci,[21] experimenting with weight lifting with two basic conditions, (1) lift with legs (as from floor) and, (2) lift with arms (as from bench), found that the time for the leg lift at low weights was 10 per cent greater than with the arm lifts.

The various effects of weight are discussed later in this section and are reflected in the table of adjustments for weight. The 10 per cent basic differential is most properly handled in the factor being discussed here and is included in the difficulty adjustments suggested for factor 1 and given in Table 20-III.

In certain industries special conditions, such as walking on slippery floors (18 per cent) or walking knee-deep in loose, granular material (38 per cent), may occur, necessitating the experimental determination of additional values for this difficulty category.

Foot pedals. Since the use of different parts of the arm changes the cycle time, it is not unreasonable to assume that this should apply in general

---

[20]Mundel and Radkins, Purdue University.

[21]Sekerci, Purdue University.

to the leg and foot as well. A study indicated that the operation of different foot pedals required different amounts of time, ranging at maximum pace from approximately 0.005 to 0.007 minute.[22] As the smallest time study element usually is 0.04 minute with stopwatches or 0.02 minute with memomotion time study, this indicates a maximum difference of 5 per cent (stopwatch) to 10 per cent (memomotion) in an element that could be attributed to the type of foot pedal. However, of five types of pedals studied, four showed a maximum range of difference of 0.0005 minute, which is approximately 1 to 2 per cent of the smallest common time study element. Realizing that this is about half of the usual element, we may deduce that the better pedals need no adjusting factor and that an increment of 5 per cent for the slowest type of pedal—fulcrum ahead of toe, and heel on rest with contact under ball of foot, as compared with the fulcrum anywhere under the foot— would not be unreasonable. If considerable resistance is offered by the pedal, then the adjustment for weight, discussed later, would also be used to compensate for this; hence, the set of adjustments in Table 20-III is suggested for factor 2.

Bimanualness. Considerable evidence exists to indicate that the simultaneous use of both hands to perform work on identical parts is highly preferable to the use of one hand. Indeed, this was listed as a basic principle with practically every motion study technique previously discussed as applicable to jobs performed at one work station. However, the use of both hands does not produce a 100 per cent increase in output but somewhat less than that. It is indicated that the use of both hands slows up the cycle.[23] In short experiments this increment was found to approximate 30 per cent. However, the two-handed work with untrained operators such as were used in the experiments would be probably subject to considerable change by the training or practice that would occur in an industrial situation. This change could well be greater than that which would be obtained with the one-hand work, which was more normal for the operators used. In addition, the change from one-hand to two-hand work increased, in some of the cases studied, the visual requirements, which are compensated for by the next factor. In actual practice, a figure considerably smaller than the one obtained from the previously referred to data was used, for some time, with apparent success.

A better estimate of a correct adjustment was obtained by Ischinger.[24] He studied eight industrial operators on five different jobs that they performed normally with both hands in a simultaneous and symmetrical fashion. He contrasted their maximum performance when doing work with both

---

[22]Barnes, Hardaway, and Pololsky, Purdue University.
[23]Mundel, Barnes, and McKenzie, University of Iowa.
[24]Ischinger, Purdue University.

hands with their maximum performance using one (preferred) hand. The jobs selected had a minimum of eye–hand coordination, and the weight of parts handled was negligible. The operators performed the one-hand version of the task 18 per cent faster than the two-hand version. (The two-hand method, of course, gave more production.) While the spread of Ischinger's data indicates that some subfactors may be involved, additional experimentation is needed to more fully ascertain this. From these data, the adjustments given in Table 20-III were computed for this factor. They have been widely used and appear to produce equitable standards.

Eye–hand coordination. Eye–hand coordination requirements primarily affect therbligs that are visually aided in performance, such as position and grasp, although other motions accompanying them, such as preceding transports, are also affected.[25]

Experimentation in this area is greatly hindered by the great effect training has on the performance of such therbligs. The time required for a not-too-difficult *position* may be reduced by 50 per cent after practice with only 3,000 cycles.[26]

Consequently, results obtained in short-run experiments are probably greatly exaggerated. MacKenzie[27] found that with one-hand work, changing the positioning requirements with $\frac{1}{2}$-inch cubes, from a general location (3-inch tolerance of placement) to placement in a slot with only $\frac{1}{64}$-inch tolerance in one direction, increased the time for *move to block, pick up, carry, place, and release* by approximately 65 per cent over the 3-inch conditions. He also found, however, that the change in tolerance from $\frac{1}{8}$ inch, a liberal tolerance, to $\frac{1}{64}$ inch accounted for only about 18 points of the increase. The difference between the $\frac{1}{8}$-inch tolerance of placement and the 3-inch tolerance was responsible for approximately 46 points of the increase. This is an unreasonable figure to accept for trained performance (MacKenzie himself places similar reservations on these data), inasmuch as no visual effort should be required at either of these last two conditions after practice, and they could well take very similar times.

Another study of a task requiring considerable eye–hand coordination showed that the pattern of eye usage shifts after practice and that a considerable reduction in cycle time takes place.[28] However, this study did not indicate whether the reduction was greater or less than that with tasks with less eye–hand coordination. Other experiments[29] showed that radically in-

---

[25]Mundel and Barnes, University of Iowa.
[26]Mundel, University of Iowa.
[27]MacKenzie, University of Iowa.
[28]Barnes, Perkins, and Juran, University of Iowa.
[29]Mundel and Barnes, University of Iowa.

creasing the eye–hand coordination increased the time for an extremely simple cycle by anywhere from 11 per cent to 25 per cent.

In short, these data indicate very little except that eye–hand coordination is a factor of apparently some magnitude affecting the pace at which work may be performed. Also, short-run experiments indicate it to be of a magnitude similar to the bimanualness adjustment deduced from short runs of one-hand versus two-hand work. Consequently, it was scaled down to obtain preliminary figures to produce the adjustments of Table 20-III, which have been used now for a long time with apparent success.

Handling requirements. An experiment was performed with operators moving small cylinders.[30] Some of these were placed on surfaces surrounded by sharp needle points; other open-ended ones were filled (as the subjects were led to believe) with an "ink" that would permanently stain their clothes. The operators were asked to move these cylinders from one electrical contact plate to another as rapidly as possible. The addition of the needle points only increased the time to *get, move, and aside* the cylinders by approximately 2 per cent, and the "ink" by 11 per cent as compared with a plain brass cylinder stored on a plain surface. This second case with the ink, however, would also qualify for approximately a 6 per cent adjustment for eye–hand coordination; consequently, the data indicate an additional adjustment of approximately 5 per cent for such a condition, which may be likened to fragility, where the exertion of muscular forces would need careful limitation. The needles surrounding the cylinders suggest a category of "control of location of hand and object but squeezing not ruled out" with a 2 per cent adjustment. In view of the short work spells and lack of training, these figures should be on the generous side. The values of Table 20-III are suggested as adjustments for this factor. They have been used in many time standards, and no difficulties have been encountered.

Weight handled or resistance encountered. Five experiments have been performed concerning this factor. Apparently conflicting data appear in the literature, but, as will be shown, these may be correlated if the correct variables are used. The setup for each experiment is described in the following paragraphs, but results are not discussed until they are summarized in Figures 39.10a, 39.10b, and 39.11.

Maass experimented with 10 college students.[31] His task consisted of moving, as fast as possible, a weighted box 18 inches across a table followed by an unloaded movement of the hand 18 inches to the point of origin, a free movement back to the box, a movement of the box, and so on, as fast as the

---

[30] Mundel and Barnes, University of Iowa.
[31] Maass, Purdue University.

operators were able to move. The operators worked roughly half of the experimental period.

A study by Solberg used 10 college students (all in excellent physical condition) to move, at their maximum pace, a loaded lever up and down for 5-minute work spells with 5-minute rest periods between each work spell.[32] The load on the lever was varied from 0.5 pound to 50.5 pounds with only one value being used for each 5-minute work spell. Solberg recorded the time per cycle for the last minute of each work spell. The subjects worked from low to high weight on one day and from high weight to low weight on another day, to minimize the effect of order presentation.

Sekerci used 10 college students to lift a bar to which weights were attached.[33] The weights used varied from 2 pounds to 44 pounds in 7-pound increments. A mechanism returned the bar to the original position (9 inches from the ground) so that lifts to 33 inches, and free movements back, could be made. Sekerci used 2-minute work spells, with the subjects working at their maximum pace with 2-minute rests between each work spell, random orders of work for each subject, and two lifting conditions: (1) with arms and trunk, and (2) with legs (by squatting and standing).

In order to relate these data with industrial practice, and in an attempt to determine the reason for the apparent differences, two additional experiments were undertaken. With the cooperation of Patrick Cudahy Co., a Milwaukee concern with an extensive shipping department, a test situation was constructed.[34] The situation required the workers to stand in one spot and lift cartons ($7\frac{1}{2}$ by $13\frac{1}{4}$ by $20\frac{1}{4}$ inches) 18 inches up and at the same time move them 2 feet to the left and deposit them on a conveyor which returned them eventually to the operator, but spaced so as to permit free lifting.

The initial position of the carton was 9 inches from the floor, and the surface of the conveyor on which they were deposited was 27 inches from the floor. The weight of the cartons varied from 2 pounds to 80 pounds in algebraic steps, although all the cartons being handled during any one run were of equal weight.

In the first experiment by Lehman, the 10 operators, *all of whom normally did work of this type*, moved five boxes from the lower platform to the conveyor, as rapidly as possible, followed by a 2-minute rest.[35] The time to lift the five boxes was recorded.

This procedure was used starting with the lightest box and proceeding up through the heavier boxes. The operators then repeated the work at each

---

[32]Solberg, Purdue University.

[33]Sekerci, Purdue University.

[34]Special thanks are due to C. J. Allen, Chief Industrial Engineer, who made the experimental facilities and workers available.

[35]Lehman, Marquette University.

weight, starting with the heaviest box and proceeding through to the lightest. The weights used in pounds were 1, 2, 2.5, 5, 8, 12, 24, 48, and 80. During the experiment, the operators worked a total of approximately 2 minutes out of 38 minutes, or 5.3 per cent of the time. The operators were given a strong incentive to work at their maximum pace by paid time off following the experimental run.

In a subsequent experiment by Kolarec,[36] the same apparatus and boxes were used, but instead of the five-box sequence, each of five workers lifted the boxes for 5 minutes followed by a 5-minute rest. During the first work period of 5 minutes, the workers worked with the 1-pound boxes, and progressed by work spells to the 80-pound boxes; then repeated the 5-minute work, 5-minute rest pattern, going from the 80-pound to the 1-pound boxes. The workers were induced to work at their maximum pace during the 3-hour experimental period by giving them the rest of the day off with full pay. All the workers studied were accustomed to handling, loading, and lifting boxes, such as were used, as a normal part of their usual job. The number of boxes handled each minute at each weight was recorded.

Figure 39.10a gives the data for each of these two, and the weight moving or lifting experiments previously referred to, in a form similar to their initial reporting. In Figure 39.10b, the data have been converted to pounds lifted or overcome per period of time, using a time value in each case that permits charting on one scale. These data from Figure 39.10b were plotted using a log-log grid as shown in Figure 39.11.

All the data exhibit an exceedingly similar slope pattern with a few exceptions. First, Solberg's is slightly flatter than the others, but it should be recalled that he recorded only the last minute of a 5-minute work spell, which may well have tended to maximize the work decrement created by the effort of lifting each weight. Second, Lehman's curve does not fall off at the higher weights as do both Sekerci's two curves, Kolarec's curve, and Maass's curve. Third, Sekerci's "leg" curve does not appear to fall off as greatly as the other three. It should be noted that Lehman's experiment did not require the workers to work as great a proportion of the time as the other experiments. Also, Sekerci's leg experiment involved a different muscle group, as compared to the other experiments.

Although a comparison of Lehman's curve with Kolarec's curve, which is plotted using the same time period, would appear to suggest a work decrement all along the scale, the data do not support such a hypothesis, inasmuch as Kolarec's workers were slower than Lehman's even during the first minute's work at the lowest weight, suggesting that this difference can be attributed to the difference in the basic abilities of the two working groups used.

---

[36]Kolarec, Marquette University.

Pounds per Pull on Lever or per Lift	Solberg: per 0.02 Minutes	Maass: per 20 Minutes	Sekerci— Leg: per 25,000 Time Units	Sekerci— Arm: per 25,000 Time Units	Lehman: per 2.5 Minutes	Kolarec: per 2.5 Minutes
0.5	50					
1.0					134	112.5
2.0			333	368	266	235
2.5					336	300
3.0	272					
4.0		615				
5.0					658	575
5.5	458					
8.0	640				1,000	880
9.0			1,430	1,540		
10.13		1,480				
10.5	795					
12.0					1,415	1,260
13.0	956					
16.0			2,420	2,530		
16.25		2,260				
18.0	1,260					
23.0		3,300	3,320			
23.13		2,480				
24.0					2,560	2,220
25.5	1,790					
30.0			4,070	3,930		
37.0			4,720	4,360		
38.0	2,440					
44.0			5,290	4,600		
48.0					4,500	3,000
50.5	3,000					
80.0					6,660	3,200

Figure 39.10a. Data from weight experiments.

The theoretical amounts of work that would be performed, if the heavier weights did not slow down the workers, can be shown by a 45° line passing through the intercept of any of the experimental curves on the Y axis (1-pound ordinate). The difference between the intersection of the same ordinate with an experimental curve would be work lost, at that weight lift, as indicated by the particular experiment, from the theoretical work capacity (based on a maintenance of the same pace as with the 1-pound lift). In all

Pounds Pull on Lever or Weight per Lift	Solberg: Time per Cycle (0.00001 minute)	Maass: Time per Cycle (minutes)	Sekerci— Leg: Relative* Time per Lift	Sekerci— Arm: Relative* Time per Lift	Lehman: Time per Cycle (minutes)	Kolarec: Boxes Lifted (minutes)
0.5	20.0					
1.0					0.0186	45
2.0			150	136	0.0188	47
2.5					0.0186	48
3.0	22.0					
4.0		0.130				
5.0					0.0190	46
5.5	24.0					
8.0	25.0				0.0200	44
9.0			157	146		
10.13		0.137				
10.5	26.4					
12.0					0.0212	42
13.0	27.2					
16.0			165	158		
16.25		0.144				
18.0	28.6					
23.0			174	173		
23.13		0.156				
24.0					0.0226	37
25.5	29.0					
30.0			184	191		
37.0			196	212		
38.0	31.1					
44.0			208	239		
48.0					0.0266	25
50.5	33.6					
80.0					0.0300	16

*In Sekerci's original report, these data were in cycles per work spell, but access to the original values could not be obtained inasmuch as the thesis was unobtainable.

Figure 39.10b. Data from Figure 39.10a converted to total pounds lifted per unit of time by selecting for each experiment a time value that permits plotting on a similar scale. (Various time periods chosen so as to make slope comparisons readily feasible on a log-log presentation.)

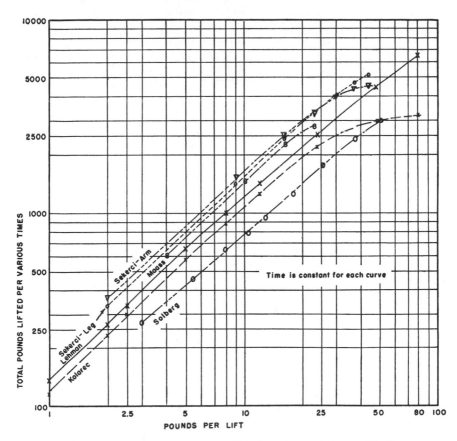

Figure 39.11. Plot of total weight lifted against weight per lift (log-log grid).

cases, this is obviously a function of the weight lifted. Trend lines were fitted to Lehman's and Kolarec's data after changing Lehman's scale so as to cause it to coincide with Kolarec's curve (taking these two as the most realistically representative of all the data). The last two experiments certainly were closest to actual work conditions and workers.

To put the data obtainable into usable form, Lehman's curve was superimposed on Kolarec's and a 45° line passed through the combined intersection of the *Y* axis at 1 pound per lift. Using this 45° line, the theoretical pounds overcome per unit of time was calculated for each lift, by 1-pound increments. This value was divided by the actual total weight lifted at each weight per lift (by pounds) using the slopes of both Lehman's and Kolarec's curves, and the excess over 1.00 used as representative of the per cent increase in time due to the weight per lift. This increase in time, in per cent, is plotted

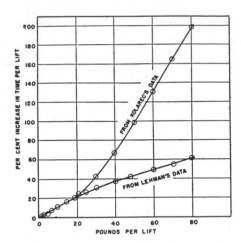

Figure 39.12. Per cent additional time as a function of pounds per lift. (Based on data of Lehman and Kolarec.)

in Figure 39.12. The slopes of Lehman's curve yielded the per cent increase in time when the load-connected-work was only 5 per cent of the cycle. The two sets of data were identical up to and including 19 pounds per lift.

Inasmuch as an objective rating applied to a work element converts the time taken to the time that would be taken with the standardized pace (and no difficulty), the time obtained after rating would be similar to the time at 1 pound (or no load) and expressed on the graph of Figure 39.13 as a 45° line. Hence, the time increment from the 45° line to Lehman's or Kolarec's curve would be the difficulty adjustment for weight expressed as a function of the pounds per lift. An assumption was made that the additional increment from 5 per cent of the time to 50 per cent of the time under load was linear, and Table 20-IV was computed.

Other special constraints on work speed. In some jobs, particularly those in heavy industry, workers may work in areas that restrict their motions, in positions that are potentially dangerous, and so forth. Special difficulty adjustments must be developed for these conditions.

Summary, difficulty adjustments. The presently available data for difficulty adjustments have been applied over an extended period of time to an extensive variety of work. They also have been instrumental in bringing peaceful solutions to situations where standards had been a source of constant arguments. They certainly offer a more consistent basis from which to work than does the conventional procedure in which the time study man subjectively evaluates the difficulty of the job in order to form a mental image of a

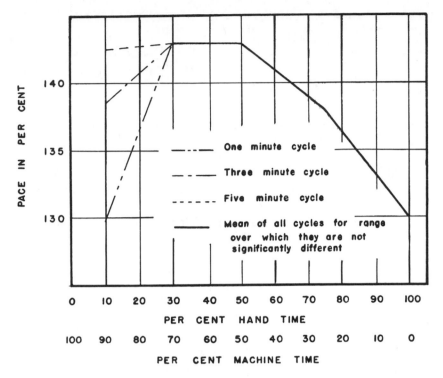

Figure 39.13. Pace and per cent of cycle that is hand time, using college students. (From R. W. Llewellyn, "An Investigation of the Effects of Machine Time on Operator Pace," M.S. thesis, Purdue University.)

suitable pace with which to compare the pace he observes. Further experimentation to extend the percentages presented in Table 20-III to unusual conditions is not difficult, provided that all parties recognize that the application of additional values may later indicate the need for, and the magnitude of, additional corrections to these new data. If rates are not consistent in ease or difficulty, the elements of sameness in the malfunctioning rates may be compared, and the corrections to the table of difficulty adjustments will be indicated.

The problem of the difficulty adjustments, however, is still a field for additional motion and time study research, but this research must be carefully performed. We must avoid sweeping statements based on short performances of laboratory tasks by students relatively unfamiliar with such work.

Finally, it is also interesting to note that the basic philosophy of the difficulty adjustments makes the time study also a motion study technique. For the most part, changes that will reduce the adjustments will also facilitate the task,[37] and consequently, these values may be used as a guide toward better methods.

Allowances for Machine Time

In Chapter 20 it was suggested that the three basic solutions to the problem of machine time allowances were:

1. Make no adjustment.
2. Adjust the standards so that the probable production increment over standard is related to the per cent of manual work in the cycle.
3. Adjust the standard so that the expectable production increment over standard is the same for all jobs.

Some basic information from experimentation. Before discussing the advantages and disadvantages of each of these three basic types of solutions (and variations thereof), some additional information should be examined.

An experimental investigation[38] of the phenomenon of increased pace on cycles containing rest gave the results shown in Figure 39.13. These data were obtained from work done by college students. They performed a job consisting of jabbing a stylus into four holes at the corners of a 4-inch square, moving around and around as rapidly as possible for various predetermined periods, then stopping and resting for predetermined periods, and so forth (to imitate tasks on which the operator worked and then rested while he waited for the machine to perform its part of the cycle). Various predetermined arrangements were tested with a series of work spells. Owing to the other obligations of the experimental operators used, the work spells were hardly extensive enough to allow the full effect of the intermittent rests to be evaluated. In other words, the full cumulative effect of the higher per cent operator-paced work was not reached. Also, the operators were not sufficiently trained to permit them to utilize the effect of the rest pauses; their skill rather than their physical capacity limited their production at the arrangements that permitted higher paces. The low skill is also suggested by the drop in production during the cycles with a small per cent hand time, which indicates an unreasonably long warm-up period for such a simple task. However, the *law of diminishing returns* undoubtedly had some effect also. Nevertheless, these data do verify that there is a pace increase possible when

---

[37]The adjustment for bimanualness is the outstanding exception.

[38]R. W. Llewellyn, "An Investigation of the Effects of Machine Time on Operator Pace," M.S. thesis, Purdue University, 1948.

each cycle contains rest, as was originally suggested by shop data given to the laboratory for study.

Similar plant data. The data obtained from earning curves at a plant where this factor was not taken into account indicate that the actual paces we may expect are shown in Figure 39.14.[39]

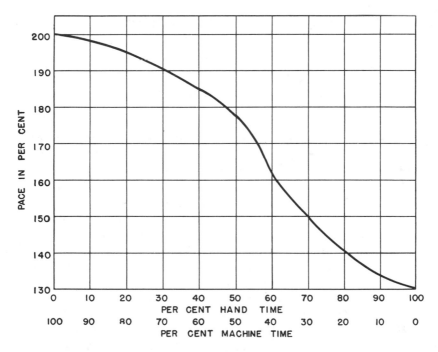

Figure 39.14. Pace and per cent of cycle that is hand time, from shop data. Per cent is based on base time.

Interpreting the plant data. It is to be noted that the intersection of either curve with the ordinate has no real significance and, also, that its actual shape as it approaches the ordinate is in doubt because the law of diminishing returns must certainly take effect some place in this region, particularly with short cycles. However, any error in this range is probably negligible because the portion of the cycle controlled by the operator, to which the error would apply, becomes exceedingly small in this region. Indeed, it would take a 100 per cent error in the curve of pace at 10 per cent hand time to affect the expected cycle performance time by as much as 5 per cent, and it is not anticipated that the errors on the curve approach such a

---

[39]This plant requested that its name not be disclosed.

figure. Below 10 per cent hand time, they are of even less consequence. It should also be noted that if the definition of standard is such that the typically expected production increment above standard is defined as other than 30 per cent, the ordinate values of Figure 39.14 will require change so that they express the expectable pace as properly related to the particular numerical designation assigned to the maximum pace typically expectable at 100 per cent hand time. For any ordinate value Y on Figure 39.14, the new Y', with the maximum pace on a 100 per cent hand job identified as X, would be

$$Y' = Y \frac{X}{130}$$

Now to return to the problem of allowances for machine time; let us examine the alternatives, one by one.

The results of alternative 1: make no adjustment. In cases where no incentives are in use, production over standard is seldom expected or realized. The production one can expect from manual, partly manual, and fully machine-controlled tasks is alike for all practical purposes, if no allowance for machine time is made. Such standards will be appropriate for all uses of standard times as given in Chapter 5. This will be true regardless of the amount of internal work, as long as the standard time for internal work does not exceed the machine-controlled time. While the manual or primarily manual job will provide the worker with greater flexibility in the use of his time in that he may voluntarily slow down and later speed up to achieve standard for the day, the machine worker may frequently have considerable rest inherent in the cycle. However, when internal work becomes extensive, such jobs require a continual, steady pace to achieve standard. Such conditions may give rise to some problem jobs in the shop. If such jobs are long-running, worker rotation, special wage rates, or special allowances may be introduced.

Interaction of alternative 1 and incentives. In shops where incentives are in use, the lack of a machine-time allowance is frequently justified by pointing to incentives as extra pay for extra work and indicating that the machine job just does not permit such activity. This creates four problem areas:

1. Production expectancy varies on different jobs, depending on the amount of machine-controlled time, making scheduling and so forth more difficult. However, the data of Figure 39.14 indicate that this does not become a problem until the machine-controlled portion of the cycle exceeds 45 per cent of the cycle, provided that there is no internal work.

2. Once the 45 per cent of cycle controlled by the machine is exceeded, earnings will vary as concerns the incentive, from job to job, even with

equal worker ability and application. This may seriously upset the established job hierarchy. It may make the manual jobs more attractive than the machine jobs, which may not be desirable.

3. The machine jobs, which, because of high overhead rates, may be well worth stimulating by incentives, may yield such poor potential that the workers do not attempt to earn such incentives. This may produce higher costs.

4. With high amounts of internal work, calling for steady application from the machine operator, real unrest may develop because of the lack of incentive potential. Further, workers may resist methods improvements on such machines. (Improvements to such tasks often involve making a maximum of the work internal.) Workers may also strenuously resist multiple machine assignments, which cut into their rest time and do not increase the potential incentive.

The results of alternative 2: adjust the standards so that the potential incentive increment over standard is related to the per cent of manual work in the cycle. One common procedure for making such an adjustment is as follows:

$$\text{allowance} = \frac{\text{rated internal manual time}}{\text{rated external manual time} + \text{machine-controlled time}}$$
$$\times \text{ per cent expected average incentive}$$

Such an allowance is applied to the whole cycle. It is to be noted that it more or less guarantees the average incentive attainment for the internal work, even if the internal work is done at standard. Such an allowance provides a considerable incentive for multiple machine assignments and increased internal work. It also tends to make the earning potential on machine-controlled operations with considerable internal work more like that of fully manual jobs. On the other hand, with cycles almost fully machine-controlled and with large amounts (but less than the machine cycle) of internal work, it tends to produce incentive pay for work at standard, creating the reverse problem of that listed as the fourth problem area under alternative 1.

Basic phenomenon is ignored with alternative 2. It should be noted that the simple allowance calculation just given neglects to take into account the effect of the increased pace possible with the rest periods inherent in the cycle. To examine how this factor may be included, let us examine a 1-minute cycle as follows:

0.3 minute; 30 per cent external manual time (rated and adjusted)
0.7 minute; 70 per cent machine-controlled time
0.2 minute; 20 per cent internal manual time (rated and adjusted)

If we consider 130 per cent the definition of expectable incentive pace, Figure 39.14 may be read directly. With the operator working 50 per cent of the cycle, we may expect a pace of 177 per cent for the external manual work, hence a performance time of approximately 0.170 minute. Now, following the policy of alternative 2, this job should have potential incentive earnings of

$$\frac{0.30 + 0.20}{1.00} \times 30 \text{ per cent} = 15 \text{ per cent}$$

Taking the rest "effect" into account. Actually, taking into account the effect of the rest, the operator can be expected, without allowance for machine time, to perform the task in $0.170 + 0.700$ minute $= 0.870$ minute. The incentive increment attained will be $(1.000 - 0.870)/1.000 = 13$ per cent. Hence, some additional allowance X will be required to achieve the policy objectives, as follows:

$$\frac{1.00\left(\dfrac{100 + X}{100}\right) - 0.870}{1.00\left(\dfrac{100 + X}{100}\right)} = 15$$

and $X = 2.35$ per cent. With this machine time allowance, the incentive earnings will be

$$\frac{1.0235 - 0.870}{1.0235} = 0.15 \quad \text{or} \quad 15 \text{ per cent}$$

A working solution. A table computed in such a manner from the data of Figure 39.14, but based on a definition of incentive performance as 120 per cent, is given in Table 39-I. This is used by the Mirro Aluminum Co. This table has been effective in increasing the equity on machine tasks and increasing the acceptability of multiple machine assignments.

The results of alternative 3: adjust the standard so that the expectable production increment over standard is the same for all jobs. While this policy makes the standards alike for scheduling, pricing, and so forth, it tends to produce incentive earnings on machine jobs without incentive effort. However, it maintains, even with incentives, the exact hierarchy of the wage scales from job to job, and may reflect the workers' concept of equity. In such cases, the introduction of multiple machine assignments and additional internal work may be resisted by the working group as "more work without more pay." On the other hand, management may feel that they have been leaning over backward to equalize incentive opportunity until such multiple machine assignments or better methods could be devised. Certainly, it is a policy whose ultimate effects depend upon the relationship between management and the working group and the understanding of the full implications

Table 39-I. MACHINE ALLOWANCE CHART, STANDARDS DEPARTMENT, MIRRO ALUMINUM COMPANY*

Per Cent of Machine Time in Total Cycle vs *Per Cent of Internal Time in Total Cycle*

Per Cent of Machine Time in Total Cycle	1 (0-5)	2 (6-10)	3 (11-15)	4 (16-20)	5 (21-25)	6 (26-30)	7 (31-35)	8 (36-40)	9 (41-45)	10 (46-50)	11 (51-55)	12 (56-60)	13 (61-65)	14 (66-70)	15 (71-75)	16 (76-80)	17 (81-85)	18 (86-90)	19 (91-95)	20 (96-100)
96-100													(apply allowances to total cycle)							
91-95			0.9	1.9	3.0	4.1	5.1	6.2	7.4	8.5	9.7	10.9	12.1	13.2	14.5	15.6	16.8	17.8	19.0	
86-90					1.2	2.3	3.5	4.7	5.9	7.4	8.8	10.1	11.5	13.0	14.2	15.6	16.8	18.0		
81-85						0.6	1.9	3.4	5.1	6.8	8.2	9.8	11.1	12.9	14.3	15.7	17.0			
76-80							0.9	2.7	4.6	6.3	8.0	9.7	11.4	13.0	14.6	16.0				
71-75							0.3	2.3	4.3	6.2	8.2	10.0	11.8	13.5	15.0					
66-70								2.2	4.3	6.4	8.5	10.4	12.3	14.0						
61-65								2.4	4.8	7.0	9.2	11.3	13.0							
56-60							0.4	2.9	5.5	7.8	10.0	12.0								
51-55							1.2	3.7	6.5	8.9	11.0									
46-50							2.1	5.0	7.6	10.0										
41-45						0.3	3.6	6.4	9.0											
36-40						1.9	5.2	8.0												
31-35					0.2	3.8	7.0													
26-30					2.3	6.0														
21-25				0.7	5.0															
16-20				4.0																
11-15			3.0																	
6-10		2.0																		
0-5	1.0																			

*Courtesy J. Ruedebusch, Director of Industrial Relations.

of subsequent changes. This may be too much to expect from some working groups.

Computing an allowance table. Following this policy, and assuming the percentages of Table 39-I to be correct, a suitable allowance may be computed for jobs rated on a basis that did not take this phenomenon into account, so that the possible production expectancy on all jobs will be 130 per cent of standard,[40] as follows:

C = per cent of base time of cycle controlled by the machine, during which the operator rests

H = per cent of base time of cycle controlled by the operator

P = possible pace for job as indicated in Figure 39.14 expressed as a fraction of 100 per cent pace

A = per cent allowance required to bring production possibility up to 130 per cent of standard

H/P = actual per cent of original base time of cycle that the hand time will actually take when performed at the pace made possible by the intermittent rest

Then

$$\frac{C + H + A}{C + H/P} = 1.3$$

and

$$A = 1.3(C + H/P) - (C + H)$$

or

$$A = 1.3(C + H/P) - 100$$

Hence, for 10 per cent machine time, we get

$$A = 1.3(10 + 90/1.34) - 100$$
$$= 1.3(77) - 100$$
$$= 100 - 100 = 0$$

For 65 per cent machine time, we get

$$A = 1.3(65 + 35/1.88) - 100$$
$$= 1.3(65 + 18.6) - 100$$
$$= 109 - 00 = 9 \text{ per cent}$$

[40] If the definition of standard involves an increment other than 30 per cent, the calculations will change accordingly, and will require first an adjustment of the ordinate in Figure 39.14.

An allowance table computed in this fashion in given in Table 39-II.

Table 39-II. MACHINE TIME ALLOWANCE* FOR 30
PER CENT PRODUCTION INCREMENT

Per Cent of Cycle *Controlled by Machine*	*Per Cent*
100	30
95	27
90	24
85	20
80	17
75	14
70	11
65	9
60	6
55	4
50	3
45 or less	00

*For a reference letter, use V followed by the per cent of the
cycle that is machine-controlled, computed on the base times.

Adjusting for internal time. There is still another group of tasks that
presents a problem requiring further manipulation of these data. Let us
assume that we have a task of which 60 per cent is controlled by the machine.
However, the operator prepares a load during part of this time and actually
works 70 per cent of the cycle. From the table given previously, an operator
with a cycle 70 per cent hand time would receive a zero allowance; the
inference is that he could go fast enough during the 70 per cent to exceed
the standard by 30 per cent without an allowance. However, in this case, his
excess productivity during 30 per cent of the cycle would show up; how fast
he went during the remainder of the hand time would be of no value, since in
this example the machine controls the production after the 40 per cent for
hand time is completed. Consequently, a new allowance, A', would have to
be computed, as follows:

$$\frac{C + H + A'}{H/P' + C} = 1.30$$

where

A' = allowance with internal work, per cent
P' = possible pace for all manual work as indicated in Figure 39.14 expressed
as a fraction of 100 per cent pace

and

$$A' = 1.30(H/P' + C) - (C + H)$$

and for the job being discussed,

$$A' = 1.30(40/1.50 + 60) - 100$$
$$= 113 - 100$$
$$= 13 \text{ per cent}$$

Note that this is larger than the allowance that would accompany a job with 60 per cent machine time and no internal work. Other situations may be evaluated in a similar manner.[41]

Summary

The materials in this chapter related to complexities of recording the method, recording the observed time, determining the number of readings required, developing skill at rating, and adjusting standards to reflect rest during machine time as well as the per cent of the cycle that is machine-controlled. While these materials may not be of interest to the casual student, they represent real areas of concern to the practitioner and his supervisor. Indeed, they should be of concern to all operating management.

PROBLEMS

39.1. Research the literature (periodical or text) and report on a reading on:
   (a) Use of motion-picture camera or tapes for recording methods and time.
   (b) Determining the number of readings required for a DTSIS.
   (c) An evaluation of ratings.
   (d) The use of machine allowances, other than delay allowances.

39.2. Same as above, but in lieu of the literature search, interview a practicing or supervisory time study observer.

39.3. A rating session with an industrial film should be held, including an evaluation of the results. Films are available from a variety of sources.

39.4. A plant defines standard as 100/125 of what the typical worker can achieve. For each of the following jobs, compute the machine allowance for each of the four possible alternatives.

---

[41]If a different definition of standard is in use, Figure 39.14 must be recalculated before this type of adjustment can be computed.

   (a) 90 per cent machine-controlled; 10 per cent internal time

   (b) 90 per cent machine-controlled; 20 per cent internal time

   (c) 80 per cent machine-controlled; 20 per cent internal time

   (d) 70 per cent machine-controlled; 10 per cent internal time.

39.5. The instructor will supply data from a rating session with eight raters. Make a complete statistical analysis of the data and prepare your critique for each rater.

39.6. What allowance would have to be added to compensate for the machine-time effect on a job where the machine controlled 75 per cent of the cycle but the operator worked during 45 per cent of the cycle? (130 per cent typical production expectancy.)

   (a) Where the policy is to equalize the potential incentive on machine and manual activities.

   (b) Where the policy is to make incentive potential proportional to per cent of time worked with no attention given to the pace increase due to intermittent rest.

   (c) Same as part (b), but taking the effect of rest into account.

39.7. What allowance would have to be added to a job to compensate for the machine-time effect if the machine controls 65 per cent of the cycle, during which time the operator rests, and standards in this shop are set so that the average operator is expected to exceed them by 25 per cent?

   (a) Where the policy is to equalize the potential incentive on machine and manual activities.

   (b) Where the policy is to make incentive potential proportional to per cent of time worked with no attention given to the pace increase due to intermittent rest.

   (c) Same as part (b), but taking the effect of rest into account.

39.8. What allowance would have to be added to a job to compensate for the machine-time effect if the machine controls 75 per cent of the cycle, during which time the operator rests, and standards in this shop are set so that the average operator is expected to exceed them by 50 per cent?

   (a) Where the policy is to equalize the potential incentive on machine and manual activities.

   (b) Where the policy is to make incentive potential proportional to per cent of time worked, with no attention given to the pace increase due to intermittent rest.

   (c) Same as part (b), but taking the effect of rest into account.

40

NOTES ON PREDETERMINED
TIME SYSTEMS

Short History

Predetermined motion time data has been in existence since the 1920s. However, the early developers of these data did not disclose either the method of developing the data or the data itself; it was closely held for proprietary use. The first general disclosure of such data occurred with the publication of *Methods-Time Measurement*, by H. B. Maynard, G. J. Stegemerten, and J. L. Schwab (McGraw-Hill, 1948). Subsequently, other sets of data were released. New systems were developed with the aim of overcoming what the authors felt were limitations or defects in existing systems. Some new systems were developed to produce freely available data so as to overcome restrictions placed on some of the data systems by the originators. Some systems were, without doubt, produced to "author a system." The number of such systems has lead some critics to refer to these as "... . the alphabet soup

660

of predetermined motion time systems: BMT, DMT, ISM, MTA, MTS, RMD, etc."[1]

HPT

The HPT data presented in this book were developed to provide data that could be used in conjunction with the difficulty adjustments of objective rating. The effects of the factors for which difficulty adjustments are made are inextricably built into all the other available data tables. Since the release of the HPT data in 1952, the data have been given extensive use in a variety of industrial plants and government installations. It should be realized that the data of HPT, when multiplied by all the combinations of difficulty adjustments, is really a table containing millions of values. When the time study men are familiar with objective rating, these HPT data may be used with a minimum of additional training. Carefully applied, the standards produced appear to have an inherent variability of about the same order of magnitude as direct time study—intensive sampling with objective rating. Occasionally, unpredicted and sizable discrepanicies do occur. See Chapter 21, footnote 4 (p. 396).

Some Problems

Without doubt, the real utility of some of the data systems has been adversely affected by the unfortunately extravagant claims—with respect to both the applicability and accuracy of these data systems—made by some of the early proponents of such data. The claims with respect to applicability have caused waste of large amounts of manpower in futile attempts to apply the PTS approach in areas more suitably handled as FIAT- or FPE-type situations. The original claims with respect to accuracy later became obviously extravagant when revisions and changes to the original data tables were published. Such changes would hardly have been needed had the original data possessed the properties claimed for them. Further, most of the systems do not produce fully compatible standards. Some of the difference can be correctly attributed to the different concept of *normal* built into the various tables; other discrepancies are difficult to explain unless one accepts the conclusion that they cannot be as accurate as many of them claim to be.

Some of the claims for accuracy have been accompanied by the attitude that with such accuracy no labor group has any grounds for or right to object

---

[1]B. Gottlieb, "A Critical Analysis of Predetermined Motion Time Systems," *Proceedings, 16th Annual Conference and Convention, American Institute of Industrial Engineers*, 1965, pp. 75–84.

to the results obtained with such data. It is extremely doubtful whether any group of organized labor would accept such a view. The right to "grieve" over a particular time standard is a right commonly sought and held by such groups; a view with which this author is sympathetic. Further, it seems possible to hold such views without being partisan; the data associated with any of the time study or work measurement procedures, including predetermined time systems, is such that it seems relatively impossible to logically defend any other point of view.

Sources of more details. The argument over these data, begun with their introduction, continues, and will probably do so in the future. For those who wish to follow the argument in detail, reference should be made to:

1. Books on:
   a. Predetermined times.
   b. Methods-time measurement.
   c. Work factor.
   d. Basic motion times.
   e. Modaps.
2. Periodical literature, such as:
   a. *Transactions*, American Institute of Industrial Engineers, Inc.
   b. *Industrial Engineering*, Journal of the American Institute of Industrial Engineers, Inc.
   c. *Journal of Methods-Time Measurement*, MTM Association
3. Conference proceedings, such as:
   a. *Proceedings, Annual Conference and Convention*, American Institute of Industrial Engineers, Inc.

Summary

A short history of PTS systems has been given. Some basic problems have been indicated.

PROBLEMS

40.1. Using HPT, develop a standard for the job shown in:
(a) Figure 27.1a.
(b) Figure 27.1b.
(c) Figure 27.2.
(d) Figure 27.3.
If additional information is needed, make and record a reasonable assumption.

40.2. Same as Problem 40.1, using any job and PTS table assigned by the instructor.

41

NOTES ON STANDARD
DATA SYSTEMS

Introduction

This chapter deals with three topics related to standard data systems which were not covered in the initial chapter on the subject, Chapter 22. First, notes are included on a basic and commonly used statistical technique, the fitting of straight lines to data by the method of least squares so as to provide an introduction to statistical problems. Second, notes are included relating to the use of work-unit analysis prior to the development of standard data for service activities. Third, the development and use of standard data in a job-type machine shop is examined.

Fitting Straight Lines to Data by the Method of Least Squares

Introduction. It is true that many small calculators will perform the necessary computations as a hard-wired operation. However, the analyst should understand what is being done so that the analyst controls the computations rather than vice versa. This understanding is an even more important requirement when the statistical analysis of raw data (prior to the development of standard data) is assisted by the sophisticated statistical software packages available with large computers.

Definition. The method of least squares is used to find the line $Y = a + bX$ such that the sum of the squares of the deviations of the data points about the line are at a minimum. The two equations

$$Na + b \sum X = \sum Y \quad \text{and} \quad a \sum X + b \sum X^2 = \sum XY$$

are solved simultaneously to obtain a and b. Y is the time and X is the variable affecting the element; hence, a value for each of these is available for each of the N sets of data used for the determination of standard elemental times.

Goodness of fit. The *goodness of fit* is determined by means of a *coefficient of correlation*. The coefficient of correlation, r, is based on two values, as follows. Let

$$\sum d^2 = \text{sum of the squares of the deviations of the data points about a line: } \bar{Y} = X$$
$$\sum d'^2 = \text{sum of the squares of the deviations of the data points about the line of least squares}$$

Then the correlation coefficient, r, is

$$r = \sqrt{\frac{\sum d^2 - \sum d'^2}{\sum d^2}}$$

Evaluation of the results. The statistical reliability[1] of r may be estimated. However, both r and its evaluation may still be affected by:

1. Chance causes.
2. Limitations on the range of the data.

---

[1] A text on statistics should be consulted. Computations should never be used blindly.

An example. A set of time study data has been plotted in Figure 41.1. Both the lines about which $\sum d^2$ and $\sum d'^2$ are computed are shown. Were computations made concerning the reliability, one would have been computationally assured that the line is a good fit. However, there is some considerable evidence that the correct line is a curve, falling off sharply as the work variable X increases. This potential relationship is indicated with a broken line. Obviously, before the standard data tables are constructed, some additional studies with much larger values of X should be obtained. Such conditions are frequently encountered. Care should be exercised in accepting the results of initial computations.

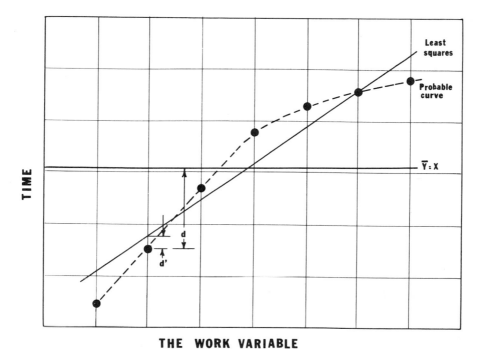

THE WORK VARIABLE

Figure 41.1. Curve fitted to time study data after line of least squares was computed and added to plot.

Work-Unit Analysis and Standard Data

Introduction. With maintenance work, or other similar activities, the tasks are usually numerous and the repetition relatively low. Standard data at the 3rd-order work-unit or 4th-order work-unit level is of great assistance in achieving managerial control. The work of setting standard times from such data is relatively small. The 3rd- or 4th-order work-unit data may be obtained in any of the three following ways:

1. Built from standard elemental data.
2. Accumulated by appropriate work measurement techniques.
3. Acquired from outside sources.

The first method, building from standard elemental data, is the most laborious, particularly if the jobs are numerous and if the standard elemental data are to be synthesized from predetermined time systems. In most cases, other time study techniques may be employed to gather the necessary data much more rapidly and economically. The third approach offers only speed of acquisition as its main advantage; the methods used in the original setting may differ from the methods in use so much that the validity of the use of such data is dubious. In most cases such borrowed data are applied with "craft allowances" ranging from 30 to 100 per cent, to allow for these unevaluated, unknown differences.

The work-unit problem. Whatever approach is chosen, the most basic problem is to organize the data for ease of access into what is called a work-unit *dictionary*. For instance, the 3rd-order work-units for aircraft maintenance could be indexed under a five-part, compound work-unit index as:

1. Model number.
2. Major category.
3. Item number.
4. Part number.
5. Action code (e.g., inspect; remove, inspect, and replace; remove and replace; remove, repair, and replace).

Note: The first four items are similar to the use of a noun and a number of adjectives to fully identify an object. The fifth part provides the verb.

In a similar fashion, plant maintenance work could be indexed as:

1. Plant component: structure; production equipment; service equipment; and so forth.
2. Specific type: wall; floor; window; lathe; shafting; truck; and so forth.
3. Model number.
4. Type.
5. Part.
6. Action code.

In any case, as a first step in constructing standard data, the manner of intended use should be fully investigated so that the work-unit dictionary can be compiled and used as a guide for the development of the actual time data.

Standard Elemental Data in Job-Type Machine Shop

Introduction. The shop in which these data were developed and used was a job shop. They did custom work for other manufacturers. Runs were usually short. Each lot of product was different. Direct time study–intensive sampling or preparing standards for each job directly from predetermined times would have taken too long and been too expensive.

Written standard practices. A written standard practice was devised for each operation performed in the shop. The operations included cutting keyways in shafts, splining shafts, slotting shaft ends, hobbing gears on shafts, and so forth.

Predetermined times. The operation of keyway milling on a Kent-Owens milling machine is shown in Figure 41.2. The HPT data buildup for the manual part of this operation is shown in Figure 41.3. The computation of the allowed time is shown in Figure 41.4.

Figure 41.2. Milling keyway on Kent–Owens mill. (Photo Courtesy of Modern Machine Works, Cudahy, Wis.)

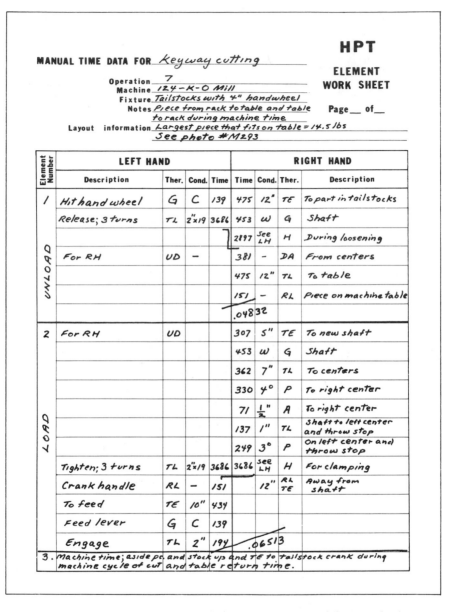

HPT

MANUAL TIME DATA FOR *Keyway cutting*

Operation __7__
Machine __124-K-O Mill__
Fixture __Tailstocks with 4" handwheel__
Notes __Piece from rack to table and table to rack during machine time__
Layout information __Largest piece that fits on table = 14.5 lbs__
__See photo #M293__

ELEMENT WORK SHEET

Page ___ of ___

Element Number	LEFT HAND Description	Ther.	Cond.	Time	Time	Cond.	Ther.	RIGHT HAND Description
1 (UNLOAD)	Hit hand wheel	G	C	139	475	12"	TE	To part in tailstocks
	Release; 3 turns	TL	2"x19	3686	453	W	G	Shaft
					2897	see LH	H	During loosening
	For RH	UD	—		381	—	DA	From centers
					475	12"	TL	To table
					151	—	RL	Piece on machine table
				.048	32			
2 (LOAD)	For RH	UD			307	5"	TE	To new shaft
					453	W	G	Shaft
					362	7"	TL	To centers
					330	4°	P	To right center
					71	1/2"	A	To right center
					137	1"	TL	Shaft to left center and throw stop
					249	3°	P	On left center and throw stop
	Tighten; 3 turns	TL	2"x19	3686	3686	see LH	H	For clamping
	Crank handle	RL	—	151		12"	RL TE	Away from shaft
	To feed	TE	10"	434				
	Feed lever	G	C	139				
	Engage	TL	2"	194	.065	13		

3. Machine time; aside pc. and stock up and TE to tailstock crank during machine cycle of cut and table return time.

Figure 41.3. First step in use of HPT data for determining standard time for manual elements for job shown in Figure 41.2.

RECAPITULATION

ELEMENTS	1	2	3	4	5	6	7	8	15
Amount of body	D-5	D-5							
Foot pedals	F-0	F-0							
Bimanualness	H-0	H-0							
Eye-hand coordination	J-2	K-4							
Handling requirements	P-2	P-2							
Weight or resistance									
TOTAL TIME IN _MIN_									
NUMBER OF OBS.									
PRO-RATE DIVISOR									
AVERAGE PER CYCLE									
RATING									
RATED TIME	0482	0653							
PER CENT OF CYCLE	—	—							
1+ADJUSTMENTS	1.09	1.11							
BASE TIME	0527	0723							
1+ALLOWANCES									
ALLOWED TIME									

(HPT VALUES written across NUMBER OF OBS./PRO-RATE DIVISOR/AVERAGE PER CYCLE)

SYMBOLS USED

C - Extra unnecessary motion
D-A.D. Dropped part
F - Unnecessary fumble
H - Unnecessary hesitation
P - Includes personal time
R - Made reject by improper work
X - Deviation from std. routine

ALLOWANCES

Personal	5 %
Mat. Hdlg.	3
TOTAL	**8** %

LOOP VALUES

Figure 41.4. Use of recapitulation section of objective time study sheet to complete determination of standard time for manual elements of job shown in Figure 41.3.

Note that the weight of the piece, which would be a variable from job to job, was not included in the standard data buildup. Allowances were also excluded.

The application form. The form used to set standards, both for making bids on potential work and for use in the shop, for keyway milling jobs is shown in Figure 41.5. Note that the weight of the workpiece is considered on this sheet. Also, the allowances are added as a final computation.

Results. Using these standard data, the shop, which had been short of work (overbid jobs) and losing money (underbid jobs), returned to steady, profitable operation.

Summary

Additional statistical work-unit identification and application problems, with respect to standard data, have been examined. Much additional material could be written. Standard data development is an extensive and important aspect of time study activity.

PROBLEMS

41.1. Fit a straight line, by the method of least squares, to the following data sets, the first number of which is the X value; the second, Y:
(a) 0.5, 1.0; 1.5, 1.0; 2.0., 2.5; 3.0, 2.5; 4.5, 3.5; 5.0, 6.0; 6.5, 6.0; 6.5, 7.0; 8.0, 6.5; 8.0, 9.5; 9.0, 9.5; 10.0, 9.5.

KEYWAY CUTTING CALCULATION SHEET

Date _____   Type of keyway _____
Customer _____ Dimensions _____
Part No. _____ Name _____ Cutter diameter _____inches
Material _____
Machine _____
Fixture _____
Weight of part _____ lbs. at operation Weight adjustment from Table I ____ = W

NOTE: Center alignment inspection by foreman, go - no-go gage during machine cycle.

1. LOAD FROM TABLE TOP (USE APPROPRIATE LINE AND MARK THIS LINE) TIME IN MINUTES
 A. Kent-Owens, middleclamp,3" handle diameter .0701 + .0632W
 B. Kent-Owens, tailstock,4" handle diameter .0723 + .0651W
 C.
 D. K & T, tailstock, 5" handle diameter   .0903 + .0813W
 E.
 F. Cinn., tailstock, 3 1/2" handle diameter .0631 + .0568W
 G. Cinn., middleclamp,3 1/2" handle diameter .0756 + .0681W
 H.

2. CUT KEYWAY, USE TABLE II FOR CORRECT MACHINE
 Cutter size _____; Feed _____; Speed _____

3. UNLOAD TO TABLE TOP (USE APPROPRIATE LINE AND MARK THIS LINE)
 A. Kent-Owens, middleclamp,3" handle diameter .0684 + .0639W
 B. Kent-Owens, tailstock,4" handle diameter .0527 + .0483W
 C.
 D. K & T tailstock, 5" handle diameter   .0704 + .0640W
 E.
 F. Cinn., tailstock 3 1/2" handle diameter .0476 + .0445W
 G. Cinn., middleclamp,3 1/2" handle diameter .0631 + .0590W

4. HANDLING TO AND FROM RACKS (CALCULATE APPROPRIATE LINE)
 A. Kent-Owens (Extend when piece is more than 14.5 lbs)
     .0329 + .0289W _____
 B. Cinn. (Extend when piece is more than 25 lbs)
     .0296 + .0260W _____
 C. K & T (No extension to total on this line)
     .0332 + .0291W _____

5. CYCLE TIME CORRECTION (Disregard this item of an extension to
         total cycle time was made in item 4)
   Time for A, B or C of item 4 + _____
   Subtract machine time, item 2 - _____
  If difference is positive extend to
  total cycle column   = _____

CALCULATIONS BY _____
       TOTAL BASE TIME, MINUTES _____

CHECKED BY _____
     ALLOWANCE: 5% Pers.; 3% MH _____

    TOTAL STANDARD TIME IN MINUTES PER UNIT _____

    STANDARD TIME IN HOURS PER 100 UNITS _____

    PRODUCTION PER HOUR AT STANDARD _____

Figure 41.5. Calculation sheet for computing standard time from standard elemental data for cutting keyways on milling machines. (The symbol w indicates the difficulty adjustment for the weight of the piece of stock.)

670

(b) 1.0, 0.5; 2.0, 1.5; 2.5, 3.0; 3.0, 4.5; 4.5, 6.0; 6.0, 6.5; 8.0, 7.5; 9.0, 8.0; 10.0, 8.5.

41.2. Plot the data of:
(a) Problem 41.1(a).
(b) Problem 41.1(b).

41.3. Comment on the desirability of using the straight line for the data of:
(a) Problem 41.1(a).
(b) Problem 41.1(b).

41.4. Write a work-unit dictionary for the maintenance of:
(a) A shop.
(b) A school.
(c) A supermarket.
(d) A standardized company gas station.
(e) A motor vehicle.
(f) A chain of motels.
(g) Any assigned facility.

41.5. Suggest a work measurement method for the part of Problem 41.4 worked on.

41.6. Prepare an application sheet, without actual time values, for use in the facility worked on in Problem 41.4.

42

NOTES ON DIRECT TIME STUDY—EXTENSIVE SAMPLING

Introduction

This chapter treats three additional aspects of direct time study—extensive sampling. First, a procedure will be examined for making observations at other-than-random-or stratified-random intervals. Second, a situation wherein the work-unit is such that it cannot be reported by a singular work count will be dealt with. A single example will be used to illustrate these first two subjects. Third, additional statistical control techniques, designed to more fully evaluate the reliability of sampling data, will be described. In the first discussion, to continue to broaden the reader's view of the range of applicability of the techniques, the illustration is drawn from a service activity.

Sampling at Nonrandom Intervals
and with Multifactor Work-Units

Introduction. A technique of some special type is needed with work-units such as *Sausage manufacturing maintained in compliance in-plant*, a work-unit described in Chapter 30 as being produced by the Meat Inspection Service of the U.S. Department of Agriculture. This work-unit presents a work measurement problem quite different from the work-units described in previous chapters. However, the Meat Inspection Service has 16 work-units with similar characteristics produced in over 3,000 locations.

Many other government regulatory agencies have similar work. The work of roving in-plant inspectors in private industry, salesmen, and similar job-holders have many aspects in common with the work of the government regulatory inspectors. There is much need for an adequate technique for determining and allocating required staff resources for such activities.

The tasks performed by the sausage (or other process plant) inspector are neither cyclic nor repetitive. The inspector of the process-type plant travels in a random fashion throughout the plant, responding only at times to specific work-generating systems, such as the before-start-up sanitation check, checking incoming shipments of raw materials, or placing government seals on products moving under federal control. Most of the time is spent maintaining surveillance, random spot-checking of meat ingredients, spices, labels, net weights, continuing plant sanitation, employee hygiene, and so forth. A motion-picture record of such an activity would be difficult to make, although it could be made. However, how one would analyze it leaves much open to question. The predetermined time system approach is clearly inappropriate. Certainly, some other technique needs to be employed.

The standard time must fit all places where the work is done. It must be remembered, in the situation being discussed, that we are not seeking a work measurement coefficient that will tell us only how much inspection capability is required for the situation which is the subject of the work measurement study, although this is the usual condition for work measurement. Rather, we are seeking to relate various dimensions (a multiplicity of factors) which might characterize any establishment to the need for inspection capability. To restate this, rather than a work measurement coefficient for a *specific* plant inspected, we are seeking the standard data coefficients for converting *any* plant of the general type serviced to a statement of the required inspection capability resource. We are seeking a coefficient for converting a quantity of workload (the pertinent dimension of each and all of the plants of the type under consideration) to a quantitative expression of required resources (inspection capability). This type of work measurement problem is akin to many other diverse government activities.

Detailed logging was unsatisfactory. It was found that it was not possible to record, by manual means or even by voice tape, a running account of what any inspector did, together with the time for each part of his activity. The work pattern was so varied that while one set of acts was being recorded (together with their time and appraisal of pace) another set of events had occurred and was missed in the recording. Further, the quantity of records that could result from the observation of one inspector would have been enormous, and from the group needed to eliminate the effect of a few individual's personal habits, the quantity preposterous. Obviously, some type of sampling was needed.

Sampling problems and their solution. Inasmuch as the inspectors worked by themselves, or completely independently from all the other inspectors in a multi-inspector establishment, it seemed clear that random trips to observe the inspector would either be a hopeless task or would produce bias in the data. The observer needed to stay with the inspector. If the observer was with the inspector at all times, a periodic type of observation would make better use of the observer's time. The observer also, as with the normal time study method, could evaluate the pace whenever the activity that was being performed, at the time an observation was made, was of such a nature as to permit this to be done. Hence, bias by the inspector's work pace could be removed.

Work-count problems and their solution. When the matter of a work count is considered, new problems arise. In contrast with most situations subjected to work measurement, the variables that affect the requisite capability of process inspection in a plant have a much more tenuous effect. Also, they could not be ascertained a priori with any certainty. This is another way of saying that the proper method of counting the work-unit could not be determined before the studies were completed. Hence, data concerning all possible ways of counting the work-unit were collected. A number of sets of complete observed time-use data, together with data relating to the possible work counts, had to be obtained in different plants so as to permit examination by mathematical techniques to determine which, if any, ways of counting had any real relationship and, if so, in what manner.

Multifactor work-unit, defined. With indirect work, situations like the sausage inspector's job are frequently encountered. The statistical interface (see Chapter 30) between the 4th-order work-units and the 5th-order work-units is either so complex, or the relationship between the two orders of work-units is so tenuous, that the necessary work time cannot be attributed to a singular work count. The work-units produced have a multifactor relationship with the time required: hence the term "multifactor work-unit."

Preparing for the study. Four veterinarians were trained to carry on the work measurement study. Three particular features are worth noting. First, the four veterinarians trained (and assigned for 6 months to this activity) were normally supervisors of inspectors of processed meat. They had also done the work themselves. They had considerable professional pride in seeing a job done correctly and had great insight into the work. Second, Donald Houston, the veterinarian who trained them, fully understood the objectives, the work-unit structure of the agency, and "stayed with" the work measurement activity to assist in any difficulties encountered. Thus, the work was a regular managerial activity rather than a nuisance project. Third, the group, as a group, made several practice studies to detect any defects in the procedure and to see if all eventualities had been anticipated. Subsequently, the procedure was reduced to its final form. These "dry runs" eliminated much potential backtracking after the study started and should be considered as an almost absolute necessity when this or any sampling technique is used.

Making observation and collecting work counts. The actual routine of the study was relatively simple. For 6 months, each of the four observers went each week to a different meat-processing plant. The plants were selected so as to give a wide variety of types and combinations of processing, and so as to cover all sections of the country.

Travel and initial observations of how effectively the inspector-to-be-observed maintained the plant in conformance with federal standards occupied each Monday. On Tuesday the observer followed the inspector all day so as to habituate the inspector to being observed. Also, values of some of the work counts were collected and arrangements made to obtain the work counts for the days during which actual sampling studies were to be made. On Wednesday and Thursday the actual sampling studies were made and values of the work counts appropriate to those days were collected. Friday was devoted to reducing the data to summary form for submission to the central office and for return travel to the observer's home station.

In this fashion 96 sets of sampling observations and related values of work counts were collected. Each set consisted of the data concerning the work counts and summarized data concerning the performance, together with a narrative description of the plant and its management so as to provide a basis for further analyzing the data in case they did not yield to routine approaches.

The following data concerning the work counts was collected separately for each 4th-order work-unit, that is, for each type of meat processing over which surveillance was maintained (e.g., sausage manufacture, meat smoking, meat curing, edible oil rendering, and so forth). (The listing also contains the instructions given the observers in order to make sure that the work counts were collected in a consistent fashion.) It will be noted that the items appear as variables unless one realizes that had they been phrased fully, the items would have appeared as: *square feet maintained under surveillance, number of rooms maintained under surveillance*, and so forth.

- a. *Square feet*. Review blueprints or step-off area when small enough. Include only the area used for the major categories to be studied (e.g., sausage, curing, etc.).
- b. *Number of rooms*. A room is a walled-off enclosure. If the room is used for more than one major category, count under each category.
- c. *Pieces of equipment*. Use the special form prepared for this purpose. Pieces of equipment, if used for more than one major category, should be counted under each category of use. Do not include small pieces of equipment such as smoke trees.
- d. *Floors*. This refers to multilevel operation. Count only the number of floors used for the major category.
- e. *Positions*. Count the number of house employees working in each major category. If the employee works in more than one major category, count in each category.
- f. *Base items*. A major processed item such as fully cooked ham, sweet pickle ham, all-meat bologna, belly bacon, jowl bacon, and so on.
- g. *Products*. The number of separately labeled outputs made during the observation period: for example, $\frac{1}{2}$-lb bacon, 1-lb bacon, hotel-pack bacon, etc.; Cryovac ham, boneless ham, sliced ham.
- h. *Batches*. In sausage, count the number of chopper loads; in curing, count the lots pumped; in smoking, count the smokehouse loads. If possible, estimate the size of the batches.
- i. *Lines*. The number of automated, mechanized, or manually coordinated packaging lines set up. For example, if six employees are packaging bacon from one slicer, count as one line.
- j. *Pounds*. Total pounds of product processed or held under the supervision of the meat inspector being logged, exclusive of pounds sliced or packaged. Product held includes ham in cure, sausage in dry room or cooler, dry salt products in storage, and so on.
- k. *Pounds sliced*. Number of pounds of products sliced under the major category.
- l. *Pounds packaged*. Number of pounds, under the major category, packaged into self-service packs: for example, 8-oz franks, 1-lb franks, and so on.

m. *Samples taken*. Samples taken for submission to government laboratory.

n. *Required samples per week*. Number of samples that must be taken
per week for submission to the government laboratory to conform to
required inspection procedures. This may vary, depending upon
station policy.

o. *Seals*. Number of government seals placed.

p. *Shipments*. Number of incoming government-sealed shipments
accepted.

q. *Export certificates*. Number of export certificates issued.

r. *Work spell*. Length of work spell of the inspector.

The data on performance were recorded over a two-day period during
which each observation made was classified both as to major category (or
4th-order work-unit), such as "smoked product inspection provided" or
"cured product inspection provided," and so forth, as well as by subcategory,
or purpose, such as with respect to sanitation, ingredients, spices, net weight
of packages, labelling, laboratory sampling, or temperature check, as appropriate. (Not all subcategories were pertinent to all 4th-order work-units.)
Further, each observation was evaluated as to necessity and pace. If the
observer noted that the inspection activity was not necessary, it was categorized as "excess." If the observation was recognized as being of a necessary
activity and was manual, it was rated. If not manual but cognitive, it was
accepted at face value, that is, given a 100 per cent rating. This part of the
procedure was akin to the handling of observations with normal direct time
study–extensive sampling.

To continue, in addition to the categories described, the observer also
had spaces on his recording form for indicating observations of the following
categories:

1. Idle or personal time.
2. Travel inside the plant when such travel could not be ascribed to
   a 4th-order work-unit (the usual case as reflection will confirm).
3. Travel external to the plants, when an inspector had a multiplant
   assignment.
4. Operational supervision; the inspector was obtaining conformance
   to federal standards by his visible presence, although he was not
   physically active.[1]

---

[1] Real insight into activity is necessary to separate this from "idle," a further reason for
the selection of the type of observer used. The observational procedure was designed to
remove any bias caused by inspector method.

Initial computations. The recorded performance observation data were summarized for each 4th-order work-unit (and its subcategories) as well as for the additional categories. A time attributable to each separate category was computed by the following formula, which is the mathematical statement of the basic sampling theorem:

$$T_a = T_t \times \frac{n}{N} \times \bar{R}(1 + \text{Adj.}) \times (1 + A)$$

where

$T_a =$ time attributable to a proper performance of the work-unit
$T_t =$ total time covered by the observations
$n =$ number of observations of the particular category
$N =$ total number of observations of all categories
$\bar{R} =$ average rating of all the observations of the category
$(1 + \text{Adj.}) =$ factor to include the difficulty adjustments (objective rating was used)
$(1 + A) =$ additive to provide a predetermined amount of personal time, wherein A is a decimal expression of the portion of time so allotted by policy

Adjusting the data for length of work spell. When all the sets of data had been collected, the analysts further adjusted the performance data so as to make all the data comparable. To achieve this adjustment, the time per category was converted to the per cent of the day taken by the category. Inasmuch as each inspector had only one day of capability available per day, this percentage is a measure of the per cent of the inspector's capability required to produce the work-unit. If the work-unit is further recognized as requiring the provision of the capability daily for a year, the per cent thus computed for a data set may be thought of as the per cent MY (man-year) required under the configuration of work-units associated with the particular data set.

Summary of kinds of data available from the study. The analysts had for each different 4th-order work-unit, and for each establishment, a number of sets of data that gave the per cent MY capability and the values for all work counts that constituted the dimensions describing the establishment.

Analysis of the data. By various mathematical techniques[2] the pertinent work counts were determined and evaluated with respect to their singular and combined relationship with the required staff capability.

---

[2]Correlation, multiple regression, and linear programming, as well as curve fitting, were used.

Evaluation also included an examination of the credibility of the results. For instance, the required MY capability for *Sausage manufacturing maintained in compliance in-plant* was finally determined to be a function of:

$$E + P + W(10)$$

where (with respect to sausage manufacture)

$E =$ number of pieces of major equipment
$P =$ number of plant direct-labor employees
$W =$ average hourly weight of product, thousands of pounds

There is no aspect of the equation above which strained the credulity of the loggers (who had experience at this work). It seemed reasonable that the number of pieces of major equipment (a measure of the sanitation-problem-including area), the number of employees (a measure of the number of places violations could occur), and the production (a measure of the amount of product that had to be observed) all affected the required capability. Further, the plotting of the values represented by the equation $E + P + W(10)$ against the MY of required capability for each data set indicated a good fit; there were no data points that suggested the existence of sizable unexplained variation. The plot of the data sets for the sausage inspection activity appears in Figure 42.1.

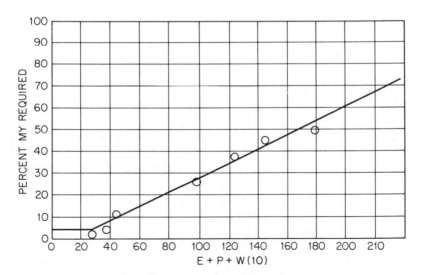

Figure 42.1. Weighted values of work counts plotted against MY of required inspection capability for "Sausage manufacturing maintained in compliance in-plant."

In a similar fashion, other 4th-order work-units were found to be determinable as follows (the values relate to equipment, personnel, and so forth, in the plant, assigned to the specific product group):

Product Group	*Factors and Relationship with MY*
Canning	$MY = f(BI + PR + 3W)$
Smoked product	$MY = f(0.5E + P + W)$
Curing	$MY = f(A + E + 2P + 5R)$

and so forth, where
f = function of
BI = number of base items
PR = number of products made during average day
W = average hourly weight of product, thousands of pounds
E = number of pieces of major equipment
P = number of direct-labor personnel in plant
R = number of rooms

Putting the results into a format for application. The final form of the data for *Sausage manufacturing maintained in compliance in-plant* is shown in Figure 42.2. Similar tables were developed for all 4th-order work-units related to processed product. Internal travel was found to be a function of the area of the plant devoted to each different type of 4th-order work-unit (e.g., a sausage plant generated an internal travel workload of 0.12 per cent of a man-year per 1,000 square feet, canning 0.04 per cent, smoking 0.18 per cent, and so forth).

Results. The data described to this point cover all activities performed within federally inspected processing establishment.[3] These standard data permit the dimensions of each establishment served (the quantity of workload) to be converted to a statement of the quantity of staff resources required for each establishment. Previous staff allocations were made on the basis of pure judgment. Work measurement was used to introduce a rational, consistent, and equitable distribution of work to individuals.

---

[3]There are some other minor activities which are performed and for which work measurement coefficients have been developed but which have not been discussed. The discussion of these would only add length to this presentation.

```
PROCESSED MEAT INSPECTION DIVISION
NON-REPETITIVE ASSIGNMENT STANDARD : #1

                    ACTIVITY : Sausage

Variables included: (1) Equipment (E) per standard list
                    (2) Plant employees (P)
                    (3) Average hourly production (W)
                    (4)
                    (5)

Equation:  Base computed value (BVC) = E + P + W(10)
```

BVC	Basic % M.Y.	BVC	Basic % M.Y.	BVC	Basic % M.Y.	BVC	Basic % M.Y.
22	1	97	26	172	51	247	76
25	2	100	27	175	52	250	77
28	3	103	28	178	53	253	78
31	4	106	29	181	54	256	79
34	5	109	30	184	55	259	80
37	6	112	31	187	56	262	81
40	7	115	32	190	57	265	82
43	8	118	33	193	58	268	83
46	9	121	34	196	59	271	84
49	10	124	35	199	60	274	85
52	11	127	36	202	61	277	86
55	12	130	37	205	62	280	87
58	13	133	38	208	63	283	88
61	14	136	39	211	64	286	89
64	15	139	40	214	65	289	90
67	16	142	41	217	66	292	91
70	17	145	42	220	67	295	92
73	18	148	43	223	68	298	93
76	19	151	44	226	69	301	94
79	20	154	45	229	70	304	95
82	21	157	46	232	71	307	96
85	22	160	47	235	72	310	97
88	23	163	48	238	73	313	98
91	24	166	49	241	74	316	99
94	25	169	50	244	75	319	100

Figure 42.2. Standard data for "Sausage manufacturing maintained in compliance in-plant."

Additional Statistical Controls for Direct Time Study—Extensive Sampling

Introduction.   In Chapters 8 and 23 data were given with respect to the number of observations needed to obtain reliable samples. Additional statistical controls may be used to check on:

1. The acceptability of each day's sample.
2. The reliability of the average $\bar{p}$, or per cent of time spent on each category, as indicated by all the samples available at any particular point in the study.

681

How the controls are employed. The manner of employing these statistical controls is as follows. Let us assume that our p_1 from the first day was 24, or 24 per cent, or 0.24 of the total time. Let us further assume that during the second to fifth day p_1 was 0.26, 0.24, 0.22, and 0.26, respectively. The cumulative average $\bar{p}$ can be found by averaging these, provided that the total number of observations each day was identical. If this condition is not met, then $\bar{p}$ may be found by

$$\bar{p} = \frac{\sum n}{\sum N}$$

where

$\sum n$ = sum of the number of observations of p on all days through the date for the calculation

$\sum N$ = sum of the number of all observations through the date for the calculation

In the case being discussed,

$$\bar{p}_1 = \frac{0.24 + 0.26 + 0.24 + 0.22 + 0.25}{5} = 0.24$$

A control chart for p may also be plotted, as with normal work sampling observations, to check on the possible occurrence of nonrandom, assignable causes which may affect the validity of a day's sample, or for trends or changes in the total situation. Likewise, the reliability of $\bar{p}$ may be checked to determine when the study has reached the desired level of accuracy.

The limits for the daily control chart for p may be plotted around the first $\bar{p}$ computed from the first five values of p. Subsequent calculations will give the reliability of $\bar{p}$.

The upper and lower control limits for p are based on $\bar{p} \pm 2\sigma_p$, where $\sigma_p = \sqrt{\bar{p}(1 - \bar{p})/N}$ and N equals the number of observations made per day. The reliability of $\bar{p}$ is based on $\bar{p} \pm 2\sigma_{\bar{p}}$, where $\sigma_{\bar{p}} = \sqrt{\bar{p}(1 - p)/\sum N}$ and $\sum N$ is the grand-total number of all observations. The 2σ limits were chosen to maintain the same reliabilities as with readings for direct time study–intensive sampling.

An example. Figure 42.3 is a control chart from the S & C study of Chapter 23 for the work unit *door installed*. It will be noted that the daily value is frequently *out of control*. Each out-of-control point has two possible causes:

1. The sample was biased by poor methods of observation.
2. The work-unit occurred an unusual number of times.

CONTROL CHART
FOR DTSES

ASSEMBLY	Area observed
DOOR	Work-unit
DOORS	Category
A.M. & M.O.	Observer
4-18 — 5-15	Dates of observations

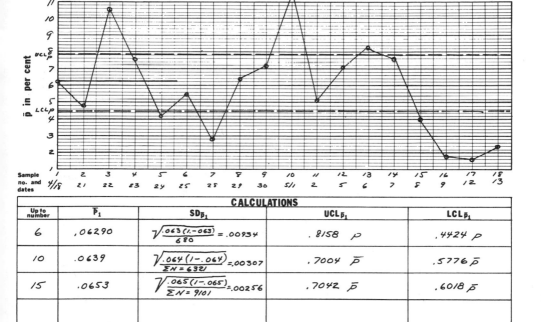

CALCULATIONS				
Up to number	$\bar{P}_1$	$SD_{\bar{p}_1}$	$UCL_{\bar{p}_1}$	$LCL_{\bar{p}_1}$
6	.06290	$\sqrt{\dfrac{.063(1-.063)}{680}} = .00934$	.8158 p	.4424 p
10	.0639	$\sqrt{\dfrac{.064(1-.064)}{\Sigma N=6321}} = .00307$	.7004 $\bar{p}$	.5776 $\bar{p}$
15	.0653	$\sqrt{\dfrac{.065(1-.065)}{\Sigma N=9101}} = .00256$	.7042 $\bar{p}$	.6018 $\bar{p}$

Figure 42.3. Control chart for daily p_1 assemble doors. (Courtesy of Industrial Engineering Dept., S & C Electric Co., Chicago, Ill.)

Each out-of-control point on Figure 42.3 was separately investigated. Each out-of-control point was traced to a fluctuation in the number of doors installed that day. However, the time standard sought was for *door installed* rather than for *average doors per day installed*. Hence, the data did not need to be rejected as would have been done had the cause been of the first type indicated.

Summary

More complex studies and more refined statistical controls were examined. The aim was to demonstrate how the technique can be altered to fit problems; thus, great applicability may be achieved. Further, it was shown how the techniques of statistical quality control may be of assistance in evaluating DTSES data.

PROBLEMS

42.1. Five telephone repairers were observed by means of systematic sampling, by five observers, during a 4-week (160-hour) period. A daily lunch break of 1 hour was excluded. Observers, accompanying the repairers, recorded their activity category at 2-minute intervals. Whenever possible the observations were given an objective rating. The data obtained, when summarized, were as given in the table (which has been simplified for this problem).

Activity Category	Number of Observa- tions	Rating (Average)	Difficulty Adjust- ment
1. Drive repair truck	3,118	100	N/A
2. Make minor vehicle repairs	958	85	12
3. Make call; no one home	719	75	12
4. Climb pole, select pair, connect	1,197	72	45
5. Replace defective instrument	3,118	90	17
6. Repair broken leads in house	2,551	60	60
7. Run in new pair for second phone	1,042	64	60
8. Idle, rest, personal	1,918	100	N/A
9. Lunch	720	100	N/A
10. Discuss problem with householder	2,397	100	N/A
11. Avoid dog	1,680	145	40
12. Make circuit check after work	960	100	N/A
13. Wrong-address call	1,680	75	12
14. Write up unit job report	1,920	100	N/A

Work-count data
  1. Phone faults corrected (10 on pole)   224
  2. Second phones installed   102
  3. Wrong-address calls made   32

(a) What are the theoretical number of observations that should have been recorded?

(b) What is the standard time per phone fault corrected?

(c) What is the standard time to connect a new phone?

(d) Comment on potential savings in the system.

42.2. The daily per cent of time spent checking stock in a warehouse and the line items checked, for a 20-day sampling study, was as shown in the table.

Date	Per Cent	Line Items	Number of Observations
12/9	42	1,024	1,200
12/10	45	1,080	1,190
12/11	43	1,060	1,204
12/13	48	1,100	1,185
12/14	41	1,020	1,200
12/17	46	1,130	1,180
12/18	43	1,020	1,090
12/19	47	1,190	1,190
12/20	42	980	1,200
12/21	43	1,075	1,186
12/24	21	475	685
12/26	15	380	528
12/27	45	1,010	1,190
12/28	47	1,200	1,200
12/29	49	1,090	1,204
12/30	20	500	684
12/31	18	450	484
1/4	17	380	668
1/5	43	1,120	1,020
1/6	42	960	1,200

(a) Plot the values of p by days.
(b) Add control limits to the plot of part (a) on the basis of the first 5 days' data.
(c) Comment on possible reasons for out-of-control points, if any.
(d) If any out-of-control points exist, should they be discarded?
(e) Compute the final $\bar{p}$.
(f) Compute the range of the probable true value of:
    1) $\bar{p}$ based on first 5 days' data.
    2) Same, 10 days.
    3) Same, 15 days.
    4) Final $\bar{p}$.

43

NOTES ON THE FRACTIONED PROFESSIONAL ESTIMATE

Introduction

In Chapter 24, in which the fractioned professional estimate was introduced, two illustrations were given. Both of these were, in the broad sense of the word, repetitive. The FPE was shown to be extremely useful (and rapid of application) with the type of work shown in those examples. However, the FPE approach has a much broader field of application, as will be seen in the three examples that follow. The first deals with the use of the FPE for a unique, one-time project; the second deals with standard times for supervisors. The third deals with an additional way of making an estimate.

Case I—The Use of the FPE
for a Project-Type Work-Unit

Project, defined

"The word *project* is used to describe all of the activity associated with the attainment of a goal when the activity necessitates some extensive, unique responses, different, in many details, from the previous performance of tasks employing the same area of knowledge or different in that considerable ingenuity may be required in devising or developing an answer.

"A project exists when the goal is to produce an advance in the state of an art, to develop a design with unique features, to select a preferable course of action from among many alternatives, and so forth.

"The meaning of the word *project* will include the performance of the design activities within an industrial enterprise, such as physical plant design; product—process design (including research and development); man—job design (including the relationships between man—jobs); control-system design; or the design of a method of employing a specific technology to handle a problem within a control system, e.g., the design of a mathematical model and the development of the method of solving such a model. A pure research activity would also be described as a project or a series of projects, depending upon the singularity or multiplicity of the separable goals, as well as upon the level of control being examined. A project, therefore, may be considered as a unique 5th-order work-unit, a one-of-a-kind output, containing highly individualized problems."[1]

Projects and network diagrams. A project, in some cases, may be represented by a network diagram, as in Figure 12.1, or each step of such a network may be considered a project, depending upon the level on control being examined. Whichever is the case, the FPE used to develop the time values shown in Figure 12.2 would be made separately for each activity in the network.

Each activity in such a network, or any separable project that is not part of a network, should have a *project plan*, whether an FPE is to be made or not. A project plan is merely a listing of the separable, discrete, homogeneous steps contemplated as necessary to achieve the goal. The conversion of a project plan to an FPE requires only the use of experience to estimate the time for each step.

---

[1] M. E. Mundel, *A Conceptual Framework for the Management Sciences.* New York: McGraw-Hill Book Company, 1967, pp. 243–244.

An example of a project plan. Part of a project plan converted to an FPE is shown in Figure 43.1. Note that each separated step begins with an action verb. Such a format assists in keeping the steps homogeneous. However, it does not seem appropriate to make this too rigid a rule. Too many small steps make a project plan confusing; steps that are too large for estimates defeat the purpose. Common sense must be employed. However, in most cases, literature search, work performed on travel status, experimental setup design, experimentation, computation, consultation, and so forth, usually should be recognized as separate steps.

TECHNICAL PLAN — LIVESTOCK SLAUGHTER INSPECTION STAFF PROJECT 67

Project Description and Objective:

To investigate present meat industry handling practices and methods, prior to and after freezing, of meat byproducts (offal) and to determine and recommend, if needed, adequate handling practices, freezing methods and controls to assure wholesomeness of meat byproducts intended for both domestic and export markets. This project will consist of three (3) subprojects.

Subproject 1

This subproject will consist of a study of current industry offal handling practices and freezing methods. This study will be limited to livers, kidneys, hearts, and tongues. This subproject will consist of four (4) phases.

PHASE I

Review literature and consult scientific experts in government, industry, and trade associations to determine recommended chilling, freezing, and holding temperatures for meat byproducts.

Man Hours	Step	
40	A.	Search published scientific sources for chilling and freezing standards and time-temperature relationships for meat byproducts.
24	B.	Search published trade association (American Society of Heating, Refrigeration & Airconditioning Engineers, etc.) sources for recommended product freezer and freezing standards.
24	C.	Consult with USDA scientific experts (microbiologists, food technologists, etc.) for unpublished scientific opinions and recommendations for freezing of meat byproducts.
20	D.	Search for and solicit industry opinions on sound industry procedures for packing and freezing meat byproducts.

PHASE II

This phase will consist of the design of a study.

Man Hours	Step	
6	A.	Consult with statistical staff to design statistically sound survey plan of industry meat byproduct handling practices and freezing methods.
12	B.	Determine type of temperature measuring devices needed to conduct survey.

Figure 43.1. Part of a project plan converted to an FPE.

Advantages of a project plan. The making of a project plan has several advantages. First, it provides a basis for a discussion between the person undertaking the project and his supervisor concerning the feasibility of the contemplated sequence. Second, a file of such project plans (after their completion) provides a record of the "know-how" for performing such work. This record is invaluable in orienting new members of project group in the manner in which work is done. Third, it facilitates the project workers proceeding in an orderly fashion. When the FPE's are added, the value of the project plan is enhanced. These FPEs provide a basis of agreement between the person undertaking the project and his supervisor concerning the depth to which the project is to be undertaken. Second, the "know-how," when the project is completed, includes the knowledge of the amount of manpower associated with steps of such projects. Third, it also orients the new employee with respect to the depth to which the typical project is pursued.

Additional values when the plan has FPEs. The project plan with its FPEs has still additional values related to managerial control. When the time dimensions of each project are known, the person undertaking the project can examine his current calendar of work and determine when he can complete each step, phase, or total project. The apparently feasible schedule may be determined. If such a schedule is satisfactory, the subsequent performance may be monitored by the person on the project and assistance sought when schedule nonconformance appears likely, as well as when subject-matter assistance is needed. A better managerial climate with less "expediting" can be obtained. If, on the other hand, the feasible schedule is not suitable, the FPEs of the project as well as other work on the calendar will be examined. The alternatives of changing project schedules, dropping projects, altering the various FPEs to change the depths to which various steps of various projects are to be pursued, and so forth, will be examined. Such a procedure makes it much more likely that the most suitable alternative will be selected than in a situation where neither project plans nor FPEs exist and chance determines the alternative that occurs.

Use with reporting systems. Simple reporting systems comparing the FPEs to actual times are frequently used to assist each person on project work to aid in maintaining managerial control; guidance is sought when nonconformance to schedules appears to be in the offing.

It can be seen that if the FPEs are unrealistic, much corrective action will take place, although attention will be focused on the cause. However, managerial action to minimize the effect of schedule failure can be taken. Also, subsequent FPEs will reflect the details of experience more accurately and will more closely approach reality.

Case II—The Use of the FPE
to Set Supervisory Standards

A variety of advantages. The application of the technique to supervisory activities has other benefits besides the determination of how much supervisory work can be given to a supervisor. The development of an FPE for a work-unit consisting of supervision is, in most cases, also the first detailing of how the task of supervision is to be performed, although certainly such detail would seem to be a prerequisite for obtaining a known quality of supervision. Of course, if how supervision is to be performed has already been documented, the development of the FPE is much easier; unfortunately, this is seldom the case. Further, the FPE represents, to a large extent, an agreement between the upper manager and the supervisor concerning how time will be spent; hence, much less time is needed for what is otherwise normal routine discussion concerning who should do what. In addition, if effective supervision is not obtained under the pattern of the FPE, then alternative strategies can be evolved and tried.

An example. An examination of the method of using an FPE with supervisory work seems a particularly useful example in that such activity is found in all organizations, both industrial and governmental. For instance, in the meat-inspection activity described in Chapter 30, there is a 4th-order work-unit, *Inspection supervision provided.* The illustration that follows consists of the data developed by means of the FPE approach, by line managers, staff, and experienced circuit supervisors for the assignment of "Officer in Charge" (OIC). The OIC is the supervisor of a group of plants and inspectors in an assigned area in the meat-inspection activity. Although this supervisory task is performed by a government employee, the procedure with an industrial job would be exactly the same. The details, of course, will vary with every supervisory position.

Uses of the FPE. It was anticipated that the data from the FPE would be used in the following manner:

1. The existing OIC assignments would be measured to determine the equity of the assignments.[2]
2. The assignments would be adjusted to provide the proper workload.[3]

---

[2] If this was not an ongoing activity, these assignments could have been synthesized. The procedure would work in the same manner except for this difference.

[3] If adequate manpower is not available, the FPE provides the basis for the OIC and his boss agreeing on what corrective action (e.g., extra time or some work slighted) is most appropriate.

3. The forecasts of new plants by size (number of inspectors) and type for future time periods would be used to determine the increment to the workload of the OICs and to assist in determining when changes in manpower resources would be needed.[4]

Starting to make the FPE. The group[5] making the FPE decided that the work of an OIC could be meaningfully divided into major subgroups:

1. Office activities.
2. District meetings.
3. Administrative travel.
4. Travel between plants.
5. Review and evaluation of program and personnel.

Making the detailed FPE. Each of these subgroups was individually examined and work-units and substeps identified. An FPE was made for each substep. The individual FPEs were added for each work-unit and, with a man-year taken as 2,080 hours of work, the hour values were converted to man-years (MY) to facilitate use. For instance, office duties, meetings, travel, and review were listed and the FPEs made as follows:

	Hours/Month
1. *Office duties*	
a. Assignment lists.	1
b. Daily assignment adjustments.	1
c. Review periodic reports.	10
d. Counsel on construction.	6
e. Correspondence and reports to District Director's Office.	20
f. Telephone contacts (all).	8
g. Arrange and promote training programs.	4
h. Appeals (packer).	4
i. Study memos, letter, regulations, etc., to keep current.	10

---

[4]With an industrial supervisor the items used to structure the FPE would probably be products, lots, machines, workers, and so forth, rather than plants and inspectors. This difference would affect only the details of making and using the FPE. The general procedures of developing and using the FPE would not change.

[5]This is from the meat-inspection activity and represents the results of the coordinating efforts of Donald Houston, a veterinarian assigned to lead the work measurement activity, with a group following only the general procedure set forth by the author; the actual work is by Houston and his group, although I have taken certain liberties to shorten the example.

	Hours/Month

j. Public relations (meetings, etc.).   2
k. Investigate and report on employee actions.   1
l. Evaluation and promotion.   2
m. Contacts within other state and federal agencies.   2
n. Industrial relations.   2
o. Union relations (AFGE).   1
p. Special projects and visitors.   12
q. Interview new employees.   2

  Total = 88 hours/month = 1,056 hours/year   51.00% MY

The details of the other work-units and their FPEs were[6]:

2. *District OIC meetings*
   One meeting per month, all day
   8 hours/month = 96 hours/year   4.60% MY

3. *Administrative travel*
   Two five-hour trips per month
   10 hours/month = 120 hours/year   5.70% MY

4. *Between-plant travel time* (based on dispersion of plants in the circuit)
   a. *All plants in circuit in one metropolitan area*
      1. For plants reporting directly to the OIC
         One trip per month; 1 hour per trip
         1 hour/plant/month = 12 hours/year/plant   0.57% MY/plant
      2. For plants reporting to a subordinate supervisor
         One trip per 3 months; 1 hour per trip
         $\frac{1}{3}$ hour/plant/month = 4 hours/plant/year   0.20% MY/plant
   b. *Plants scattered in one metropolitan area and in surrounding territory*
      1. For plants reporting directly to the OIC
         One trip per month, two hours per trip
         2 hours/plant/month = 24 hours/plant/year   1.14% MY/plant
      2. For plants reporting to a subordinate supervisor
         One trip per 3 months; 2 hours per trip
         $\frac{2}{3}$ hour/plant/month = 8 hours/year   0.39% MY/plant
   c. *Plants scattered all over the territory with no cluster in any metropolitan area*
      1. For plants reporting directly to the OIC
         One trip per month at 3 hours per trip
         3 hours/plant/month = 36 hours/plant/year   1.71 MY/plant
      2. For plants reporting to a subordinate supervisor
         One trip per 3 months; 3 hours per trip
         1 hour/plant/month = 12 hours/plant/year   0.57% MY/plant

5. *Review and evaluation of program and personnel*
   a. *Standard for VS review[7]*
      Review use of procedures
      Antemortem and humane slaughter   1

	Hours/Month
Postmortem	$\frac{5}{6}$
Carcass dispositions	$\frac{1}{2}$
Product review	
Offal: coolers, condemned inedible control	1
Departimental review	
Sanitation, facilities, general operation	1
Administrative review	1
Direct supervision, training, planning	$1\frac{1}{2}$
Habituation and internal plant travel	$\frac{1}{2}$
Total = 82 hours/year =	3.84% MY/inspector
b. *Standard for VI review*	
Review use of procedures	
Antemortem and humane slaughter	$\frac{1}{2}$
Postmortem	1
Carcass dispositions	$\frac{1}{2}$
Product review	
Offal, cooler, condemned inedible control	$\frac{1}{2}$
Departmental review	
Sanitation, facilities, general operations	$\frac{3}{4}$
Administrative review	$\frac{1}{4}$
Direct supervision; training, planning	2
Habituation, internal plant travel	$\frac{1}{4}$
Total = 69 hours/year =	3.31% MY/inspector

[6]The list given here is somewhat shorter than the actual list so as to shorten the presentation. The part shown has been chosen so as to show all the kinds of work-units encountered and how they were used to set standards.

[7]A category of inspector who is also a subordinate supervisor. In the review of a VS the work accomplished by his subordinates is also reviewed. This is why the VS review time is somewhat greater than the VI review time shown later.

Applying the FPE. The following formula was used, together with the FPE data, to compute the supervisory workload associated with any circuit:

$$SWL = 61.30 + T_1 N_1 + T_2 N_2 + VS \times 3.84 + VI \times 3.31$$

where

SWL = supervisory workload, percent of man-years/year
61.30 = percent of man-years for the constant part of the work: (1) office duties; (2) district OIC meetings; (3) administrative travel
T_1 = travel time based on how the plants are dispersed
N_1 = number of plants reporting directly to an OIC

T_2 = travel time based on how the plants are dispersed
N_2 = number of plants reporting to a subordinate supervisor
VS = number of subordinate supervisors reporting to OIC
VI = number of inspectors reporting directly to the OIC

Using the FPEs and the formula above, the Meat Inspection Service determines its supervisory manpower needs. The same data are used to assign work to individuals, constantly appraise the changes in workload, and redistribute the work as feasible. Further, the standards constitute a plan for supervision; if difficulties are encountered, the strategy may be changed in a known fashion.

A Different Way of Making Estimates

Introduction. Multiplant operations, decentralized government agencies, and public utilities have many service activities for which standard times have not been established. There may be a sudden need such standard times to respond to a demand for labor cost information from higher levels of the organization, those in the budget process or rate-setting regulatory agencies, as applicable to the types of organizations cited. Although such standard times could be established by mathematical techniques, sufficient data may not be available. Although such data could be developed, time may be so short that some more rapid technique is desired to serve the immediate need.

What resources the organization has available. In almost all cases, such organizations (as have been mentioned) have experienced supervisors, of the work under scrutiny, who know, but in an unorganized manner, how long such work takes. It is this "data bank" that must be tapped.

The first estimate. One approach is to gather a group of supervisors and focus their attention on the jobs in question, one job at a time, in the following manner:

1. Individually, without consulting each other, they record whether the average occurrence of the job in question takes a matter of:
   a. Months.
   b. Weeks.
   c. Days.
   d. Hours.
   e. Minutes.
   f. Seconds.

2. Having made their decisions individually, they pool their information and discuss the typical response.
3. They agree, after discussion (it can sometimes be heated unless the person chairing the meeting keeps the discussion on a substantive level) on the time scope of the job (1.a to 1.f).

The Delphi method. The procedure of individually estimating, receiving pooled responses, and reestimating is referred to as the *Delphi method.*

Finer estimates. After the general time frame has been agreed upon, a more precise estimate may be made in one of two ways:

1. The supervisors may attempt to individually specify an average time for the work under scrutiny and reprocess it, as with the first estimate, until a value is agreed upon.
2. Each supervisor may attempt to develop a five (or more)-part distribution of times, prior to pooling. With a job which has been agreed upon as taking hours, for example, a supervisor may estimate that:
   a. 10 per cent take 1 hour, on the average.
   b. 20 per cent take 2 hours, on the average.
   c. 65 per cent take 3 hours, on the average.
   d. 3 per cent take 5 hours, on the average.
   e. 2 per cent take 15 hours, on the average.

The use of the distribution seems to force a more ready recognition of the few exceptions, leading to more rapid agreement on a final weighted average.

Summary. These methods have been used with good results. The standard times set can be roughly evaluated by taking a past period of work counts and comparing the earned hours (using the estimated standard times) to the payroll hours worked. There should not be sizable discrepancies between these two values.

Summary

The application of the FPE approach to staff and supervisory work has been shown. When these examples are considered, in addition to those shown in Chapter 24, the broad range of application of the FPE can be appreciated.

PROBLEMS

43.1. (a) Make a project plan for a term paper.
(b) Add FPEs to each step of part (a).

43.2. Make a project plan, together with FPEs, to investigate a recent fire reported in a newspaper.

43.3. Same as Problem 43.2, but with respect to a traffic accident.

43.4. Make a project plan, together with FPEs, for a research project (selected or assigned, as designated by the instructor).

43.5. By observation or discussion, establish an FPE of a supervisor's job in a local plant.

44 NOTES ON STANDARD TIMES BY FIAT

As shown in Chapter 25, a standard time by fiat is a simple concept. However, the most important problem with fiat standards is the problem of properly identifying a situation calling for such standards *when encountered*. Failure to do so can cause, at times, enormous amounts of wasted effort. One such misdirected exercise will be discussed in this chapter.

Ward Secretaries in Veterans' Administration Hospitals[1]

Introduction. In order to reduce the workload of the critically short nursing and medical staff, as well as to improve the atmosphere in the wards,

---

[1]Adapted from M. E. Mundel, *Measuring and Enhancing the Productivity of Service and Government Organizations.* Tokyo: The Asian Productivity Organization, 1975.

697

the Veterans' Administration (VA) created a new position called *Ward Secretary.*

Ward Secretary. Each Ward Secretary worked only in one ward. A Ward Secretary had the following duties:

1. Keep the ward roster, by beds.
2. Post patients' records.
3. Order medication (administering excluded).
4. Distribute mail in the ward.
5. Distribute laundry in the ward.
6. Type patients' VA or other insurance forms, as requested by patients.
7. Obtain newspapers, writing materials, clothing, and so forth, as appropriate, when needed.
8. "Keep an eye" on the ward; be there to do work when needed; provide a "presence" in the ward; and so forth.

The problem. The productivity of each VA hospital, computed on the basis of the total of employees' earned hours divided by the hours worked, was incredibly low. Managers were certain that some large source of error existed.

The work measurement error. Standard times had been set, using predetermined time system data (PTS), for the Ward Secretary's duties, which were numbered 1 through 5 of the list previously given. These work-units had been chosen because work counts could be obtained for these five work-units. An allowance was provided to give time for the duties numbered 6 and 7; 8 was completely neglected.

The development of the standard times by PTS required the listing of all the steps required to do the work, such as for duty 1:

1. Push back chair.
2. Stand up.
3. Walk six steps to patient roster.
4. Reach 14 inches for roster with right hand.
5. Grasp roster.
And so forth.

Subsequently, times for each step were taken from tables of motion times for such steps. The work was laborious, voluminous, and tedious. Much time was used.

What should have been done. As an alternative approach, a short period of observation would have revealed that a Ward Secretary spent a large portion of the available time on duty 8. The assignment of a Ward Secretary was essentially a fiat assignment. The detailed data were unnecessary and erroneous; the important and very time-consuming duty, number 8, had been neglected.

The consequences of the work measurement error. Using the detailed standard times set by PTS, the typical Ward Secretary, during an 8-hour tour of duty, was credited with approximately 4 earned hours. If the earned and worked hours, computed in such a manner for 40 Ward Secretaries, are combined with other employee categories, a low productivity should come as no surprise. In the total VA hospital system, this error source was erroneously indicating an overmanning of approximately 1,600 persons.

Correcting the error. The assignment of a Ward Secretary should have been recognized as a job with a fiat time standard. The work spent at developing detailed standards was unnecessary. A Ward Secretary spending an 8-hour tour of duty in a ward should be thought of as earning 8 hours. Like many fiat assignments, the internal labor productivity of such a job is rather inflexible. The VA, rather than being overmanned by 1,600 people, was really undermanned. Not all wards had Ward Secretaries. In many places where two shifts of Ward Secretaries would have been desirable, staff for two shifts was not available.

The erroneous reports were corrected.

Summary

Fiat time standards are a simple concept. Much effort can be saved if a situation calling for a fiat standard is so identified early in the work measurement process. Situations with fiat standards are not susceptible to productivity improvement by increased physical effort; changes in method or trade-offs between staff and equipment are required.

PROBLEMS

44.1. Would the elimination of the Ward Secretaries in the VA increase hospital productivity? Defend your position.

44.2. From commonly observed experience, indicate at least four common situations where the labor productivity of what would otherwise require fiat manning has been improved by an equipment change. (The instructor may choose to rule out certain examples.)

45

NOTES ON TIME STANDARDS
BY MATHEMATICAL ANALYSIS

Introduction

In Chapter 26, in the section on curve fitting, it was indicated that situations might be encountered where the use of appropriate work-units would cause the number of kinds of work-units to exceed the number of data sets available from past records. A data set, it may be recalled, consisted of information concerning a number of work counts of a variety of work-units, together with the aggregated work time for the total outputs in the data set. In such cases, the normal use of simple simultaneous equations is not feasible. Also, curves cannot be fitted until each work-unit is given a weight. Such a problem did not arise in the use of self-reporting for the work measurement study in the Office of the Solicitor (Chapter 26) because the data were collected so that a work time was found separately for the total work count of each work-unit. It is worth noting, however, that the number of work-units was very large. If the total work time in each location had not been so separated, it is doubtful whether even a mathematical "unscrambling" or attributing of the work time to each separate work count would have been feasible.

This chapter will examine techniques for setting standard times in situations where the records of work time are only available as an aggregated value and the number of work-units exceeds the number of data sets. In such cases either of two mathematical techniques, linear programming or multiple regression, must be used. In addition to the mathematical techniques, a variety of methods for obtaining work-time and work-count data will be discussed.

The Use of Linear Programming and Multiple Regression for Deriving Standards

Field of use. These mathematical techniques are employed when the four following conditions describe the available data:

1. The output data and the associated time data are available only as a series of data sets, with each data set containing work counts of more than one work-unit.
2. The work count of each work-unit in each data set is separated from the work counts of other work-units.
3. The time for each kind of work-unit is not separated in each data set; the work time is available only as an aggregate for each data set.
4. The number of different work-units for which work counts are available exceeds the number of data sets available; the reverse may also be true.

Such a set of conditions is encountered frequently when using historical production and work-time data. The reason for the conditions existing is simple. The data are taken from past records. At the time the events occurred, the work-unit structure was not defined. Hence, no division was made of the total time for the data set.

Examples. For instance, the work-time and work-count data available may concern the installation of ship boilers. The total time attributable to each boiler installation may be a matter of record. So are the drawings of each boiler, but each drawing was not described as a series of work counts of the variety of 4th-order work-units which comprise the 5th-order work-unit, *the particular boiler installed.* Hence, time was not allocated by the 4th-order work-units. The 4th-order work-units and their associated work count may be such as given in the partial list that follows:

(square feet of)	boiler foundation installed
(number of)	boiler tubes installed
(volume of)	header installed
(weight of)	header installed
(number of)	large valves installed

As an additional example, the data may relate to a foundry making very large castings, such as steel rolling-mill frames, or main ship engines. The historical data may be identical in nature to the data for the installation of boilers.

The nature of such data is different from that of the example given in Chapter 23, the metal-clad assembly department of the S & C Electric Company. In that case it was noted that: "The employees work as a group on a considerable number of units simultaneously, shifting back and forth and helping each other as needed. It is difficult to get accurate timekeeping charges against each unit."

The four conditions describing a situation for which the techniques to be described will be applicable will also be found when the relationship between the 3rd-order work-units and the 4th-order work-units is complex and hard to define. Reference should be made to the section on *inspectors of processed meat*, Chapter 42, where multifactor work-units were encountered.

A similar problem will also be encountered when making a work measurement study of some industrial sales forces, public relations groups, press services, or Market News Services or Crop Reporting Services of the U.S. government. It is worth noting that in these situations the development of standard times by means of other work measurement approaches may appear to be (and may actually be) a hopeless task. However, with the techniques that will be described in this section, the development of standard times is a relatively simple task, even in such situations.

The basic mathematical model. In all the preceding situations (with the exception of the S & C Electric Company) the N number of data sets may be represented by a series of N equations, one for each data set, constituting a partial mathematical model of the situation, as follows:

$$\mathrm{BV} + \mathrm{WC}_1 1 \times \mathrm{ST}_1 + \mathrm{WC}_2 1 \times \mathrm{ST}_2 \cdots + \mathrm{WC}_i 1 \times \mathrm{ST}_i + S_1$$
$$= \sum \mathrm{MH}_1$$

$$\mathrm{BV} + \mathrm{WC}_1 2 \times \mathrm{ST}_1 + \mathrm{WC}_2 2 \times \mathrm{ST}_2 \cdots + \mathrm{WC}_i 2 \times \mathrm{ST}_i + S_2$$
$$= \sum \mathrm{MH}_2$$

$$\mathrm{BV} + \mathrm{WC}_1 N \times \mathrm{ST}_1 + \mathrm{WC}_2 N \times \mathrm{ST}_2 \cdots + \mathrm{WC}_i N \times \mathrm{ST}_i + S_N$$
$$= \sum \mathrm{MH}_N$$

where

$\quad$ BV = basic value associated with any level of operation (unknown)

$\quad$ $WC_1 1$ = work count of work-unit 1 in data set 1 (known)

$\quad$ $WC_2 1$ = work count of work-unit 2 in data set 1 (known)

$\quad$ $WC_i 1$ = work count of work-unit i in data set 1 (known)

$\quad\;$ ST_1 = standard time for work-unit 1 (unknown)

$\quad\;$ ST_2 = standard time for work-unit 2 (unknown)

$\quad\;$ ST_i = standard time for work-unit i (unknown)

$\qquad S_1$ = some amount of time (called *slack time*), contained in data set 1, which cannot be explained by the indicated relationships (unknown).

$\sum MH_1$ = sum of the man-hours of work time associated with data set 1 (known) (in some cases it may be more convenient to use man-weeks, man-months, or man-years; whichever unit of time is used, the standards derived by mathematical methods will have the same dimensions as used in the sum)

$\quad$ $WC_1 2$ = work count of work-unit 1 in data set 2, and so forth, down to data set N

Linear programming and *multiple regression* are the names associated with two mathematical routines for finding values for the unknown in such systems of equations.

Linear programming. The use of linear programming provides a solution wherein:

1. $S_1 + S_2 + \cdots + S_N$ = minimum value; referred to as *residual slack*.
2. BV, $S_1, S_2, \ldots, S_N \geqq 0$
3. $ST_1, ST_2, \ldots, ST_i \geqq 0$
4. It is not necessary that all work-units be assigned a standard time other than zero.
5. Further, it is assumed that any work count multiplied by the appropriate standard time is part of a linear function (e.g., a work count of 30 of work-unit 1 takes three times as long as a work count of 10 of work-unit 1, one-half the time of a work count of 60, and so forth).

Multiple regression. The use of multiple regression provides a solution wherein:

1. $S_1{}^2 + S_2{}^2 + \cdots + S_N{}^2$ = minimum value; referred to as *residual variance*.
2. BV, $S_1, S_2, \ldots, S_N \geqq 0$
3. $ST_1, ST_2, \ldots, ST_i \geqq 0$

4. It is not necessary that all work-units be assigned a standard time other than 0.
5. Further, as with linear programming, it is also assumed that any work count multiplied by the appropriate standard time is part of a linear function.

Logic needed in addition to mathematics. An examination of the conditions under which a solution is obtained by either method will indicate that the techniques must not be employed blindly. The solutions reached must be carefully examined and the validity checked with other criteria. For instance, Figure 45.1 shows the plot of a series of data sets and the solutions arrived at by linear programming (LP) and multiple regression (MR).

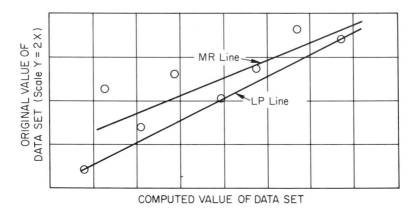

Figure 45.1. Lines fitted to data sets by LP and MR.

The solutions for ST_1, ST_2, ..., ST_i are obviously different for the LP and MR solutions.

Features of the LP solution. The use of LP to obtain a solution assumed that:

1. The average difficulty of the work-units contained within any one work count does not vary significantly from data set to data set.
2. The best performance was reasonable; the other data sets represent siutuations where more work could be done, or unique unmeasured conditions exist.
3. The work counts are parts of linear functions.
4. The relative values assigned to ST_1, ST_2, ..., ST_i are reasonable.
5. The work-units assigned a standard time of zero are not related to the use of time.

Not all of these assumptions are acceptable unless more facts are obtained. The partial mathematical model of the situation wherein only N equations are used is not a complete model. Note also that assumptions 1 and 2 above may be somewhat difficult to verify without a trial application of the answers derived by the use of LP. Assumptions 3, 4, and 5 are different in that they may be avoided by adding supplementary equations to the system of N equations or by converting work counts which are parts of nonlinear functions to become parts of linear functions.

Features of the MR solution. The use of MR to obtain a solution assumed that:

1. The average difficulty of the work-units contained within any one work count could vary significantly from data set to data set.
2. The data sets represent performances of unknown quality; only the average has any significance; it is a performance which, it is reasonable to expect, will be repeated in the future. (This statement can be altered if use is made of an M arrived at by judgment and which is used to modify $\sum MH_1, \sum MH_2, \ldots, \sum MH_N$. In such cases, additional information is needed to develop a value for M.)
3. The work counts are parts of linear functions.
4. The relative values assigned $ST_1, ST_2, \ldots, ST_i$ are reasonable even if they are negative.
5. The work-units assigned a standard time of zero are not related to the use of time.

As with LP solutions, not all of these assumptions would be acceptable without more facts. More information is needed. Also, as with LP, assumptions 1 and 2 are difficult to verify without a trial application of the answers derived by the use of MR. Also, as with LP, additional equations or conversions of work counts may be added to the system of N equations to avoid assumptions 3, 4, and 5. Obviously, the possibility of a negative standard time (part of assumption 4) must be eliminated if the answers are to be believable.[1]

Choosing between LP and MR. From the foregoing comparison, we see that assumptions 1 and 2 should be used as criteria for choosing one or the other of the two techniques. Obviously, each situation must be examined carefully to see which assumptions are most appropriate. Therefore, the general guides that follow, resulting from a variety of applications, are not to be taken as inflexible criteria.

---

[1] In general, a negative standard time indicates some deficiency in the system of work-units used in the N equations, or an incorrect method of counting.

When LP is better. LP appears more appropriate when the work situation has either of the following two sets of characteristics (in addition to the four general conditions given at the start of this chapter):

1. The relationship between the 3rd- and 4th-order work-units is not direct, e.g., the "sausage manufacture surveillance provided" described in Chapter 21, or in Market News Services, sales offices, and so forth. (An example will be given later.)

2. The relationship between the 3rd- and 4th-order work-units is direct and the work count of each kind of work-unit in each data set is so high that a different mix of difficulty from data set to data set is not a tenable assumption.

When MR is better. MR appears more appropriate when the work situation contains one of the following three sets of characteristics (in addition to the four general conditions given at the start of this chapter):

1. The relationship between the 3rd- and 4th-order work-units is direct, as with ship boiler erection or with the making of large castings, and when, in addition, the work count of the work units is so low that a different mix of difficulty from data set to data set is a tenable assumption.

2. The relationship between the 3rd- and 4th-order work-units is direct and the work count from data set to data set is high, but there is reason to believe that the work-units are not really the same from data set to data set for reasons beyond the control of the producing group.

3. The relationship between the 3rd- and 4th-order work-units is direct and the work count from data set to data set is high, but there is reason to believe that there are random errors in most of the work counts.

When LP and MR appear equally appropriate. When work situations not clearly meeting any of the five types given above for LP and MR are encountered, the only reasonable approach seems to be to try both and examine the answers and the situation before making a choice.[2]

---

[2]In addition to the criteria affecting the choice of LP or MR, a few additional criteria have been suggested at various times. These additional criteria are of a somewhat different nature. For instance, LP has been suggested as preferable because it is easier to explain to a group. This may be so, but an explanation of either technique assumes a considerable educational level. If a group can understand one technique, they should be able to understand the other. Further, the time to run an MR-type solution on a computer has been shown to be less than the time to run an LP solution. However, this seems a frivolous criterion; a problem with eight data sets and 10 different work-units per data set may take 12 seconds to run by MR, 24 seconds by LP. In addition, it has been shown that if data are generated

LP or MR, by hand computation. With either MR or LP, hand computations with problems containing many work-units and many data sets can be long and arduous. Some shortcut methods are available for both LP and MR. For large problems, access to a computer and appropriate programs is highly desirable. However, even large computers have limitations concerning problem size. Common limits range from 20 data sets of 30 work counts each (medium computers) to 50 data sets with 200 work counts (large computers). Problems whose size is in excess of these limits are conceivable but not common.[3]

Work Measurement Standards for the U.S. Livestock Market News Service

Introduction. As an illustration, let us examine an application of LP to a service-type activity. This illustration represents type 1 of those situations listed as appropriate to handle by LP. The particular illustration has been chosen for a variety of reasons. First, it represents a type of work for which work measurement has often been thought of as infeasible. Second, the application to a problem of this type requires that many additional types of facts, not contained in the N basic equations, be brought into the problem. Thus, the solution of this problem facilitates the demonstration of many things that could not be demonstrated with a simpler example. The illustration concerns the development and use of time standards for assisting managerial control in the Livestock Market News Service of the Consumer and Marketing Service, U.S. Department of Agriculture.

The objective of the Livestock Market News Service is to facilitate the maintenance of an orderly market by gathering, collating, and distributing timely information concerning the supply and movement of market cattle, swine, sheep, and wool to buyers and sellers. To this end they gather information concerning daily transactions with respect to quantity and price, evaluate the reliability of the information, observe incoming shipments with regard to quantity and quality, and distribute information concerning volume,

---

at random to produce data sets meeting exactly condition 1 of the conditions for which MR is suggested as a technique, MR predicts the place such future data sets would fall on a plot of actual versus computed work time better than LP. This should surprise no one. Figure 45.1 demonstrates that this is the nature of the techniques. Finally, it has been suggested that if a computer is available and only an MR or LP program is available, the available program should be used. As an expedient, this may be so, but if the other program would fit the situation better, it would seem advisable to obtain a program; they are available. The criteria given in the text are suggested as more appropriate than these peripheral considerations.

[3] However, I have not yet encountered problems larger than this.

quality, and price to their local area and to a central location for nationwide distribution.

Previously discussed work measurement methods. It would not appear that any of the work measurement procedures discussed to this point would be feasible in such a situation; the work is nonroutine. Even an FPE would be difficult except on a day-to-day basis; this does not appear as a reasonable solution. Hence, some new approach is needed.

More details about the Service. The Livestock Market News Service maintains offices in 42 cities in the United States. These are cities in which there is an important market in cattle, swine, sheep, or wool. It is important to be able to determine the amount of manpower needed to man each existing office and to periodically evaluate the changes in markets so as to reallocate staff. Also, when a new market develops, work measurement data of the standard data type is needed to determine the required manpower. What is needed is standard data for all 4th-order work-units associated with any 5th-order work-unit, *A (specific) market reported.*

Getting a list of 4th-order work units. People knowledgeable in Market News work were asked to assist in preparing a list of potential 4th-order work-units. (These are the types of "components" from which "a market reported" can be said to be "assembled.")

The 4th-order work-units listed contained the following:

1. Classes of animals reported.
2. Number of terminal market days reported.
3. Number of buyers and sellers associated with the market reported.
4. Auction market days reported by federal personnel.
5. Carload lots of cattle reported at terminal markets (by 1,000's).
6. Carload lots of cattle reported at auction markets (by 1,000's).
7. Carload lots of cattle sold direct reported (by 1,000's).
8. Carload lots of swine reported at terminal markets (by 1,000's).
9. Carload lots of swine reported at auction markets (by 1,000's).
10. Carload lots of swine sold direct reported (by 1,000's).
11. Carload lots of sheep reported at terminal markets (by 1,000's).
12. Carload lots of sheep reported at auction markets (by 1,000's).
13. Carload lots of sheep sold direct reported (by 1,000's).
14. Carload lots of wool reported (by 1,000's).
15. Auction market days reported by state personnel supervised.
16. Number of state reporters supervised.
17. Number of miles to make market rounds per week driven.

It should be noted that, as is appropriate to 4th-order work-units of a service activity, the outputs have been described with a past-tense verb to reduce confusion with activities. Also, all the figures are "dimensions" of the market reported; they are not changeable by choice. Further, the tasks associated with something "reported" are not pinned down; the relationship between the 3rd- and 4th-order work-units is not clear. The situation is clearly identifiable as the first type given as appropriate for LP.[4]

Selecting the data sources. As the next step in the problem situation being examined, the people familiar with the Livestock Market News Service were asked to pick 12 offices meeting certain criteria. The reasons for this selection of 12 from the 42 offices were as follows:

1. As will be noted when the criteria are given, this increased the knowledge concerning that which the data represented.
2. It set aside 30 other data sets to test the credibility of the data obtained.
3. This would reduce the size of the problem to fit the available computer.

The 12 offices were selected to meet the following criteria:

1. The 12 included various sizes of offices, ranging from the smallest to the largest.
2. Only offices were selected that were known to require good reporters and where the workload was felt to be such that the reporters were kept relatively busy at all times. Hence, the N equations would be realistic if the work-units had been chosen well.
3. Only offices were selected that were considered to be doing an effective job of reporting. There seemed to be no point in developing standards for work improperly done.
4. The mix of offices selected included offices such that each work-unit appeared in one or more of the N (12) equations. In this way all work-units would be considered.

Obtaining the work-count and work-time data. Subsequently, the work-count for each of the 4th-order work-units was obtained for each of the 42 offices. The work time for each office was determined from payroll records. The 12 (N) equations were constructed where the WC_1 through WC_{17}

---

[4]There are also market reporting groups for poultry and dairy products, cotton, grain, vegetables, and so forth. The approach described here can be used with all of these and similar information-gathering groups. A press service seems no different.

represented the work counts of the 17 different work-units; S_1 through S_{12} represented the 12 unexplained times; the $\sum MH_1$ through $\sum MH_{12}$ represented the payroll time in the 12 offices; ST_1 through ST_{17} represented the 17 potential standard data values sought.

Processing the data. The data were fed to a computer under control of an LP program. Answers were returned after approximately 2 minutes of computer mainframe time.[5]

Defects in the initial solution. The answers were examined by the Market News people for credibility. For instance, the free operation of the computer produced a value for work-unit 15, *Auction market days reported by state personnel supervised*, but a 0.0 value for work-unit 16, *Number of state reporters supervised.* Discussion among the Market News group led to the conclusion that this was an error; market days reported by state personnel could increase without adding to the workload of federal personnel, but increasing the number of state reporters would increase the workload. Hence, 4th-order work-unit 15 was removed from the equations.

Reprocessing the data and new defects in the answer. A new solution was obtained and work-unit 16 then had a value. However, other difficulties remained. *Terminal swine reported* (work-unit 8) had a value, but *Auction swine reported* (work-unit 9) and *Direct swine reported* (unit 10) did not; the computer gave the standard time for work-units 9 and 10 as 0.0. That work-unit 9 was 0.0 was believable; the state reporters covered the swine auction markets. A 0.0 standard time for *Direct swine reported* was not acceptable.

Correcting for the new defects. The consensus of the Market News people was that the work associated with an increment to a market of *1,000 carload lots of direct swine reported* was less than that associated with *1,000 carload lots of terminal swine reported* but more than that associated with an increment of *one buyer or seller associated with the market reported.* Hence, two new equations were written:

$$T_{10} \leq T_8$$
$$T_{10} \geq T_3$$

Additional corrections. In a similar fashion, many additional equations (called *restrictions*) were added. However, one work-unit that was considered important either failed to enter the solution, or when driven in by restrictions, produced intolerably large slack values in offices that were considered to be efficient. The work-unit was 17, *Number of miles to make market rounds per week driven.* Examination of this variable showed that high values were

---

[5]A terminal having access to a time-shared computer was used.

associated with offices still being shown as having excessive slack, but all offices whose slack had approached a 0.0 value showed about 300 miles for a work count for this work-unit. To remove this nonlinear effect, work-unit 17 was altered to *Number of miles in excess of 300 per week to make market rounds driven.*

The final solution. The final solution had the following properties:

1. It fitted all N equations.
2. It fitted all restricting equations.
3. It produced values for all the variables people knowledgeable in the work felt were important. (The list was somewhat shorter than the original list.)
4. The relative values of the time standards for the different work-units, and so forth, seemed appropriate to those who knew the work.
5. It left only a small amount of residual slack at locations where its existence seemed readily explainable as extra capability. The people who had worked in those offices felt more work could be done if necessary. (However, typical was an office with five people and 4.80 man-years of workload, and so forth.)

Validating the answers with additional facts. Subsequently, the standard data derived from the computer were applied to the other 30 offices to compare the evaluation of the manning thus obtained with all other information available concerning each office. The data were used to determine the man-years of effort per year that appeared necessary to staff each office; this was compared to current manning.

The data indicated slightly more than one extra man-year at one office (slack = 1.1 man-years); this turned out to be the home office of the "relief" man, who substituted for other personnel when they were on vacation. The data also indicated an office where 2 man-years were assigned but only 0.75 man-year of work appeared. This was found to be the office used for training new men; the 0.75 man-year of workload enabled an experienced man to report the market and still have time for training; the additional man-year was the trainee. Several offices were found with excessive manpower; later, site visits confirmed this and the employees were transferred to other types of work in the department. In no case did there appear any conflict between the workload as measured by the standard data and all available facts.

Using the standard time values. The data are currently used to determine annual budgets, and to periodically assess the change in markets and the advisability of reallocating personnel. The data can be used to determine the resources required when a new market reporting office is considered.

Notes on the time and effort spent using this work measurement approach. It is worth noting that the largest amount of time spent by those engaged on the project was in determining the list of potential 4th-order work-units. Work on this was carried on sporadically over a 3-month period. The work counts were determined by each office in about 1 week. Four hours at the computer terminal were used during the actual development of the standard times.

When the organization does not have multiple offices. In the case of a single office at a single location, the necessary N sets of data may be obtained by taking each year or each month as a data set. Provided that the mix of work-units varies from data set to data set, the computational methods, discussed to this point, may be applied.

Work Measurement Standards When Linear Programming or Multiple Regression Are Used and the Work Counts Display Nonlinearity

The basic problem. In some cases, using either LP or MR, the plot of the man-hour data originally associated with the data sets versus the man-hour values for the data sets computed from the derived weights may resemble Figure 45.2.

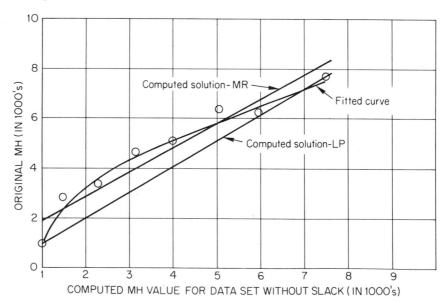

Figure 45.2. Curve fitted to data-set points after inspection revealed nonlinearity.

The solution. Obviously, the linear answer is wrong; a curve must be fitted to the data points as shown in Figure 45.2. (*Note:* The curve was the last item drawn in Figure 45.2; it was added after the nonlinearity was detected.) If the shape of the curve is credible and the fit reasonable, the curve may now be used to construct a table of standard times that does not assume linearity of the work counts. For each computed value of the abscissa, a man-hour time may be determined. For instance, in Figure 45.2, for a weighted value of the work counts of 4,000, a man-hour time of 5,100 is indicated. A table may be constructed for all possible values of the sum of the weighted work counts to produce a standard data table.

Work Measurement Standards by Linear Programming or Multiple Regression When Some of the Work Counts Are Known to Come from Nonlinear Functions

When determining the work units to use in LP or MR computations, information may be available which indicates that the work counts of one or more of the variables will not be part of a linear system. For instance, the time spent pouring molten steel into a large casting will not be a function of the amount of steel. The time for the work-unit, *steel poured*, will consist of a constant time for the tasks of set up and prepare, plus a variable depending upon the amount of steel poured. In a similar fashion, the work count of almost any work-unit produced, in different amounts at different times, may belong to a nonlinear function. The work counts associated with the data sets need conversion to a linear function before the LP or MR solution is attempted. Figure 45.3 is the conversion graph for such a work-unit. If the actual work count is used to enter the abscissa, the "converted work count" which is part of a linear system may be read off the ordinate. The lines added to the curve indicate how an original work count of 19 is altered to a "converted work count" of 22, which is then part of a linear function.

Reducing the Possiblity That "Rationalization" Is Used to Explain Residual Slacks in an LP or Residual Variance in an MR Problem

In the application of LP to the Livestock Market News Service, the slacks obtained in the original solution were carefully examined for credibility after the solution had been obtained. Some may feel that this offers

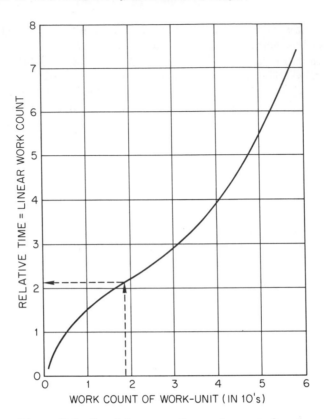

Figure 45.3. Graph for converting work count of work-unit that is part of a nonlinear function to a linear work count.

too much opportunity to "bend" the facts to fit the solution. Hence, in some applications, the anticipated slacks are removed from the $\sum$ MH terms before the solution is attempted in order to reduce the chance of biasing the analysis of the results. Instead of using actual payroll man-hours for each data set, an estimate of the required man-hours is used. If the estimates are correct and rational, the residual slacks in the solution will approach 0.0. This presumes, if the answers are credible, that the basis of the estimates was good. However, the use of the mathematical procedure reduces the basis of these estimates to a quantitative statement of the weights for each component part. These quantitative statements make possible the subsequent computation of work standards without the use of the skill previously employed in making the estimate. Like the FPE, the technique makes it possible to change from a "visceral standard" to documented, repeatable information.

Other Methods of Gathering Work-Count and Work-Time Information

In the case of the office of the Solicitor (Chapter 26) the work-count and work-time data were obtained by having the lawyers record all the work they did and all the time they worked. In some applications, such an approach may produce an overwhelming amount of data. Further, such an approach requires a considerable period of time; in some instances such time may not be available.

Hence, the next two topics will concern alternative methods of gathering work-time and work-count data.

Recording with a sampling "traveler." A special form for recording work time may be attached to every fifth or tenth document. For instance, in a purchasing office, every incoming requisition whose number ends in 0 or 5 may have such a form attached. This form would be used to log the work time of every person who worked on this requisition and the resultant purchase order. The result would be a random 20 per cent work-time sample, matching 20 per cent of the work count.

A reconstructed log. In some organizations, such as the National Institute of Allergies and Infectious Diseases (see Chapter 30), the work flow may be such as to require a whole year to record all events and appropriate times. In some cases the time for logging may not be available; the need for data may be too immediate. Further, the year of logging may be burdensome.

A work log of time used, by work-unit, can be reconstructed if the work is such that it:

1. For the most part, takes at least a half a day per task.
2. Leads to reports that are on file.
3. In general, is scheduled and planned on a calendar.[6]

In such cases it has been possible to develop a work count of work-units and a work time by work-units, for a sizable professional group, in a matter of days. The standard times developed from such data represent an unknown level of diligence, but they do provide a basis for budgeting and for measuring changes in labor productivity in subsequent years.

---

[6]Proposed by Anne Kaufman, National Institutes of Health.

Summary

Methods of handling work-time and work-count data, when the number of kinds of work-units exceeds the number of available data sets, have been discussed. Additional methods of obtaining work-time and work-count data have been examined.

PROBLEMS

45.1. Three direct time study—extensive sampling data sets have been obtained. These provide the following information:

Set	Allowed Time (minutes)	Work Counts of 4th-Order Work-Units			
		Work-Unit 1	Work-Unit 2	Work-Unit 3	Work-Unit 4
1	51	4	3	6	2
2	92	8	1	7	9
3	52	2	3	6	5

Determine the standard time for each work-unit.

45.2. Same as Problem 45.2, but the data are as follows:

Set	Allowed Time (hours)	Work Counts of 4th-Order Work-Uints			
		Work-Unit 1	Work-Unit 2	Work-Unit 3	Work-Unit 4
1	56	2	2	6	6
2	51	6	6	2	1
3	29	3	1	5	2

Determine the standard time for each work unit.

45.3. Same as Problem 45.1, but the data are

		Work Counts of 4th-Order Work-Units				
	Allowed					
	Time	*Work-*	*Work-*	*Work-*	*Work-*	*Work-*
Set	*(hours)*	*Unit 1*	*Unit 2*	*Unit 3*	*Unit 4*	*Unit 5*
1	98	6	3	5	4	7
2	74	4	9	6	8	4
3	169	8	1	12	2	11

Determine a standard time for each work-unit.

45.4. A series of jobs exist, each consisting of the production of a different mix of work-units 1 through 4 of Problem 45.1, but each job having a fixed mix. Using the values obtained from Problem 45.1 (or values assigned by the instructor) and assuming an 8-hour day:

(a) Develop a table showing the relationship between the mix of work-units 1 through 4 to the per cent of the worker's capability required to produce the assigned mix of work-units, from 1 to 100 per cent of a worker's capability.

(b) How reliable do you think this table is? Explain your position.

(c) If you do not think the table is reliable, describe in detail the steps necessary to obtain a reliable table. Be quantitative.

45.5. The data available for the operation of a particular kind of an office of the government are as follows:

		Work Counts, by Work-Units							
	Time Used								
Location	*(man-hours)*	1	2	3	4	5	6	7	8
1	15,440	120	250	180	200	500	150	800	100
2	15,440	140	200	200	300	100	50	1,000	50
3	27,020	130	100	280	150	100	500	50	2,000
4	21,320	160	100	300	100	800	200	1,000	500

Further, the personnel in the office work an actual average of 1,930 hours per year.

(a) Determine the standard time for each of the work-units by linear programming.

(b) Same as part (a), but the smallest standard time is to be 1 man-hour.

(c) Determine the standard time for each of the work-units by multiple regression.

(d) Same as part (c), but the smallest standard time is to be 1 man-hour.

(e) Compare the results obtained from parts (a) and (c).

46

WORK MEASUREMENT PROCEDURE MANUAL AND REPORTING FORMS

Introduction

The work of a motion and time study department should follow formal procedures. The procedures should, of course, be changed when needs so dictate, but the changes should be formalized. This chapter describes a formal time study manual and formal procedures for reporting on work done. These materials are offered as a guide for an organization; as a point of departure in creating a manual and reporting procedures to meet specific needs.

Manual of Time Study Procedure[1]

Basic policy. The head of time study shall review all actual time studies and accept responsibility for their reliability. He shall be responsible for

---

[1]This manual, concerned primarily with direct time study–intensive sampling, is sufficiently important that an outline manual is reproduced here, almost in full. Various versions

proposing changes in the time study procedure but must obtain the chief manufacturing engineer's consent to all major revisions of the procedure. The head of time study shall see that all time studies follow standard time study practice.

1.00 Summary

1.01. Notify operator and foreman that study is to be made and why.[2]

1.02. The time study man is responsible for the standardization of the work station, etc. Check method of performance with foreman. Do not issue orders directly to operator. Deal through the foreman unless he, in your presence, instructs the operator to do the job in the manner you request.

1.03. Identify the work-unit, the unit in which production is measured. Break the job into the smallest elements which are practical and possible to accurately time and which conform with standard elements. Write these elements in detail on the front of the time study sheet and number them. Abbreviate these descriptions on the back of the sheet for actual timing.

1.04. Fill in on the time study sheet the department, the operation, the operator's name and clock number, the details of tools and such information as make, model, and serial number of equipment or machines, and all other applicable items. Make a written note of any condition of equipment or workplace affecting the operation.

1.05. Enter the time of starting the actual recordings on the back of time study sheet and start the stopwatch. Write the time that each element is completed in the space provided at the side of each element description.

1.06. Continue to leave the watch running, and repeatedly record the time required to complete each element. Write down the watch reading and description for any foreign element (nonroutine occurrence) that happens during the course of the study. Make notes concerning foreign elements in the lower right corner of the time study sheet. Use standard symbols where possible.

1.07. Before leaving the job, rate the performance of the operator and list the difficulties inherent in each element of the job.

---

of this manual are in use by numerous companies. If other methods of work measurement are to be used, a similar manual should be developed. This illustration will then serve as a guide. In order to conserve space, where material normally included in the manual is reproduced elsewhere in this book, the location of the material will be cited instead.

[2]The procedure items are numbered so as to facilitate reference to them. In some shops union participation is obtained by training union time-study stewards in the procedure. In actual practice, each detail item appears on a separate sheet, so as to facilitate adding notes or issuing revised detail sheets, one at a time, as needed.

1.08. The next step (done in office) is to subtract each element clock reading from the preceding reading to get the actual time required for that element.

1.09. Calculate the observed average time for each element. This is the sum of the individual time values divided by the number of values. If the element does not occur per each work-unit, indicate the proper pro rata divisor.

1.10. The average time for each element is then multiplied by the rating. The answer is the rated time for the element.

1.11. When weights or resistances over 20 lb are involved in the job, the per cent of cycle taken by each rated element time is computed. The difficulty adjustments are entered. The rated time for each element is multiplied by 1.00 plus the total difficulty correction for the element, expressed as a decimal. This product is the base time.

1.12. The allowances are added to give the total standard time per time study element.

1.13. The standard time per unit is in minutes. It is necessary to convert it into standard hours per 100 or per 1,000 units. This is done by multiplying by 100 or 1,000 and dividing by 60.

1.14. The production per hour or hourly requirements are found by dividing the standard hours per 100 units into 100 or per 1,000 units into 1,000.

1.15. The piece rate per 100 or per 1,000 pieces is obtained by multiplying the hours per 100 or per 1,000 pieces by the base rate applicable to the operations. The standard time for standard-time plans is found by multiplying the unit time by appropriate values.

1.16. The time study must be approved by the Time Study Supervisor prior to issue.

1.17. The time study will be used to issue the:
a. Instruction and layout (see Figures 27.1 through 27.4).
b. Notice of piece-work rate.
1.18. In cases of failure to meet standard production:
a. A "Production Check Study" will be made and a report issued.
b. A meeting may be held with the departmental foreman and steward to discuss the problem.

2.00 Detail sheets

2.10. SPECIFIC RESPONSIBILITY OF THE TIME STUDY SECTION

Time study is a service activity. The time study man works primarily in an advisory capacity and should not tell production supervisors what to do. The time study man can and should advise supervisors whenever possible and work closely with them. The time study man may also make recommendations to the plant superintendent.

The time study man is responsible for establishing the method of operation for the job to be studied. The time study man, together with the foreman, will make the necessary adjustments and see that the operator is following the required method before beginning the study. The time study man is responsible for recording all the conditions surrounding the job which he feels will help ensure that the resulting production standard will be for a particular standardized operation.

The time study section is responsible for:

1. Establishing and issuing time standards, after management's approval, for production supervisors, and payroll and production control. Unless the supervisor has a factual basis for rejection, these standards are valid.

2. Establishing and recording acceptable work methods on operations, including speeds, feeds, motions, hand and crew balance, etc., in cooperation with production supervisors.

3. The acceptability of the operator available for the time study.

4. Notifying supervisors of changes and improvements that will facilitate production and reduce costs, and for reporting on such changes.

5. Notifying production supervision and timekeeping of changes in standards.

6. Keeping records of production standards up to date.

7. Assisting in determining causes of failures to achieve standard production.

8. Periodically auditing standards for correctness and applicability and taking the necessary steps to correct for any changes introduced since the setting of the standard.

2.20. SERVICE OPERATORS

In order to study the required time, the time study man will (after obtaining the foreman's help or permission) instruct the service operators in following a prescribed routine designed to accomplish the necessary work with the least effort.

2.30. SPECIFIC RESPONSIBILITY OF PRODUCTION SUPERVISORS

The production supervisor's duties include "properly understanding and using production standards." The production supervisor is responsible for:

1. Instructing employees in standard methods.

2. Cooperating with time study in the development of standard methods. The foreman must define the work assignment of the employees under study.

3. Assisting in the elimination of unnecessary walking, handling, motions, and delays.

4. Controlling delay time. Seeing that allowable delays are accurately reported.

5. Knowing what is included in the standard method when instructing a worker.

6. Reporting all changes in methods to the time study department.

7. Eliminating any work designed to overcome a temporary difficulty.

8. Letting the operators know the piece price or standard time or standard production expected per hour.

9. Seeing that time, production, and delays are accurately reported.

10. Seeing that the proper management attitude is maintained during all discussions of time standards.

3.10. DESCRIPTION

The written description of the job, in the form of elemental descriptions, should be sufficiently complete with respect to motions, tools and equipment, workplace, workplace layout, flow of work, working conditions, and material so that the job could be reproduced from this description in only one manner.

3.11. ELEMENTS

The job will be described in terms of elements chosen so as to meet the following criteria:

a. Easily detected and definite end-points.
b. As small as is convenient to time.
c. As unified as possible.
d. Hand time separate from machine time.
e. Constant elements separated from variable elements (as concerns a series of similar jobs).
f. Regular elements separated from irregular elements.
g. Internal hand time kept separate from external time.

3.20. WORK-UNITS

The work-unit or unit of production should be identified and defined.

3.31. FILMS

On jobs that are difficult or laborious to describe, a motion-picture record of the method should be made.

3.32. STILL PICTURES

Polaroid pictures of workplaces should be used instead of sketches in all possible cases. A boldly marked scale should be included in the picture to give dimensions. (Use sticks with alternate black-and-white bands 1 inch in length.)

4.10. MATERIALS HANDLING

Indicate all pertinent information relative to delivery and disposal of direct and indirect materials.

4.20. ASSISTANCE GIVEN

Indicate all service performed by others for the operator and the relationship of these services to productivity.

6.10. NUMBER OF READINGS

An adequate sample of readings shall be taken so as to be reasonably representative of the performance observed. (See Table 20-I.)

7.10. PACE DEFINED

The pace of the operator (or speed or rate of activity) is defined as the rate at which muscular force is applied to the body, arm, hand, or finger movements, disregarding the effect of job difficulty, and hence, as the observed rate of *acceleration of the body member controlling the speed of doing the work*. (See Chapters 20 and 39.)

7.11. PACE SCALE

The 100 per cent pace and speeds above and below 100 per cent are specifically defined in the standard film loop when it is projected at the proper speed. (See Chapters 20 and 39.)

7.12. ASSIGNING A RATING

The time study man shall be familiar with the appearance of the standard scale of paces and shall appraise the job while he is studying it so that when he has completed his readings of the element times, he is able to enter a pace rating for each element. This pace rating shall represent the average speed at which that element was performed as compared with the standard loop, and no attention shall be paid to the difficulty of the element. Any numerical value which the time study man deems appropriate may be used, although in most cases extreme performance should be scrutinized carefully. (It is to be noted that extremely difficult elements may be performed slowly, particularly where heavy weights are involved.)

7.13. MOTION PICTURES

In cases where the time study man is not certain of the rating, a film record may be made. Use the constant-speed drive on the camera and take pictures during the time study.

7.20. PROCESS-CONTROLLED ELEMENTS

The time for some elements in controlled by the requirements of what is being worked with rather than by the diligence of the operator. In such cases, no matter how hard the operator tried to do the task, he could not reduce the time. If the time observed is the time required by the material rather than the operator, a rating of —[the value representing the typical incentive pace] will be used without any adjustments for difficulty.

9.10. PRORATING

Prorating is reducing the allowed time per time study element to a per work-unit value. For example, if the time study element was for "closing box of two toy engines," then only one-half of the time study element is allowable per engine; hence, divide by 2.

11.10. DIFFICULTY ADJUSTMENTS

The difficulty on each element will be described in terms of the proper classification from each of the six categories of difficulty. The selected values will be added for each element, to obtain the total difficulty correction for the element. If an element has a weight adjustment, the per cent of the rated cycle time taken by that element will be computed, to determine the correct weight adjustment. (See Tables 20-III and 20-IV.)

11.20. DIFFICULTY-ADJUSTMENT NOTES; GENERAL

The difficulty-adjustment value for an element for any category will be the maximum applicable value for any reasonable portion of the element. (If weight or resistance to overcome is involved, the hand motions toward the weight should be in the same element as those with the weight.)

11.21. DIFFICULTY-ADJUSTMENT NOTES; WHEN PEDALS ARE HELD DOWN DURING THE JOB

The pedal correction is only for those elements in which the pedal is actually manipulated. Such elements, if similar to subsequent ones except for this factor, may have to be prorated.

11.22. DIFFICULTY-ADJUSTMENT NOTES; INTERPRETING DIFFICULTY CLASSIFICATIONS

Elements corresponding to the various categories are listed next to them as a guide. (These notes should refer to elements, not jobs. A specially annotated table of difficulty adjustments should be prepared in each plant.)

12.10. PERSONAL ALLOWANCES

A personal allowance of 5 per cent will be added to the difficulty-adjusted time to allow for personal needs.

12.20. ALLOWANCES, MACHINE-CONTROLLED OPERATIONS

The standard time will be based on the standard time for manual work during the machine downtime plus the allowance-adjusted time for the machine-controlled part of the cycle. (See the material on machine allowance, Chapter 39.)

17.10. GENERAL POLICY

In general, the standard will be developed for the smallest group practical; the individual standard will be given whenever economically and practically possible.

17.20. CREW STANDARDS, LINE ACTIVITY

Where a crew works in sequence, the standard will be based on the longest operation in the line. The standard will also give the efficiency of the line setup so as to indicate the possibilities of future rate changes by a better distribution of work. The efficiency will be computed in the following manner:

$$\frac{\dfrac{\text{total standard time of all operations}}{\text{number of operators}}}{\text{standard time of longest operation}} \times 100 = \text{per cent efficiency}$$

When the individual instruction cards are issued, the per cent of full operation will be indicated for each position. Delay time will be shown and line balance noted as the cause when appropriate.

Reports

Introduction. Reports from staff departments, such as an industrial engineering department, usually fall into two categories: (1) reports on specific proposals, and (2) reports on total accomplishment, usually on a periodic basis.

Although staff reports are often informal, varying in style from project to project and period to period, they are usually a recurring type of report. In such cases, considerable executive time may be saved by adopting a standard format. The use of a standard format becomes increasingly important as the organization becomes large, although in the small organization, where each man may have a variety of duties, the time-saving aspects of standard formats may well be worthwhile.

Project reports. When the report concerns the subjects covered in this book, the individual report on a specific project normally forms the basis for acceptance or rejection by operating management. If the individual reports are adequately formulated, the problem of submitting adequate periodic reports concerning total activity (in these areas) is greatly facilitated. The periodic report may be merely a summary of the individual reports.

A form suitable for summarizing the content of a method proposal is shown in Figure 46.1. A form for summarizing a procedure proposal is shown

METHOD PROPOSAL SUMMARY

1. _____ Date
2. _____ To
3. _____ From
4. _____ Subject
   _____
   _____

1. Fewer people
2. Fewer steps
3. Less time on a step or steps
4. Less time in production
5. Less space
6. Less time for critical skills
7. Less time on critical equipment
8. Increased quality
9. Less cost
10. Less skill on step or steps
11. Better control

Improvement will (insert proper numbers or describe if not classifiable):
5. _____

6. If this proposal is approved, it will be necessary to (summarize):
_____
_____
_____
_____
_____

ATTACHMENTS: (Insert number of sheets in boxes, follow with page numbers on lines.

7. ☐ _____ Cost of change estimate detail sheet
8. ☐ _____ Original charts
9. ☐ _____ Proposed charts, including summary and comparison
10. ☐ _____ Proposed equipment list and details of placement
11. ☐ _____ Jig, fixture, workplace or layout sketches or drawings
12. ☐ _____ Job instruction sheets as required for proposal
13. ☐ _____ Additional attachments list

14. FINANCIAL ASPECTS OF CHANGE AND EVALUATION OF CHANGE (summary)

a. _____ (who) estimated the annual volume
b. _____ (who) estimated the fixed cost of tools , etc.
c. _____ Time new method would take to pay for itself out of savings.
d. _____ Original labor cost and hours, annual
e. _____ Proposed labor cost and hours, estimated annual
f. _____ Cost of change
g. _____ Net savings, annual (estimated)

Figure 46.1. Form for methods improvement proposal.

PROCEDURE PROPOSAL SUMMARY

1. _____ Date   1. Fewer people
                                2. Fewer steps

2. _____ To   3. Less time on a step or steps
                                4. Less time to complete procedure

3. _____ From   5. Less space
                                6. Less time for critical personnel

4. _____ Subject   7. Less time on critical equipment
                                8. Increased accuracy

5. $ _____ Annual saving   9. Cheaper procedure
                                10. Less skill on step or steps

6. ___ New Procedure   11. Better control

7. ___ Improved Procedure

8. ___ Improved Forms

9. Improvement will (insert proper numbers or describe if not classifiable):

10. If this proposal is approved it will be necessary to (summarize):

_____

_____

_____

_____

_____

11. Financial aspects of change and evaluation of change. (summary)

_____

_____

_____

_____

_____

ATTACHMENTS (insert number of sheets before comma, follow with page numbers

12. ___ , _____ Original procedure chart

13. ___ , _____ Proposed procedure chart

14. ___ , _____ Proposed equipment list and details of placement

15. ___ , _____ Form designs and instructions for use, by forms

16. ___ , _____ Job instructions, by persons

17. ___ , _____ Additional attachments list

Figure 46.2a. Form for procedures improvement proposal.

in Figure 46.2a. Inasmuch as it it difficult to summarize a process chart—combined analysis directly on the chart, when the procedure proposal concerns two process charts—combined analysis, Figure 46.2a is often accompanied by a form such as shown in Figure 46.2b, which summarizes and contrasts the details of the two procedures involved: the original and the pro-

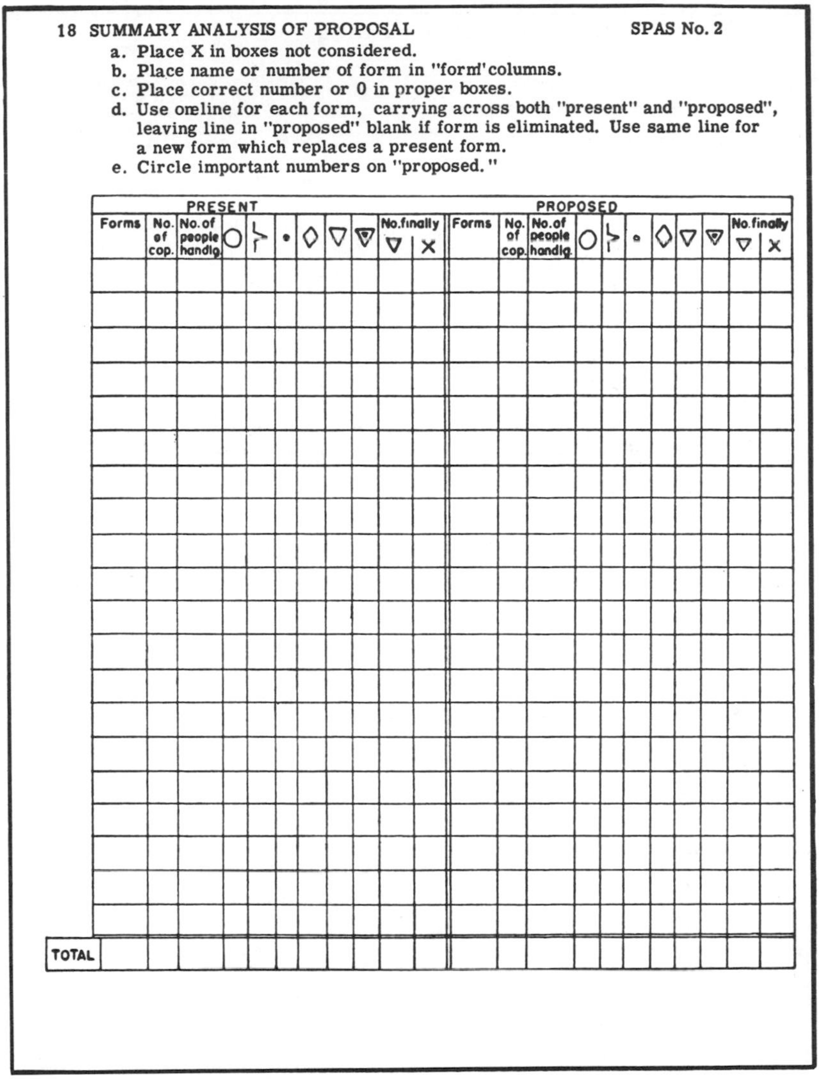
18 SUMMARY ANALYSIS OF PROPOSAL SPAS No. 2

a. Place X in boxes not considered.
b. Place name or number of form in "form" columns.
c. Place correct number or 0 in proper boxes.
d. Use oneline for each form, carrying across both "present" and "proposed", leaving line in "proposed" blank if form is eliminated. Use same line for a new form which replaces a present form.
e. Circle important numbers on "proposed."

	PRESENT											PROPOSED										
	Forms	No. of cop.	No. of people handlg.	○	⇒	•	◇	▽	▽	No. finally ▽	No. finally X	Forms	No. of cop.	No. of people handlg.	○	⇒	•	◇	▽	▽	No. finally ▽	No. finally X
TOTAL																						

Figure 46.2b. Supplementary form for Figure 46.2a for summarizing and comparing two process charts—combined analysis.

posed. A form suitable for summarizing the content of a work measurement proposal is shown in Figure 46.3.

Performance and audit reports. In addition to reporting on performance, the industrial engineering group should audit the status of the aspects

WORK MEASUREMENT PROPOSAL SUMMARY

REASONS FOR STDS.

1. _____ Date   LIST A

                                      A. Correct clerical error

2. _____ To   B. Correct old standard

                                        C. Improve quality or yield

3. _____ From   D. Change in work unit

4. Subject _____   E. Other. Specify in item 7

   _____ Revised standard

                                         LIST B

   _____ Reason for revision from list A   F. New product

                                        G. New equipment

   _____ New standard   H. New method

                                        I. Other. Specify in item 7

   _____ Reason for setting from list B

5. _____ Check if method proposal gives details

6. Operation or operations and department or departments involved:

   _____

   _____

   _____

7. If this standard is installed, it will be necessary to (summarize, including any unusual features) :

   _____

   _____

   _____

   _____

8. _____ Job description sheets have been prepared.

9. ATTACHMENTS. Insert number of sheets before comma; give page numbers and, except for preprinted item, titles, after comma.
   _____ , Production standards notice

   _____ , _____

   _____ , _____

10. FINANCIAL ASPECTS OF CHANGE

   a. _____ (who) estimated annual volume

   b. _____ Current or original annual labor cost and hours

   c. _____ Proposed annual labor cost and hours

   d. _____ Net annual estimated cost reduction, dollars

Figure 46.3. Form for work measurement proposal.

of the plant to which they contribute. A series of forms for reporting such audits appear as Figures 46.4 through 46.6.

Periodic reports from staff departments serve several purposes. The first is to inform management concerning the results of the activity in order to answer the ever-present question, "Is it worth continuing?" The second

METHODS IMPROVEMENT AUDIT

_____ INSTALLATION
_____ DEPARTMENT
_____ DATE ENDING, PERIOD COVERED

I   METHODS PROPOSALS
   a. _____ Total proposals of method changes made to date.
   b. _____ Cumulative changes proposed prior to period and awaiting action.
   c. _____ Number of method changes proposed in period.
   d. _____ Number of changes acted upon:
                   1. From prior periods.
      _____   2. From reporting period.
   e. _____ Number of changes awaiting action (b + c - d).
   f. _____ Number of changes approved and approval %; -- $\frac{\text{Approved}}{\text{Reviewed}}$ x 100.
                   1. For prior period
      _____   2. For reporting period.
   g. _____ 1. Total savings of accepted proposals from Method Proposal
                   Summary Sheets, cumulative for a year.
      _____ 2. Average increase in production for reporting period.
      _____ 3. List highest three.
      _____
      _____

II TALLY OF CATEGORIES OF SAVINGS ACCEPTED DURING PERIOD
   a. _____ Fewer people
   b. _____ Fewer steps.
   c. _____ Less time on a step or steps.
   d. _____ Less time in production.
   e. _____ Less space.
   f. _____ Less time for critical skills.
   g. _____ Less time on critical equipment.
   h. _____ Increased quality.
   i. _____ Less cost.
   j. _____ Less skill on step or steps.
   k. _____ Better control.
   l. _____ Miscellaneous.

III NUMBER OF SOP'S PREPARED FOR SHOP METHODS
   a. _____ Revised.
   b. _____ New.

IV SUGGESTION PROGRAM
   a. _____ Number of suggestions submitted for period and cumulative.
   b. _____ Percent of employees submitting suggestions.
   c. _____ Percent and number of ideas approved.
   d. _____ Number of suggestions held over one month.
   e. _____ Cost savings resulting from suggestion program:
                   From reporting period.
      _____   Cumulative yearly.
   f. _____ Cost of administering program.

V   TRAINING PROGRAMS
   a. _____ Number of supervisors trained in Methods Improvement:
                   1. In reporting period
      _____   2. Total, to date.
   b. _____ Number of employees trained in Methods Improvement:
                   1. In reporting period
      _____   2. Total, to date.

Figure 46.4. Form for periodic reporting or auditing of methods improvement activity. (Developed by I. Lazarus.)

purpose is to create a record indicating rate of performance, to assist in planning future activity or programming the workload of individuals responsible for the function reported on. The reports may also serve the industrial engineering administrator in operating his department, and in planning budgets.

PROCEDURE ANALYSIS ACTIVITY AUDIT SHEET

_____ INSTALLATION

_____ DEPARTMENT

_____ DATE ENDING, PERIOD COVERED

I   ACTIVITY

    a. _____   Total number of man hours spent in procedure and forms design studies during report period.

    b. _____   Reduction in man hours of work load per month as result of procedure and forms design changes during report period.

    c. _____   Cumulative savings in man hours, 12 month running total.

II   PROCEDURE STUDIES

    a. _____   Approximate number of procedures in installation.

    b. _____   Total number of procedures recorded graphically at end of last report period.

    c. _____   Number of procedures recorded graphically during this report period.

    d. _____   Total number of procedures recorded graphically at end of this report period.

    e. _____   Number of procedures proposals submitted during report period.

    f. _____   Number of procedure proposals accepted during report period.

III   FORMS DESIGN

    a. _____   Total number of forms in use at end of last report period.

    b. _____   New forms created not replacing old forms.

    c. _____   Number of forms combined.

    d. _____   Number of forms eliminated.

    e. _____   Net change in number of forms in use.

    f. _____   Total number of forms in use at end of present report period.

    g. _____   Number of forms simplified.

Figure 46.5. Form for periodic reporting or auditing of procedure improvement activity. (Developed by J. Moquin.)

An alternative form of a methods improvement audit appears in Figure 46.7; an alternative form of a work measurement audit appears in Figure 46.8. By evaluating these activities in terms of the contribution to plant profitability, the economics of the operation of the motion and time study activity are readily evaluated by line management. These periodic reports

WORK MEASUREMENT DEVELOPMENT AND APPLICATION AUDIT

_____ INSTALLATION

_____ DEPARTMENT

_____ DATE ENDING, PERIOD COVERED

I LABOR STANDARD DEVELOPMENT
a. _____ Total number labor standards, present.
b. _____ Total number labor standards, last report.
c. _____ Number increase over last report.
d. _____ Total number labor standards revised; method change.
e. _____ Total number labor standards revised; other reasons.

II STANDARD HOURS COVERAGE
a. _____ Total elapsed man hours during report period covered by standards.
b. _____ Total daywork man hours during report period.
c. _____ Total elapsed man hours during report period.
d. _____ Percentage of hours covered by labor standards.
e. _____ Total number of operations performed during report period covered by labor standards.
f. _____ Total number of operations performed during report period not covered by labor standards.
g. _____ Percentage of operations performed covered by labor standards.

III PRODUCTIVITY ANALYSIS
a. _____ Total standard man hours produced on work covered by standards.
b. _____ Total elapsed man hours on work covered by standards.
c. _____ Percentage productivity.

IV DELAY HOURS ANALYSIS
a. _____ Total machine downtime delay man hours.
b. _____ Percentage of machine downtime to total elapsed man hours.
c. _____ Total set-up delay man hours.
d. _____ Percentage of set-up time to total elapsed man hours.
e. _____ Total miscellaneous delay man hours.
f. _____ Percentage of miscellaneous delay to total elapsed man hours.
g. _____ Total delay man hours.
h. _____ Percentage total delay to total elapsed man hours.

Figure 46.6. Form for periodic reporting or auditing of work measurement activity. (Developed by C. S. Schneider.)

are readily assembled. Each motion and time study analyst, after completing a task, turns in two copies of a form similar to Figure 46.1 or 46.3, as appropriate. One copy is filed in a monthly folder, from which the monthly audit is compiled. The other copy is filed by the analyst, to facilitate periodic review of his activity by the chief industrial engineer.

C.B.C.   CODE 1464 - B   DATE March 30

METHODS IMPROVEMENT AUDIT - PERIOD   Feb. 14   TO Mar. 13

1. METHODS PROPOSALS

a. _____ 64 _ Total Proposals of Method Changes made this fiscal year to date.

b. _____ 11 _ Number of method changes proposed in period.

c. $110,247 1. Total savings of accepted proposals from Method Proposal
                summary sheet made to date.

d. $ 22,703 2. Total savings of accepted proposals from Method Proposal
                summary sheet cumulative for period.
                Less cost of Equipment if any and IE expense $1,900 per period.

   $ 10,851 List highest three.   Pickle Curing*

     7,088   Casing Shipping*

     4,772   Hog Kill*

2. TALLY OF CATEGORIES OF SAVINGS ACCEPTED DURING PERIOD

       a. _6_ Fewer people

       b. _-_ Fewer steps

       c. _6_ Less time on a step or steps

       d. _-_ Less time in production

       e. _-_ Less space

       f. _2_ Less time for critical skills

       g. _-_ Less time on critical equipment

       h. _-_ Increased quality

       i. _6_ Less cost

       j. _-_ Less skill on step or steps

       k. _-_ Better control

       l. _-_ Miscellaneous

                              Submitted by _____

       * See attached sheet for details

Figure 46.7. Plant methods improvement audit. (Courtesy of C. J. Allen, Chief Industrial Engineer, Patrick Cudahy, Inc., Cudaly, Wis.)

P.C. Inc 1463-B

DATE **March 30**

Mr. M. F. Cudahy:

WORK MEASUREMENT PROGRESS REPORT - PERIOD _____ **Feb. 14** _____ TO **Mar. 13** _____

1. STANDARDS ESTABLISHMENT

    A. Number of Standards at beginning of period.   **3,772**

    B. Number of New Standards Established. ("Established" includes re-organizing job, writing-up method, reviewing method with supervision and steward).   **11**

    C. Total number of Standards at end of period.   **3,783**

2. STANDARDS REVISIONS

    A. Number of Standards indicated as needing revision during period.   **131**

    B. Number of Standards revised.   **131**

    C. Net result of those revised.

        A. Total estimated annual man hours of work affected.   **313,477**

        B. Total number of workers affected.   **344**

        C. Estimated reduction in annual earned direct labor hours.   **9,722**

        D. Estimated annual payroll savings for month.   **$26,220**
    Includes cost of equipment, if any plus IE expense $4,212/period

        E. Estimated annual payroll savings this fiscal year to date.*   **$133,350**

RECAP		PLANT EVALUATIONS
Total plant	**1,372**	Hog Kill - 1
Total incentive paid workers	**1,091**	Hog Offal - 5
Total production workers	**1,222**	
% Total production workers	**90.1**	RECAP
% Total Plant	**78.5**	Standards $ **3,517**
% Jobs on Standard	**82.3**	Methods $ **22,703**
% Hours on Standard	**73.7**	Total $ **26,220**

* Total represents decreased labor minus increased costs due operations changed for quality or yield reasons.

cc: HH Holcomb, JK Stark, E Scheidenhelm, N Sherman, MS Hungness

Submitted by _____

Figure 46.8. Plant work measurement audit. (Courtesy of C. J. Allen, Chief Industrial Engineer, Patrick Cudahy, Inc., Cudahy, Wis.)

Summary

This book has described various ways by which the productivity of various types of work may be increased. The effort toward improvement requires managerial support. This chapter has been concerned with the problems of carring on the measurement of labor performance in a formalized manner and reporting on the accomplishments to obtain such support.

PROBLEMS

46.1. Write a procedure manual for:
   (a) Direct time study—extensive sampling.
   (b) Predetermined time system.
   (c) Fractioned professional estimates.
   (d) Time standards by mathematical analysis.

46.2. Obtain a copy of a time study manual, and critique it.

46.3. Obtain a copy of U.S. Air Force MIL-STD-1567, and critique it.

SUGGESTED TOPICS FOR FURTHER STUDY

The library resources vary in the many places and lands in which it is hoped this book will be found useful. Unfortunately, books, related magazines, bulletins, and research reports are frequently unavailable a few years after their issuance. Hence, instead of a frustrating list of relatively unavailable publications, the following list of related or impinging topics is given to aid in searching available library resources.

Allowances, time study
Bar charting
Clerical work measurement
Critical path planning
Effectiveness measurement
Ergonometrics
Flow process charts
Forms design

Gantt charts
Group technology
Industrial engineering
Industrial relations
Information flow analysis
Leveling of time studies
Linear programming
Load charts

Management science
Man and machine analysis
Materials handling
Memomotion study
Methods study
Micromotion study
Motion economy
Motion study
Network analysis
Office work simplification
Operation charts
Operations research
Payments by results
Performance measurement
PERT
Plant layout
Predetermined time standards
Probability guides
Process design
Production control
Production planning

Productivity
Project planning
Queuing
Rating
Ratio–delay study
Standard time
System analysis
Therblig analysis
Time-lapse photography
Time standards
Time study
Value analysis
Value engineering
Wage incentives
Work design
Work measurement
Work sampling
Work simplification
Work standards
Work study
Work-unit analysis

INDEX